A SECOND COURSE IN
MATHEMATICAL ANALYSIS

A SECOND COURSE IN
MATHEMATICAL ANALYSIS

BY

J. C. BURKILL, F.R.S.
Master of Peterhouse, Cambridge

AND

H. BURKILL
Senior Lecturer
University of Sheffield

CAMBRIDGE
AT THE UNIVERSITY PRESS
1970

Published by the Syndics of the Cambridge University Press
Bentley House, 200 Euston Road, London N.W.1
American Branch: 32 East 57th Street, New York, N.Y.10022

Library of Congress Catalogue Card Number: 69–16278

Standard Book Numbers:
521 04381 6 Vol. 1
521 07519 X Vol. 2

Printed in Great Britain
at the University Printing House, Cambridge
(Brooke Crutchley, University Printer)

CONTENTS

PREFACE

This course of analysis is intended for mathematical specialists in their second and third years at Universities. We assume familiarity with the concept of a limit and its applications to infinite series and to the differential and integral calculus. The contents of *A First Course in Mathematical Analysis* by one of us (J.C.B.), which do not include Cauchy sequences, upper and lower limits or uniform convergence, form a suitable foundation for this book. From time to time we shall need to refer to a basic work on analysis and for simplicity we shall refer to the *First Course*, shortening its title to C1.

An undergraduate after his first year should be ready for a more abstract setting and prepared to think in metric spaces instead of the Euclidean line or plane. The study of metric spaces provides not only a means of unifying different topics in analysis but also a natural link with topology.

Chapters 1–9 concentrate on general analysis and real functions and 10–14 on complex functions. These last five chapters are largely independent of 6–9, and the reader who wishes to reach Cauchy's theorem quickly needs little more than §8.1 and §8.2.

After careful thought we decided to treat the Riemann and Riemann–Stieltjes integrals fully (chapters 6 and 8) and to leave the Lebesgue integral out. For many purposes the Riemann integral is sufficient, and the inclusion of an adequate account of Lebesgue measure and integration would impair the balance of the book.

We are very grateful to Dr L. Mirsky and to Professor G. E. H. Reuter, each of whom has read the entire manuscript. The final form of it owes a great deal to their care and vigilance and to the experience which they gained from their own teaching.

J.C.B.
H.B.

August 1968

1

SETS AND FUNCTIONS

1.1. Sets and numbers

It has been recognized since the latter part of the nineteenth century that the idea of number (real and complex), and therefore all analysis, is based on the theory of sets. In modern analysis the dependence is explicit, for the language and algebra of sets are in constant use.

The reader is likely to be familiar with the intuitive notion of a set and with the basic operations on sets. In this section we therefore confine ourselves to fixing the terminology and recapitulating the results that will be used subsequently. The notes at the end of the chapter refer to books which develop set theory systematically from explicitly stated axioms.

There are synonyms for the word set such as *collection* and *space*. The members of a set are also called its *elements* or *points*. The statement that a belongs to (or is a member of) the set A is written

$$a \in A.$$

If a does not belong to A we write

$$a \notin A.$$

We denote by $\varnothing$ *the empty set*, namely the set which has no members.

Inclusion of sets. Suppose that every member of the set A also belongs to the set B, i.e. that

$$x \in A \Rightarrow x \in B. \tag{1.11}$$

Then we say that A is contained in B and write

$$A \subset B \quad \text{or} \quad B \supset A;$$

the set A is said to be a *subset* of B. Note that any set is a subset of itself. Also the empty set is a subset of every set; for (1.11) is logically equivalent to

$$x \notin B \Rightarrow x \notin A$$

and, when A is the empty set, this implication clearly holds for any set B.

If $A \subset B$ and $B \subset A$, i.e.

$$x \in A \Leftrightarrow x \in B,$$

then A and B have the same members and we write

$$A = B.$$

If $A \subset B$ and $A \neq B$, we can call A a *proper* subset of B.

Union and intersection. Given any two sets A, B, the set which consists of all the elements belonging to A or to B (or to both) is called the *union* of A and B and is denoted by

$$A \cup B.$$

The set consisting of the elements which belong to both A and B is called the *intersection* of A and B and is denoted by

$$A \cap B.$$

The sets A, B are said to be *disjoint* if they have no elements in common, i.e. if $A \cap B = \varnothing$.

Theorem 1.11. *The following identities hold for any sets A, B, C.*

(i) $A \cup B = B \cup A$, $\quad A \cap B = B \cap A$;

(ii) $A \cup (B \cup C) = (A \cup B) \cup C$, $\quad A \cap (B \cap C) = (A \cap B) \cap C$;

(iii) $A \cup (B \cap C) = (A \cup B) \cap (A \cup C)$,

$\quad A \cap (B \cup C) = (A \cap B) \cup (A \cap C)$.

Proof. We illustrate the argument used for establishing these identities by proving the first result in (iii).

Let x be any element of $A \cup (B \cap C)$. Then $x \in A$ or $x \in B \cap C$. If $x \in A$, then $x \in A \cup B$ and $x \in A \cup C$, so that $x \in (A \cup B) \cap (A \cup C)$. If $x \in B \cap C$, then $x \in B$ and $x \in C$ so that again $x \in (A \cup B) \cap (A \cup C)$. We have therefore shown that

$$A \cup (B \cap C) \subset (A \cup B) \cap (A \cup C). \qquad (1.12)$$

Now let $y \in (A \cup B) \cap (A \cup C)$. Then $y \in A \cup B$ and $y \in A \cup C$. It follows that $y \in A$ or that $y \in B$ and $y \in C$, i.e. that $y \in A$ or $y \in B \cap C$. Hence $y \in A \cup (B \cap C)$ and so

$$(A \cup B) \cap (A \cup C) \subset A \cup (B \cap C). \qquad (1.13)$$

The inclusion relations (1.12) and (1.13) now yield the required identity. |

The laws of operation with $\cup$ and $\cap$ on sets have some likeness to those with $+$ and $\times$ on numbers. In fact $\cup$ and $\cap$ have replaced the signs for sum and product formerly used in the algebra of sets. The likeness is only partial: the identity $A \cup A = A$ has no analogue in numbers; the second law in theorem 1.11 (iii) has an analogue, but not the first.

The definitions of union and intersection previously given may be extended. If $\mathscr{C}$ is an arbitrary collection of sets, the union of these sets, denoted by

$$\bigcup_{A \in \mathscr{C}} A,$$

is the set consisting of all those elements which belong to at least one of the sets A. The intersection of the sets of $\mathscr{C}$, denoted by

$$\bigcap_{A \in \mathscr{C}} A,$$

is the set consisting of the elements which belong to all the sets A.

We require one more operation with sets. If A, B are any two sets, the *difference*

$$A - B$$

is the set consisting of those elements which belong to A but not to B. Note that the definition does not require B to be a subset of A. However a particularly important case occurs when all sets under consideration are subsets of a given set X. Then the set $X - A$ is called the *complement* of A (relative to X) and we shall denote it by A'. Clearly $(A')' = A$.

Theorem 1.12. *For any collection $\mathscr{C}$ of subsets of a set X,*

$$\left(\bigcup_{A \in \mathscr{C}} A \right)' = \bigcap_{A \in \mathscr{C}} A' \quad \text{and} \quad \left(\bigcap_{A \in \mathscr{C}} A \right)' = \bigcup_{A \in \mathscr{C}} A'.$$

The proof is left to the reader.

Theorem 1.12 shows that the operations $\cup$ and $\cap$ are complementary and that an algebraic identity involving $\cup$ and $\cap$ will admit of a *dual* identity obtained by interchanging $\cup$, $\cap$. For example, in each pair of identities in theorem 1.11 the second follows from the first.

We shall take for granted that the reader is familiar with the systems of real and complex numbers. There are well known and simple methods for obtaining the complex numbers from the real numbers. For our purposes it is only necessary to postulate the existence of the system of real numbers as one satisfying certain axioms which are given in C1 (11–12) and are restated in the notes at the end of the chapter. However, it is possible to construct the real numbers from much more primitive objects. Again the notes contain some remarks on the subject.

The set of real numbers will be denoted by R^1, the set of complex numbers by Z. Our notation for the *open interval* in R^1 which consists of the points x such that $a < x < b$ is (a, b). The *closed interval* of points x such that $a \leqslant x \leqslant b$ is written $[a, b]$. The intervals defined by $a \leqslant x < b$ and $a < x \leqslant b$ are written $[a, b)$ and $(a, b]$. The infinite interval $[a, \infty)$ consists of the points x such that $x \geqslant a$. Corresponding interpretations are put on the expressions (a, ∞), $(-\infty, a)$, $(-\infty, a]$. Finally, $(-\infty, \infty)$ is R^1.

Exercises 1(*a*)

1. Prove that

(i) $A \cup B = A \Leftrightarrow A \supset B$;   (ii) $A \cap B = A \Leftrightarrow A \subset B$.

2. Prove that

$$(B \cup C) \cap (C \cup A) \cap (A \cup B) = (B \cap C) \cup (C \cap A) \cup (A \cap B).$$

3. Show that $A - B = A \cap B'$. Hence or otherwise prove that

(i) $(A - B) \cap (A - C) = A - (B \cup C) = (A - B) - C$;

(ii) $(A - B) \cup (A - C) = A - (B \cap C)$;

(iii) $(A - C) \cap (B - C) = (A \cap B) - C$;

(iv) $(A - C) \cup (B - C) = (A \cup B) - C$.

4. Prove that

(i) $(A - B) \cap C = (A \cap C) - (B \cap C)$;

(ii) $(A - B) \cup C = (A \cup C) - (B \cup C) \Leftrightarrow C = \varnothing$.

5. Prove that

(i) $(A \cup B) - B = A \Leftrightarrow A \cap B = \varnothing$;

(ii) $(A - B) \cup B = A \Leftrightarrow A \supset B$.

6. The *symmetric difference $A \triangle B$* of the sets A, B is defined as the set of elements in A or in B, but not in both. Show that

$$A \triangle B = (A - B) \cup (B - A) = (A \cup B) - (A \cap B) = (A \cup B) \cap (A' \cup B').$$

Deduce that

(i) $A \triangle B = \varnothing \Leftrightarrow A = B$;   (ii) $A' \triangle B' = A \triangle B$.

7. Prove that

(i) $A \triangle B = B \triangle A$;   (ii) $(A \triangle B) \triangle C = A \triangle (B \triangle C)$.

8. Prove that

(i) $(A \triangle B) \cap C = (A \cap C) \triangle (B \cap C)$;

(ii) $(A \triangle B) \cup C = (A \cup C) \triangle (B \cup C) \Leftrightarrow C = \varnothing$;

(iii) $(A \triangle B) \cup (A \triangle C) = (A \cup B \cup C) \cap (A' \cup B' \cup C')$.

9. Show that $\quad A \triangle B = C \triangle D \Leftrightarrow A \triangle C = B \triangle D.$

10. When E is a finite set (i.e. one with finitely many elements), denote by $|E|$ the number of its elements.

Show that, if the sets A, B are finite, then

$$|A| + |B| = |A \cup B| + |A \cap B|.$$

1.2. Ordered pairs and Cartesian products

A set with elements $a, b, c, \ldots$ is often denoted by the symbol

$$\{a, b, c, \ldots\}. \tag{1.21}$$

The notation calls for a number of comments.

(i) In such a listing of elements it is immaterial whether a particular element appears once or several times. This convention is purely a matter of convenience. For instance it allows us to denote the set of roots of a complex quadratic equation by $\{a, b\}$, whether $a \neq b$ or $a = b$.

(ii) The order in which $a, b, c, \ldots$ are written in $\{a, b, c, \ldots\}$ has no significance. For example $\{a, b\} = \{b, a\}$.

(iii) It is important to distinguish between the object a and the set $\{a\}$, i.e. the set whose only member is a. Thus the set $\{\varnothing\}$ has one element, while $\varnothing$ has none.

The notation (1.21) has a useful variant. Given a set X, denote by $P(x)$ a statement, which is either true or false, about the element x of X. The set of all $x \in X$ for which $P(x)$ is true is written

$$\{x \in X | P(x)\},$$

or simply $\qquad \{x | P(x)\}$

if the identity of the set X is clear from the context.

Illustrations

(i) $\{x \in Z | x^2 + 1 = 0\} = \{i, -i\}$;

(ii) $\{x \in R^1 | x^2 + 1 = 0\} = \varnothing$;

(iii) $\{x \in R^1 | a \leqslant x \leqslant b\} = [a, b]$.

The notion of a set does not involve any ordering among the elements. For instance in the set $\{a, b\}$, a and b have the same status.

The *ordered pair* (a, b) is the set $\{a, b\}$ together with the ordering 'first a, then b'. Thus (a, b) and (b, a) are different unless $a = b$. (The context will determine whether the expression (a, b) stands for an ordered pair or an open interval.)

The intuitive concept of an ordered pair (a, b) may be formalized by the definition
$$(a, b) = \{\{a\}, \{a, b\}\}.$$
(See exercise 1(b), 1.)

An ordered set $(a_1, ..., a_n)$ of any finite number n (> 1) of elements is defined in a similar way.

Definition. *Given the non-empty sets* $X_1, ..., X_n$, *their Cartesian product*
$$X_1 \times ... \times X_n \qquad (1.22)$$
is the collection of all ordered sets $(x_1, ..., x_n)$ *such that*
$$x_1 \in X_1, ..., x_n \in X_n.$$
If $X_1 = ... = X_n = X$, *the set* (1.22) *is denoted by* X^n; X^1 *is taken to be* X.

For $n > 1$, $R^n = R^1 \times ... \times R^1$ (n factors) is the set of all ordered sets $(x_1, ..., x_n)$ of real numbers. We define an *interval* in R^n as a set
$$I_1 \times ... \times I_n,$$
where $I_1, ..., I_n$ are intervals in R^1. For example $[a, b] \times [c, d]$ is the set of points $(x, y) \in R^2$ such that $a \leqslant x \leqslant b$ and $c \leqslant y \leqslant d$.

1.3. Functions

Let X, Y be two non-empty sets. A *relation from X to Y* is a subset of $X \times Y$. If a relation f, i.e. a subset of $X \times Y$, is such that, *for every* $x \in X$, *there is one and only one member* (x, y) *of* f, then f is said to be a *function on X to* (or *into*) Y; we express this symbolically by writing
$$f : X \to Y.$$
The set X is called the *domain* of f.

In some contexts the terms *mapping, transformation,* or *operator* are often used as synonyms for function. Let (x, y) be an element of the function $f : X \to Y$. Then we say that y is the *value* of f at x (or the *image* of x under f) and we write $y = f(x)$. If E is any subset of X, the subset of Y
$$\{y \in Y | y = f(x) \quad \text{and} \quad x \in E\}$$
is called the image of E under f; it is denoted by $f(E)$. The set $f(X)$ is called the *range* of f. Note that $f(X)$ may be a proper subset of Y.

If a function f has its domain in X and its range in Y, we say that f is a function *from* X to Y. A function from R^1 to R^1 will be called a *real function*, a function from Z to Z a *complex function*.

Illustrations

(i) We obtain a function on X to Y whenever we have a rule for assigning a single value y in Y to each x of X. Thus the equation $x^2 + y^3 = 1$ defines a function on R^1 to R^1 whose range is $(-\infty, 1]$.

(ii) The equation $x^2 + y^2 = 1$ defines a relation but not a function on $[-1, 1]$ to $[-1, 1]$ because, whenever $-1 < x < 1$, it is satisfied by two values of y.

(iii) The logarithmic function is the function on $(0, \infty)$ to R^1 which assigns the value $\log x$ to the positive number x.

(iv) The sequence $a_1, a_2, a_3, \ldots$ is a function whose domain is the set of positive integers exhibited as suffixes.

More generally, given two arbitrary non-empty sets A, I and a function on I to A, we sometimes find it convenient to write the image of a member i of I in the form a_i.

(v) Denote by Φ the set of all real functions ϕ defined by an equation of the form

$$\phi(x) = \alpha x + \beta \quad (x \in R^1),$$

where α, β are real numbers. A function $f: R^2 \to \Phi$ is obtained by assigning to each point $(a, b) \in R^2$ the function $\phi_{a,b}$ given by $\phi_{a,b}(x) = ax + b \quad (x \in R^1)$.

Let f be a function on X into Y. If $f(X) = Y$, i.e. if every $y \in Y$ is the image under f of at least one $x \in X$, then we say that f is *surjective*, or that f maps X *onto* Y.

If different elements of X have different images, i.e. if $f(x) = f(x')$ implies $x = x'$, then f is called *injective*.

If f is both surjective and injective, so that every value y in Y is assumed once and once only, then f is said to be *bijective*. A bijective function is also called a *bijection*; it effects a *one-to-one correspondence* between its domain and range.

Consider again the illustrations above.

(i) The function maps R^1 onto $(-\infty, 1]$; it is not injective.

(iii) The logarithmic function is a bijection on $(0, \infty)$ to R^1.

(v) This function is a bijection on R^2 to Φ.

Inverse functions. Suppose that the function f has domain X and range Y. To f, which is a subset of $X \times Y$, there corresponds the subset of $Y \times X$ defined as

$$\{(y, x) | (x, y) \in f\}. \tag{1.31}$$

Plainly this is a function with domain Y and range X if and only if f is bijective. When f is bijective, (1.31) is called the *inverse function* of f and is denoted by f^{-1}. For instance the logarithmic function has an inverse with domain $(-\infty, \infty)$ and range $(0, \infty)$; the inverse is the exponential function.

If the function $f: X \to Y$ is injective, then $f: X \to f(X)$ is bijective and $f^{-1}: f(X) \to X$ is a function *from* Y to X. For example the function $f: R^1 \to R^1$ defined by

$$f(x) = \frac{1}{1+e^x}$$

has the inverse $f^{-1}: (0, 1) \to R^1$ given by

$$f^{-1}(y) = \log\left(\frac{1}{y}-1\right).$$

It is clear that, if f has an inverse, then f^{-1} has an inverse and $(f^{-1})^{-1} = f$.

Inverse images. Let $f: X \to Y$ be an arbitrary function and let E be any subset of Y. We denote by $f^{-1}(E)$ the set

$$\{x \in X | f(x) \in E\}.$$

This set is called the *inverse image* of E. It is important to note that the definition of $f^{-1}(E)$ does not presuppose the existence of the inverse function f^{-1}. However, when f^{-1} does exist, then $f^{-1}(E)$ is the image of E under f^{-1}.

If f is the function in (i) on p. 7 then, for example,

$$f^{-1}((0, 1]) = (-1, 1);$$

the *function f^{-1}* does not exist.

Composition of functions. Given the functions $f: X \to Y$ and $g: Y \to W$, the function $h: X \to W$ defined by

$$h(x) = g\{f(x)\} \quad (x \in X)$$

is called the *composition* of g and f and is written $g \circ f$.

For three functions $f_1: X_1 \to X_2$, $f_2: X_2 \to X_3$, $f_3: X_3 \to X_4$, we clearly have
$$f_3 \circ (f_2 \circ f_1) = (f_3 \circ f_2) \circ f_1$$
so that the expression $f_3 \circ f_2 \circ f_1$ has a meaning. The process of composition may be extended to any number of functions.

For a non-empty set A, denote by i_A the *identity function* on A defined by
$$i_A(x) = x \quad (x \in A).$$

If, now, $f: X \to Y$ is any function,
$$f \circ i_X = f, \quad i_Y \circ f = f;$$
and, when f is bijective,
$$f^{-1} \circ f = i_X, \quad f \circ f^{-1} = i_Y.$$

Theorem 1.31.

(i) *If the function* $f: X \to Y$ *is surjective and if there is a function* $g: Y \to X$ *such that* $g \circ f = i_X$, *then* f *is bijective and* $g = f^{-1}$.

(ii) *If the function* $f: X \to Y$ *is bijective and if the function* $g: Y \to X$ *is such that* $f \circ g = i_Y$, *then* $g = f^{-1}$.

Proof.

(i) If $f(x) = f(x')$, then

$$x = g\{f(x)\} = g\{f(x')\} = x'$$

and so f is injective as well as surjective. We now have

$$g = g \circ i_Y = g \circ (f \circ f^{-1}) = (g \circ f) \circ f^{-1} = i_X \circ f^{-1} = f^{-1}.$$

(ii) $g = i_X \circ g = (f^{-1} \circ f) \circ g = f^{-1} \circ (f \circ g) = f^{-1} \circ i_Y = f^{-1}.$ |

In theorem 1.31 (ii) the bijectiveness of f is not a consequence of surjectiveness, as it is in part (i). For instance let $X = \{a, b\}$ $(a \neq b)$, $Y = \{c\}$ and let f, g be defined by $f(a) = f(b) = c$, $g(c) = a$. Then f is surjective and $f \circ g = i_Y$, but f is not bijective.

If the functions $f: X \to Y$, $g: Y \to W$ are bijective, then $g \circ f: X \to W$ is bijective. It follows from theorem 1.31, or it may be proved directly, that $(g \circ f)^{-1} = f^{-1} \circ g^{-1}.$

Restriction and extension of functions. Suppose that $X_1 \supset X_2$ and that $f: X_1 \to Y$, $g: X_2 \to Y$ are functions such that $f(x) = g(x)$ for all $x \in X_2$. We then say that g is *the restriction* of f to X_2 and that f is *an extension* of g to X_1. For example let $f_1: R^1 \to R^1$ and $f_2: R^1 \to R^1$ be given by $\quad f_1(x) = x, \quad f_2(x) = |x| \quad (x \in R^1).$

The function $g: [0, \infty) \to R^1$ defined by

$$g(x) = x \quad (x \geqslant 0)$$

is the restriction to $[0, \infty)$ of both f_1 and f_2; and f_1, f_2 are both extensions of g to R^1.

Exercises 1(*b*)

1. Show that $\{\{a\}, \{a, b\}\} = \{\{c\}, \{c, d\}\}$ if and only if $a = c$ and $b = d$.

2. Prove that

    (i) $(A \cup B) \times C = (A \times C) \cup (B \times C)$,

    (ii) $(A \cap B) \times C = (A \times C) \cap (B \times C)$,

    (iii) $(A - B) \times C = (A \times C) - (B \times C)$.

3. Prove that $A \times B = B \times A$ if and only if $A = B$. Is $(A \times A) \times A$ necessarily the same set as $A \times (A \times A)$?

4. Show that $\{a\} \times \{a\} = \{\{\{a\}\}\}$.

5. Let $\mathscr{C}$ be a collection of subsets of a set X. Prove that, for any function $f\colon X \to Y$,

$$\text{(i) } f(\bigcup_{A \in \mathscr{C}} A) = \bigcup_{A \in \mathscr{C}} f(A), \quad \text{(ii) } f(\bigcap_{A \in \mathscr{C}} A) \subseteq \bigcap_{A \in \mathscr{C}} f(A).$$

Show that, if f is injective, then identity holds in (ii); and that f is injective if

$$f(A \cap B) = f(A) \cap f(B)$$

for all subsets A, B of X.

6. Show that, if $f\colon X \to Y$ is any function, then

$$f(A - B) \supset f(A) - f(B)$$

for all subsets A, B of X; and that identity holds for all A, B if and only if f is injective.

7. Let $\mathscr{C}$ be a collection of subsets of a set Y. Show that, for any function $f\colon X \to Y$,

$$\text{(i) } f^{-1}(\bigcup_{B \in \mathscr{C}} B) = \bigcup_{B \in \mathscr{C}} f^{-1}(B), \quad \text{(ii) } f^{-1}(\bigcap_{B \in \mathscr{C}} B) = \bigcap_{B \in \mathscr{C}} f^{-1}(B).$$

8. Show that, if $f\colon X \to Y$ is any function, then

$$f^{-1}(C - D) = f^{-1}(C) - f^{-1}(D)$$

for all subsets C, D of Y.

9. Show that, if $f\colon X \to Y$ is any function, then

$$\text{(i) } f^{-1}(f(A)) \supset A, \quad \text{(ii) } f(f^{-1}(B)) \subset B$$

for all subsets A of X and all subsets B of Y. Prove also that in (i) identity holds for all A if and only if f is injective; and in (ii) identity holds for all B if and only if f is surjective.

10. The functions $f_1\colon X_1 \to X_2$, $f_2\colon X_2 \to X_3$, $f_3\colon X_3 \to X_4$ are such that the compositions $f_2 \circ f_1\colon X_1 \to X_3$ and $f_3 \circ f_2\colon X_2 \to X_4$ are both bijective. Show that f_1, f_2, f_3 are all bijective.

11. Let R be a relation from X to X (i.e. a subset of $X \times X$) and write xRy when $(x, y) \in R$. The relation is said to be
   (i) *reflexive* if $x \in X \Rightarrow xRx$;
   (ii) *symmetric* if $xRy \Rightarrow yRx$;
   (iii) *transitive* if xRy and $yRz \Rightarrow xRz$.
   Criticize the following 'argument'. The relation R is known to be symmetric and transitive. Writing x for z in (iii) and using (ii) we obtain $xRy \Rightarrow xRx$. Hence R is reflexive.

1.4. Similarity of sets

If there is a one-to-one correspondence between two sets A, B, i.e. if there exists a bijection $f\colon A \to B$, then the two sets are said to be *similar* and we write $\qquad A \sim B.$

It is easy to see that similarity has the following three properties:

(i) $A \sim A$ (reflexivity);

(ii) if $A \sim B$, then $B \sim A$ (symmetry);

(iii) if $A \sim B$ and $B \sim C$, then $A \sim C$ (transitivity).

If we confine ourselves to a given collection $\mathscr{C}$ of sets, then similarity is a relation from $\mathscr{C}$ to $\mathscr{C}$ and, in view of (i)–(iii) above, it is an equivalence relation. The reader has probably met such relations in other branches of mathematics, particularly algebra.

Finite sets are similar if and only if they have the same number of elements. This implies, in particular, that a finite set cannot be similar to a proper subset of itself. We shall see that infinite sets always contain proper subsets to which they are similar (theorem 1.42, corollary). A simple example is the following. Let A be the set of all integers and let B be the set of all even integers. Then $A \sim B$, since the function given by $y = 2x$ $(x \in A, y \in B)$ establishes the required one-to-one correspondence.

A set which is similar to the set $\{1, 2, ...\}$ of positive integers is said to be *countably* (or *enumerably*) *infinite*. Such a set is therefore one that can be arranged as a sequence. A set is called *countable* if it is either finite or countably infinite.

We now show that there are infinite sets which are not countable.

Theorem 1.41. *The set of real numbers in the interval* $(0, 1)$ *is not countable.*

Proof. Suppose that the real numbers in the interval $(0, 1)$ do form a countable set. This set is certainly infinite and can therefore be arranged as a sequence

$$a_1, a_2, a_3, ...$$

Now represent each number a_n as an infinite decimal

$$a_n = 0.\alpha_{n1}\alpha_{n2}\alpha_{n3} ...$$

in which recurring 9's are not used. Such a representation is unique. Define

$$\beta_n = \begin{cases} 1 & \text{when } \alpha_{nn} \neq 1, \\ 2 & \text{when } \alpha_{nn} = 1. \end{cases}$$

Then

$$b = 0.\beta_1\beta_2\beta_3 ...$$

is a real number between 0 and 1 which differs from a_n in the nth decimal place. Hence b is not one of the numbers $a_1, a_2, a_3, ...$ and this contradicts the original assumption that all real numbers in the interval $(0, 1)$ appear in the sequence (a_n). |

Countably infinite sets are the 'smallest' infinite sets in the sense of the theorem below.

Theorem 1.42. *Every infinite set contains a countably infinite subset.*

Proof. Let A be an infinite set. Take some element of A and call it a_1. The set $A - \{a_1\}$ is not empty; denote one of its elements by a_2. This process may be continued indefinitely: at the nth stage the set

$$A - \{a_1, ..., a_{n-1}\}$$

cannot be empty since A is infinite. The set $\{a_1, a_2, ...\}$ is a countably infinite subset of A. |

Corollary. *Every infinite set contains a proper subset to which it is similar.*

Proof. Let A be an infinite set and let $S = \{a_1, a_2, ...\}$ be a countably infinite subset of A. The function $f: A \to A - \{a_1\}$ defined by

$$f(x) = \begin{cases} a_{n+1} & \text{if } x = a_n \quad (n = 1, 2, ...), \\ x & \text{if } x \in A - S \end{cases}$$

is a bijection. Thus $A \sim A - \{a_1\}$. |

The next theorem is frequently used.

Theorem 1.43. *Every subset of a countable set is countable.*

Proof. Let A be a countable set and let B be a subset of A. When B is finite there is nothing to prove. Assume therefore that B is infinite, so that A is infinite and, being countable, can be arranged as a sequence (a_n). Since every set of positive integers has a least member, there is an element a_{n_1} of B with least suffix n_1. Next let a_{n_2} be the element of $B - \{a_{n_1}\}$ with least suffix n_2. The procedure may be repeated indefinitely since B is not finite. The sequence

$$a_{n_1}, a_{n_2}, a_{n_3}, ... \tag{1.41}$$

consists of elements of B and, in fact, exhausts B. For take any element of B. It is of the form a_k, since it is a member of A, and it must therefore occur among the first k elements of the sequence (1.41). Hence B is countable. |

Theorem 1.44. *The set P^2 of all ordered pairs (p, q) of positive integers is countable.*

Proof. The elements of P^2, all of which appear in the array

$$(1, 1), (1, 2), (1, 3), ...,$$
$$(2, 1), (2, 2), (2, 3), ...,$$
$$(3, 1), (3, 2), (3, 3), ...,$$
$$......................................,$$

may be arranged as a sequence in several simple ways. Perhaps the simplest is the enumeration by diagonals:

$$(1, 1); \ (2, 1), (1, 2); \ (3, 1), (2, 2), (1, 3); \ ...;$$
$$(n, 1), (n-1, 2), ... (1, n); \ .... \ |$$

Exercise. Show that the set P^n ($n \geqslant 2$) of all ordered sets $(p_1, ..., p_n)$ of n positive integers is countable.

Theorem 1.45. *The union of a countable collection of countable sets is countable.*

Proof. Let $\{A_1, A_2, ...\}$ be a finite or countably infinite collection of countable sets. If

$$B_1 = A_1, \qquad B_n = A_n - \bigcup_{k=1}^{n-1} A_k \quad (n > 1),$$

then the B_n are mutually disjoint and

$$\bigcup_n A_n = \bigcup_n B_n = C,$$

say. (The non-committal notation $\bigcup_n$ is used since the number of sets may be finite or infinite.) By theorem 1.43 each of the sets B_n is countable. Therefore, if B_n is not empty, its elements may be arranged in a possibly terminating sequence

$$b_{n1}, b_{n2}, ...$$

Now let S be the subset of P^2 which contains a pair (p, q) if and only if B_p is either infinite or has at least q members. By theorems 1.43 and 1.44, S is countable. Since there is an obvious one-to-one correspondence between S and C, it follows that C is countable. $|$

Theorem 1.46.

   (i) *The set of all rational numbers is countable.*
   (ii) *The set of all irrational numbers is not countable.*

Proof. A given positive rational number is uniquely representable in the form p/q, where p, q are coprime positive integers. With this number we now associate the ordered pair (p, q). Thus the set of positive rational numbers is similar to a subset of P^2 and is therefore countable. Part (i) now follows easily by use of theorem 1.45.

Since the set of real numbers in $(0, 1)$ is not countable, by theorem 1.43 the set R^1 of all real numbers cannot be countable. If now the set of irrational numbers were countable, (i) and theorem 1.45 would show R^1 to be countable. |

Instead of considering all rational and irrational numbers we may restrict ourselves to those in any given interval. It follows immediately from theorem 1.43 that the analogue of (i) holds. The analogue of (ii) can then be proved as before by using exercise 2 below.

Exercises 1 (c)

1. Let C be a countable set. Show that, if A is any infinite set, then

$$A \cup C \sim A;$$

and that, if B is an uncountable set, then

$$B - C \sim B.$$

2. (i) Show that all finite open intervals in R^1 are similar.
(ii) Show that all open intervals in R^1 are similar.
(iii) Show that all intervals in R^1 are similar.

3. Let $\mathscr{C}$ be a countable collection of disjoint sets such that, for all $A \in \mathscr{C}$, $A \sim R^1$. Show that

$$\bigcup_{A \in \mathscr{C}} A \sim R^1.$$

4. Show that, if the sets $A_1, ..., A_n$ are countable, then so is $A_1 \times ... \times A_n$. (In particular, if Q is the set of rational numbers, then Q^n is countable.)

5. Let $\mathscr{P}$ be the set of all sequences of 0's and 1's. Using the binary representation of real numbers prove that $\mathscr{P}$ is similar to the interval $(0, 1)$.
Show also that $\mathscr{P}^2 \sim \mathscr{P}$ and deduce that, if the sets $A_1, ..., A_n$ are similar to R^1, then $A_1 \times ... \times A_n \sim R^1$. (In particular, $R^n \sim R^1$.)

6. Prove that the set of all polynomials

$$a_0 x^n + a_1 x^{n-1} + ... + a_{n-1} x + a_n$$

with integral coefficients is countable. Deduce that the set of algebraic numbers is countable. (An algebraic number is a number which is a root of an algebraic equation with integral coefficients.)

7. Let R_i ($0 \leqslant i \leqslant 9$) be the set of those real numbers in $(0, 1)$ which do not use the numeral i in their decimal representation. Prove that $R_i \sim (0, 1)$. (Suppose that recurring 9's are not used.)

8. A set E of real numbers is such that every series

$$\sum_{n=1}^{\infty} x_n,$$

whose terms are distinct elements of E, converges. Show that E is countable.

9. Given a set A, denote by $\mathscr{S}_A$ the set of subsets of A (including $\varnothing$ and A). Show that
 (i) if A has n elements, then $\mathscr{S}_A$ has 2^n elements;
 (ii) if A is countably infinite, then $\mathscr{S}_A \sim R^1$;
 (iii) if A is any set, then $A \not\sim \mathscr{S}_A$ (though clearly A is similar to a subset of $\mathscr{S}_A$).

10. Show that any collection of disjoint open intervals of R^1 is countable. Deduce that, if J is any interval in R^1 and the function $f: J \to R^1$ is monotonic, then the set of points of discontinuity of f is countable.
 (If $c \in J$ and c is not an end point, then

$$f(c+) = \lim_{x \to c+} f(x), \quad f(c-) = \lim_{x \to c-} f(x)$$

exist; see C1, 32.)

NOTES ON CHAPTER 1

§ **1.1.** *Axiomatic set theory.* The intuitive notion of a set, though perfectly adequate for everyday use by analysts, is, in fact, self-contradictory. To demonstrate this we describe the famous Russell paradox. A set may or may not be a member of itself: most sets are not and we call these *normal*; an example of a set which *is* a member of itself is the set consisting of all infinite sets (for clearly there are infinitely many of these). Consider the set N of all normal sets. Is N normal or not? If N is normal, then, by definition of normality, $N \notin N$; but, by definition of N, a set which does not belong to N is not normal and we have a contradiction. The supposition that N is not normal is equally untenable. The mathematician's way out of this dilemma is to construct an axiomatic system designed to mimic the desirable features of the intuitive approach and to avoid its pitfalls. The axiomatic theory of sets is described in *Theory of Sets and Transfinite Numbers* by B. Rotman and G. T. Kneebone. A more informal attitude is adopted by P. R. Halmos in *Naive Set Theory*.
 The theory of sets is the starting point of the monumental *Éléments de Mathématique*, which is designed to present the whole of pure mathematics in strictly logical order. This work, by a group of French mathematicians writing under the collective name of N. Bourbaki, has been appearing in sections since 1939 and is not yet (in 1969) complete.
 The real number system. The set R^1 of real numbers, together with the operations of addition and multiplication and an ordering relation, is an ordered field: it satisfies the following six algebraic and three ordinal axioms.
 A1. To every pair of real numbers a, b correspond a real number $a+b$, the *sum* of a and b, and a real number ab, the *product* of a and b.
 A2. $a+b = b+a$; $ab = ba$.
 A3. $(a+b)+c = a+(b+c)$; $(ab)c = a(bc)$.
 A4. There are real numbers 0, 1 such that

$$0+a = a \quad \text{and} \quad 1a = a$$

for all real numbers a.

A5. For every real number a, there is a real number x such that $a+x = 0$.

For every real number a, other than 0, there is a real number y such that $ay = 1$.

A6. $(a+b)c = ac+bc$.

O1. For every two real numbers a, b, one and only one of

$$a > b, \quad a = b, \quad b > a$$

is true.

O2. If $a > b$ and $b > c$, then $a > c$.

O3. If $a > b$, then $a+c > b+c$; if also $c > 0$, then $ac > bc$.

What distinguishes the real number system from other ordered fields is *Dedekind's axiom*:

If L, R are two non-empty sets of real numbers such that $L \cup R = R^1$ and every member of L is less than every member of R, then either L has a largest member or R has a least member.

The construction of such a richly endowed system is, not surprisingly, a major undertaking. The first step is to obtain the natural numbers $0, 1, 2, \ldots$; and the most satisfying way of doing this is to use only the machinery of set theory (see, for instance, Halmos's *Naive Set Theory*). Another possibility is to begin with the set of natural numbers, but to assume no more than a few simple properties (Peano's axioms). This approach is adopted by L. W. Cohen and G. Ehrlich (*The Real Number System*), E. Landau (*Foundations of Analysis*) and H. A. Thurston (*The Number System*). There are now standard algebraic methods for proceeding first to the integers $(0, \pm 1, \pm 2, \ldots)$ and then to the rational numbers. At this stage we have arrived at an ordered field which does not, however, satisfy Dedekind's axiom. There are two principal constructions for the final step to the real numbers. The method of cuts is mentioned in C1 and is given in detail by Landau. Cauchy (or fundamental) sequences are used by Cohen and Ehrlich and by Thurston. This process is also described in the notes at the end of chapter 3, since it is closely related to some of the contents of that chapter.

2

METRIC SPACES

2.1. Metrics

Classical analysis deals with sets of points in and functions defined on Euclidean space or the complex plane, while in more recent developments prominent parts are played also by other types of spaces. These spaces lead to theories with many similar features. The reason is that the spaces share the same underlying structure. It is therefore both illuminating and also economical of effort to develop many aspects of analysis in a general setting which includes as special cases the various spaces of particular importance. A fundamental property which these spaces have in common is that in all of them there is a distance between any two points. We now lay down the conditions which such a distance function or metric must satisfy.

Definition. Suppose that X is a (non-empty) set and that ρ is a real valued function on $X \times X$ with the following three properties.

M 1. $\rho(x, y) \geqslant 0$ *for all* $x, y \in X$ *and* $\rho(x, y) = 0$ *if and only if* $x = y$;

M 2. $\rho(x, y) = \rho(y, x)$ *for all* $x, y \in X$;

M 3. $\rho(x, z) \leqslant \rho(x, y) + \rho(y, z)$ *for all* $x, y, z \in X$ *(the triangle inequality).*

The function ρ is called a metric *or* distance *on X; and X, taken together with the metric ρ, is called a* metric space *which we denote by (X, ρ).*

The notation (X, ρ) emphasizes the fact that the set X and the metric ρ have equal shares in the construction of the metric space. A given set may give rise to many different metric spaces by having different metrics associated with it. Nevertheless, when it is clear which metric is being used we may, for brevity, write X rather than (X, ρ) for the metric space.

Illustrations

(i) Any set X can be equipped with a metric;

$$\rho(x, y) = \begin{cases} 0 & \text{if } x = y, \\ 1 & \text{if } x \neq y \end{cases}$$

is such a metric, but not a very interesting one. (This metric is called the *discrete metric*.)

(ii) On R^1 the *usual metric* is given by

$$\rho(x, y) = |x-y|.$$

The *usual metric* on R^n is defined as follows. If

$$x = (x_1, \ldots, x_n), \quad y = (y_1, \ldots, y_n) \in R^n,$$

$$\rho(x, y) = \sqrt{\{(x_1-y_1)^2 + \ldots + (x_n-y_n)^2\}}. \tag{2.11}$$

The case $n = 1$ gives the earlier definition of the usual metric on R^1.

Plainly ρ satisfies M1 and M2. Putting $a_i = x_i-y_i$, $b_i = y_i-z_i$ we see that M3 is equivalent to

$$\left(\sum_{i=1}^{n} (a_i+b_i)^2\right)^{\frac{1}{2}} \leqslant \left(\sum_{i=1}^{n} a_i^2\right)^{\frac{1}{2}} + \left(\sum_{i=1}^{n} b_i^2\right)^{\frac{1}{2}}. \tag{2.12}$$

By squaring both sides of (2.12) we obtain yet another equivalent form, namely

$$\sum_{i=1}^{n} a_i b_i \leqslant \left(\sum_{i=1}^{n} a_i^2\right)^{\frac{1}{2}} \left(\sum_{i=1}^{n} b_i^2\right)^{\frac{1}{2}}. \tag{2.13}$$

This now follows from the fact that, for all real ξ,

$$\sum_{i=1}^{n} (a_i\xi+b_i)^2 = A\xi^2 + 2H\xi + B \geqslant 0,$$

so that $$H^2 \leqslant AB,$$

where $$A = \sum_{i=1}^{n} a_i^2, \quad H = \sum_{i=1}^{n} a_i b_i, \quad B = \sum_{i=1}^{n} b_i^2.$$

Therefore ρ, defined by (2.11) is a metric.

The metric space (R^n, ρ), where ρ is the usual metric, is called *n-dimensional Euclidean space*.

(iii) On Z, the complex plane, the usual metric is of the same form as that on R^2. If $z = x+iy$ and $z' = x'+iy'$ (x, y, x', y' real),

$$\rho(z, z') = |z-z'| = \sqrt{\{(x-x')^2 + (y-y')^2\}}.$$

In fact, when equipped with their usual metrics, Z and R^2 are essentially the same metric spaces in which only the notations differ.

(iv) Let $B[a, b]$ be the set of bounded real functions on the interval $[a, b]$ and, for $f, g \in B[a, b]$, let
$$\rho(f, g) = \sup_{a \leqslant x \leqslant b} |f(x)-g(x)|.$$

Clearly ρ has the properties M1 and M2. M3 follows from the inequalities

$$\sup |\phi(x)+\psi(x)| \leqslant \sup (|\phi(x)| + |\psi(x)|) \leqslant \sup |\phi(x)| + \sup |\psi(x)|.$$

(v) Let S be the space of sequences $x = (x_1, x_2, \ldots)$ of complex numbers such that Σx_n converges absolutely. For $x, y \in S$ put

$$\rho(x, y) = \sum_{n=1}^{\infty} |x_n-y_n|.$$

It is easy to show that ρ is a metric on S.

Let (X, ρ) be a metric space and let Y be a subset of X. If σ is the restriction of ρ to $Y \times Y$, i.e. $\sigma(x, y) = \rho(x, y)$ for all $x, y \in Y$, then

σ is a metric on Y; it is said to be the metric *induced* by (X, ρ) on Y. The metric space (Y, σ) is called a *(metric) subspace* of (X, ρ). For instance let R^3 have its usual metric ρ given by

$$\rho(x, y) = \sqrt{\{(x_1 - y_1)^2 + (x_2 - y_2)^2 + (x_3 - y_3)^2\}}.$$

If Y is the set of points $x \in R^3$ with $x_3 = c$, the metric σ induced on Y is given by

$$\sigma(x, y) = \sqrt{\{(x_1 - y_1)^2 + (x_2 - y_2)^2\}}.$$

In future, unless there is a statement to the contrary, the metrics on R^n and Z will be taken to be the usual metrics; and subsets of R^n and Z will be taken to have the metrics induced by the usual ones.

Sequences of points. The following definition extends the notion of convergence of sequences of real numbers to sequences of points in a metric space.

Definition. *The sequence* (x_n) *of points in the metric space* (X, ρ) *is said to* converge *to the point* $x(\in X)$ *if*

$$\rho(x_n, x) \to 0 \quad as \quad n \to \infty. \tag{2.14}$$

We write, as in classical analysis, $x_n \to x$ *as* $n \to \infty$ *or* $\lim_{n \to \infty} x_n = x$.

Since $\rho(x, y)$ is a real number, we are able to use the concept of convergence in R^1. We could replace (2.14) by the phrase 'given $\epsilon > 0$, there is an n_0 such that $\rho(x_n, x) < \epsilon$ for $n > n_0$'. This form of the definition reduces to the familiar definition of a convergent sequence of real numbers when (X, ρ) is R^1 with its usual metric.

Exercise. Prove that a sequence cannot converge to two distinct limits.

Illustrations

(i) In R^2 let

$$x_n = \left(\frac{n}{2n+1}, \frac{2n^2}{n^2-2}\right).$$

Then, as $n \to \infty$,

$$x_n \to (\tfrac{1}{2}, 2),$$

since

$$\rho(x_n, x) = \sqrt{\left\{\left(\frac{1}{4n+2}\right)^2 + \left(\frac{4}{n^2-2}\right)^2\right\}} \to 0.$$

(ii) In $B[1, 2]$ with $\rho(f, g) = \sup_{1 \leqslant x \leqslant 2} |f(x) - g(x)|$, let f_n be given by

$$f_n(x) = (1 + x^n)^{1/n} \quad (1 \leqslant x \leqslant 2).$$

Then

$$f_n \to f,$$

where

$$f(x) = x \quad (1 \leqslant x \leqslant 2).$$

For, when $1 \leqslant x \leqslant 2$,

$$0 \leqslant f_n(x) - f(x) = x(x^{-n} + 1)^{1/n} - x < x(2^{1/n} - 1) \leqslant 2(2^{1/n} - 1)$$

and so

$$\rho(f_n, f) \leqslant 2(2^{1/n} - 1) \to 0 \quad as \quad n \to \infty.$$

When the metrics ρ, σ on a set X are such that a sequence converges in (X, ρ) if and only if it converges in (X, σ), then ρ and σ are said to be *equivalent metrics*. We shall see, as the theory develops, that a change to an equivalent metric preserves many of the properties of a metric space.

A sufficient, though not necessary condition for the metrics ρ, σ to be equivalent is the existence of strictly positive constants λ, μ such that
$$\lambda\rho(x, y) \leqslant \sigma(x, y) \leqslant \mu\rho(x, y)$$

for all $x, y \in X$. (See exercise 2(a), 7.)

Exercises 2(a)

1. Is the function ρ given by $\rho(x, y) = |x^2 - y^2|$ a metric on (a) $(-\infty, \infty)$, (b) $[0, \infty)$?

2. Let C be the set of bounded and continuous real functions on R^1 and let ρ be defined by
$$\rho(\phi, \psi) = \sup_{\substack{-\infty < x < \infty \\ 0 < h \leqslant 1}} \left| \int_x^{x+h} \{\phi(t) - \psi(t)\}\, dt \right|.$$

Show that ρ is a metric on C.

3. Let X be a non-empty set and let ρ be a real-valued function on $X \times X$ such that

　(i) $\rho(x, y) = 0$ if and only if $x = y$; and
　(ii) $\rho(x, y) \leqslant \rho(x, z) + \rho(y, z)$ for all $x, y, z \in X$.

Prove that ρ is a metric on X.

4. In R^n let $x_k = (\xi_{1k}, \ldots, \xi_{nk})$. Show that
$$x_k \to x = (\xi_1, \ldots, \xi_n)$$
as $k \to \infty$ if and only if, for $1 \leqslant i \leqslant n$, $\xi_{ik} \to \xi_i$.

5. In a metric space (X, ρ), $x_n \to x$ and $y_n \to y$ as $n \to \infty$. Prove that
$$\rho(x_n, y_n) \to \rho(x, y).$$

6. Show that, if ρ is a metric on X, then so is σ given by
$$\sigma(x, y) = \frac{\rho(x, y)}{1 + \rho(x, y)}$$
and that ρ, σ are equivalent metrics.

7. Prove that the metrics ρ, σ on X are equivalent if there are constants λ, $\mu > 0$ such that
$$\lambda\rho(x, y) \leqslant \sigma(x, y) \leqslant \mu\rho(x, y)$$

for all $x, y \in X$. Give an example to show that the converse is false.

8. On R^2, σ and τ are defined by

$$\sigma(x, y) = \max\left(|x_1 - y_1|, |x_2 - y_2|\right), \quad \tau(x, y) = |x_1 - y_1| + |x_2 - y_2|.$$

Show that σ and τ are metrics which are equivalent to one another and to the usual metric ρ.

Compare the sets of points x in R^2 given by $\rho(x, a) \leqslant 1$, $\sigma(x, a) \leqslant 1$, $\tau(x, a) \leqslant 1$ respectively.

9. Denote by σ the discrete metric on R^1 (p. 17, illustration (i)). Prove that σ and the usual metric ρ on R^1 are not equivalent.

10. Let C be the set of real functions continuous on the interval $[0, 1]$ and define ρ, σ, τ by

$$\rho(f, g) = \sup_{0 \leqslant x \leqslant 1} |f(x) - g(x)|,$$

$$\sigma(f, g) = \left(\int_0^1 |f(x) - g(x)|^2 \, dx\right)^{\frac{1}{2}},$$

$$\tau(f, g) = \int_0^1 |f(x) - g(x)| \, dx.$$

(A function continuous on a closed interval is bounded and integrable.) Prove that σ and τ are metrics on C. Show also that, for all $f, g \in C$,

$$\rho(f, g) \geqslant \sigma(f, g) \geqslant \tau(f, g).$$

For $n = 1, 2, \ldots$, the functions $f_n, g_n \in C$ are defined by

$$f_n(x) = \begin{cases} 1 - nx & (0 \leqslant x \leqslant n^{-1}), \\ 0 & (n^{-1} < x \leqslant 1); \end{cases} \quad g_n(x) = \begin{cases} n^{\frac{1}{2}}(1 - nx) & (0 \leqslant x \leqslant n^{-1}), \\ 0 & (n^{-1} < x \leqslant 1). \end{cases}$$

Prove that the sequence (f_n) converges in (C, σ) and in (C, τ), but not in (C, ρ); and that (g_n) converges in (C, τ), but not in (C, ρ) or in (C, σ).

11. Let $(X_1, \rho_1), \ldots, (X_n, \rho_n)$ be any metric spaces. Verify that metrics on $X_1 \times \ldots \times X_n$ are given by

$$\sigma_1(x, y) = \max\{\rho_1(x_1, y_1), \ldots, \rho_n(x_n, y_n)\},$$

$$\sigma_2(x, y) = \sqrt{\{\rho_1^2(x_1, y_1) + \ldots + \rho_n^2(x_n, y_n)\}},$$

$$\sigma_3(x, y) = \rho_1(x_1, y_1) + \ldots + \rho_n(x_n, y_n),$$

where $x = (x_1, \ldots, x_n)$, $y = (y_1, \ldots, y_n)$ and $x_i, y_i \in X_i$ $(i = 1, \ldots, n)$. Show also that $\sigma_1, \sigma_2, \sigma_3$ are all equivalent.

2.2. Norms

The theory of vector spaces is partly algebraic and partly analytical. We need only the most basic algebraic ideas and these are, no doubt, known to the reader. But, largely to fix notation and terminology, we formally define a vector space.

Definition. Let V be a non-empty set and suppose that (i) *to any two elements x, y of V there corresponds an element in V, called their sum,* *which is denoted by* $x+y$; *and* (ii) *to any real number* α *and any* x *in* V *there corresponds an element in* V, *called the* scalar product *of* α *and* x, *which is denoted by* αx. (Thus addition is a function on $V \times V$ into V and scalar multiplication is a function on $R^1 \times V$ into V.) *Suppose also that the following conditions are satisfied.*

(1) $x+y = y+x$;

(2) $x+(y+z) = (x+y)+z$;

(3) *there is an element* $\theta \in V$ (*the zero element*) *such that* $x+\theta = x$ *for all* $x \in V$;

(4) *to every element* $x \in V$ *there corresponds an element* $-x \in V$ *such that* $x+(-x) = \theta$;

(5) $\alpha(\beta x) = (\alpha\beta)x$;

(6) $1x = x$;

(7) $(\alpha+\beta)x = \alpha x+\beta x$;

(8) $\alpha(x+y) = \alpha x+\alpha y$.

Then V (*together with the operations of addition and scalar multiplication defined in* V) *is called a* vector (*or* linear) *space over* R^1, *or a* real vector space.

If the scalar product is formed with complex numbers α *and the same properties* (1)–(8) *hold, then* V *is said to be a* vector space over Z, *or a* complex vector space.

The simplest example of a real vector space is R^n with addition and scalar multiplication defined by

$$(x_1, \ldots, x_n)+(y_1, \ldots, y_n) = (x_1+y_1, \ldots, x_n+y_n),$$

$$\alpha(x_1, \ldots, x_n) = (\alpha x_1, \ldots, \alpha x_n).$$

A more sophisticated example is $B[a, b]$; here $f+g$ and αf are defined by

$$(f+g)(x) = f(x)+g(x), \quad (\alpha f)(x) = \alpha f(x) \qquad (a \leqslant x \leqslant b).$$

We shall assume the elementary algebraic properties of vector spaces. Some of these are given in exercise 2(*b*), 1.

Definition. *Let* V *be a* vector space. *The real valued function* $\|.\|$ *on* V *is called a* norm *on* V, *and* V *is called a* normed vector space, *if*

N 1. $\|x\| \geqslant 0$ *for all* $x \in V$ *and* $\|x\| = 0$ *if and only if* $x = \theta$;

N 2. $\|x+y\| \leqslant \|x\|+\|y\|$ *for all* $x, y \in V$;

N 3. $\|\alpha x\| = |\alpha|\|x\|$ *for every number* α *and all* $x \in V$.

It is easy to see that, in a normed vector space, a metric ρ may be defined by

$$\rho(x, y) = \|x-y\|.$$

In fact, the metrics in illustrations (ii)–(v) of §2.1 all arise from norms. For instance in R^n the *usual norm* is given by

$$\|x\| = \sqrt{(x_1^2 + \ldots + x_n^2)}$$

and in $B[a, b]$ $$\|f\| = \sup_{a \leqslant x \leqslant b} |f(x)|.$$

However, a metric on a vector space need not be derivable from a norm (see exercise 2(b), 2).

More specialized than normed vector spaces are *inner product spaces*.

Definition. *Let V be a complex normed vector space and suppose that, with every ordered pair (x, y) of elements in V there is associated a complex number, denoted by $x.y$, and that the following conditions are satisfied.*

I1. $x.y = \overline{y.x}$ *for all $x, y \in V$;*

I2. $(\alpha x + \beta y).z = \alpha(x.z) + \beta(y.z)$ *for all numbers α, β and all $x, y, z \in V$;*

I3. $x.x = \|x\|^2$ *for all $x \in V$.*

Then $x.y$ is called the inner product *of x and y and V is called a* complex inner product space.

If V is a real normed vector space, $x.y$ is real and I1–I3 *hold, then V is called a* real inner product space. *Note that, in this case,* I1 *reduces to $x.y = y.x$.*

It is shown in exercise 2(b), 4 that no more than one inner product can be associated with a given norm.

Once again R^n provides the most obvious illustration. The inner product for the usual norm is defined by

$$x.y = x_1 y_1 + \ldots + x_n y_n.$$

The inequality (2.13) can now be written as

$$x.y \leqslant \|x\| \, \|y\|$$

and is a particular case of the following result.

Theorem 2.21. *(Cauchy's inequality.) If V is a real or complex inner product space, then, for all $x, y \in V$,*

$$|x.y| \leqslant \|x\| \, \|y\|.$$

Equality holds if and only if one of x, y is a scalar multiple of the other.

Proof. Since $\theta . y = 0$ (see exercise 2(b), 6), we may assume that $x \neq \theta$. Take

$$\beta = -\frac{y.x}{\|x\|^2}.$$

Then
$$0 \leqslant \|\beta x + y\|^2 = (\beta x + y).(\beta x + y)$$
$$= |\beta|^2 \|x\|^2 + \beta(x.y) + \bar{\beta}(y.x) + \|y\|^2$$
$$= \frac{|x.y|^2}{\|x\|^4} \|x\|^2 - 2\frac{|x.y|^2}{\|x\|^2} + \|y\|^2.$$

Moreover $\|\beta x + y\| = 0$, i.e. $\beta x + y = \theta$, if and only if $y = \alpha x$ for some number α. |

Exercises 2(b)

1. Show that the following identities hold in a vector space:
   (i) $0x = \theta$;   (ii) $(-1)x = -x$;   (iii) $\alpha\theta = \theta$;
   (iv) if $x - y$ denotes $x + (-y)$, then

$$-(x+y) = -x-y, \quad -(x-y) = y-x.$$

2. Prove that the discrete metric σ on R^1 (see illustration (i) of § 2.1) cannot be derived from a norm.

3. Two norms on a vector space V are *equivalent* if they give rise to equivalent metrics. Prove that the norms $\|.\|_1$ and $\|.\|_2$ are equivalent if and only if there exist strictly positive constants λ, μ such that

$$\lambda\|x\|_1 \leqslant \|x\|_2 \leqslant \mu\|x\|_1$$

for all $x \in V$. (Contrast exercise 2(a), 7.)

4. Show that if, in a normed vector space, an inner product exists, then it is unique.

5. The space T consists of the sequences $x = (x_1, x_2, \ldots)$ of complex numbers such that $\Sigma|x_n|^2$ converges. Verify that ρ given by

$$\rho(x, y) = \left(\sum_{n=1}^{\infty} |x_n - y_n|^2\right)^{\frac{1}{2}} \quad (x, y \in T)$$

is a metric on T; that the metric arises from a norm; and that to this norm there corresponds the inner product

$$x.y = \sum_{n=1}^{\infty} x_n \bar{y}_n.$$

6. Show that in an inner product space $x.\theta = \theta.x = 0$.

7. Prove that, if V is an inner product space, then, for all $x, y \in V$,

$$\|x+y\|^2 + \|x-y\|^2 = 2(\|x\|^2 + \|y\|^2).$$

Hence give an example of a normed vector space which is not an inner product space.

8. The sequences (x_n), (y_n) in the normed vector space V converge to x, y respectively, and the numerical sequences (α_n), (β_n) converge to α, β respectively. Show that

$$\alpha_n x_n + \beta_n y_n \to \alpha x + \beta y.$$

Prove also that, if V possesses an inner product, then

$$x_n . y_n \to x . y.$$

9. *Alternative definition of inner product space.* Let V be a complex (real) vector space and suppose that, with every ordered pair (x, y) of elements in V, there is associated a complex (real) number $x . y$ such that I1, I2 are satisfied and
   I3′. for all $x \in V$, $x . x \geqslant 0$ and $x . x = 0$ if and only if $x = \theta$.
Show that $\|x\| = \sqrt{(x . x)}$ is a norm on V.

2.3. Open and closed sets

In this section we consider subsets of a fixed metric space (X, ρ). We shall use the symbol X for the metric space as well as the set without reiterating that ρ is specified.

Let $a \in X$ and let r be a positive real number. The set of points x such that $\rho(a, x) < r$ is called the *open ball* with *centre* a and *radius* r; we shall denote it by $B(a; r)$. In R^1, $B(a; r)$ is the open interval $(a-r, a+r)$. In R^2 and in Z an open ball is usually called an *open disc*.

Definition. *Given a subset E of X (which may be X itself), the point $c \in X$ is said to be a* limit point *of E if, for every $\epsilon > 0$, there is an $x \in E$ such that* $0 < \rho(c, x) < \epsilon,$

i.e. $B(c; \epsilon) - \{c\}$ contains a point of E.

A limit point c of E can also be defined by either of the following two conditions:

   (i) Every open ball $B(c; \epsilon)$ contains infinitely many points of E.

   (ii) There is a sequence of points $x_n \in E$ such that $x_n \neq c$ and $x_n \to c$ as $n \to \infty$.

It is clear that a point c with properties (i) or (ii) is a limit point of E. Conversely, if c is a limit point of E, then, for every n, there is a point $x_n \in E$ such that $0 < \rho(c, x_n) < 1/n$

and so (ii) is satisfied. Also (ii) implies (i).

The conditions (i) or (ii) show that a finite set cannot have a limit point. An infinite set may or may not have a limit point. For example the subset $\{1, 2, 3, \ldots\}$ of R^1 has no limit point.

It is important to note that a limit point of a set E may or may not belong to E. For instance if, in R^1, E is the interval $(0, 1)$, then every point in the interval $[0, 1]$ is a limit point of E.

A point c of E is said to be an *isolated point* of E if it is not a limit point of E. Thus, if c is an isolated point of E, there is a $\delta > 0$ such that $B(c; \delta)$ contains no point of E other than c.

Definition. *Given a set $E \subset X$, the point $c \in E$ is said to be an* interior point *of E if there is a $\delta > 0$ such that $B(c; \delta) \subset E$.*

Illustrations

(i) All points of an open ball $B(a; r)$ are interior points of $B(a; r)$. For suppose that $c \in B(a; r)$, so that $\rho(a, c) = s < r$. If $\delta = r - s$ and $\rho(c, x) < \delta$, then

$$\rho(a, x) \leqslant \rho(a, c) + \rho(c, x) < s + \delta = r$$

and so $x \in B(a; r)$. Therefore $B(c; \delta) \subset B(a; r)$, i.e. c is an interior point of $B(a; r)$.

(ii) $X = R^1$.

Let $E = [0, 1)$. The set of limit points of E is $[0, 1]$; the set of interior points of E is $(0, 1)$.

Let E be the set of rational points of R^1. Every point of R^1 is a limit point of E. E has no interior points.

(iii) $X = Z$.

Let E be the annulus consisting of the points z such that $1 < |z| < 2$. The set of limit points is the set of z such that $1 \leqslant |z| \leqslant 2$. The set of interior points is E.

Let E be the set of points $x + iy$ such that $y = -1$ and $0 \leqslant x \leqslant 1$. The set of limit points is E. There are no interior points.

When the space X has no isolated points (as is the case with R^n and Z) then an interior point of a subset E of X is also a limit point of E. But if X is, for instance, the space consisting of all the integers (and ρ is the usual metric in R^1), then every subset E of X is without limit points although all the points of E are interior points of E.

Definition. (i) *A set F in X is said to be* closed *if F contains all its limit points.*

(ii) *A set G in X is said to be* open *if every point of G is an interior point of G.*

Illustrations

(i) Every open ball in a metric space is open. A set $\{x \mid \rho(a, x) \leqslant r\}$ is closed (see exercise 2(c), 2) and is consequently called the *closed ball* with centre a and radius r.

Any set without limit points is closed; in particular, every finite set is closed.

(ii) In R^n, every open interval is open and every closed interval is closed.

(iii) In R^1, the interval $[0, 1)$ is neither open nor closed; the set of rational points is neither open nor closed.

(iv) In Z, the annulus defined by $1 < |z| < 2$ is open; the real axis (im $z = 0$) is closed; the set of points $z = (1 + i)/n$ $(n = 1, 2, \ldots)$ is neither open nor closed.

(v) Let $(C[a, b], \rho)$ be the metric space of real functions continuous (and so bounded) on the interval $[a, b]$, in which

$$\rho(f, g) = \sup_{a \leqslant x \leqslant b} |f(x) - g(x)|.$$

Then the set E_1 of f such that

$$\inf_{a \leqslant x \leqslant b} f(x) > 0$$

is open and the set E_2 of f such that

$$f(a) = 1$$

is closed.

(vi) In any metric space the empty set and the whole space are both open and closed. In R^n and Z these are the only sets that are both open and closed (theorem 3.32, corollary). In an arbitrary metric space there may be others. For instance in the space of integers (with the usual metric of R^1) every set is both open and closed. The question is taken up again in §3.3.

(vii) The property of being open or closed (or neither) is not intrinsic to a given set E; it is relative to the set X of which E is regarded as a subset and the metric ρ with which X is endowed. For instance $[0, 1)$ is neither open nor closed as a subset of R^1 equipped with the usual metric, but $[0, 1)$ is closed as a subset of $(-1, 1)$ with the same metric; and as a subset of (R^1, σ), where σ is the discrete metric, $[0, 1)$ is both open and closed.

A convenient characterization of closed sets is embodied in the following result.

Theorem 2.31. *A necessary and sufficient condition for the set F to be closed is that, whenever (x_n) is a convergent sequence of points in F,* $\lim_{n \to \infty} x_n \in F$.

Proof. First suppose that F is closed and that $x_n \to x$, where $x_n \in F$ for every n. If $x_n = x$ for all sufficiently large n, then, *a fortiori*, $x \in F$; otherwise x is a limit point of F and again $x \in F$.

Next, let F have the property described in the theorem. If F has no limit points, then F is closed. If F has a limit point x, then there is a sequence (x_n) of points in F such that $x_n \to x$, and so $x \in F$. |

Theorem 2.32. (i) *The set G in X is open if and only if G' $(= X - G)$ is closed.*

(ii) *The set F in X is closed if and only if F' $(= X - F)$ is open.*

Proof. We need only show that (a) if G is open, then G' is closed and (b) if F is closed, then F' is open.

(a) Let G be open. If G' has no limit points, then G' is closed. If G' has a limit point c, then every open ball $B(c; \epsilon)$ contains points of G'. Hence c cannot be an interior point of G and, since G is open, $c \notin G$. Thus $c \in G'$. We have therefore shown that G' contains all its limit points.

(*b*) Let F be closed. If F' is empty, then F' is open. If F' is not empty, let x be any point of F'. Since F is closed and $x \notin F$, it follows that x is not a limit point of F. Hence there is a $B(x; \delta)$ free of points of F, i.e. there is a $B(x; \delta) \subset F'$. Thus the arbitrary point x of F' is an interior point of F'. |

We have stressed the fact that a given set is open or closed (or neither) only with respect to the metric space in which it is embedded. It is therefore of interest to note that, if ρ and σ are equivalent metrics on X, then the metric spaces (X, ρ) and (X, σ) have the same open and the same closed sets. The converse also holds. Both statements follow easily from the last two theorems.

The next theorem concerns collections of open or closed sets. These collections may be finite or infinite and need not be countable.

Theorem 2.33. (i) *The union of any collection of open sets is open.*
(ii) *The intersection of any collection of closed sets is closed.*

Proof. (i) Let $\mathcal{G}$ be a collection of open sets G. If $x \in \bigcup_{G \in \mathcal{G}} G$, then there is a member of $\mathcal{G}$, say G^*, such that $x \in G^*$. Since G^* is open, x is an interior point of G^*, i.e. there is a δ such that

$$B(x; \delta) \subset G^*$$

and so also
$$B(x; \delta) \subset \bigcup_{G \in \mathcal{G}} G.$$

Hence x is an interior point of $\bigcup_{G \in \mathcal{G}} G$. This shows that $\bigcup_{G \in \mathcal{G}} G$ is open.

(ii) Let $\mathcal{F}$ be a collection of closed sets F. We could show directly that $\bigcap_{F \in \mathcal{F}} F$ contains all its limit points; but, to illustrate the use of complements in proving 'dual' results for open and closed sets, we use (i) as follows.

Since F' is open, whenever $F \in \mathcal{F}$, by (i),

$$\left(\bigcap_{F \in \mathcal{F}} F \right)' = \bigcup_{F \in \mathcal{F}} F'$$

is open. Hence, by theorem 2.32, $\bigcap_{F \in \mathcal{F}} F$ is closed. |

Theorem 2.34. (i) *The intersection of a* finite *collection of open sets is open.*
(ii) *The union of a* finite *collection of closed sets is closed.*
The proof is left to the reader.

Exercise. Give examples of (i) a sequence of open sets whose intersection is not open and (ii) a sequence of closed sets whose union is not closed.

Let (Y, σ) be a subspace of the metric space (X, ρ). Then (Y, σ) has open and closed sets and, although a set which is open (closed) in Y is, of course, generally not open (closed) in X, the relationship between open and closed sets in X and in Y is very simple.

Theorem 2.35. *Let (Y, σ) be a metric subspace of the metric space (X, ρ). Then a set is open in (Y, σ) if and only if it is of the form $Y \cap G$, where G is open in (X, ρ). A similar result holds for closed sets.*

Proof. First, if G is open in X, then clearly every point of $Y \cap G$ is an interior point of $Y \cap G$ in Y.

Next, let E be open in Y. Then, for every $x \in E$ we can find a δ_x such that $y \in E$ whenever $y \in Y$ and $\sigma(x, y) < \delta_x$. If now $B(x; \delta_x)$ denotes an open ball in X, the set

$$G = \bigcup_{x \in E} B(x; \delta_x)$$

is open in X and $E = Y \cap G$.

The result for closed sets may be deduced by the method of taking complements. |

Corollary. *If Y is open in X, then a subset of Y is open in Y if and only if it is open in X. If Y is closed in X, then a subset of Y is closed in Y if and only if it is closed in X.*

Definition. *Let E be any subset of X.*
   (i) *The* interior *of E, denoted by $E°$, is the set of interior points of E.*
   (ii) *The* closure *of E, denoted by $\bar{E}$, is the union of E and the set of limit points of E.*

Thus $\bar{E}$ is the set of x such that every $B(x; \epsilon)$ contains at least one point of E or, equivalently, $x \in \bar{E}$ if and only if x is the limit of a sequence of points in E.

For alternative definitions of the notions of interior and closure see exercise 2(*c*), 6.

Theorem 2.36. *For any set E in X, $E°$ is open and $\bar{E}$ is closed.*

Proof. If $E°$ is empty, it is open. If $E°$ is not empty, let x be any point of $E°$. Then there is a $B(x; \delta) \subset E$. Every $y \in B(x; \delta)$ is an interior point of $B(x; \delta)$ and so of E. Hence $B(x; \delta) \subset E°$ and therefore x is an interior point of $E°$.

It may be shown directly that $\bar{E}$ is closed. The result also follows from the identity
$$(\bar{E})' = (E')^{\circ}$$
which is easily proved. (See exercise 2(c), 7.) |

Open and closed sets could also have been defined in terms of interiors and closures: G is open if and only if $G^{\circ} = G$; F is closed if and only if $\bar{F} = F$.

Definition. *A* frontier point *of a set E in X is a point c such that every open ball $B(c; \epsilon)$ contains at least one point of E and at least one point of E'. The set of all frontier points of E is called the* frontier *of E and is denoted by* fr E.

Clearly E and E' have the same frontier.

Since a frontier point of E is a point of E which is a limit point of E' or a point of E' which is a limit point of E, it follows that
$$\text{fr } E = \bar{E} - E^{\circ}. \tag{2.31}$$

Illustrations

(i) Let E be the interval $[0, 1)$ in R^1. Then $E^{\circ} = (0, 1)$, $\bar{E} = [0, 1]$ and fr$E = \{0, 1\}$.

(ii) Let (X, ρ) be the space of integers with the metric induced by R^1. If $E = B(c; 1) = \{c\}$, then $E^{\circ} = \bar{E} = E$. Note that $\bar{E}$ is not the closed ball
$$\{x | \rho(x, c) \leqslant 1\} = \{c-1, c, c+1\}.$$

Exercises 2(c)

1. Show that every infinite set X may be equipped with a metric ρ which is such that X has a limit point in (X, ρ).

2. Prove that, in any metric space (X, ρ), the closed ball $\{x | \rho(a, x) \leqslant r\}$ is closed.

3. Prove that a bounded, closed set of real numbers contains its supremum and infimum.

4. Let E be an arbitrary set in a metric space. Show that the set E^* of limit points of E is closed.

5. Show that in R^n with the usual metric any collection of disjoint open sets is countable. Is this true for an arbitrary metric space?

6. (i) Show that E° is the union of all open sets contained in E and so is the 'largest' open set contained in E.

(ii) Show that $\bar{E}$ is the intersection of all closed sets containing E and so is the 'smallest' closed set containing E.

7. Prove that, for any subset E of X,
$$E^{\circ} \cup \text{fr } E \cup (E')^{\circ} = X$$
and
$$(\bar{E})' = (E')^{\circ}.$$
(The set $(E')^{\circ}$ is called the *exterior* of E.)

8. Prove that
$$\operatorname{fr} E = \bar{E} \cap \bar{E'}$$
and that fr E is closed.

9. (i) Prove that
$$\overline{E_1 \cup \dots \cup E_n} = \bar{E}_1 \cup \dots \cup \bar{E}_n,$$
but that, for an infinite collection $\mathscr{E}$ of sets E, generally the relation
$$\overline{\bigcup_{E \in \mathscr{E}} E} \supset \bigcup_{E \in \mathscr{E}} \bar{E}$$
only holds.

(ii) Prove that, for any collection $\mathscr{E}$ of sets E,
$$\overline{\bigcap_{E \in \mathscr{E}} E} \subset \bigcap_{E \in \mathscr{E}} \bar{E}.$$

Construct an example to show that identity need not hold even for a finite collection.

(iii) Deduce the results for interiors corresponding to (i) and (ii).

10. The metric spaces (X, ρ), (X, σ) have equivalent metrics. Show that every subset E of X has the same limit points and interior points in (X, ρ) and (X, σ).

11. If E is a set in a metric space and the set D is such that
$$D \subset E \subset \bar{D},$$
then D is said to be *dense* in E. (For instance the set of rational numbers is dense in R^1.) Show that, if C is dense in D and D is dense in E, then C is dense in E.

NOTES ON CHAPTER 2

§**2.1.** The idea of a metric space is due to M. Fréchet. In his doctoral thesis, published in 1906, he examined various sets of axioms for a metric. The set $M1 - M3$ is the one which experience showed to be the most fruitful.

The inequality (2.12), which is equivalent to the triangle inequality for the usual metric on R^n, is a particular case of Minkowski's inequality
$$\left(\sum_{i=1}^{n} (a_i + b_i)^r \right)^{1/r} \leqslant \left(\sum_{i=1}^{n} a_i^r \right)^{1/r} + \left(\sum_{i=1}^{n} b_i^r \right)^{1/r}$$
in which $a_i, b_i \geqslant 0$ $(i = 1, \dots n)$ and $r \geqslant 1$. (In (2.12) the a_i, b_i may be negative, but the step from non-negative a_i, b_i to arbitrary ones is trivial.) The inequality (2.13), which we call Cauchy's inequality in theorem 2.21, is also known as Schwarz's inequality (or even the Cauchy–Schwarz inequality). It is a particular case of Hölder's inequality.
$$\sum_{i=1}^{n} a_i b_i \leqslant \left(\sum_{i=1}^{n} a_i^p \right)^{1/p} \left(\sum_{i=1}^{n} a_i^q \right)^{1/q},$$
where $a_i, b_i \geqslant 0$ $(i = 1, \dots, n)$, $p > 1$ and $p^{-1} + q^{-1} = 1$. Both Hölder's and Minkowski's inequalities may be generalized in a variety of ways. (See Hardy, Littlewood and Pólya, *Inequalities*.)

3

CONTINUOUS FUNCTIONS ON
METRIC SPACES

3.1. Limits

The notion of continuity is based on that of a limit and we first
define the limit of a function whose domain and range lie in arbitrary
metric spaces.

Definition. *Let (X, ρ) and (Y, σ) be metric spaces. Let f be a function
with domain $E \subset X$ and with range in Y. Suppose also that x_0 is a
limit point of E and $y_0 \in Y$. We then say*

$$f(x) \to y_0 \quad as \quad x \to x_0 \quad or \quad \lim_{x \to x_.} f(x) = y_0$$

if, given $\epsilon > 0$, there is a $\delta > 0$ such that

$$\sigma(f(x), y_0) < \epsilon \quad whenever \quad x \in E \text{ and } 0 < \rho(x, x_0) < \delta.$$

Notes. (i) It is irrelevant whether x_0 belongs to the domain E of f
or not.

(ii) When $X = Y = R^1$ (or Z) the above definition reduces to the
standard definition of the limit of a real (or complex) function.

Exercise. Show that a function cannot converge at a point to more than one
limit.

Illustrations

(i) $X = Y = R^1$
   (a) With $E = R^1$, $\lim_{x \to x_0} x^k = x_0^k$ ($k = 1, 2, \ldots$).
   (b) With $E = R^1 - \{0\}$, $\lim_{x \to 0} x \sin(1/x) = 0$, $\lim_{x \to 0} \sin(1/x)$ does not exist.
   (c) Let E be the set of rational numbers and let f be defined on E by the
equation
$$f(p/q) = 1/q$$
when p, q are coprime integers and $q \geqslant 1$. We shall show that, for every $c \in R^1$,
$\lim_{x \to c} f(x) = 0$.

Given $\epsilon > 0$, let q_0 be an integer such that $1/q_0 < \epsilon$. Denote by Q_0 the set of
rational numbers p/q with $q < q_0$. We note that, in any finite interval, there is
only a finite number of members of Q_0. Hence there is a $\delta > 0$ such that the
intervals $(c - \delta, c)$ and $(c, c + \delta)$ contain no members of Q_0. It follows that,
whenever $0 < |c - p/q| < \delta$,

$$0 \leqslant f\left(\frac{p}{q}\right) = \frac{1}{q} \leqslant \frac{1}{q_0} < \epsilon.$$

(ii) $X = R^2$, $E = R^2 - \{(0, 0)\}$, $Y = R^1$.

Let g_1 be given by
$$g_1(x, y) = \frac{x^3}{x^2+y^2}.$$

Then
$$|g_1(x, y)| = |x| \frac{x^2}{x^2+y^2} \leqslant |x|$$

and
$$\lim_{(x,\, y)\to(0,\, 0)} g_1(x, y) = 0.$$

Let g_2 be given by
$$g_2(x, y) = \frac{x^2}{x^2+y^2}.$$

Then $\lim\limits_{(x,\, y)\to(0,\, 0)} g_2(x, y)$ does not exist. For, if $y = mx$, we have (for all $x \neq 0$),
$$g_2(x, mx) = \frac{x^2}{x^2+m^2x^2} = \frac{1}{1+m^2};$$

and so g_2 takes all values between 0 and 1 in any annulus $0 < x^2+y^2 < \delta^2$.

(iii) Let $E = X = Y = C[0, 1]$ (the set of real functions continuous on $[0, 1]$). The metric ρ is, as is usual on $C[0, 1]$, given by
$$\rho(\phi_1, \phi_2) = \sup_{0\leqslant t\leqslant 1} |\phi_1(t) - \phi_2(t)|.$$

The function h from $C[0, 1]$ to $C[0, 1]$ is given by $h(\phi) = \psi$, where
$$\psi(t) = \int_0^t \phi(\tau)\, d\tau.$$

If, for instance, ϕ_0 is given by $\phi_0(t) = t^2$, then $\lim\limits_{\phi\to\phi_0} h(\phi) = \psi_0$, where $\psi_0(t) = \frac{1}{3}t^3$. For, if $\rho(\phi_0, \phi) < \epsilon$, then, for $0 \leqslant t \leqslant 1$,

$$|\psi_0(t) - \psi(t)| = \left|\tfrac{1}{3}t^3 - \int_0^t \phi(\tau)\, d\tau\right| = \left|\int_0^t \{\tau^2 - \phi(\tau)\}\, d\tau\right|$$
$$\leqslant \int_0^t |\tau^2 - \phi(\tau)|\, d\tau \leqslant \epsilon$$

and therefore $\rho(\psi_0, \psi) < \epsilon$.

A sequence a_1, a_2, a_3 ... may be regarded as a function on the set $\{1, \frac{1}{2}, \frac{1}{3}, \ldots\}$. It follows that the notion of a sequential limit may be expressed by means of the notion of a functional limit. What is less obvious, and more interesting, is that the opposite statement is also true.

Theorem 3.1. *Let (X, ρ) and (Y, σ) be metric spaces. Let f be a function with domain $E \subset X$ and with range in Y and suppose that x_0 is a limit point of E. Then $f(x) \to y_0$ as $x \to x_0$ if and only if, for every sequence (x_n) in $E - \{x_0\}$ such that $x_n \to x_0$, $f(x_n) \to y_0$ as $n \to \infty$.*

Proof. (i) Suppose that $f(x) \to y_0$ as $x \to x_0$. Then, given $\epsilon > 0$, there is a $\delta > 0$ such that
$$\sigma(f(x), y_0) < \epsilon \quad \text{whenever } x \in E \text{ and } 0 < \rho(x, x_0) < \delta.$$

Now, if (x_n) is any sequence in $E - \{x_0\}$ such that $x_n \to x_0$, there is an integer n_0 such that

$$0 < \rho(x_n, x_0) < \delta \quad \text{for} \quad n > n_0.$$

Therefore $\qquad\qquad \sigma(f(x_n), y_0) < \epsilon \quad \text{for} \quad n > n_0,$

i.e. $\qquad\qquad\qquad f(x_n) \to y_0 \quad \text{as} \quad n \to \infty.$

(ii) Suppose that, whenever (x_n) is in $E - \{x_0\}$ and $x_n \to x_0$, $f(x_n) \to y_0$.

If $f(x) \not\to y_0$ as $x \to x_0$, then there is an $\epsilon > 0$ with the property that there is no $\delta > 0$ such that $\sigma(f(x), y_0) < \epsilon$ whenever $x \in E$ and $0 < \rho(x, x_0) < \delta$. Therefore, given any n, there is an $x_n \in E$ such that

$$0 < \rho(x_n, x_0) < 1/n \quad \text{and} \quad \sigma(f(x_n), y_0) \geqslant \epsilon.$$

Thus there is a sequence (x_n) in $E - \{x_0\}$ such that $x_n \to x_0$ and $f(x_n) \not\to y_0$. This contradicts our original hypothesis and so $f(x) \to y_0$ as $x \to x_0$. ∎

For functions with values in a normed vector space V there is, as might be expected, an algebra of limits. Let f, g be such functions defined on a subset E of X. Also suppose that E has a limit point x_0 and that

$$f(x) \to y_0, \quad g(x) \to z_0 \quad \text{as} \quad x \to x_0.$$

When ϕ, ψ are real or complex valued functions on E and

$$\phi(x) \to \alpha, \quad \psi(x) \to \beta \quad \text{as} \quad x \to x_0,$$

then $\qquad\qquad \phi(x)f(x) + \psi(x)g(x) \to \alpha y_0 + \beta z_0. \qquad\qquad (3.11)$

If V possesses an inner product, then also

$$f(x) . g(x) \to y_0 . z_0; \qquad\qquad (3.12)$$

and if V is R^1 or Z, $\qquad f(x)/g(x) \to y_0/z_0 \qquad\qquad (3.13)$

provided that $z_0 \neq 0$.

These results may be proved directly or they may be deduced, by means of theorem 3.1, from the corresponding results on sequences (exercise 2(b), 8).

Exercises 3(a)

1. The real valued function f on $R^2 - \{(0, 0)\}$ is defined by

$$f(x, y) = \frac{x^2 y}{x^4 + y^2}.$$

Show that $\lim\limits_{(x,\, y) \to (0,\, 0)} f(x, y)$ does not exist.

2. Let (X, ρ), (Y, σ) be metric spaces. Let f be a function with domain $E \subset X$ and with range in Y, and suppose that x_0 is a limit point of E. Show that, if $\lim_{n \to \infty} f(x_n)$ exists for every sequence (x_n) in $E - \{x_0\}$ such that $x_n \to x_0$, then $\lim_{x \to x_0} f(x)$ exists.

3. Let (X, ρ), (Y, σ) be metric spaces, and let E be a subset of X, x_0 a limit point of E, y_0 a point of Y. Show that, if the function $f : E \to Y$ is such that $f(x) \to y_0$ as $x \to x_0$, then $y_0 \in \overline{f(E)}$; and that if, in addition, f is injective, then y_0 is a limit point of $f(E)$.

4. Let (X, ρ), (Y, σ), (W, τ) be metric spaces and let D, E be subsets of X, Y respectively. Let $h = g \circ f : D \to W$ be the composition of the functions $f : D \to Y$, where $f(D) \subset E$, and $g : E \to W$. Also (i) x_0 is a limit point of D and $f(x) \to y_0$ as $x \to x_0$; (ii) y_0 is a limit point of E and $g(y) \to w_0$ as $y \to y_0$. Show that, when f is injective (so that, by **3**, y_0 is automatically a limit point of E), then $h(x) \to w_0$ as $x \to x_0$. Construct an example in which $h(x) \not\to w_0$.

5. Prove that the statements (3.11)–(3.13) hold under the conditions given at the end of the section.

3.2. Continuous functions

The definition of continuity which we shall now give is a natural generalization of the concept familiar in elementary analysis.

Definition. Let (X, ρ) and (Y, σ) be metric spaces. Then the function $f : X \to Y$ is said to be continuous *at the point x_0 of X if, given $\epsilon > 0$, there is a $\delta > 0$ such that*

$$\sigma(f(x), f(x_0)) < \epsilon \quad whenever \quad \rho(x, x_0) < \delta$$

(or $f(B(x_0; \delta)) \subset B(f(x_0); \epsilon)$).

If f is continuous at all points of a set, it is said to be continuous on *that set.*

The definition implies that f is continuous at a limit point x_0 of X if

$$f(x) \to f(x_0) \quad as \quad x \to x_0;$$

and that f is continuous at all isolated points of X (for if x_0 is an isolated point and δ is sufficiently small, the only point x such that $\rho(x, x_0) < \delta$ is x_0 itself).

Let f be a function whose domain is a subset X_1 of a metric space (X, ρ). When considering the limits of f we are interested in the limit points of X_1 and these need not belong to X_1. However, for questions of continuity, only the points of the domain X_1 of f are relevant. Since X_1, endowed with the metric induced by (X, ρ), is itself a metric space, it is therefore simplest in this context to treat X_1 as the fundamental space. This is, in fact, what we did in the definition of continuity.

Illustrations. The function f of illustration (i)(c) on p. 32 whose domain is the set of rational numbers, is discontinuous everywhere. If f_1, with domain R^1 is defined by

$$f_1(x) = \begin{cases} f(x) & \text{when } x \text{ is rational,} \\ 0 & \text{when } x \text{ is irrational,} \end{cases}$$

then f_1 is continuous at all irrational points and discontinuous at all rational points.

The function $h: C[0, 1] \to C[0, 1]$ of illustration (iii) on p. 33 is continuous on $C[0, 1]$. This follows from the inequality

$$\rho(h(\phi_1), h(\phi_2)) \leqslant \rho(\phi_1, \phi_2)$$

which holds for all ϕ_1, ϕ_2 of $C[0, 1]$.

Let (X, ρ) be any metric space, let a be a fixed point of X and let the function $g: X \to R^1$ be defined by the equation

$$g(x) = \rho(a, x).$$

Since
$$\rho(a, x_1) - \rho(a, x_2) \leqslant \rho(x_2, x_1)$$

and
$$\rho(a, x_2) - \rho(a, x_1) \leqslant \rho(x_1, x_2),$$

$$|g(x_1) - g(x_2)| = |\rho(a, x_1) - \rho(a, x_2)| \leqslant \rho(x_1, x_2)$$

for all x_1, x_2 in X. Hence g is continuous on X.

Algebraic operations on continuous functions with values in a normed vector space again yield continuous functions. If the functions $f, g: X \to V$ and the functions $\phi, \psi: X \to R^1$ (or Z) are continuous at x_0, then $\phi f + \psi g$ is continuous at x_0; so is $f.g$ if V is an inner product space, and f/g is continuous at x_0 if V is R^1 or Z and $g(x_0) \neq 0$. All this is obvious when x_0 is an isolated point of X and otherwise follows from the corresponding results on limits. In the next theorem we return to functions with arbitrary ranges. The result contrasts with exercise 3(a), 4.

Theorem 3.21. *Let (X, ρ), (Y, σ), (W, τ) be metric spaces and let $h = g \circ f: X \to W$ be the composition of the functions $f: X \to Y$ and $g: Y \to W$. If f is continuous at x_0 and g is continuous at $y_0 = f(x_0)$, then h is continuous at x_0.*

Proof. Given ϵ, there is an η such that

$$\tau(g(y), g(y_0)) < \epsilon \quad \text{whenever} \quad \sigma(y, y_0) < \eta.$$

There is now a δ such that

$$\sigma(f(x), f(x_0)) < \eta \quad \text{whenever} \quad \rho(x, x_0) < \delta$$

and so δ is also such that

$$\tau(h(x), h(x_0)) < \epsilon \quad \text{whenever} \quad \rho(x, x_0) < \delta. \ |$$

It is important to note that the inverse function of a bijective continuous function is generally not continuous. A simple example is the function f with domain $X = [0, 1) \cup [2, 3]$ and range $Y = [0, 2]$ defined by

$$f(x) = \begin{cases} x & \text{for } 0 \leqslant x < 1, \\ x-1 & \text{for } 2 \leqslant x \leqslant 3. \end{cases}$$

If X and Y have the metrics induced by R^1, then f is continuous on X. However f^{-1}, which is given by

$$f^{-1}(y) = \begin{cases} y & \text{for } 0 \leqslant y < 1, \\ y+1 & \text{for } 1 \leqslant y \leqslant 2, \end{cases}$$

is discontinuous at the point 1. For other examples see exercises 3(b), 4 and 5.

The concept of continuity can be formulated in a variety of ways. We shall next consider equivalent definitions, first for continuity at a point and then for continuity on the domain of the function.

Theorem 3.22. *Let (X, ρ) and (Y, σ) be metric spaces. A necessary and sufficient condition for the function $f: X \to Y$ to be continuous at the point x_0 is that $x_n \to x_0$ implies $f(x_n) \to f(x_0)$.*

The proof is similar to that of theorem 3.1. The theorem may also be deduced from theorem 3.1, but slight complications arise, since members of the sequence (x_n) may now be x_0.

Theorem 3.22 shows that the introduction of equivalent metrics in the domain or the range of a function does not affect the property of continuity: if ρ, ρ_1 are equivalent metrics on X and σ, σ_1 are equivalent metrics on Y, and if $f: X \to Y$ is continuous at x_0 with respect to ρ and σ, then f is also continuous at x_0 with respect to ρ_1 and σ_1.

Theorem 3.23. *Let (X, ρ) and (Y, σ) be metric spaces. Each of the following conditions is necessary and sufficient for the function $f: X \to Y$ to be continuous on X:*

(i) *Whenever G is open in Y, then $f^{-1}(G)$ is open in X.*
(ii) *Whenever F is closed in Y, then $f^{-1}(F)$ is closed in X.*

Proof. (i) First suppose that f is continuous on X and that the set G is open in Y.

If the set $f^{-1}(G)$ is empty then it is also open. If $f^{-1}(G)$ is not empty, let x_0 be any one of its points. Then $f(x_0) \in G$ and, since G is open, there is an ϵ such that

$$y \in G \quad \text{whenever} \quad \sigma(y, f(x_0)) < \epsilon.$$

But, since f is continuous at x_0, there is a δ such that

$$\sigma(f(x), f(x_0)) < \epsilon \quad \text{whenever} \quad \rho(x, x_0) < \delta.$$

Hence $\qquad f(x) \in G \quad \text{whenever} \quad \rho(x, x_0) < \delta,$

i.e. $\qquad x \in f^{-1}(G) \quad \text{whenever} \quad \rho(x, x_0) < \delta.$

This shows that x_0 is an interior point of $f^{-1}(G)$. It follows that $f^{-1}(G)$ is open in X.

Next suppose that, whenever G is open in Y, then $f^{-1}(G)$ is open in X.

Let x_0 be any point of X and take any positive ϵ. The open ball $B(f(x_0); \epsilon)$ is open in Y and so, by hypothesis, the set $f^{-1}\{B(f(x_0); \epsilon)\}$ is open in X. This set also contains x_0 and therefore there is a $\delta > 0$ such that $\qquad x \in f^{-1}\{B(f(x_0); \epsilon)\} \quad \text{whenever} \quad \rho(x, x_0) < \delta,$

i.e. $\qquad f(x) \in B(f(x_0); \epsilon) \quad \text{whenever} \quad \rho(x, x_0) < \delta,$

i.e. $\qquad \sigma(f(x), f(x_0)) < \epsilon \quad \text{whenever} \quad \rho(x, x_0) < \delta.$

Thus f is continuous at x_0 and so on X.

(ii) This part follows from (i) by the method of complements since, if E is any subset of Y,

$$f^{-1}(E) \cap f^{-1}(E') = \varnothing \quad \text{and} \quad f^{-1}(E) \cup f^{-1}(E') = X. \;|$$

In the last theorem the inverse images may not be replaced by direct images. It is not true that $f(S)$ is open (closed) in Y whenever S is open (closed) in X. The function defined on p. 37 will illustrate this (exercise 3(b), 6) but it is instructive to employ also more commonplace examples. For instance the function f with domain R^1 and range R^1 given by $\qquad f(x) = x(x-1)(x-2)$

maps the open interval $(0, 1)$ onto $(0, \frac{3}{8}]$. Next, let g, with domain $(-\infty, \infty)$ and range $[-\frac{1}{2}, \frac{1}{2}]$ be defined by

$$g(x) = \frac{x}{1+x^2}.$$

This function maps the closed interval $[1, \infty)$ on the interval $(0, \frac{1}{2}]$ which is not closed in $[-\frac{1}{2}, \frac{1}{2}]$.

We have previously mentioned that a bijective continuous function $f : (X, \rho) \to (Y, \sigma)$ need not be such that $f^{-1} : (Y, \sigma) \to (X, \rho)$ is continuous. When f^{-1} *is* continuous, f is called a *homeomorphism*. (Note that then f^{-1} is also a homeomorphism.) Two metric spaces are said to be *homeomorphic* if there exists a homeomorphism on one into the other. If $f : (X, \rho) \to (Y, \sigma)$ is a homeomorphism, then, by theorem 3.23, f maps open (closed) sets in X into open (closed) sets in Y and there is a bijection between the open (respectively closed) sets in the two spaces.

A bijective function $f : (X, \rho) \to (Y, \sigma)$ such that

$$\sigma(f(x_1), f(x_2)) = \rho(x_1, x_2)$$

for all $x_1, x_2 \in X$ is called an *isometry*. It is plain that an isometry is a homeomorphism. Two metric spaces are called *isometric* if there exists an isometry on one into the other. Such spaces have identical metric properties although their elements may be of entirely different kinds. For instance R^1 is isometric with the subspace Λ of $C[0, 1]$ consisting of the functions ϕ_λ $(-\infty < \lambda < \infty)$ defined by

$$\phi_\lambda(x) = \lambda x \quad (0 \leqslant x \leqslant 1).$$

Another example of an isometric pair is provided by R^2, Z.

Linear functions. We briefly consider a class of functions which plays an important part in modern analysis.

Definition. *Let V, W be two real (complex) normed vector spaces. A function $f : V \to W$ is called* linear *if*

$$f(\alpha_1 x_1 + \alpha_2 x_2) = \alpha_1 f(x_1) + \alpha_2 f(x_2)$$

for all $x_1, x_2 \in V$ and all real (complex) numbers α_1, α_2.

Clearly $f(\theta) = \theta$, where the letter θ on the left stands for the zero element in V and that on the right for the zero element in W. (It is not customary to use distinctive notations for the zero element and the norm in the two spaces.)

Illustrations

(1) The function $f : C[a, b] \to R^1$ defined by the relation

$$f(\phi) = \int_a^b \phi(t)\, dt \quad (\phi \in C[a, b])$$

is linear.

(2) The reader is probably acquainted with linear functions on one Euclidean space into another. We shall use three properties of these functions which are

proved in most books on linear algebra. (See, for instance, L. Mirsky, *Introduction to Linear Algebra*, chapters IV and V.)

(i) The function $f: R^m \to R^n$ is linear if and only if it is of the form

$$f(x_1, ..., x_m) = (y_1, ..., y_n),$$

where

$$\left.\begin{aligned} y_1 &= a_{11}x_1 + ... + a_{1m}x_m, \\ &\cdots\cdots\cdots\cdots\cdots \\ y_n &= a_{n1}x_1 + ... + a_{nm}x_m \end{aligned}\right\} \tag{3.21}$$

or

$$\begin{pmatrix} y_1 \\ \vdots \\ y_n \end{pmatrix} = \begin{pmatrix} a_{11}...a_{1m} \\ \cdots\cdots \\ a_{n1}...a_{nm} \end{pmatrix} \begin{pmatrix} x_1 \\ \vdots \\ x_m \end{pmatrix}.$$

The $n \times m$ matrix (a_{ji}) is said to *represent* f.

(ii) If the linear functions $f: R^m \to R^n$ and $g: R^n \to R^p$ have matrices A, B, then the linear function $g \circ f: R^m \to R^p$ is represented by the matrix BA.

(iii) Let $f: R^m \to R^n$ be a linear function with matrix A. Then f is injective if and only if $n \geqslant m$ and A has (maximum) rank m. Also f is surjective (i.e. the range of f is the whole of R^n) if and only if $n \leqslant m$ and A has (maximum) rank n. Thus, a necessary and sufficient condition for f to be bijective (so that f^{-1} exists and is again a linear function) is that $m = n$ and A has maximum rank, i.e. $\det A \neq 0$. When f is bijective, f^{-1} has matrix A^{-1}.

It is easy to see that a linear function on R^m into R^n is continuous at all points of its domain (see exercise 3(b), 8). A consequence of the theorem below is that any linear function either is everywhere continuous or is nowhere continuous.

Theorem 3.24. *Let V, W be normed vector spaces. If the function $f: V \to W$ is linear, then the following three statements are equivalent.*

(i) *f is continuous on V.*

(ii) *There is a point $x_0 \in V$ at which f is continuous.*

(iii) *$\|f(x)\|/\|x\|$ is bounded for $x \in V - \{\theta\}$.*

Proof. (ii) $\Rightarrow$ (i). Take any $\epsilon > 0$. Since f is continuous at x_0, there is a $\delta > 0$ such that $\|x - x_0\| < \delta$ implies $\|f(x) - f(x_0)\| < \epsilon$.

Let c be any point of V and take any x such that $\|x - c\| < \delta$. Then $\|(x - c + x_0) - x_0\| < \delta$ and therefore $\|f(x - c + x_0) - f(x_0)\| < \epsilon$, i.e. $\|f(x) - f(c)\| < \epsilon$. Thus f is continuous at c.

(i) $\Rightarrow$ (iii). Suppose that $\|f(x)\|/\|x\|$ is not bounded. Then, given any integer n, there is a point $x_n \in V - \{\theta\}$ such that $\|f(x_n)\|/\|x_n\| > n$. If $v_n = (n\|x_n\|)^{-1}x_n$, then $\|v_n\| = 1/n$ and

$$\|f(v_n)\| = \frac{\|f(x_n)\|}{n\|x_n\|} > 1.$$

Hence f is not continuous at θ.

(iii) $\Rightarrow$ (ii). Let $N = \sup \|f(x)\|/\|x\|$. Since $f(\theta) = \theta$, the inequality

$$\|f(x)\| \leqslant N\|x\|$$

shows that f is continuous at θ. $\blacksquare$

Given the normed vector spaces V, W, denote by $L(V, W)$ the set of continuous linear functions on V into W. With the usual interpretation of addition and scalar multiplication of vector valued functions, $L(V, W)$ is a vector space. Moreover it is easily verified that the equation

$$\|f\| = \sup_{x \in V - \{\theta\}} \|f(x)\|/\|x\| \quad (f \in L(V, W)) \tag{3.22}$$

defines a norm on $L(V, W)$.

Notes. (i) In (3.22) the symbol $\|.\|$ is used in three different senses.

(ii) $\|f\| = \sup\limits_{\|x\|=1} \|f(x)\|$.

(iii) For all $x \in V$, $\|f(x)\| \leqslant \|f\|\,\|x\|$.

Theorem 3.25. *If (a_{ji}) is the matrix of the function $f \in L(R^m, R^n)$, then*

$$\max_{i,j} |a_{ji}| \leqslant \|f\| \leqslant \left(\sum_{i=1}^m \sum_{j=1}^n a_{ji}^2\right)^{\frac{1}{2}}. \tag{3.23}$$

Proof. Let x be any point of R^m and let $y = f(x)$. Applying Cauchy's inequality (2.13) to (3.21) we have

$$y_1^2 + \ldots + y_n^2 \leqslant \left(\sum_{i=1}^m a_{1i}^2\right)\left(\sum_{i=1}^m x_i^2\right) + \ldots + \left(\sum_{i=1}^m a_{ni}^2\right)\left(\sum_{i=1}^m x_i^2\right)$$

i.e.
$$\|y\|^2 \leqslant \left(\sum_{i=1}^m \sum_{j=1}^n a_{ji}^2\right)\|x\|^2.$$

This proves the right-hand inequality in (3.23).

Next, if $|a_{qp}| = \max\limits_{i,j} |a_{ji}|$, let x^* be the vector with pth component 1 and all other components 0. Then $\|x^*\| = 1$ and

$$\|f(x^*)\| = (a_{1p}^2 + \ldots + a_{np}^2)^{\frac{1}{2}} \geqslant |a_{qp}|. \quad \blacksquare$$

Exercises 3(*b*)

1. Prove the second part of theorem 3.23 without using the first part.

2. Let (X, ρ), (Y, σ) be metric spaces. Show that $f: X \to Y$ is continuous at the point x_0 if and only if, given any open set N in Y which contains $f(x_0)$, there is an open set M in X such that $x_0 \in M$ and $f(M) \subset N$.

3. Construct a function f on a metric space (X, ρ) which is discontinuous at all points of a dense subset E of X, but is such that the restriction of f to E is continuous on E.

4. Let X, Y be similar, infinite sets and suppose that their metrics ρ, σ are such that (X, ρ) has no limit points while (Y, σ) has at least one limit point. Show that, if the function $f: X \to Y$ is bijective, then f is continuous on X, but f^{-1} is not continuous on Y. (Thus (X, ρ), (Y, σ) are not homeomorphic.) Give an example of such a function f.

5. The metric spaces (X, ρ), (X, σ) have the same underlying set X. Let

$$I: (X, \rho) \to (X, \sigma)$$

be the identity function given by $I(x) = x$. Show that I is a homeomorphism if and only if ρ and σ are equivalent metrics. Give an example to show that, when ρ, σ are not equivalent, I may be continuous without I^{-1} being continuous.

6. Let f be the function with domain $X = [0, 1) \cup [2, 3]$ and range $Y = [0, 2]$ defined on p. 37. Find a set G open in X and a set F closed in X such that $f(G)$ is not open in Y and $f(F)$ is not closed in Y.

7. The function g with domain $(-\infty, \infty)$ and range $[0, \infty)$ is given by

$$g(x) = (x-1)^2 e^x.$$

Show that there are open (closed) sets in $(-\infty, \infty)$ which are mapped by g into sets that are not open (closed) in $[0, \infty)$.

8. Let (X, ρ) be a metric space and let $f_1, \ldots, f_n$ be functions on X to R^1. The function $f = (f_1, \ldots, f_n): X \to R^n$ is given by

$$f(x) = (f_1(x), \ldots, f_n(x)).$$

Prove that f is continuous at the point $x_0 \in X$ if and only if, $f_1, \ldots, f_n$ are continuous at x_0.

Hence, or otherwise, show that every linear function on R^m into R^n is everywhere continuous.

9. Let (X, ρ), (Y, σ), (W, τ) be metric spaces. To define continuity of a function on $X \times Y$ into W we use one of the equivalent metrics on $X \times Y$ of exercise $2(a)$, 11.

Given $f: X \times Y \to W$ and $x_0 \in X$, $y_0 \in Y$, define the functions g on X and h on W by

$$g(x) = f(x, y_0) \quad (x \in X), \qquad h(y) = f(x_0, y) \quad (y \in Y).$$

Show that, if f is continuous at (x_0, y_0), then g, h are continuous at x_0, y_0 respectively. Construct an example to show that g, h may be continuous at x_0, y_0 respectively without f being continuous at (x_0, y_0).

10. Let D be the vector space of real differentiable functions on the interval $[0, 1]$ and define the norm on D by

$$\|\phi\| = \sup_{0 \leqslant x \leqslant 1} |\phi(x)| \quad (\phi \in D).$$

Let $f: D \to R^1$ be the linear function defined by

$$f(\phi) = \phi'(0) \quad (\phi \in D).$$

Show that f is not continuous.

11. The linear function $f: R^n \to R^1$ is given by

$$f(x_1, \ldots, x_n) = a_1 x_1 + \ldots + a_n x_n^{\circ}.$$

Find $\|f\|$.

12. Let U, V, W be normed vector spaces and let $f: U \to V$, $g: V \to W$ be continuous linear functions. Show that

$$\|g \circ f\| \leqslant \|g\| \|f\|.$$

Deduce that, if f has a continuous inverse f^{-1}, then

$$\|f\| \|f^{-1}\| \geqslant 1.$$

Give an example of a linear function f such that

$$\|f\| \|f^{-1}\| > 1.$$

3.3. Connected metric spaces

In the classical account of continuity the more striking results, such as those relating to bounds and intermediate values are proved for continuous functions defined on finite closed intervals. These results do not hold for arbitrary continuous functions, but they can be regained by suitably restricting the domain. The theme of the remainder of this chapter is the relation between the structure of a metric space and the properties of the continuous functions defined on it.

We remarked in chapter 2 that there exist metric spaces (X, ρ) in which sets other than $\varnothing$ and X are both open and closed. A space in which there are no such sets is called *connected*. Since a set is both open and closed if and only if its complement is both closed and open, the condition for a space not to be connected is that it is the union of two non-empty, disjoint open sets. The notion of connectedness can also be applied to subsets of a space. The subset E of X is said to be connected if the metric space (E, σ) is connected, where σ is the metric induced by (X, ρ). In view of theorem 2.35 we may therefore formulate the definition of connectedness in the following way.

Definition. Let (X, ρ) be a metric space.

(i) *The open subset G of X (which may be X itself) is said to be* connected *if there do not exist two non-empty open sets G_1, G_2 such that* $$G_1 \cap G_2 = \varnothing \quad and \quad G_1 \cup G_2 = G.$$

(ii) *An arbitrary subset E of X is said to be* connected *if there do not exist open sets G_1, G_2, such that*

$$G_1 \cap E \neq \varnothing, \quad G_2 \cap E \neq \varnothing, \quad G_1 \cap G_2 \cap E = \varnothing, \quad G_1 \cup G_2 \supset E.$$

It is clear that part (i) is a particular case of part (ii). Note that, by definition, the empty set is connected.

The appropriateness of the word 'connected' is shown by the next two theorems.

Theorem 3.31. *A necessary and sufficient condition for a non-empty set in R^1 to be connected is that it is an interval.*

Proof.

Necessity. Let E be a non-empty set in R^1 which is not an interval. Then there are three real numbers a, q, b such that $a < q < b$ and $a \in E$, $b \in E$ while $q \notin E$. The sets $G_1 = (-\infty, q)$ and $G_2 = (q, \infty)$ are open and interset E since they contain a, b respectively. Also clearly

$$G_1 \cap G_2 \cap E = \varnothing \quad \text{and} \quad G_1 \cup G_2 \supset E.$$

Therefore E is not connected.

Sufficiency. Let I be an interval and let G_1, G_2 be open sets which intersect I, but are such that

$$G_1 \cap G_2 \cap I = \varnothing.$$

Take a point $a \in G_1 \cap I$ and a point $b \in G_2 \cap I$. Since $a \neq b$, we may suppose that $a < b$. Then $[a, b]$ is a subinterval of I. Let u be the upper bound of the set $G_1 \cap [a, b]$. Since a, b are interior points of G_1 and G_2 respectively and $G_1 \cap G_2 \cap [a, b]$ is empty, it follows that

$$a < u < b.$$

If now δ is any positive number such that

$$a < u - \delta < u + \delta < b,$$

$(u - \delta, u]$ contains at least one point of G_1 and $(u, u + \delta)$ contains no points of G_1. Hence $(u - \delta, u + \delta)$ cannot be a subset of either G_1 or G_2. Since G_1, G_2 are open, it follows that $u \notin G_1$ and $u \notin G_2$, so that $G_1 \cup G_2 \not\supset I$. Therefore I is connected. |

Theorem 3.32. *A necessary and sufficient condition for a (non-empty) open set in R^n (or Z) to be connected is that any two of its points may be joined by a polygon lying entirely in the set. The polygon may be chosen so that its sides are parallel to the coordinate axes.*

Proof.

Necessity. Let G be open and connected. Take a point c in G and let G_1 be the set of points in G which can be joined to c by a polygon in G. Let $G_2 = G - G_1$ so that $G_1 \cap G_2 = \varnothing$ and $G_1 \cup G_2 = G$. We shall see that G_1 and G_2 are also open.

If x is any point of G_1, there is a $B(x; \epsilon) \subset G$. All points in $B(x; \epsilon)$ can be joined to x by a line segment in $B(x; \epsilon)$ and so in G. Since x can be joined to c by a polygon in G, so therefore can all points in $B(x; \epsilon)$. Hence $B(x; \epsilon) \subset G_1$. Thus G_1 is open. The same kind of argument shows that G_2 is open.

Since G is connected and is the union of the two disjoint open sets G_1, G_2, at least one of these sets is empty. But $c \in G_1$ and so G_2 is empty. Therefore $G = G_1$, i.e. all points of G can be joined to c by means of a polygon in G. Finally, any two points of G can be joined by way of c.

The polygon could be one with sides parallel to the axes, for a point in an open ball of R^n may be joined to the centre by n segments parallel to the axes.

Sufficiency. Suppose that G is open and that any two points of G may be joined by a polygon lying entirely in G. Let G_1, G_2 be non-empty, disjoint open sets contained in G. We have to show that $G_1 \cup G_2 \neq G$.

Take a point $c_1 \in G_1$ and a point $c_2 \in G_2$. There is a polygon in G which joins c_1 and c_2. This polygon must have a segment with end points p, q such that $p \in G_1$ and $q \in G_2$. The segment may be represented by the vector equation

$$x = x(t) = p + (q - p)t \quad (0 \leqslant t \leqslant 1).$$

Let u be the upper bound of the numbers t such that $x(t) \in G_1$. Arguing in much the same way as in the proof of theorem 3.31 we see that $0 < u < 1$ and $x(u)$ does not belong to G_1 or to G_2. Therefore $G_1 \cup G_2 \neq G$. |

Corollary. *The spaces R^n and Z are connected.*

An open, connected subset of a metric space is called a *region*. Regions in Z will prove to be the natural domains of complex functions (chapter 10).

Definition. *Let (X, ρ) be a metric space and let $E \subset X$. A set C is called a* component *of E if it is a maximal connected subset of E (i.e. if C is a connected subset of E and if no other connected subset of E contains C).*

It is an immediate consequence of the lemma below that any two distinct components of a set are disjoint.

Lemma. *Let $\mathscr{E}$ be a collection of connected sets E in a metric space. If $\bigcap_{E \in \mathscr{E}} E$ is not empty, then $\bigcup_{E \in \mathscr{E}} E$ is connected.*

Proof. Suppose that $H = \bigcup_{E \in \mathscr{E}} E$ is not connected. Then there are open sets G_1, G_2 such that

$$G_1 \cap H \neq \varnothing, \quad G_2 \cap H \neq \varnothing, \quad G_1 \cap G_2 \cap H = \varnothing, \quad G_1 \cup G_2 \supset H.$$

Let $a \in \bigcap_{E \in \mathscr{E}} E$. Since $a \in H$, we may assume that $a \in G_1$. If E is now any set of $\mathscr{E}$, $a \in E$ and $E \subset H$ so that

$$G_1 \cap E \neq \varnothing, \quad G_1 \cap G_2 \cap E = \varnothing, \quad G_1 \cup G_2 \supset E.$$

As E is connected, we must also have $G_2 \cap E = \varnothing$. It follows that $G_2 \cap H = \varnothing$, which is a contradiction. |

Theorem 3.33. *Any set in a metric space has a unique decomposition into components.*

Proof. Let E be a set in a metric space. For any $x \in E$, denote by C_x the union of all the connected subsets of E which contain x. Since $\{x\}$ is connected (see exercise 3(c), 2), C_x is not empty and, by the lemma, C_x is connected. C_x is also a component. For if C is a connected subset of E which contains C_x, then C contains x and so, by definition of C_x, $C_x \supset C$. Finally it is clear that any component of E is a set C_x. |

Corollary 1. *An open set in R^n or Z is the union of countably many disjoint regions.*

Proof. Let G be an open set in R^n or Z and let C be a component of G. If $x \in C$, then $C_x = C$. Also there is a δ such that $B(x; \delta) \subset G$ and, since $B(x; \delta)$ is connected (theorem 3.32); $B(x; \delta) \subset C_x = C$. Thus C is open, i.e. C is a region. It now follows from exercise 2(c), 5 that the set of components of G is countable. |

Corollary 2. *An open set in R^1 is the union of countably many open intervals.*

Theorem 3.34. *If the domain of a continuous function is connected, then so is the range.*

Proof. Let (X, ρ), (Y, σ) be metric spaces and suppose that the function $f: X \to Y$ is continuous. If $f(X)$ is not connected, then Y contains open subsets G_1, G_2 which intersect $f(X)$ and are such that

$$G_1 \cap G_2 \cap f(X) = \varnothing \quad \text{and} \quad G_1 \cup G_2 \supset f(X). \tag{3.31}$$

By theorem 3.23, $f^{-1}(G_1)$ and $f^{-1}(G_2)$ are open. These sets are also non-empty, since $G_1 \cap f(X)$ and $G_2 \cap f(X)$ are non-empty. Moreover, by (3.31),

$$f^{-1}(G_1) \cap f^{-1}(G_2) = \varnothing \quad \text{and} \quad f^{-1}(G_1) \cup f^{-1}(G_2) = X$$

and so X is not connected. Hence if X is connected, then so is $f(X)$. |

Corollary. *If the domain of a real-valued, continuous function is connected, then the range is an interval.*

Exercise. Prove that, if the real function f is continuous on the interval $[a, b]$, then f assumes in this interval every value between $f(a)$ and $f(b)$. (The intermediate value theorem of classical analysis, theorem 3.6 in C1.)

Exercises 3(c)

1. Suppose that $E \subseteq X \cap Y$ and that the metrics ρ, σ on X, Y respectively are identical on E. Show that E is a connected set in (X, ρ) if and only if it is a connected set in (Y, σ).

2. Show that, in any metric space, a set consisting of a single point is connected; and that a set which consists of more than one point and contains an isolated point is not connected.

3. Prove that a metric space (X, ρ) is connected if and only if its only subsets with empty frontier are $\varnothing$ and X.

4. Show that in the space of irrational numbers the only connected sets are those consisting of single points.

5. Prove that, if a subset E of a metric space is connected, then $\bar{E}$ is also connected. If $\bar{E}$ is connected, is E necessarily connected?

6. Show that the components of a closed subset of a metric space are closed. Are the components of an open subset necessarily open?

7. Show that the subset

$$\{(x, y) | x = 0, -1 \leqslant y \leqslant 1\} \cup \{(x, y) | 0 < x \leqslant 1, y = \sin(1/x)\}$$

of R^2 is connected.

3.4. Complete metric spaces

Readers are familiar with the idea of a convergent sequence. We now introduce a closely related concept.

Definition. *The sequence* (x_n) *of points in the metric space* (X, ρ) *is said to be a* Cauchy sequence *if, given* $\epsilon > 0$, *there is an* n_0 *such that*

$$\rho(x_m, x_n) < \epsilon \quad \text{whenever} \quad m, n > n_0.$$

Theorem 3.41. *If (x_n) is a convergent sequence in a metric space (X, ρ), then it is a Cauchy sequence.*

Proof. Let $\lim\limits_{n\to\infty} x_n = x$. Given $\epsilon > 0$, there is an n_0 such that $\rho(x_n, x) < \epsilon$ for $n > n_0$. If now $m, n > n_0$,

$$\rho(x_m, x_n) \leqslant \rho(x_m, x) + \rho(x_n, x) < 2\epsilon. \mid$$

The converse of theorem 3.41 is not true in every metric space. For instance if X is the interval $(0, 1)$ of R^1, then the sequence $(1/n)$ is a Cauchy sequence, for, when $\epsilon > 0$,

$$\left|\frac{1}{m} - \frac{1}{n}\right| < \epsilon \quad \text{if} \quad m, n > \frac{1}{\epsilon}.$$

However $(1/n)$ does not converge in $(0, 1)$, since there is no point x in $(0, 1)$ such that $1/n - x \to 0$ as $n \to \infty$. Another subspace of R^1 in which the converse of theorem 3.41 fails is the metric space of rational numbers. (See exercise $3(d), 2$.)

Definition. *Let (X, ρ) be a metric space. A subset E of X (which may be X itself) is said to be* complete *if every Cauchy sequence in E has a limit in E. If X is complete, we say that the metric space (X, ρ) is* complete.

Notes. (i) In any metric space the empty set is complete (since it contains no Cauchy sequences).

(ii) The non-empty set E in the metric space (X, ρ) is complete if and only if the metric subspace (E, σ) of (X, ρ) is complete.

We shall establish the vital result that R^1 is complete. First we prove a lemma which has independent interest.

Lemma. *Every sequence of real numbers has a monotonic subsequence.*

Proof. Let (x_n) be any sequence of real numbers. We call x_p a *terrace point* if $x_n \leqslant x_p$ for all $n \geqslant p$.

If there are infinitely many terrace points, let x_{ν_1} be the first, x_{ν_2} the second, and so on. The sequence (x_{ν_k}) is clearly decreasing.

If there are finitely many terrace points (perhaps none), take an integer ν_1 so that no x_n, for $n \geqslant \nu_1$, is a terrace point. There is now a $\nu_2 > \nu_1$ such that $x_{\nu_2} > x_{\nu_1}$. Again there is a $\nu_3 > \nu_2$ such that $x_{\nu_3} > x_{\nu_2}$; and continuing in this way we obtain an increasing subsequence (x_{ν_k}). $\mid$

We do not need the full strength of the lemma, but only the special case that a bounded sequence has a convergent subsequence.

Theorem 3.42. *The space R^1 (with its usual metric) is complete.*

Proof. Let (x_n) be a Cauchy sequence in R^1. We first show that (x_n) is bounded. Since there is an integer p such that

$$|x_n - x_m| < 1 \quad \text{whenever} \quad m, n \geqslant p,$$

it follows that $\quad |x_n| \leqslant |x_n - x_p| + |x_p| < |x_p| + 1$

whenever $n \geqslant p$. Thus, for all n,

$$|x_n| \leqslant \max (|x_1|, |x_2|, ..., |x_{p-1}|, |x_p| + 1).$$

It now follows from the lemma that (x_n) has a bounded, monotonic and therefore convergent subsequence, say (x_{ν_k}). Suppose that

$$x_{\nu_k} \to x \quad \text{as} \quad k \to \infty.$$

We shall show that $x_n \to x$ as $n \to \infty$.

Take any positive ϵ. Since (x_n) is a Cauchy sequence, there is an integer n_0 such that

$$|x_m - x_n| < \epsilon \quad \text{for} \quad m, n \geqslant n_0.$$

Also since $x_{\nu_k} \to x$, there is an integer k_0 such that

$$|x_{\nu_k} - x| < \epsilon \quad \text{for} \quad k \geqslant k_0.$$

As $\nu_k \geqslant k$, we then have, for $n \geqslant \max (n_0, k_0)$,

$$|x_n - x| \leqslant |x_n - x_{\nu_n}| + |x_{\nu_n} - x| < 2\epsilon.$$

Thus $x_n \to x$ as $n \to \infty$. |

Corollary. *The spaces R^n and Z (with their usual metrics) are complete.*

Proof. We need only consider R^n. Let (x_r) be a Cauchy sequence in R^n. If $x_r = (\xi_{1r}, ..., \xi_{nr})$, then

$$\left. \begin{matrix} |\xi_{1r} - \xi_{1s}| \\ \cdots\cdots\cdots \\ |\xi_{nr} - \xi_{ns}| \end{matrix} \right\} \leqslant \rho(x_r, x_s)$$

and so each of the sequences $(\xi_{1r})_{r \geqslant 1}, ..., (\xi_{nr})_{r \geqslant 1}$ is a Cauchy sequence in R^1. Therefore, by the theorem, there are real numbers $\xi_1, ..., \xi_n$ such that

$$\xi_{1r} \to \xi_1, ..., \xi_{nr} \to \xi_n$$

as $r \to \infty$. If $x = (\xi_1, ..., \xi_n)$, we then have

$$\rho(x_r, x) \leqslant |\xi_{1r} - \xi_1| + ... + |\xi_{nr} - \xi_n|$$

and, since the right-hand side tends to 0 as $r \to \infty$, it follows that $x_r \to x$ as $r \to \infty$. |

Theorem 3.41 and the fact that R^1 and Z are complete may be combined in the following statement. *A necessary and sufficient condition for the sequence (x_n) of real or complex numbers to converge is that, given $\epsilon > 0$, there is an integer n_0 such that $|x_m - x_n| < \epsilon$ whenever $m, n > n_0$.* In classical analysis this is referred to as the *General Principle of Convergence*.

For subsets of metric spaces there is an intimate relation between the properties of being complete and of being closed.

Theorem 3.43. *Let (X, ρ) be a metric space and let E be a subset of X.*
  (i) *If E is complete, then E is closed.*
  (ii) *If X is complete and if E is closed, then E is complete.*

Proof. (i) Let (x_n) be a sequence of points in E which converges to a point $x \in X$. By theorem 3.41, (x_n) is a Cauchy sequence in X and so in E. Since E is complete, it follows that (x_n) converges in E, i.e. that $x \in E$. Theorem 2.31 now shows that E is closed.

  (ii) Let (x_n) be a Cauchy sequence in E. Since (x_n) is a Cauchy sequence in X and X is complete, there is an $x \in X$ such that $x_n \to x$. By theorem 2.31, $x \in E$. |

Several important metric spaces have functions as their elements. We shall now consider two such spaces which we have previously met in specialized form.

Let (Y, τ) be a metric space. The subset E of Y is said to be *bounded* if there is an open ball that contains E; and, if X is an arbitrary set, the function $\phi : X \to Y$ is called bounded if the subset $\phi(X)$ of Y is bounded. We denote by $B(X, Y)$ the set of bounded functions on X into Y and we define the usual metric ρ on $B(X, Y)$ by

$$\rho(\phi, \psi) = \sup_{x \in X} \tau(\phi(x), \psi(x)) \quad (\phi, \psi \in B(X, Y)).$$

When X is also equipped with a metric, so that continuity on X has a meaning, we denote by $C(X, Y)$ the subspace of $B(X, Y)$ consisting of the bounded, continuous functions on X into Y. The usual metric on $C(X, Y)$ is the one induced by $(B(X, Y), \rho)$. When $Y = R^1$, $B(X, Y)$ and $C(X, Y)$ are usually written $B(X)$ and $C(X)$ respectively. We have also previously used, and we retain, the notation $B[a, b]$ and $C[a, b]$ for $B(X)$ and $C(X)$ when X is the interval $[a, b]$.

Theorem 3.44. *If X is any set and (Y, τ) is a complete metric space, then $B(X, Y)$ is complete.*

Proof. Let (ϕ_n) be a Cauchy sequence in $B(X, Y)$ and let x be any point of X. Since

$$\tau(\phi_m(x), \phi_n(x)) \leqslant \sup_{\xi \in X} \tau(\phi_m(\xi), \phi_n(\xi)) = \rho(\phi_m, \phi_n),$$

it follows that $(\phi_n(x))$ is a Cauchy sequence in the complete metric space (Y, τ). Therefore $\lim_{n \to \infty} \phi_n(x)$ exists for every $x \in X$. Let $\phi : X \to Y$ be the function given by

$$\phi(x) = \lim_{n \to \infty} \phi_n(x) \quad (x \in X).$$

The next step is to show that $\phi \in B(X, Y)$. There is an integer p such that

$$\rho(\phi_m, \phi_n) < 1 \quad \text{whenever} \quad m, n \geqslant p.$$

Therefore, for any $x \in X$,

$$\tau(\phi_m(x), \phi_n(x)) < 1 \quad \text{whenever} \quad m, n \geqslant p. \tag{3.41}$$

Taking $n = p$ and letting $m \to \infty$ in (3.41) we get (see exercise $2(a), 5$)

$$\tau(\phi(x), \phi_p(x)) \leqslant 1.$$

Since this holds for all $x \in X$ and since ϕ_p is bounded (i.e. $\phi_p(X)$ is contained in an open ball), it follows that ϕ is bounded.

We must still prove that $\phi_n \to \phi$, i.e. that $\rho(\phi_n, \phi) \to 0$ (which is a stronger assertion than the statement that $\phi_n(x) \to \phi(x)$ for every x —see exercise $3(d), 7$). Given $\epsilon > 0$, there is an n_0 such that

$$\rho(\phi_m, \phi_n) < \epsilon \quad \text{whenever} \quad m, n > n_0$$

and so, by an argument similar to the one we have just used, we see that, for every $x \in X$,

$$\tau(\phi(x), \phi_n(x)) \leqslant \epsilon \quad \text{whenever} \quad n > n_0.$$

Thus

$$\rho(\phi, \phi_n) = \sup_{x \in X} \tau(\phi(x), \phi_n(x)) \leqslant \epsilon \quad \text{whenever} \quad n > n_0$$

and so $\phi_n \to \phi$. |

Corollary. *If X is any set, the space $B(X)$ is complete.*

Theorem 3.45. *If (X, σ) is any metric space and (Y, τ) is a complete metric space, then $C(X, Y)$ is complete.*

Proof. If (ϕ_n) is a Cauchy sequence in $C(X, Y)$ and so also in $B(X, Y)$, then, by theorem 3.44, there is a $\phi \in B(X, Y)$ such that $\phi_n \to \phi$ as $n \to \infty$. We have to show that ϕ is continuous on X.

Let x_0 be an arbitrary point of X and take $\epsilon > 0$. There is an n such that

$$\rho(\phi, \phi_n) = \sup_{x \in X} \tau(\phi(x), \phi_n(x)) < \epsilon;$$

and, since ϕ_n is continuous, there is a $\delta > 0$ such that

$$\tau(\phi_n(x_0), \phi_n(x)) < \epsilon \quad \text{whenever} \quad \sigma(x_0, x) < \delta.$$

Hence, when $\sigma(x_0, x) < \delta$,

$$\tau(\phi(x_0), \phi(x)) \leqslant \tau(\phi(x_0), \phi_n(x_0)) + \tau(\phi_n(x_0), \phi_n(x)) + \tau(\phi_n(x), \phi(x))$$
$$< 3\epsilon.$$

Thus ϕ is continuous at x_0 and, since x_0 was an arbitrary point of X, ϕ is continuous on X. |

Corollary 1. *If (X, σ) is any metric space and (Y, τ) is a complete metric space, then $C(X, Y)$ is closed in $B(X, Y)$.*

Corollary 2. *If (X, σ) is any metric space, the space $C(X)$ is complete.*

We end this section with a result that has many applications in analysis.

Definition. *Let (X, ρ) be a metric space and let $\Omega: X \to X$ be a function mapping X into itself. If now there is a non-negative number $k < 1$ such that* $$\rho(\Omega(x), \Omega(x')) \leqslant k\rho(x, x')$$

for all $x, x' \in X$, then Ω is called a contraction mapping.

Clearly a contraction mapping is continuous.

Theorem 3.46. (Banach's fixed point principle.) *If Ω is a contraction mapping on a complete metric space (X, ρ), then the equation $\Omega(x) = x$ has one and only one solution (i.e. the mapping $\Omega: X \to X$ leaves one and only one point unchanged).*

Proof. Let x_0 be an arbitary point of X and define the members of the sequence (x_n) by the recurrence relation

$$x_{n+1} = \Omega(x_n) \quad (n \geqslant 0).$$

Then
$$\rho(x_m, x_{m+1}) = \rho(\Omega(x_{m-1}), \Omega(x_m))$$

$$\leqslant k\rho(x_{m-1}, x_m)$$

$$\dots \dots \dots \dots \dots$$

$$\leqslant k^m \rho(x_0, x_1)$$

and

$$\rho(x_n, x_{n+r}) \leqslant k^n \rho(x_0, x_r)$$
$$\leqslant k^n \{\rho(x_0, x_1) + \rho(x_1, x_2) + \ldots + \rho(x_{r-1}, x_r)\}$$
$$\leqslant k^n \rho(x_0, x_1)(1 + k + \ldots + k^{r-1})$$
$$< \frac{k^n}{1-k} \rho(x_0, x_1).$$

Since $k^n \to 0$ as $n \to \infty$, (x_n) is therefore a Cauchy sequence and X, being complete, contains a point x such that $x_n \to x$. Now, since Ω is continuous, by theorem 3.22,

$$\Omega(x_n) \to \Omega(x).$$

But also
$$\Omega(x_n) = x_{n+1} \to x$$

and therefore $\Omega(x) = x$.

To prove the uniqueness of the fixed point, suppose that

$$\Omega(x) = x \quad \text{and} \quad \Omega(x') = x'.$$

Then
$$\rho(x, x') = \rho(\Omega(x), \Omega(x')) \leqslant k\rho(x, x')$$

and so, since $k < 1$, $\rho(x, x') = 0$, i.e. $x = x'$. |

Many theorems on the existence and uniqueness of the solutions of differential, integral and other equations are traditionally proved by various devices involving successive approximations. The essence of these procedures is embodied in Banach's fixed point principle whose use therefore greatly simplifies the classical proofs. We illustrate the method by proving a theorem on implicitly defined functions.

Theorem 3.47. *Let (x_0, y_0) be an interior point of a set E in R^2 and suppose that the function $f: E \to R^1$ satisfies the following conditions.*
    (i) *$f(x_0, y_0) = 0$;*
    (ii) *f is continuous in an open set G containing (x_0, y_0);*
    (iii) *f_y exists in G and is continuous at (x_0, y_0), and $f_y(x_0, y_0) \neq 0$.*
Then there exist a rectangle

$$M \times N = [x_0 - \alpha, x_0 + \alpha] \times [y_0 - \beta, y_0 + \beta]$$

and a continuous function $\phi: M \to N$ such that $y = \phi(x)$ is the only solution lying in $M \times N$ of the equation $f(x, y) = 0$.

Proof. Put
$$q = \frac{1}{f_y(x_0, y_0)}.$$

Since f_y is continuous at (x_0, y_0), there is a rectangle

$$[x_0 - \delta, x_0 + \delta] \times [y_0 - \beta, y_0 + \beta]$$

contained in G in which

$$|1 - qf_y(x, y)| < \tfrac{1}{2}. \tag{3.42}$$

Using the conditions (i) and (ii) we can now find a positive number $\alpha \leqslant \delta$ such that, for $x \in M = [x_0 - \alpha, x_0 + \alpha]$,

$$|qf(x, y_0)| < \tfrac{1}{2}\beta. \tag{3.43}$$

By theorem 3.43, $N = [y_0 - \beta, y_0 + \beta]$ is complete. Hence, by theorems 3.44 and 3.45, the spaces $B(M, N)$, $C(M, N)$, equipped with their usual metrics, are complete. Let Ω be the mapping on $B(M, N)$ defined by $\Omega(\psi) = \chi$, where

$$\chi(x) = \psi(x) - qf(x, \psi(x)) \quad (x \in M).$$

Note that, if ψ is continuous, then χ is also continuous.

We first show that Ω maps $B(M, N)$ into itself. Let $\psi \in B(M, N)$. Then, for $x \in M$, we have, by the mean value theorem,

$$\begin{aligned}
\Omega(\psi)(x) - y_0 &= \psi(x) - qf(x, \psi(x)) - y_0 \\
&= [\psi(x) - qf(x, \psi(x))] - [y_0 - qf(x, y_0)] + qf(x, y_0) \\
&= [\psi(x) - y_0][1 - qf_y(x, u)] + qf(x, y_0),
\end{aligned}$$

where u lies between y_0 and $\psi(x)$ and so in N. Hence, by (3.42) and (3.43),

$$|\Omega(\psi)(x) - y_0| < \tfrac{1}{2}|\psi(x) - y_0| + \tfrac{1}{2}\beta < \beta.$$

We use a similar argument to prove that Ω is a contraction mapping. If $\psi_1, \psi_2 \in B(M, N)$ and $x \in M$,

$$\begin{aligned}
\Omega(\psi_1)(x) - \Omega(\psi_2)(x) &= [\psi_1(x) - qf(x, \psi_1(x))] - [\psi_2(x) - qf(x, \psi_2(x))] \\
&= [\psi_1(x) - \psi_2(x)][1 - qf_y(x, v)],
\end{aligned}$$

where v lies between $\psi_1(x)$ and $\psi_2(x)$. Therefore, by (3.42),

$$\rho(\Omega(\psi_1), \Omega(\psi_2)) \leqslant \tfrac{1}{2}\rho(\psi_1, \psi_2).$$

It now follows from theorem 3.46 that Ω has a unique fixed point, which means that there is a unique function $\phi : M \to N$ such that $\Omega(\phi) = \phi$, i.e. $f(x, \phi(x)) = 0$ for $x \in M$. (The uniqueness implies, in particular, that $\phi(x_0) = y_0$.) We still have to prove that the function ϕ is continuous. But we have remarked that, if ψ is continuous, then so is $\Omega(\psi)$. Hence, if Ω^* is the restriction of Ω to $C(M, N)$, then Ω^* is a contraction mapping on $C(M, N)$. Therefore Ω^* has a fixed point and, since a fixed point of Ω^* is also a fixed point of Ω, it follows that ϕ is the fixed point of Ω^*. Thus ϕ belongs to $C(M, N)$ and so is continuous. |

A result similar to theorem 3.47 was proved in C1 (167) by entirely elementary means. But the present method of proof can be used for the general implicit function theorem involving systems of equations. (See theorem 7.43.)

Exercises 3(d)

1. Let (X, ρ) be a metric space and let E be a subset of X which has a limit point x_0. Suppose also that (Y, σ) is a complete metric space and that the function $f : E \to Y$ is such that, given $\epsilon > 0$, there exists $\delta > 0$ such that

$$\sigma(f(x), f(x')) < \epsilon \quad \text{whenever} \quad x, x' \in E \cap B(x_0; \delta).$$

Prove that $\lim_{x \to x_0} f(x)$ exists.

2. For $n = 1, 2, \ldots,$ let $\quad s_n = 1 + \dfrac{1}{1!} + \dfrac{1}{2!} + \ldots + \dfrac{1}{n!},$

Prove that (s_n) is a Cauchy sequence in Q, the space of rational numbers. Show also that, for every integer q, $q! s_q$ is an integer and $0 < q!(e - s_q) < 1$ (where $e = \sum_0^\infty 1/n!$). Deduce that e is not rational and that Q is not complete.

3. Let D be the space of real functions differentiable on $[0, 2]$ and let ρ be the metric on D induced by $B[0, 2]$. Show that (D, ρ) is not complete.

4. The metric σ on R^1 is given by

$$\sigma(x, y) = \left| \frac{x}{1 + |x|} - \frac{y}{1 + |y|} \right| \quad (x, y \in R^1).$$

Show that (R^1, σ) is not complete.

5. The metrics ρ, σ on a set X are equivalent. Are (X, ρ) and (X, σ) necessarily both complete or both not complete?

6. (i) Prove that the intersection of any collection of complete subsets of a metric space is complete.

(ii) Prove that the union of a finite number of complete subsets of a metric space is complete.

7. If (X, ρ) is a metric space, we define the *diameter* $\rho(E)$ of a bounded subset E of X by

$$\rho(E) = \sup_{x_1, x_2 \in E} \rho(x_1, x_2).$$

Prove the following result.

Cantor's intersection theorem. If (X, ρ) is a complete metric space and (F_n) is a contracting sequence of non-empty, closed sets in X such that $\rho(F_n) \to 0$ as $n \to \infty$, then $\bigcap_{n=1}^{\infty} F_n$ consists of exactly one point.

Show that, if any one of the conditions

$$\text{(i) } (X, \rho) \text{ is complete,} \quad \text{(ii) } F_n \text{ is closed,} \quad \text{(iii) } \rho(F_n) \to 0$$

is omitted, then $\bigcap_{n=1}^{\infty} F_n$ may be empty.

8. Give an example of a sequence (ϕ_n) and a function ϕ, all in $B[0, 1]$, such that $\phi_n(x) \to \phi(x)$ for every $x \in [0, 1]$, but $\phi_n \not\to \phi$.

9. Let V, W be normed vector spaces. Prove that, if W is complete, then $L(V, W)$ is complete.

(Note that the metrics on $L(V, W)$ and on $C(V, W)$ are defined quite differently. The only element common to the two spaces is the zero function $\Theta : V \to W$ such that $\Theta(x) = \theta$ for all $x \in V$.)

10. The real function f is continuous on the interval $[a, b]$ and differentiable in (a, b). Also $a \leqslant f(x) \leqslant b$ for $a \leqslant x \leqslant b$ and $|f'(x)| \leqslant k < 1$ for $a < x < b$. Prove that equation $f(x) = x$ has one and only one solution in $[a, b]$.

11. Prove the following implicit function theorem.

Let S be the strip $[a, b] \times (-\infty, \infty)$ of R^2 and suppose that the function $f : S \to R^1$ satisfies the following conditions:
  (i) f is continuous;
  (ii) f_y exists in S and there are constants m, M such that

$$0 < m \leqslant f_y(x, y) \leqslant M$$

for all $(x, y) \in S$.

Then there exists a continuous function ϕ on $[a, b]$ such that, for $a \leqslant x \leqslant b$, $y = \phi(x)$ is the only solution of the equation $f(x, y) = 0$.

12. Prove the following theorem.

Let S be the rectangle $[x_0 - a, x_0 + a] \times [y_0 - b, y_0 + b]$ of R^2. The function $f : S \to R^1$ is continuous and satisfies the following *Lipschitz condition*: There is a constant A such that

$$|f(x, y_1) - f(x, y_2)| \leqslant A |y_1 - y_2| \text{ for all } (x, y_1), (x, y_2) \in S.$$

Also let $|f(x, y)| \leqslant B$ in S; and let α in $[0, a]$ be such that

$$\alpha < \min (1/A, b/B).$$

Then the differential equation

$$\frac{dy}{dx} = f(x, y),$$

subject to the initial condition $y = y_0$ when $x = x_0$, has a unique solution in the interval $[x_0 - \alpha, x_0 + \alpha]$.

(First verify that $y = \phi(x)$ is a solution of the differential equation, subject to the given initial condition, if and only if ϕ satisfies the integral equation

$$\phi(x) = y_0 + \int_{x_0}^{x} f(t, \phi(t)) \, dt.)$$

13. Let $\Omega : R^1 \to R^1$ be defined by

$$\Omega(x) = x + 1/(1 + e^x).$$

Show that (i) Ω has no fixed point, and (ii) for all distinct $x_1, x_2 \in R^1$,

$$|\Omega(x_1) - \Omega(x_2)| < |x_1 - x_2|.$$

14. Given $\Omega : X \to X$, define $\Omega^{(1)} = \Omega$ and $\Omega^{(n)} = \Omega \circ \Omega^{(n-1)}$ ($n \geqslant 2$). Show that, if the metric space (X, ρ) is complete and there is a positive integer r such that $\Omega^{(r)}$ is a contraction mapping, then Ω has a unique fixed point.

15. The function $\Omega : C[0, 1] \to C[0, 1]$ is defined by

$$\Omega(\phi)(x) = \int_0^x \phi(t)\, dt \quad (\phi \in C[0, 1], \ x \in [0, 1]).$$

Prove that Ω is not a contraction mapping, but that $\Omega^{(2)}$ is a contraction mapping.

3.5. Completion of metric spaces

In this section the notion of isometry, defined on p. 39 is essential. We also recall (see exercise 2(c), 11) that, if (Y, σ) is a metric space, the subset D of Y is *dense* in Y if $\overline{D} = Y$.

Definition. *Let (X, ρ) be a metric space. The complete metric space (Y, σ) is said to be a* completion of *(X, ρ) if it contains a metric subspace (X_0, σ_0) such that*
  (i) *(X, ρ), (X_0, σ_0) are isometric, and*
  (ii) *X_0 is dense in Y.*

Theorem 3.5. *Every metric space has a completion and all the completions of the space are isometric.*

Proof. Let (X, ρ) be a metric space. We begin by constructing a metric space (X^*, ρ^*) which will be seen to be a completion of (X, ρ).

(i) *Definition of X^*.* We call two Cauchy sequences (x_n), (x_n') in (X, ρ) *equivalent* and write $(x_n) \sim (x_n')$ if $\rho(x_n, x_n') \to 0$ as $n \to \infty$. It is easily checked that we have, in fact, defined an equivalence relation. We denote by X^* the set whose elements are the equivalence classes generated by this relation.

(ii) *Definition of ρ^*.* Let x^*, y^* be two elements of X^* and let (x_n), (y_n) be arbitrary members of x^*, y^* respectively. Then (x_n), (y_n) are Cauchy sequences in (X, ρ) and

$$|\rho(x_m, y_m) - \rho(x_n, y_n)| \leqslant \rho(x_m, x_n) + \rho(y_m, y_n).$$

Therefore $(\rho(x_n, y_n))$ is a Cauchy sequence in R^1 and, since R^1 is complete, $\lim\limits_{n\to\infty} \rho(x_n, y_n)$ exists. Moreover, if (x_n'), (y_n') are any other members of x^* and y^*, i.e. $(x_n') \sim (x_n)$ and $(y_n') \sim (y_n)$, then

$$|\rho(x_n, y_n) - \rho(x_n', y_n')| \leqslant \rho(x_n, x_n') + \rho(y_n, y_n')$$

and so

$$\lim_{n\to\infty} \rho(x_n', y_n') = \lim_{n\to\infty} \rho(x_n, y_n).$$

We are therefore justified in defining the function ρ^* on $X^* \times X^*$ by

$$\rho^*(x^*, y^*) = \lim_{n\to\infty} \rho(x_n, y_n),$$

but we must still verify that ρ^* is a metric on X^*.

Clearly $\rho^*(x^*, y^*) \geqslant 0$ and $\rho^*(x^*, x^*) = 0$. Now suppose that $\rho^*(x^*, y^*) = 0$. If $(x_n) \in x^*$, $(y_n) \in y^*$, then $\lim \rho(x_n, y_n) = 0$, i.e. $(x_n) \sim (y_n)$, and $x^* = y^*$. Thus M1 is satisfied. It is clear that M2 holds. To prove M3 we take $(x_n) \in x^*$, $(y_n) \in y^*$, $(z_n) \in z^*$ and let $n \to \infty$ in the inequality

$$\rho(x_n, z_n) \leqslant \rho(x_n, y_n) + \rho(y_n, z_n).$$

(iii) (X^*, ρ^*) *contains a subspace isometric with* (X, ρ). The subset X_0 of X^* is defined as follows: $x^* \in X_0$ if and only if there is an $x \in X$ such that the (Cauchy) sequence $(x, x, x, \dots)$ belongs to x^*. This x^* therefore consists of all the sequences in X which converge to x. Evidently different points x, y in X determine distinct members of X_0. Thus the function $f : X \to X_0$ defined by

$$f(x) = x^*,$$

where x^* is the equivalence class containing $(x, x, \dots)$, is bijective.

If $x, y \in X$ and $f(x) = x^*$, $f(y) = y^*$, so that $(x, x, \dots) \in x^*$ and $(y, y, \dots) \in y^*$, then, by definition of ρ^*,

$$\rho^*(x^*, y^*) = \rho(x, y).$$

Hence f is an isometry.

(iv) X_0 *is dense in* X^*. Take an arbitrary element x^* of X^*. Let (x_n) be a member of x^* and, for each k, denote by x_k^* the element of X_0 which contains the sequence $(x_k, x_k, \dots)$. Then

$$\rho^*(x^*, x_k^*) = \lim_{n\to\infty} \rho(x_n, x_k)$$

and, since (x_n) is a Cauchy sequence, the right-hand side tends to 0 as $k \to \infty$. Thus $x_k^* \to x^*$. This means that $\overline{X_0} = X^*$, i.e. that X_0 is dense in X^*.

(v) (X^*, ρ^*) *is complete.* Let (x^*_n) be an arbitrary Cauchy sequence in X^*. Since X_0 is dense in X^*, for each n there is a $y_n^* \in X_0$ such that
$$\rho^*(x^*_n, y_n^*) < 1/n.$$

The equivalence class y_n^* contains a sequence $(y_n, y_n, ...)$ of points in X. Now

$$
\begin{aligned}
\rho(y_m, y_n) &= \rho^*(y_m^*, y_n^*) \\
&\leqslant \rho^*(y_m^*, x^*_m) + \rho^*(x^*_m, x^*_n) + \rho^*(x^*_n, y_n^*) \\
&< \rho^*(x^*_m, x^*_n) + \frac{1}{m} + \frac{1}{n}.
\end{aligned}
$$

Since (x^*_n) is a Cauchy sequence in X^*, it follows that (y_n) is a Cauchy sequence in X. Let $y^* \in X^*$ be the equivalence class containing (y_n). Then

$$\rho^*(y^*, x^*_k) \leqslant \rho^*(y^*, y_k^*) + \rho^*(y_k^*, x^*_k) < \lim_{n \to \infty} \rho(y_n, y_k) + 1/k.$$

As (y_n) is a Cauchy sequence, the right-hand side tends to 0 as $k \to \infty$, i.e. $x^*_k \to y^*$. Hence (X^*, ρ^*) is complete.

(vi) *Any two completions of* (X, ρ) *are isometric.* Let (Y, σ) be a completion of (X, ρ) and let (Y_0, σ_0) be a subspace of (Y, σ) isometric with (X, ρ). Proceeding as in (i) we group the Cauchy sequences (y_n) in Y_0 into equivalence classes y^*.

Since Y is complete, every Cauchy sequence in Y_0 has a limit in Y and clearly all Cauchy sequences belonging to the same equivalence class y^* have the same limit y. Conversely, to any $y \in Y$, there corresponds an equivalence class y^*. For, since Y_0 is dense in Y, there is a sequence (y_n) in Y_0 which converges to y and y^* is the equivalence class determined by (y_n). We have therefore set up a bijection between the equivalence classes y^* and the points y of Y.

Let (W, τ) be another completion of (X, ρ) with subspace (W_0, τ_0) isometric to (X, ρ). As before, we have a bijection between the points w of W and the equivalence classes w^* of Cauchy sequences in W_0.

The spaces (Y_0, σ_0) and (W_0, τ_0) are isometric, since each is isometric with (X, ρ). If (y_n), (w_n) are corresponding sequences in Y_0 and W_0 respectively, then

$$\sigma(y_m, y_n) = \tau(w_m, w_n)$$

and therefore, when one is a Cauchy sequence, so is the other. Moreover, if (y'_n), (w'_n) also correspond, then $(y_n) \sim (y'_n)$ if and only if $(w_n) \sim (w'_n)$. Hence the isometry between (Y_0, σ_0) and (W_0, τ_0)

induces a bijection between the set of y^* and the set of w^* and thence leads to a bijection between Y and W (which preserves the original bijection between Y_0 and W_0).

Now let y, $y' \in Y$ and w, $w' \in W$ be any corresponding pairs of points. Let (y_n), (y_n') and (w_n), (w_n') be corresponding sequences from Y_0 and W_0 respectively such that $y_n \to y$, $y_n' \to y'$ and $w_n \to w$, $w_n' \to w'$. In view of the isometry between (Y_0, σ_0) and (W_0, τ_0) we have

$$\sigma(y, y') = \lim_{n \to \infty} \sigma(y_n, y_n') = \lim_{n \to \infty} \tau(w_n, w_n') = \tau(w, w').$$

Thus the bijection between (Y, σ) and (W, τ) is, in fact, an isometry. |

It is now easy to construct a completion of (X, ρ) which actually contains (X, ρ) as a subspace. Let

$$X^+ = X \cup (X^* - X_0).$$

We obtain a natural correspondence between the points x^+ of X^+ and the points x^* of X^* by using the bijection between X and X_0 and by letting the points of $X^* - X_0$ in X^+ and in X^* correspond to themselves. The metric ρ^+ on X^+ is then defined by

$$\rho^+(x^+, y^+) = \rho^*(x^*, y^*).$$

(X^+, ρ^+) is the desired completion of (X, ρ). We have proved that *any metric space can be embedded in a complete metric space which is the closure of that space.*

Exercises 3(e)

1. Let (X, ρ) be a metric space. Show that, if (Y, σ) is a complete metric space containing a subspace isometric with (X, ρ), then (Y, σ) contains a completion of (X, ρ).

2. Show, by means of an example, that a metric space (X, ρ) may have completions (Y, σ) and (Y_1, σ_1) such that (Y_1, σ_1) is a proper subspace of (Y, σ) (i.e. $Y - Y_1 \neq \varnothing$).

3. Given the metric space (X, ρ), let (X^+, ρ^+) be a completion which contains (X, ρ) as a subspace. Prove that a contraction mapping Ω on (X, ρ) has a unique continuous extension $\Omega^+ : X^+ \to X^+$, and that Ω^+ is a contraction mapping on (X^+, ρ^+).

3.6. Compact metric spaces

Among the properties of metric spaces considered by us the one with the most far-reaching consequences is compactness.

Definition. *Let (X, ρ) be a metric space. A subset E of X (which may be X itself) is said to be* compact *if every sequence in E has subsequence which converges in E. If X is compact, we say that the metric space (X, ρ) is compact.*

Notes. (i) In any metric space the empty set is compact. (ii) The non-empty set E in the metric space (X, ρ) is compact if and only if the subspace (E, σ) of (X, ρ) is compact.

Illustrations

(i) In any metric space, a set of only a finite number of points is compact.

To prove this we need only remark that the empty set is compact and that, if (x_n) is a sequence in the set $\{\xi_1, ..., \xi_k\}$, at least one point ξ_i appears infinitely often in (x_n).

(ii) Every finite closed interval of R^n is compact.

First consider an interval $[a, b]$ of R^1. If (x_n) is a sequence in $[a, b]$, then (x_n) is bounded and, by the lemma to theorem 3.42, contains a subsequence (x_{ν_k}) which converges to a point $x \in R^1$. But, for all k, $a \leqslant x_{\nu_k} \leqslant b$ and so

$$a \leqslant \lim_{k \to \infty} x_{\nu_k} \leqslant b.$$

Thus $x_{\nu_k} \to x \in [a, b]$ and so $[a, b]$ is compact.

Next, let $I = [a, b] \times [c, d]$ be a finite closed interval in R^2 and let (x_n) be a sequence in I. If $x_n = (\xi_n, \eta_n)$, then, since (ξ_n) lies in $[a, b]$, (ξ_n) has a subsequence (ξ_{μ_j}) which converges to a point ξ in $[a, b]$. Put $\xi_{\mu_j} = \xi_j'$, $\eta_{\mu_j} = \eta_j'$. As η_j' lies in $[c, d]$, there is a subsequence (η_{λ_i}') which converges to a point η in $[c, d]$. Also $\xi_{\lambda_i}' \to \xi$, since $\xi_j' \to \xi$. But (ξ_{λ_i}') and (η_{λ_i}') are subsequences (ξ_{ν_k}) and (η_{ν_k}) of (ξ_n) and (η_n) respectively. Therefore the subsequence (x_{ν_k}) converges to the point $x = (\xi, \eta)$ in I. Hence I is compact.

The argument leading from R^1 to R^2 may be adapted to make the inductive step from R^{n-1} to R^n.

(iii) R^n and Z are not compact. For instance in R^1 the sequence of positive integers has no convergent subsequence.

(iv) $C(X)$ (with its usual metric) is not compact. For, whatever metric X has, the functions ϕ_n $(n = 1, 2, ...)$ given by

$$\phi_n(x) = n \quad (x \in X)$$

are continuous on X and the sequence (ϕ_n) clearly has no convergent subsequence.

The last two illustrations show that a complete metric space need not be compact. On the other hand, compactness implies completeness.

Theorem 3.61. *A compact set in a metric space is also complete.*

Proof. Let (X, ρ) be a metric space and let E be a compact set in X. A Cauchy sequence (x_n) in E has a subsequence (x_{ν_k}) which converges to a point $x \in E$. Then

$$\rho(x_n, x) \leqslant \rho(x_n, x_{\nu_n}) + \rho(x_{\nu_n}, x) \to 0,$$

i.e. $x_n \to x$. (Essentially the same argument, in an expanded version, forms the last part of the proof of theorem 3.42.) |

The property of compactness may also be expressed slightly differently.

Theorem 3.62. *A set E in a metric space is compact if and only if every infinite subset of E has at least one limit point in E.*

Proof.

(i) Suppose that the set E in a given metric space is compact. Let A be an infinite subset of E. Then A contains a countably infinite subset the points of which may be arranged in a sequence (x_n). Since E is compact, (x_n) has a subsequence (x_{v_k}) which converges to a point $x \in E$. Clearly x is a limit point of the set $\{x_{v_1}, x_{v_2}, ...\}$ and so of the set A.

(ii) Suppose that every infinite subset of E has at least one limit point in E. Let (x_n) be any sequence in E. The set of points $\{x_1, x_2, ...\}$ may then be finite or infinite. If $\{x_1, x_2, ...\}$ is finite, then (as in illustration (i) on p. 61) at least one of its points occurs infinitely many times in the sequence (x_n). Thus (x_n) has a convergent subsequence. If the set $\{x_1, x_2, ...\}$ is infinite, then it has a limit point x in E and this is clearly the limit of a subsequence of (x_n). |

A bounded set in R^n lies in a closed interval. As we have seen that such an interval is compact, we now have the following result: *A bounded, infinite set in R^n has at least one limit point.* This is the Bolzano–Weierstrass theorem, one of the earliest results in point set theory.

The next result is an analogue of theorem 3.43.

Theorem 3.63. *Let (X, ρ) be a metric space and let E be a subset of X.*
(i) *If E is compact, then E is bounded and closed.*
(ii) *If X is compact and if E is closed, then E is compact.*

Proof.

(i) If E is compact then it is also complete and so closed by theorem 3.43 (i). The direct proof is also quite simple.

Let a be a fixed point of X. If E is not bounded, there is a sequence (x_n) in E such that
$$\rho(x_n, a) > n \quad (n = 1, 2, ...).$$

Thus $\rho(x_n, a) \to \infty$ as $n \to \infty$. If (x_{v_k}) is any subsequence of (x_n), $\rho(x_{v_k}, a) \to \infty$ as $k \to \infty$ and so (x_{v_k}) cannot converge. Hence E is not compact.

(ii) Let (x_n) be any sequence in E and so in X. Since X is compact, there is a subsequence (x_{v_k}) with limit x in X; and, since E is closed, theorem 2.31 shows that $x \in E$. Therefore E is compact. |

Corollary. *In R^n a set is compact if and only if it is bounded and closed.*

Although in the space R^n the converse of (i) is true, in general it is false. A counter example is given in exercise $3(f)$, 3.

In the rest of this section we investigate the properties of continuous functions with compact domains.

Theorem 3.64. *If the domain of a continuous function is compact, then the range is also compact.*

Proof. Let (X, ρ) be a compact metric space, let (Y, σ) be any metric space and suppose that the function $f : X \to Y$ is continuous on X. We wish to show that the subset $f(X)$ of Y is compact.

Take any sequence (y_n) in $f(X)$. For each n choose an x_n such that $f(x_n) = y_n$. (There may be several x for which $f(x) = y_n$.) Since X is compact, the sequence (x_n) contains a subsequence (x_{ν_k}) which converges to a point $x \in X$. Then, by theorem 3.22, the continuity of f ensures that $f(x_{\nu_k}) \to f(x)$. Hence, if $f(x) = y$, $y_{\nu_k} \to y$, where $y \in f(X)$. $\vert$

Corollary 1. *If the domain of a continuous function is compact, then the range is bounded and closed.*

Corollary 2. *If the domain of a real-valued continuous function is compact, then the function is bounded and attains its upper and lower bounds.*

The second corollary follows from the fact that a bounded, closed set in R^1 contains its supremum and infimum (exercise $2(c)$, 3). It generalizes the corresponding elementary result in which the domain of the function is a finite, closed interval (C1, theorem 3.71).

Theorem 3.65. *If the domain of a bijective, continuous function is compact, then the inverse function is also continuous (i.e. the function is a homeomorphism).*

First proof. Let $f : X \to Y$ be a bijective, continuous function with compact domain (X, ρ) and with range (Y, σ).

Let (y_n) be a sequence in Y which converges to a point $y \in Y$. We wish to show that $f^{-1}(y_n) \to f^{-1}(y)$. Put $f^{-1}(y_n) = x_n$. Since X is compact, (x_n) contains at least one convergent subsequence. Let (x_{ν_k}) be such a sequence with limit x, say. Then, as f is continuous,

$$y_{\nu_k} = f(x_{\nu_k}) \to f(x).$$

But $y_{\nu_k} \to y$ and so $f(x) = y$, i.e. $x = f^{-1}(y)$. Therefore every convergent subsequence of (x_n) converges to $x = f^{-1}(y)$. It then follows by the compactness of X (see exercise $3(f), 4$) that $x_n \to x$, or $f^{-1}(y_n) \to f^{-1}(y)$.

Second proof. This is shorter, but more sophisticated.

Let F be a closed set in X. By theorem 3.63 (ii), F is compact and, by theorem 3.64, $f(F)$ is compact. Then, by theorem 3.63 (i), $f(F)$ is closed in Y. Thus $(f^{-1})^{-1}(F)$ (i.e. $f(F)$) is closed in Y whenever F is closed in X. Therefore, by theorem 3.23 (ii), f^{-1} is continuous on Y. |

An immediate corollary is the well known theorem (C1, theorem 3.9) that *a strictly monotonic, continuous, real function on an interval has a continuous inverse.*

The property of uniform continuity, possessed by real functions continuous on a closed interval, was briefly discussed in C1 (§3.8). It was used to prove that a continuous function is integrable. We now consider uniform continuity in the general setting of metric spaces.

We recall that, when (X, ρ) and (Y, σ) are metric spaces, the function $f: X \to Y$ is continuous at the point $c \in X$ if, given ϵ, there is a δ such that

$$\sigma(f(x), f(c)) < \epsilon \quad \text{whenever} \quad \rho(x, c) < \delta.$$

Here the number δ depends on c (as well as on ϵ) and generally one cannot expect one δ to suit all c's. When this *is* the case, i.e. when, given ϵ, there is a $\delta = \delta(\epsilon)$, *independent of c*, such that, for *all c in X,*

$$\sigma(f(x), f(c)) < \epsilon \quad \text{whenever} \quad \rho(x, c) < \delta,$$

then f is said to be uniformly continuous on X. Actually a slightly different, but equivalent form of the definition is more useful.

Definition. *When (X, ρ) and (Y, σ) are metric spaces the function $f: X \to Y$ is said to be* uniformly continuous *on X if, given ϵ, there is a δ such that*

$$\sigma(f(x_1), f(x_2)) < \epsilon \quad \text{whenever} \quad \rho(x_1, x_2) < \delta.$$

The condition is clearly equivalent to the set

$$E(\delta) = \{\sigma(f(x_1), f(x_2)) | \rho(x_1, x_2) < \delta\}$$

being bounded and its supremum $\omega(\delta)$ tending to 0 as $\delta \to 0$.

Uniform continuity of course implies ordinary continuity. That continuity does not generally imply uniform continuity is easily shown by examples.

(i) The real function f on $(0, 1)$ given by $f(x) = 1/x$ is continuous, but not uniformly continuous. For

$$E(\delta) = \left\{ \left\| \frac{1}{x_1} - \frac{1}{x_2} \right\| \, 0 < x_1, x_2 < 1; \, |x_1 - x_2| < \delta \right\}$$

$$\supset \left\{ \left| \frac{1}{\frac{1}{2}x} - \frac{1}{x} \right| \, 0 < x < \delta \right\} = \left\{ \left| \frac{1}{x} \right| \, 0 < x < \delta \right\}$$

and so $E(\delta)$ is not bounded.

(ii) The real function g on R^1 given by $g(x) = \cos(x^2)$ is also continuous, but not uniformly continuous since

$$\omega(\delta) = \sup_{|x_1 - x_2| < \delta} |\cos(x_1^2) - \cos(x_2^2)| = 2.$$

(For if $x_1 = [(n+1)\pi]^{\frac{1}{2}}$, $x_2 = (n\pi)^{\frac{1}{2}}$, then $x_1 - x_2 < n^{-\frac{1}{2}}$.)

With the help of the next theorem it is also easy to exhibit uniformly continuous functions.

Theorem 3.66. *A continuous function with a compact domain is uniformly continuous.*

Proof. Let $f: X \to Y$ be a continuous function whose domain (X, ρ) is compact and whose range lies in an arbitrary metric space (Y, σ).

Suppose that f is not uniformly continuous. Then there exists an $\epsilon > 0$ with the property that, to every $\delta > 0$ there correspond points $x, y \in X$ such that

$$\rho(x, y) < \delta \quad \text{and} \quad \sigma(f(x), f(y)) \geq \epsilon.$$

Thus there are sequences (x_n), (y_n) of points in X such that

$$\rho(x_n, y_n) < 1/n \quad \text{and} \quad \sigma(f(x_n), f(y_n)) \geq \epsilon.$$

Since X is compact, the sequence (x_n) contains a convergent subsequence (x_{ν_k}) with limit α, say. We have

$$\rho(y_{\nu_k}, \alpha) \leq \rho(y_{\nu_k}, x_{\nu_k}) + \rho(x_{\nu_k}, \alpha) < 1/\nu_k + \rho(x_{\nu_k}, \alpha) \to 0$$

as $k \to \infty$. Hence $\quad x_{\nu_k} \to \alpha \quad$ and $\quad y_{\nu_k} \to \alpha;$

and, since f is continuous,

$$f(x_{\nu_k}) \to f(\alpha) \quad \text{and} \quad f(y_{\nu_k}) \to f(\alpha).$$

Thus $$\sigma(f(x_{\nu_k}), f(y_{\nu_k})) \to 0. \tag{3.61}$$

3

However $((x_{\nu_k}, y_{\nu_k}))$ is a subsequence of the sequence $((x_n, y_n))$ and therefore

$$\sigma(f(x_{\nu_k}), f(y_{\nu_k})) \geqslant \epsilon \quad \text{for every } k. \tag{3.62}$$

As (3.61) and (3.62) are incompatible, f is uniformly continuous. |

A simple example (for instance a constant function on a non-compact set) shows that the condition of the last theorem is not necessary for uniform continuity.

Exercises 3(f)

1. The metrics ρ, σ on a set X are equivalent. Prove that (X, ρ) is compact if and only if (X, σ) is compact.

2. Prove, without using theorem 3.43, that a compact subset of a metric space is closed.

3. Let $B[0, 1]$ have its usual metric and let ϕ be the function defined by $\phi(x) = 0$ for $0 \leqslant x \leqslant 1$. Show that the bounded and closed set $\overline{B(\phi; 1)}$ is not compact.

4. Let (x_n) be a sequence in a compact metric space. Show that, if every convergent subsequence (there is at least one) converges to the point x, then $x_n \to x$.

5. (i) Prove that the intersection of any collection of compact subsets of a metric space is compact. (Show first that the intersection of a compact set and a closed set is compact.)

(ii) Prove that the union of a finite number of compact subsets of a metric space is compact.

6. Show that, in a compact metric space, every contracting sequence of non-empty, closed sets has a non-empty intersection (cf. exercise 3(d), 7).

7. The function $f: [a, \infty) \to R^1$ is continuous, and $f(x) \to l$ as $x \to \infty$. Prove that f is uniformly continuous on $[a, \infty)$.

8. The real functions f, g, h, k on $[0, \infty)$ are defined by $f(x) = x^2$, $g(x) = \sin x$, $h(x) = (x+1)^{-1} \cos (x^2)$, $k(x) = x^{\frac{3}{2}}$. Which are uniformly continuous on $[0, \infty)$?

9. A real function on an interval $[a, b]$ is called *piecewise linear* if there are points $x_i (i = 0, 1, ..., n)$ such that $a = x_0 < x_1 < ... < x_{n-1} < x_n = b$ and the function is linear on each interval $[x_{i-1}, x_i]$ $(i = 1, ..., n)$.

Show that, given any function $f \in C[a, b]$ and any $\epsilon > 0$, there is a piecewise linear function $g \in C[a, b]$ such that $\rho(f, g) < \epsilon$ (where ρ is the usual metric on $C[a, b]$).

Note. In classical analysis and algebra a real function f is called *linear* if it is of the form $f(x) = ax + b$ $(x \in R^1)$, i.e. if it has a straight line graph. The nomenclature is carried over to functions on R^n to R^1 and to complex functions.

This usage clashes with the definition of a linear function on p. 39 unless $b = 0$. It is, however, long-established and expressive and we occasionally employ it. The context will make clear the interpretation intended. (The vector space terminology for the function f above is *affine*.)

10. The functions $f, g : C[0, 1] \to C[0, 1]$ are given by

$$f(\phi)(x) = \int_0^x \phi(t)\, dt, \quad g(\phi)(x) = \int_0^x \phi^2(t)\, dt \quad (\phi \in C[0, 1], \quad x \in [0, 1]).$$

Show that both functions are continuous on $C[0, 1]$. Are they uniformly continuous?

11. Let E be a compact set in a metric space (X, ρ). Prove that there are points $x, y \in E$ such that
$$\rho(E) = \rho(x, y).$$

(For the definition of the diameter $\rho(E)$ of E see exercise 3(d), 7.) Show also that, if E is assumed to be merely bounded and closed, then such points x, y need not exist.

12. In a metric space (X, ρ), the *distance* between the point c and the subset E of X, denoted by $\rho(c, E)$, is defined by

$$\rho(c, E) = \inf_{x \in E} \rho(c, x).$$

Prove that the function $f : X \to R^1$ given by

$$f(x) = \rho(x, E)$$

is uniformly continuous on X. Prove also that $\rho(x, E) = 0$ if and only if $x \in \bar{E}$.

13. In a metric space (X, ρ), the *distance* between the subsets A, B of X, denoted by $\rho(A, B)$, is defined by
$$\rho(A, B) = \inf_{x \in A,\ y \in B} \rho(x, y).$$

Prove that, if A, B are compact, there are points $a \in A$, $b \in B$ such that

$$\rho(A, B) = \rho(a, b).$$

Prove also that in R^n (with its usual metric) $\rho(A, B) = \rho(a, b)$ for some $a \in A$, $b \in B$ if one of the sets A, B is compact and the other is closed.

14. Let (X, ρ) be a metric space, let A, B be disjoint subsets of X and suppose that one of A, B is compact, the other closed. Prove that
(i) $\rho(A, B) > 0$,
(ii) there need not exist points $a \in A$, $b \in B$ such that $\rho(A, B) = \rho(a, b)$.

15. Give an example of a metric space (X, ρ) with disjoint, closed subsets A, B such that $\rho(A, B) = 0$.

3.7. The Heine–Borel theorem

In theorem 3.62 we considered a property which was evidently closely related to that of compactness and we proved that the two are, in fact, equivalent. In this section we shall establish a characterization of an entirely different kind.

Definition. A collection $\mathscr{A}$ of sets A is said to be a covering of a set E if
$$E \subset \bigcup_{A \in \mathscr{A}} A.$$

The covering is finite if $\mathscr{A}$ has a finite number of members only. When E and the sets A of $\mathscr{A}$ all lie in a metric space, the covering is said to be open if every set A of $\mathscr{A}$ is open.

If $\mathscr{A}$ and $\mathscr{A}_1$ are both coverings of E and $\mathscr{A}_1 \subset \mathscr{A}$, we say that $\mathscr{A}_1$ is, relatively to $\mathscr{A}$, a subcovering of E.

Illustrations

(i) In Z let $E = \overline{B(0; 1)}$ and let $A_\zeta = B(\zeta; 1)$. Then E is covered by the sets

$$A_\zeta \quad (\zeta \in E).$$

(ii) The interval (0, 1) is covered by the open intervals

$$(x^2, x) \quad (0 < x < 1).$$

(iii) The set $C[0, 1]$ is covered by the sets

$$A_n = \{\phi | \phi \in B[0, 1]; n-1 < \phi(0) < n+1\} \quad (n = 0, \pm 1, \pm 2, \ldots).$$

The covering (i) clearly contains finite subcoverings. The covering in (ii) contains only infinite subcoverings, while that in (iii) has no proper subcoverings.

We shall show that a necessary and sufficient condition for a set in a metric space to be compact is that every open covering of the set should contain a finite subcovering. The sufficiency is easily verified; the proof of the necessity requires some preparation.

Lemma 1. *Let E be a compact set in a metric space (X, ρ) and let ϵ be a given positive number. Then there is a finite number of points $c_1, \ldots, c_p$ in E such that*

$$E \subset \bigcup_{n=1}^{p} B(c_n; \epsilon).$$

Proof. Take c_1 to be any point of E. If $E_1 = E - B(c_1; \epsilon)$ is not empty, let c_2 be any point of E_1. Generally, if it has been possible to choose $c_1, \ldots, c_k$ and if

$$E_k = E - \bigcup_{n=1}^{k} B(c_n; \epsilon)$$

is not empty, c_{k+1} is taken to be some point of E_k. We wish to show that, after a finite number of steps, we arrive at an empty E_k.

Suppose that no E_k is empty, so that we obtain an infinite sequence (c_n). When $n > m$, $c_n \notin B(c_m; \epsilon)$, i.e.

$$\rho(c_m, c_n) \geqslant \epsilon.$$

Hence (c_n) cannot have a convergent subsequence (for such a subsequence would have to be a Cauchy sequence). This contradicts the hypothesis that E is compact. |

Lemma 2. *Let E be a compact set in a metric space (X, ρ) and let $\mathscr{G}$ be an open covering of E. Then there is a positive number α such that, for every $x \in E$, the open ball $B(x; \alpha)$ is contained in some member G of $\mathscr{G}$.*

Proof. Suppose that the lemma is false. Then, for every n, there is an $x_n \in E$ such that $B(x_n; 1/n)$ is not contained in any G of $\mathscr{G}$. Since E is compact, (x_n) has a subsequence (x_{v_k}) which converges to a point $x^* \in E$. Let G^* be a member of $\mathscr{G}$ which contains x^*. Since G^* is open, there is a $\delta > 0$ such that

$$B(x^*; 2\delta) \subset G^*.$$

For all sufficiently large values of k, $\rho(x_{v_k}, x^*) < \delta$. Hence there is an integer r such that

$$\rho(x_r, x^*) < \delta \quad \text{and} \quad r > 1/\delta.$$

If then $x \in B(x_r; 1/r)$,

$$\rho(x, x^*) \leqslant \rho(x, x_r) + \rho(x_r, x^*) < 2\delta$$

i.e. $x \in B(x^*; 2\delta)$. Therefore

$$B(x_r; 1/r) \subset B(x^*; 2\delta) \subset G^*.$$

This is a contradiction, since the $B(x_n; 1/n)$ were chosen to be open balls not lying in any member G of $\mathscr{G}$. |

Theorem 3.7. (Heine–Borel.) *A set in a metric space is compact if and only if every open covering of the set contains a finite subcovering.*

Proof.

(i) Let E be a compact set in a metric space and let $\mathscr{G}$ be an open covering of E. By lemma 2, there is an $\alpha > 0$ such that, for every $x \in E$, $B(x; \alpha)$ is contained in some G of $\mathscr{G}$. Having chosen α, we can, by lemma 1, find points $c_1, ..., c_p$ such that

$$E \subset \bigcup_{n=1}^{p} B(c_n; \alpha).$$

Now, for each n such that $1 \leqslant n \leqslant p$, there is a member G_n of $\mathscr{G}$ such that $B(c_n; \alpha) \subset G_n$. Then

$$E \subset \bigcup_{n=1}^{p} G_n$$

and so $\{G_1, ..., G_n\}$ is a finite subcovering.

(ii) Let E be a non-compact set in a metric space. Then, by theorem 3.62, E contains an infinite subset D which has no limit points in E. Therefore to each point $c \in D$ there corresponds an open ball $G_c = B(c; r_c)$ such that $G_c \cap D = \{c\}$. Moreover the open set $G = (\overline{D})'$ is such that $G \cup D \supset E$. Therefore

$$\mathscr{G} = \{G\} \cup \{G_c | c \in D\}$$

is an open covering of E. But the only member of $\mathscr{G}$ which contains a given point $c \in D (\subset E)$ is G_c. Hence every subcovering must contain the infinite collection $\{G_c | c \in D\}$ and so $\mathscr{G}$ contains no finite subcovering. |

Exercises 3(g)

1. A set E in a metric space is called *totally bounded* if it has the property which was proved for compact sets in lemma 1, namely that, given $\epsilon > 0$, there are points $c_1, \dots, c_p$ such that
$$E \subset \bigcup_{n=1}^{p} B(c_n; \epsilon).$$

Prove that a totally bounded set is bounded and show, by means of an example, that a bounded set need not be totally bounded.

2. Let E be a totally bounded set in a metric space. Show that, given a sequence (x_n) in E, there are sequences $(x_{1,n}), (x_{2,n}), (x_{3,n}), \dots$ such that $(x_{1,n}) = (x_n)$ and for $k \geq 2$,

    (i) $(x_{k,n})$ is a subsequence of $(x_{k-1,n})$,

    (ii) $(x_{k,n})$ lies in an open ball of radius $1/k$.

Prove that $(x_{n,n})$ is a Cauchy subsequence of (x_n).

    Deduce that a set in a metric space is compact if and only if it is totally bounded and complete.

3. Prove theorem 3.66 by means of the Heine–Borel theorem.

4. The function $f: [a, b] \to R^1$ is said to be increasing at the point $x \in [a, b]$ if there is a $\delta(x) > 0$ such that
$$f(u) \leq f(x) \leq f(v)$$
when $u \in (x - \delta(x), x] \cap [a, b], \quad v \in [x, x + \delta(x)) \cap [a, b].$

Show that, if f increases at every point of $[a, b]$, then f increases in $[a, b]$ (i.e. $f(\alpha) \leq f(\beta)$ when $a \leq \alpha \leq \beta \leq b$).

NOTES ON CHAPTER 3

§3.4. Although the fixed point principle has only recently become well known, S. Banach (1892–1945) proved it as long ago as 1920. It formed part of his doctoral thesis, which was published under the title 'Sur les opérations dans les ensembles abstraits et leur application aux équations intégrales' in *Fund. Mathematicae* 3 (1922), 133–81.

§3.6. What is now called the Bolzano–Weierstrass theorem was discovered by B. Bolzano (1781–1848), a profound Czech mathematician whose writings were largely ignored during his life-time. The theorem reached a wide public only when it figured in Weierstrass's lectures at the University of Berlin.

§3.7. E. Borel (1871–1956) proved in 1894 that a covering of a finite closed (linear) interval by a countable collection of open intervals has a finite sub-covering. The restriction of countability was removed by H. Lebesgue (1875–1941) who made extensive use of the improved result in his theory of integration (1902). By 1905, theorem 3.7 (for bounded closed sets in R^n) was essentially known. (See Borel's note in *Comptes Rendus* 140 (1905, I), 298–300.)

In 1871, H. E. Heine (1821–81) established the uniform continuity of a real function continuous on a finite closed interval. Since his argument contains the seeds of a covering theorem, the appellation 'Heine–Borel' has become attached to Borel's original theorem and its extensions; but 'Borel–Lebesgue', which is sometimes used for the post 1902 versions, is probably more appropriate.

Topological spaces. Once Fréchet had led the way to abstraction with his introduction of metric spaces, it was not long before mathematicians began to look for further generalizations. The neat definition below, which was arrived at only after a good deal of experimentation, rests on the observation that many properties of metric spaces and of continuous functions on metric spaces can be expressed by means of open sets without explicit mention of the metric. The definition of connectedness is already in this form; the Heine–Borel theorem gives an alternative definition of compactness in terms of open coverings; while in exercise 3(*b*), 2 and theorem 3.23 (i) open sets are used to formulate the property of continuity.

Given a non-empty set X, a collection $\mathcal{T}$ of subsets of X is said to be a *topology* on X if it has the following three properties:
 (i) $\varnothing$ and X belong to $\mathcal{T}$,
 (ii) the union of any collection of sets in $\mathcal{T}$ belongs to $\mathcal{T}$,
 (iii) the intersection of a finite number of sets in $\mathcal{T}$ belongs to $\mathcal{T}$.
If $\mathcal{T}$ is a topology on X, then the pair $(X, \mathcal{T})$ is called a *topological space* and the members of $\mathcal{T}$ are called the *open sets* of $(X, \mathcal{T})$. Since the collection of open sets of a metric space has the properties (i)–(iii), a metric space is a topological space. Equivalent metrics on a set generate the same topological space.

A topological space is called *metrizable* if there exists a metric on the under-lying set such that the collection of open sets produced by it coincides with the given topology. There are simple examples to show that not every topological space is metrizable. For instance, if X is an infinite set, let its topology consist of $\varnothing$ and all subsets G of X such that G' is finite. If ρ is any metric on X, let a, b be arbitrary distinct points of X and put $\rho(a, b) = r(>0)$. The open balls $B(a; \frac{1}{2}r)$, $B(b; \frac{1}{2}r)$ are both open in (X, ρ), but, since they are disjoint, they cannot both have finite complements. $B(a; \frac{1}{2}r)$ and $B(b; \frac{1}{2}r)$ therefore cannot both be members of the given topology. Thus topological spaces are genuine generalizations of metric spaces.

Although many features of metric spaces survive the transition to topological spaces, there are, inevitably, some casualties. The most notable of these is the property of completeness; for the notion of a Cauchy sequence has no interpre-tation in topological spaces. (The theory of *uniform spaces* has been created to accommodate a suitable generalization.)

Real numbers. In the notes at the end of chapter 1 we referred to the method of cuts which may be used for constructing the system of real numbers from that of rational numbers. It leads to Dedekind's fundamental theorem (p. 16) from which, in turn, follow results such as those on the existence of suprema and infima of bounded sets of real numbers, the convergence of bounded monotonic sequences (C1, theorems 1.8 and 2.6), and the completeness of R^1 (theorem 3.42).

We now sketch a construction of the real numbers which is based on the notion of a Cauchy sequence. For a detailed exposition see H. A. Thurston, *The Number System.*

Let Q be the set of rational numbers. Elementary analysis within Q employs the customary definitions with *rational* ϵ's. Thus the sequence (x_n) of rational numbers is said to converge to the rational number x, and we write $x_n \to x$, if, given any rational $\epsilon > 0$, there is an integer n_0 such that

$$|x_n - x| < \epsilon \quad \text{whenever} \quad n > n_0.$$

Again, a Cauchy sequence in Q is a rational sequence (x_n) such that, given any rational $\epsilon > 0$, there is an n_0 such that

$$|x_m - x_n| < \epsilon \quad \text{whenever} \quad m, n > n_0.$$

We define the equivalence of two Cauchy sequences (x_n), (x'_n) in Q as at the beginning of the proof of theorem 3.5: $(x_n) \sim (x'_n)$ if $x_n - x'_n \to 0$ as $n \to \infty$. The equivalence classes generated in this way are then called *real numbers.* We denote real numbers by starred letters. If x is a rational number, the real number x^* which contains the sequence $(x, x, \ldots)$ is called the corresponding real rational number. Note, however, that in general the use of the symbol x^* does not carry with it the implication that x^* is necessarily a real *rational* number.

The next step is to define the algebraic operations with real numbers. If x^*, y^* are two real numbers and (x_n), $(x'_n) \in x^*$, (y_n), $(y'_n) \in y^*$, then $(x_n + y_n) \sim (x'_n + y'_n)$ and $(x_n y_n) \sim (x'_n y'_n)$. We can therefore define $x^* + y^*$ and $x^* * y^*$ to be the real numbers containing the Cauchy sequences $(x_n + y_n)$ and $(x_n y_n)$ respectively. Also $-x^*$ is, unambiguously, defined as the real number containing $(-x_n)$. The definition of x^{*-1}, when $x^* \neq 0^*$, is not quite so simple to justify. It needs the following lemma.

(1) *If $(x_n) \in x^* \neq 0^*$, then there is a rational number $k > 0$ such that $|x_n| > k$ for all sufficiently large n.*

If (1) were false, there would exist a subsequence (x_{ν_k}) tending to 0 and this, in turn, would imply that the Cauchy sequence (x_n) tends to 0. Thus (1) holds. It follows that, if (x_n), $(x'_n) \in x^* \neq 0^*$ and $x_n \neq 0$, $x'_n \neq 0$ for all n, then (x_n^{-1}) and (x'^{-1}_n) are equivalent Cauchy sequences. We can therefore define x^{*-1} to be the real number containing (x_n^{-1}). It is now easy to prove that the real numbers form a *field* whose zero element is 0^* and whose unit element is 1^*. (See axioms A 1–A 6 on p. 15.)

A slightly stronger form of (1) is that, if $(x_n) \in x^* \neq 0^*$, then either (i) there is a rational $k > 0$ such that $x_n > k$ for all sufficiently large n, or (ii) there is a rational $k > 0$ such that $x_n < -k$ for all sufficiently large n. We then say

$$x^* > 0^* \quad \text{or} \quad x^* < 0^*$$

according as (i) or (ii) holds (for the same statement holds for all Cauchy sequences in x^*). Clearly $x^* > 0^*$ if and only if $-x^* < 0^*$. We define

$$x^* > y^* \quad (\text{or } y^* < x^*)$$

to mean that $x^* - y^* > 0^*$. With this definition of $>$ it may be shown that the field of real numbers is, in fact, *ordered*. (See axioms O 1–O 3 on p. 16.)

Finally we define the modulus of a real number by

$$|x^*| = \begin{cases} x^* & \text{if} \quad x^* \geqslant 0^*, \\ -x^* & \text{if} \quad x^* < 0^*. \end{cases}$$

It is a simple matter to verify that the bijection between the rational numbers and the corresponding real rational numbers preserves algebraic relations and order, i.e. if x, y are rational numbers, then

(2) $\quad (x+y)^* = x^*+y^*, \quad (xy)^* = x^*y^*. \quad (-x)^* = -x^*, \quad (x^{-1})^* = x^{*-1}$

and

(3) $$x^* \gtreqless y^* \text{ according as } x \gtreqless y.$$

Moreover, by (2) and (3),

(4) $$|x|^* = |x^*|$$

and so also

(5) $$|x^*| \gtreqless |y^*| \text{ according as } |x| \gtreqless |y|.$$

At this stage we could develop the elementary analysis of real numbers on the basis of the familiar definitions of convergent sequences and Cauchy sequences (which now use *real* ϵ^*'s). Taking this development for granted we next show that *the set of real rational numbers is dense in the set of real numbers*, i.e. that, given any real number x^*, there is a sequence (x_n^*) of real rational numbers such that $x_n^* \to x^*$. This is a corollary of the following assertion.

(6) *If x^*, y^* are real numbers and $x^* < y^*$, there is a real* rational *number u^* such that $x^* < u^* < y^*$.*

To prove this let $(x_n) \in x^*$, $(y_n) \in y^*$. There is a rational number $k > 0$ such that $y_n - x_n > k$ for $x \geqslant \nu_1$, say. Also $|x_m - x_n| < \frac{1}{4}k$ for $m, n \geqslant \nu_2$, say. Let $\nu = \max(\nu_1, \nu_2)$ and put $u = x_\nu + \frac{1}{2}k$. Since

$$x_n + \tfrac{1}{4}k < u < y_n - \tfrac{1}{4}k$$

for $n \geqslant \nu$, the real rational number u^* (which contains $(u, u, ...)$) is such that $x^* < u^* < y^*$.

The last theorem has three further important consequences.

(7) *If x, x_n $(n = 1, 2, ...)$ are rational and x^*, x_n^* $(n = 1, 2, ...)$ are the corresponding real rational numbers, then $x_n^* \to x^*$ if and only if $x_n \to x$.*

Suppose that $x_n \to x$. Given any real number $\epsilon^* > 0^*$, by (6) there is a real rational number η^* such that $0^* < \eta^* < \epsilon^*$. By (3), $\eta > 0$ and so $|x_n - x| < \eta$ for $n \geqslant \nu$, say. This implies, by (5), that $|x_n^* - x^*| < \eta^* < \epsilon^*$ for $n \geqslant \nu$. Thus $x_n^* \to x^*$. The opposite implication follows immediately from (5).

(8) *A real rational sequence (x_n^*) is a Cauchy sequence if and only if the corresponding rational sequence (x_n) is a Cauchy sequence.*

The proof is similar to that of (7).

(9) *If x^* is any real number, if $(x_n) \in x^*$ and x_n^* is the real rational number corresponding to x_n, then $x_n^* \to x^*$.*

Take any real number $\epsilon^* > 0^*$. By (6), there are real rational numbers η^*, $k^* > 0$ such that $0^* < \eta^* + k^* < \epsilon^*$. Since (x_n) is a Cauchy sequence, there exists ν such that
$$-\eta - k < x_m - x_n < \eta + k$$
when $m, n \geqslant \nu$. Let $m \geqslant \nu$. Then since $(\eta, \eta, ...) \in \eta^*$ and $(x_m - x_1, x_m - x_2, ...) \in x_m^* - x^*$, the definition of inequality among real numbers implies that

$$-\epsilon^* < -\eta^* < x_m^* - x^* < \eta^* < \epsilon^*.$$

Hence $x_m^* \to x^*$.

We are now able to prove the fundamental theorem of the present theory.

(10) *Every Cauchy sequence of real numbers converges.*

We adapt part (v) of the proof of theorem 3.5. Let (x^*_n) be any Cauchy sequence of real numbers. By (6), for each n there is a real rational number y_n^*, corresponding to a rational number y_n, such that

$$|x^*_n - y_n^*| < n^{*-1}.$$

Then
$$|y_m^* - y_n^*| \leqslant |y_m^* - x^*_m| + |x^*_m - x^*_n| + |x^*_n - y_n^*|$$
$$< |x^*_m - x^*_n| + m^{*-1} + n^{*-1}.$$

Since (x^*_n) is a Cauchy sequence, so therefore is (y_n^*). By (8), (y_n) is a Cauchy sequence of rational numbers which determines a real number y^*. By (9), $y_n^* \to y^*$. Since

$$|x^*_n - y^*| \leqslant |x^*_n - y_n^*| + |y_n^* - y^*| < n^{*-1} + |y_n^* - y^*|,$$

it follows that $x^*_n \to y^*$.

We have shown that, with the usual metric, the set of real numbers is a completion of the set of real rational numbers. Some of the arguments had counterparts in the proof of theorem 3.5. There could be no question of *applying* the theorem, since real numbers are employed both in its enunciation and its proof.

Rational numbers, having served their purpose, need no longer be used. We therefore abandon the * notation for real numbers. In particular, $0, \pm 1, \pm 2, \ldots$ will now denote the real numbers which were formerly written $0^*, \pm 1^*, \pm 2^*, \ldots$

One more theorem is needed to set analysis on its customary course.

(11) *If the non-empty set E of real numbers is bounded above, then it has a least upper bound (supremum).*

Let U be the set of upper bounds of E. Take $x_1 \in U'$, $x_2 \in U$, so that $x_1 < x_2$. Either $\frac{1}{2}(x_1 + x_2) \in U'$ or $\frac{1}{2}(x_1 + x_2) \in U$. In the former case put $x_3 = \frac{1}{2}(x_1 + x_2)$, $x_4 = x_2$; in the latter put $x_3 = x_1$, $x_4 = \frac{1}{2}(x_1 + x_2)$. Continue this process. When $x_1, \ldots, x_{2n}$ have been chosen, then

$$x_{2n+1} = \frac{1}{2}(x_{2n-1} + x_{2n}), \quad x_{2n+2} = x_{2n} \quad \text{or} \quad x_{2n+1} = x_{2n-1}, \quad x_{2n+2} = \frac{1}{2}(x_{2n-1} + x_{2n})$$

according as $\frac{1}{2}(x_{2n-1} + x_{2n})$ belongs to U' or to U. We have

$$x_1 \leqslant x_3 \leqslant x_5 \leqslant \ldots \leqslant x_6 \leqslant x_4 \leqslant x_2,$$

with $x_{2n-1} \in U'$, $x_{2n} \in U$ for $n = 1, 2, \ldots$. Also clearly

$$|x_m - x_n| \leqslant (x_2 - x_1)/2^\nu \text{ if } m, n \geqslant 2\nu + 1$$

and so (x_n) is a Cauchy sequence. By (10), (x_n) converges to a limit x.

If $y \in E$, $y \leqslant x_{2n}$ for all n. Hence $y \leqslant x$ for every $y \in E$, i.e. $x \in U$. But, if $z < x$, there is an n such that

$$z < x_{2n+1} \leqslant x$$

and so $z \in U'$. Hence x is the least member of U, i.e. x is the supremum of E.

Clearly Dedekind's theorem is a consequence of (11).

From this point the development of analysis follows the same lines whether cuts or Cauchy sequences are used to define real numbers. The two methods yield logically distinct objects (a cut of the rational numbers is not the same as an equivalence class of Cauchy sequences), but the effect is the same, since precisely the same theorems hold for the two systems.

LIMITS IN THE SPACES R^1 AND Z

4.1. The symbols O, o, $\sim$

In this chapter we shall develop the theory of sequences and series of real and complex numbers, building upon foundations such as are laid in C1. We start by defining three symbols which contribute to brevity of statement.

Suppose that (x_n) and (v_n) are sequences of real numbers.

(1) When $v_n > 0$, we relate the magnitude of x_n to that of v_n by two definitions.

(i) *If there is a constant K such that*

$$|x_n| \leqslant K v_n$$

for all n, we write $x_n = O(v_n)$.

(ii) *If, as $n \to \infty$,*

$$\frac{x_n}{v_n} \to 0,$$

we write $x_n = o(v_n)$.

(2) *If, as $n \to \infty$,*

$$\frac{x_n}{v_n} \to 1,$$

we write $x_n \sim v_n$.

Illustrations

$x_n = O(1)$ means that x_n is bounded.
$x_n = o(1)$ means that $x_n \to 0$ as $n \to \infty$.
If $x_n = 5n^2 - 7n + 9$, then

$$x_n = O(n^2), \quad x_n \sim 5n^2, \quad x_n = o(n^3).$$

Observe too that $x_n = O(n^3)$ and that O does not rule out o or $\sim$.

The notation lends itself to an extended usage. We can write, for instance,

$$O(v_n) + O(w_n) = O(v_n + w_n). \tag{4.11}$$

This means that, if a given sequence (x_n) satisfies $x_n = O(v_n)$ and a given sequence (y_n) satisfies $y_n = O(w_n)$, then $x_n + y_n = O(v_n + w_n)$. Note that, if $v_n = w_n$, the relation (4.11) becomes

$$O(v_n) + O(v_n) = O(v_n).$$

We are also able to express concisely a number of other steps which could occur in the manipulation of sequences, such as

$$O(v_n) + o(v_n) = O(v_n),$$

$$O(v_n)\,O(w_n) = O(v_n w_n),$$

$$O(v_n)\,o(w_n) = o(v_n w_n),$$

$$\sum_{r=1}^{n} O(1) = O(n),$$

$$v_n \sim w_n \Rightarrow v_n + o(w_n) \sim w_n.$$

To demonstrate how these symbols retain the gist of an argument and slough off what does not matter, we investigate the convergence of Σu_n, where

$$u_n = \frac{5n^2 - 7n + 9}{n^k},$$

and k is a constant.

We have $u_n = O(1/n^{k-2})$, so, if $k > 3$, Σu_n converges.

An O relation, expressing as it does only an upper bound, can never establish divergence. However here we have

$$u_n \sim \frac{5}{n^{k-2}}$$

and so Σu_n diverges if $k \leqslant 3$.

We have defined O, o, $\sim$ for functions of the integral variable n as $n \to \infty$. The definitions are adaptable to functions of the continuous variable x as x tends either to ∞, $-\infty$ or to a finite limit.

Illustrations

As $x \to \infty$, $x^{100} = o(e^x)$, $\sin x = O(1)$.

As $x \to 0+$, $x^2 = o(x)$, $e^{-1/x} = o(1)$, $\sin x = O(x)$ or $\sin x \sim x$.

As $x \to 0$, $\sin (1/x) = O(1)$, $1 - \cos x \sim \frac{1}{2}x^2$.

Exercises 4(a)

1. Find whether the relations $x_n = O(v_n)$, $x_n = o(v_n)$, $x_n \sim v_n$ imply the corresponding relations between

$$(a)\ \sum_{r=1}^{n} x_r \quad \text{and} \quad \sum_{r=1}^{n} v_r, \qquad (b)\ \prod_{r=1}^{n} x_r \quad \text{and} \quad \prod_{r=1}^{n} v_r.$$

2. Prove the following results.

(i) If $x_n = O(1/n^k)$, where $k > 1$, then

$$\sum_{r=n}^{\infty} x_r = O(1/n^{k-1})$$

and

$$\sum_{r=1}^{n} x_r = A + O(1/n^{k-1}),$$

where A is a constant.

(ii) If $x_n = O(1/n)$, then

$$\sum_{r=1}^{n} x_r = O(\log n).$$

(iii) If $x_n = O(n^k)$, where $k > -1$, then

$$\sum_{r=1}^{n} x_r = O(n^{k+1}).$$

(Use the inequality $1/r^k < \int_{r-1}^{r} (1/x^k)\,dx$ $(k > 0)$.)

3. Prove that, if $u_n \sim v_n$ and $v_n = O(1)$, then $u_n - v_n = o(1)$. Show also that the hypothesis $v_n = O(1)$ cannot be omitted.

4. Show that, if $x_n = O(1)$ and $2x_n \leqslant x_{n-1} + x_{n+1}$ for all n, then (x_n) converges.

5. Establish the following relations.
  (i) As $x \to 0$, $\log (1+x) = x + O(|x|^2)$.
  (ii) As $x \to 0$, $(1+x)^\alpha = 1 + \alpha x + O(|x|^2)$.
  (iii) As $x \to 0$, $\sin x = x - \tfrac{1}{6}x^3 + O(|x|^5)$.

6. Prove that, for $r > \tfrac{1}{2}$,

$$\frac{1}{r} < \int_{r-\frac{1}{2}}^{r+\frac{1}{2}} \frac{1}{x}\,dx < \frac{4r}{4r^2-1}$$

and deduce that, as $r \to \infty$, the value of the integral is $1/r + O(1/r^3)$. Hence show that

$$\sum_{r=n+1}^{2n} \frac{1}{r} = \log 2 - \frac{1}{4n+2} + O\left(\frac{1}{n^2}\right).$$

7. Prove that, as $x \to \infty$,

$$\sum_{n=1}^{\infty} \frac{1}{x^2+n^2} \sim \frac{\pi}{2x}.$$

(Use the fact that, if the function f on $[1, \infty)$ is positive and decreasing, then $\sum_{n=1}^{N} f(n) - \int_{1}^{N} f(x)\,dx$ tends to a limit l such that $0 \leqslant l \leqslant f(1)$ (C1, 141).)

4.2. Upper and lower limits

So far in this book (and in C1) the symbol ∞ has only occurred in the designation of an interval (such as $0 < x < \infty$) or as part of the phrase '$x_n \to \infty$'. It is now convenient to extend the usage. For instance we shall use 'lim $x_n = \infty$' to mean the same thing as '$x_n \to \infty$'; or 'sup $x_n = \infty$' to mean that (x_n) is not bounded above. Other phrases involving ∞ occur in the two definitions of this section. However, as heretofore, we shall not attempt to perform any algebraic operations with ∞ and $-\infty$.

The sequence

$$x_n = \sin \tfrac{1}{2}n\pi + \frac{1}{n}$$

does not converge, but

$$x_{4r-1} \to -1, \quad x_{2r} \to 0, \quad x_{4r+1} \to 1$$

as $r \to \infty$. Hence 1 and -1 are the greatest and least numbers which are limits of *subsequences* of (x_n). These will be called the *upper* and *lower limits* of (x_n) (see theorem 4.21, corollary 2). For the formal definitions we prefer, however, a different approach.

Given any sequence (x_n) of real numbers, consider successively the sequences

$$\left.\begin{aligned}
x_1, x_2, x_3, \ldots, x_n, x_{n+1}, \ldots, \\
x_2, x_3, \ldots, x_n, x_{n+1}, \ldots, \\
\cdots\cdots\cdots\cdots\cdots\cdots \\
x_n, x_{n+1}, \ldots, \\
\cdots\cdots\cdots\cdots
\end{aligned}\right\} \qquad (4.21)$$

If (x_n) is bounded above, these sequences have suprema

$$K_1, K_2, \ldots, K_n, \ldots.$$

For all n, $K_n \geqslant K_{n+1}$, since $\{x_n, x_{n+1}, x_{n+2}, \ldots\} \supset \{x_{n+1}, x_{n+2}, \ldots\}$ (in fact $K_n = \max(x_n, K_{n+1})$). Hence K_n tends either to a finite limit or to $-\infty$. Moreover $K_n \to -\infty$ if and only if $x_n \to -\infty$. For if $x_n \to -\infty$, then, given any number K, however large, there is an n_0 such that $x_n < -K$ for $n > n_0$. Hence, for $n > n_0$, $K_n = \sup\limits_{r \geqslant n} x_r \leqslant -K$ and so $K_n \to -\infty$. On the other hand, if $K_n \to -\infty$, then $x_n \to -\infty$, since $x_n \leqslant K_n$.

If (x_n) is not bounded above, none of the sequences (4.21) is bounded above, i.e. $\sup\limits_{r \geqslant n} x_r = \infty$ for all n.

Definition. *Let (x_n) be a sequence of real numbers.*
If (x_n) is bounded above,

$$\Lambda = \lim_{n \to \infty} (\sup_{r \geqslant n} x_r)$$

is called the upper limit *(or* limit superior*) of (x_n) and we write*

$$\overline{\lim_{n \to \infty}}\, x_n = \Lambda \quad or \quad \limsup_{n \to \infty} x_n = \Lambda.$$

If (x_n) is not bounded above, we write

$$\overline{\lim_{n \to \infty}}\, x_n = \infty \quad or \quad \limsup_{n \to \infty} x_n = \infty.$$

If (x_n) is bounded below and k_1, k_2, ... are the infima of the sequences (4.21), then $k_n \leqslant k_{n+1}$ for all n and k_n tends to a finite limit unless $x_n \to \infty$. We are therefore led to the following definition.

Definition. *If* (x_n) *is bounded below,*

$$\lambda = \lim_{n \to \infty} (\inf_{r \geqslant n} x_r)$$

is called the lower limit (*or* limit inferior) *of* (x_n) *and we write*

$$\varliminf_{n \to \infty} x_n = \lambda \quad or \quad \liminf_{n \to \infty} x_n = \lambda.$$

If (x_n) *is not bounded below, we write*

$$\varliminf_{n \to \infty} x_n = -\infty \quad or \quad \liminf_{n \to \infty} x_n = -\infty.$$

We have
$$\liminf_{n \to \infty} x_n \leqslant \limsup_{n \to \infty} x_n, \tag{4.22}$$

since $\inf_{r \geqslant n} x_r \leqslant \sup_{r \geqslant n} x_r$; and
$$\liminf (-x_n) = -\limsup x_n, \tag{4.23}$$

since $\inf_{r \geqslant n} (-x_r) = -\sup_{r \geqslant n} x_r$. Note that the upper and lower limits in (4.22) and (4.23) need not be finite.

A convenient characterization of finite upper and lower limits is given in the next theorem.

Theorem 4.21.

(i) *The number* Λ *is the upper limit of the sequence* (x_n) *if and only if, given* $\epsilon > 0$,

    (a) $x_n < \Lambda + \epsilon$ *for all sufficiently large n; and*

    (b) $x_n > \Lambda - \epsilon$ *for infinitely many n.*

(ii) *The number* λ *is the lower limit of the sequence* (x_n) *if and only if, given* $\epsilon > 0$,

    (c) $x_n > \lambda - \epsilon$ *for all sufficiently large n; and*

    (d) $x_n < \lambda + \epsilon$ *for infinitely many n.*

Proof.

(i) *Necessity.* Suppose that $\Lambda = \limsup x_n = \lim K_n$, where $K_n = \sup_{r \geqslant n} x_r$. Then, given $\epsilon > 0$, there is N such that

$$K_N < \Lambda + \epsilon$$

and so
$$x_n < \Lambda + \epsilon \quad \text{for} \quad n \geqslant N.$$

This is (a). To prove (b) we need only show that, given any integer p, there is an $n > p$ such that $x_n > \Lambda - \epsilon$.

Since (K_m) decreases, $K_m \geqslant \Lambda$ for all m. So, by the definition of K_{p+1}, there is an $n \geqslant p+1$ such that

$$x_n > K_{p+1} - \epsilon \geqslant \Lambda - \epsilon.$$

This gives (b).

Sufficiency. Suppose that (a) and (b) hold. By (a), $K_n \leqslant \Lambda + \epsilon$ for $n \geqslant N$; by (b), $K_n > \Lambda - \epsilon$ for all n. Thus $K_n \to \Lambda$, i.e. $\Lambda = \limsup x_n$.

(ii) This follows from (i) and (4.23). |

Corollary 1. *The sequence (x_n) tends to l as $n \to \infty$ if and only if*

$$\liminf x_n = \limsup x_n = l.$$

Proof.

(i) l finite.

If $x_n \to l$, then, given $\epsilon > 0$, there is N such that

$$l - \epsilon < x_n < l + \epsilon \tag{4.24}$$

for $n \geqslant N$. Hence $l = \limsup x_n$ and $l = \liminf x_n$.

If $\limsup x_n = \liminf x_n = l$, then (a) and (c) show that (4.24) holds for all sufficiently large n. Therefore $x_n \to l$ as $n \to \infty$.

(ii) l infinite.

$x_n \to \infty$ if and only if $\liminf x_n = \infty$. Also $\liminf x_n = \infty$ implies $\limsup x_n = \infty$.

$x_n \to -\infty$ if and only if $\limsup x_n = -\infty$; and $\limsup x_n = -\infty$ implies $\liminf x_n = -\infty$. |

Corollary 2. $\Lambda = \limsup x_n$, *when finite, is the largest number which is the limit of a subsequence of (x_n); $\lambda = \liminf x_n$, when finite, is the smallest such number.*

Proof. The conditions (a) and (b) show that, for every $\epsilon > 0$, there are infinitely many n such that $\Lambda - \epsilon < x_n < \Lambda + \epsilon$. From this follows that there is a subsequence (x_{n_k}) with limit Λ. Also (x_n) cannot have a subsequence converging to a number $\Lambda' > \Lambda$, for if $\epsilon < \Lambda' - \Lambda$, there is only a finite number of n for which $x_n > \Lambda' - \epsilon$.

The statement about lower limits is proved similarly or by use of (4.23). |

The analogue of corollary 2 for infinite upper limits is that $\limsup x_n = -\infty$ if and only if $x_n \to -\infty$; and $\limsup x_n = \infty$ if and only if there is a subsequence of (x_n) which diverges to ∞. The first statement has already been proved; for the proof of the second see exercise 4(b), 2. There is a similar characterization of infinite lower limits.

Illustrations

(i) $x_n = (-1)^n$; $\liminf x_n = -1$, $\limsup x_n = 1$.

(ii) $x_n = 2n/(n+1)$; $\liminf x_n = \limsup x_n = 2$.

(iii) $x_n = n[1+(-1)^n]$; $\liminf x_n = 0$, $\limsup x_n = \infty$.

(iv) $x_n = 1-n$; $\liminf x_n = \limsup x_n = -\infty$.

A proof of the completeness of R^1 was given in theorem 3.42, but a more natural proof uses upper and lower limits.

Let (x_n) be a Cauchy sequence. Given $\epsilon > 0$, there is an N such that
$$|x_m - x_n| < \epsilon \quad \text{for} \quad m, n \geqslant N.$$
Thus
$$x_N - \epsilon < x_n < x_N + \epsilon \quad \text{for} \quad n \geqslant N$$
and so
$$x_N - \epsilon \leqslant k_N \leqslant \lim_{n\to\infty} k_n \leqslant \lim_{n\to\infty} K_n \leqslant K_N \leqslant x_N + \epsilon.$$
Hence $\liminf x_n$ and $\limsup x_n$ are finite and
$$0 \leqslant \limsup x_n - \liminf x_n \leqslant 2\epsilon.$$
Since ϵ is arbitrary, $\limsup x_n = \liminf x_n$ and, by theorem 4.21 corollary 1, (x_n) converges. |

We now turn to functions of a continuous variable for which we define upper and lower limits at ∞ (or $-\infty$) and at a point.

First suppose that f is defined on an interval (c, ∞). *If f is bounded above in some interval (X, ∞), we define*
$$\limsup_{x\to\infty} f(x) = \lim_{x\to\infty} (\sup_{t \geqslant x} f(t));$$
if f is unbounded above in every interval (X, ∞) we write
$$\limsup_{x\to\infty} f(x) = \infty.$$
The definition of $\liminf_{x\to\infty} f(x)$ is similar.

Next let f be defined on $(a-h, a) \cup (a, a+h)$, say. *If f is bounded above in some set $(a-\eta, a) \cup (a, a+\eta)$, we define*
$$\limsup_{x\to a} f(x) = \lim_{\delta\to 0+} (\sup_{0 < |x-a| < \delta} f(x));$$
if f is unbounded above in every set $(a-\eta, a) \cup (a, a+\eta)$, we write
$$\limsup_{x\to a} f(x) = \infty.$$

The analogues of theorem 4.21 and corollary 1 are left to the reader.

Illustrations

(i) $\limsup\limits_{x\to\infty} \sin x = 1$, $\liminf\limits_{x\to\infty} \sin x = -1$.

(ii) $\limsup\limits_{x\to 0} \sin (1/x) = 1$, $\liminf\limits_{x\to 0} \sin (1/x) = -1$.

(iii) $\limsup\limits_{x\to 0} e^{1/x} = \infty$, $\liminf\limits_{x\to 0} e^{1/x} = 0$.

Exercises 4(b)

1. Prove that $x_n = O(v_n)$ if and only if $\liminf (x_n/v_n)$ and $\limsup (x_n/v_n)$ are finite.

2. Prove that $\limsup x_n = \infty$ if and only if a subsequence of (x_n) diverges to ∞; and that $\liminf x_n = -\infty$ if and only if a subsequence of (x_n) diverges to $-\infty$.

3. Let (x_n), (y_n) be sequences of real numbers such that $x_n \leqslant y_n$ for all n. Prove that

(i) $\liminf x_n \leqslant \liminf y_n$,   (ii) $\limsup x_n \leqslant \limsup y_n$.

4. Show that, if (x_n), (y_n) are bounded sequences, then

$$\liminf x_n + \liminf y_n \leqslant \liminf (x_n+y_n) \leqslant \liminf x_n + \limsup y_n$$

$$\leqslant \limsup (x_n+y_n) \leqslant \limsup x_n + \limsup y_n.$$

Construct sequences (x_n), (y_n) for which strict inequalities occur at every step.

5. The sequences (x_n), (y_n) are non-negative and bounded. Establish a chain of inequalities as in **4** with products in place of sums and give an example in which strict inequalities occur at every step.

6. (i) Let $y_n \to y > 0$. Prove that
    (a) if $\liminf x_n$ is finite, then

$$\liminf (x_ny_n) = y \liminf x_n;$$

(b) if $\liminf x_n = \infty\,(-\infty)$, then $\liminf (x_ny_n) = \infty\,(-\infty)$;
(c) if $\limsup x_n$ is finite, then

$$\limsup (x_ny_n) = y \limsup x_n;$$

(d) if $\limsup x_n = \infty\,(-\infty)$, then $\limsup (x_ny_n) = \infty\,(-\infty)$.
(ii) Show that (a), (b), (c), (d) need not hold if $y_n \to 0$.
(iii) Let $y_n \to y \leqslant 0$. Prove that, if (x_n) is bounded, then

$$\liminf (x_ny_n) = y \limsup x_n, \quad \limsup (x_ny_n) = y \liminf x_n.$$

7. The function $f: R^1 \to R^1$ is continuous and increasing. Prove that, if (x_n) is any bounded sequence of real numbers, then

$$f(\liminf x_n) = \liminf f(x_n), \quad f(\limsup x_n) = \limsup f(x_n).$$

8. Let
$$s_n = u_1 + \ldots + u_n, \quad t_n = v_1 + \ldots + v_n,$$

where (u_n) is an arbitrary sequence of real numbers and (v_n) is a sequence of positive terms such that $t_n \to \infty$. Prove that

$$\liminf \frac{u_n}{v_n} \leqslant \liminf \frac{s_n}{t_n} \leqslant \limsup \frac{s_n}{t_n} \leqslant \limsup \frac{u_n}{v_n}.$$

9. Let (x_n) be a sequence of real numbers. Show that

(i) $\liminf x_n \leqslant \liminf \dfrac{x_1 + \ldots + x_n}{n} \leqslant \limsup \dfrac{x_1 + \ldots + x_n}{n} \leqslant \limsup x_n;$

and, if $x_n > 0$ for all n,

(ii) $\liminf x_n \leqslant \liminf (x_1 \ldots x_n)^{1/n} \leqslant \limsup (x_1 \ldots x_n)^{1/n} \leqslant \limsup x_n;$

(iii) $\liminf \dfrac{x_{n+1}}{x_n} \leqslant \liminf x_n^{1/n} \leqslant \limsup x_n^{1/n} \leqslant \limsup \dfrac{x_{n+1}}{x_n}.$

10. Prove that, as $n \to \infty$,
   (i) if $k > -1$,
$$1^k + 2^k + \ldots + n^k \sim n^{k+1}/(k+1)$$
(cf. exercise $4(a)$, 3);
   (ii) $n^{1/n} \to 1$;
   (iii) $(n!)^{1/n} \sim n/e$.

11. *One-sided upper and lower limits.* Suppose that the real function f is defined in the interval $(a, a+h)$. If f is unbounded above in every interval $(a, a+\eta)$, we write $\limsup\limits_{x \to a+} f(x) = \infty$; otherwise we define

$$\limsup_{x \to a+} f(x) = \lim_{\delta \to 0+} (\sup_{a < x < a+\delta} f(x)).$$

$\limsup\limits_{x \to a-} f(x)$ and the two one-sided lower limits are defined similarly.

Prove that, if f is defined on $(a-h, a) \cup (a, a+h)$,

$$\limsup_{x \to a} f(x) = \max \{\limsup_{x \to a-} f(x), \limsup_{x \to a+} f(x)\},$$

$$\liminf_{x \to a} f(x) = \min \{\liminf_{x \to a-} f(x), \liminf_{x \to a+} f(x)\}.$$

12. *Semi-continuity.* The function $f: [a, b] \to R^1$ is said to be *upper semi-continuous* at the point $c \in (a, b)$ if

$$f(c) \geqslant \limsup_{x \to c} f(x);$$

and f is said to be *lower semi-continuous* at c if

$$f(c) \leqslant \liminf_{x \to c} f(x).$$

Semi-continuity at a, b is defined by use of the appropriate one-sided upper and lower limits.

Show that, if f is upper (lower) semi-continuous in $[a, b]$, then f is bounded above (below) and attains its supremum (infimum).

13. For the function $f: [a, b] \to R^1$ define

$$\underline{D}f(c) = \liminf_{x \to c} \frac{f(c) - f(x)}{c - x}, \quad \overline{D}f(c) = \limsup_{x \to c} \frac{f(c) - f(x)}{c - x}$$

when $c \in (a, b)$, and define $\underline{D}f(a)$, $\overline{D}f(a)$, $\underline{D}f(b)$, $\overline{D}f(b)$ as the corresponding one-sided upper and lower limits.

Prove that, if $\underline{D}f(x) \geq 0$ for $a \leq x \leq b$, then f increases in $[a, b]$. (First take $\underline{D}f(x) > 0$ in $[a, b]$ and use exercise 3(g), 4.) Show also that the condition $\overline{D}f(x) > 0$ in $[a, b]$ does not ensure that f increases in $[a, b]$.

4.3. Series of complex terms

The familiar tests of Cauchy and d'Alembert for series of positive terms (C1, 88) may be strengthened by the use of upper and lower limits. It is also convenient to have these tests in a form in which they apply to series of complex terms, and ensure either absolute convergence (and so convergence) or divergence.

Theorem 4.31. (The root test–Cauchy.) *Let Σu_n be a series of complex terms.*

(i) *If $\limsup |u_n|^{1/n} < 1$, then Σu_n converges absolutely.*

(ii) *If $\limsup |u_n|^{1/n} > 1$, then Σu_n diverges.*

Proof.

(i) Let $\limsup |u_n|^{1/n} = \Lambda$ and take c so that $(0 \leq) \Lambda < c < 1$. By theorem 4.21, there exists N such that

$$|u_n|^{1/n} < c \quad \text{and so} \quad |u_n| < c^n$$

for all $n \geq N$. Since $0 < c < 1$, Σc^n converges and so, by comparison, $\Sigma |u_n|$ converges, i.e. Σu_n converges absolutely.

(ii) Suppose that $\limsup |u_n|^{1/n} > 1$. If the upper limit is infinite, $|u_n|^{1/n}$ is unbounded. Otherwise we again appeal to theorem 4.21. In either case we see that there are infinitely many n for which $|u_n|^{1/n} > 1$, or $|u_n| > 1$. Therefore $u_n \not\to 0$ and so Σu_n diverges. |

Theorem 4.32. (The ratio test–d'Alembert.)

(i) *If $\limsup |u_{n+1}/u_n| < 1$, then Σu_n converges absolutely.*

(ii) *If $\liminf |u_{n+1}/u_n| > 1$, then Σu_n diverges.*

Proof.

(i) If $\limsup |u_{n+1}/u_n| < c < 1$, there is an N such that

$$|u_{n+1}/u_n| < c \quad \text{for} \quad n \geq N.$$

Then, for $n > N$,

$$|u_n| = |u_N| \left|\frac{u_{N+1}}{u_N}\right| \cdots \left|\frac{u_n}{u_{n-1}}\right| < |u_N| c^{n-N} = Ac^n$$

and therefore $\Sigma|u_n|$ converges.

(ii) There is an N such that $|u_{n+1}/u_n| > 1$ for $n \geqslant N$. Hence $|u_N| < |u_{N+1}| < \ldots$ and so $u_n \nrightarrow 0$. $\blacksquare$

Exercise. Give an example of a convergent series Σu_n with

$$\limsup |u_{n+1}/u_n| > 1.$$

By exercise 4(*b*), 9(iii),

$$\liminf |u_{n+1}/u_n| \leqslant \liminf |u_n|^{1/n}$$
$$\leqslant \limsup |u_n|^{1/n} \leqslant \limsup |u_{n+1}/u_n|.$$

Therefore theorem 4.32 follows from and is weaker than theorem 4.31. (See exercise 4(*c*), 2.)

Exercises 4(c)

1. Give examples to show that, when $\limsup |u_n|^{1/n} = 1$, the series Σu_n may converge or diverge. Show also that both convergent and divergent series may be such that $\limsup |u_{n+1}/u_n| = 1$ or $\liminf |u_{n+1}/u_n| = 1$.

2. Give an example of a series which theorem 4.31 shows to converge and to which theorem 4.32 is not applicable. Give a corresponding example of a divergent series.

3. The sequence (u_n) of real numbers is such that $\limsup u_n < 0$. Show that, if $a > 1$, then the series

$$\Sigma a^{u_1 + \cdots + u_n}$$

converges.

4. Investigate the convergence of the series whose nth terms are

$$\text{(i)} \quad \frac{[(2n)!]^2}{(4n)!} 10^n, \quad \text{(ii)} \quad \frac{(n!)(3n)!}{(4n)!} 10^n.$$

5. Show that the complex series

$$\Sigma \frac{n!}{(1 - in)^n}$$

converges absolutely.

6. Find all the real values of x for which the series $\Sigma x^n/n^x$ converges.

7. Find all the pairs of complex number a, b for which the series $\Sigma a^{n^2} b^n$ converges.

4.4. Series of positive terms

In this section all series have positive terms. Although theorems 4.31 and 4.32 are useful for such series, more delicate tests are available.

The proofs of the last two theorems use a *comparison principle*, introduced in C 1, in which the terms of a given series are compared with those of a series whose convergence or divergence is known. We may also compare *ratios* of terms.

Theorem 4.41. *If, for all $n \geqslant N$,*

$$\frac{u_{n+1}}{u_n} \leqslant \frac{v_{n+1}}{v_n}$$

and Σv_n converges, so does Σu_n. If the opposite inequality holds and Σv_n diverges, then Σu_n also diverges.

Proof. The statement about convergence follows from the relation, for $n > N$,

$$u_n = u_N \frac{u_{N+1}}{u_N} \cdots \frac{u_n}{u_{n-1}} \leqslant u_N \frac{v_{N+1}}{v_N} \cdots \frac{v_n}{v_{n-1}} = \frac{u_N}{v_N} v_n. \;|$$

To apply comparison principles we need a stock of standard series which are known either to converge or to diverge. Such a stock can be obtained from the theorem (C 1, 141) that if, on $[a, \infty)$, f is a positive decreasing function, then the integral and the series

$$\int_a^\infty f(x)\,dx, \quad \sum_a^\infty f(n)$$

both converge or both diverge.

For the next theorem only, write $\log_r x = \log(\log_{r-1} x)$ $(r \geqslant 2)$, where $\log_1 x = \log x$.

Theorem 4.42. (The logarithmic scale.) *Each of the series*

$$\Sigma \frac{1}{n^k}, \Sigma \frac{1}{n(\log n)^k}, \ldots, \Sigma \frac{1}{n \log n \log_2 n \ldots \log_{r-1} n \, (\log_r n)^k}, \ldots$$

converges if $k > 1$ and diverges if $k \leqslant 1$.

Proof. It must be supposed that the summation starts with a value of n (say a) large enough for the repeated logarithms to be defined.

The value of the associated integral

$$\int_a^X \frac{dx}{x \log x \log_2 x \ldots \log_{r-1} x \, (\log_r x)^k}$$

is

$$\left[\frac{(\log_r x)^{1-k}}{1-k}\right]_a^X \quad \text{if } k \neq 1$$

and is

$$\left[\log_{r+1} x\right]_a^X \quad \text{if } k = 1.$$

As $X \to \infty$, the integral tends to ∞ if $k \leq 1$ and to a finite limit if $k > 1$. |

The next theorem arises from the application of theorem 4.41 to the first two series of the logarithmic scale.

Theorem 4.43. (Ratio test of Gauss.) *If it is possible to write*

$$\frac{u_n}{u_{n+1}} = 1 + \frac{\mu}{n} + O\left(\frac{1}{n^2}\right),$$

where μ is a constant, then Σu_n converges if $\mu > 1$ and diverges if $\mu \leq 1$.

Proof. Suppose $\mu > 1$. Let $1 < q < \mu$. Let $v_n = n^{-q}$. Then, by exercise $4(a)$, $5(\text{ii})$,

$$\frac{v_n}{v_{n+1}} = \left(\frac{n+1}{n}\right)^q = 1 + \frac{q}{n} + O\left(\frac{1}{n^2}\right).$$

The inequality, for all sufficiently large n,

$$\frac{u_n}{u_{n+1}} \geqslant \frac{v_n}{v_{n+1}}$$

gives the convergence of Σu_n.

Similarly, if $\mu < 1$, Σu_n diverges.

If $\mu = 1$, define

$$v_n = \frac{1}{n \log n}.$$

$$\frac{v_n}{v_{n+1}} = \left(1 + \frac{1}{n}\right)\left\{1 + \frac{\log(1 + n^{-1})}{\log n}\right\}$$

$$= 1 + \frac{1}{n} + \frac{1}{n \log n} + O\left(\frac{1}{n^2}\right).$$

For all sufficiently large n,

$$\frac{u_n}{u_{n+1}} \leqslant \frac{v_n}{v_{n+1}}.$$

Therefore Σu_n diverges. |

The convergence tests of this section and the last are all based on comparison with standard series. No such test can be universal; for, given any convergent (divergent) series, there always exists a series that converges (diverges) more slowly. This fact is illustrated by exercises 7–10 below.

Exercises 4(d)

1. *Another comparison principle.* Show that, if Σu_n, Σv_n are series of positive terms and

$$0 < \liminf \frac{u_n}{v_n} \leqslant \limsup \frac{u_n}{v_n} < \infty,$$

then Σu_n and Σv_n either both converge or both diverge.

2. Investigate the convergence or divergence of the series whose nth terms are

$$\text{(i)} \quad n^{-p}(\log n)^{-q}, \qquad\qquad \text{(ii)} \quad (\log n)^{-\log n},$$

$$\text{(iii)} \quad (\log \log n)^{-\log n}, \qquad \text{(iv)} \quad (\log n)^{-\log \log n},$$

$$\text{(v)} \quad n^{-1-(1/n)}, \qquad\qquad \text{(vi)} \quad 1 - n \log \frac{2n+1}{2n-1},$$

$$\text{(vii)} \quad \frac{(n+a)^n}{n^{n+a}} \quad (a \text{ real}), \qquad \text{(viii)} \quad \left(\frac{1.3...(2n-1)}{2.4...2n} \right)^p,$$

$$\text{(ix)} \quad |a^{1/n}-1|^k \quad (a > 0), \qquad \text{(x)} \quad (n^{1/n}-1)^k.$$

3. Show that the series

$$\Sigma \left(1 - \frac{k \log n}{n} \right)^n, \quad \Sigma \left(1 - \frac{\log n + k \log \log n}{n} \right)^n$$

converge for $k > 1$ and diverge for $k \leqslant 1$. (Compare with Σn^{-k} and $\Sigma n^{-1}(\log n)^{-k}$.)

4. Find all the positive values of a for which the series

$$\Sigma \frac{(an)^n}{n!}$$

converges.

5. *The condensation test.* The sequence (u_n) is positive and decreasing and p is an integer greater than 1. Prove that the series

$$\sum_1^\infty u_n \quad \text{and} \quad \sum_1^\infty p^n u_{p^n}$$

both converge or both diverge. (Group the terms in Σu_n appropriately; it is advisable first to consider the case $p = 2$.)

Deduce the logarithmic scale.

6. Let Σu_n be a divergent series of positive terms and let $s_n = u_1 + ... + u_n$. Show that the series

$$\Sigma(u_n/s_n^k)$$

converges if $k > 1$ and diverges if $k \leqslant 1$. [For convergence prove that, when $k \leqslant 1$,

$$\frac{u_n}{s_n^k} \leqslant \frac{1}{k-1} \left(\frac{1}{s_{n-1}^{k-1}} - \frac{1}{s_n^{k-1}} \right).]$$

7. Given a convergent series Σu_n of positive terms, construct a convergent series Σv_n such that $v_n/u_n \to \infty$.

8. Given a divergent series Σu_n of positive terms, construct a divergent series Σv_n of positive terms such that $v_n/u_n \to 0$.

9. Given an increasing sequence (q_n) with $\lim q_n = \infty$, construct a convergent series Σu_n such that $\Sigma u_n q_n$ diverges.

10. Let
$$u_{11} + u_{12} + u_{13} + \ldots,$$
$$u_{21} + u_{22} + u_{23} + \ldots,$$
$$u_{31} + u_{32} + u_{33} + \ldots,$$
$$\ldots \ldots \ldots \ldots \ldots$$

be an infinite scale of divergent (convergent) series of positive terms each diverging (converging) more slowly than the one before in the sense that, for every r, $u_{r+1, n}/u_{r, n} \to 0 \ (\infty)$ as $n \to \infty$. Construct a series which diverges (converges) more slowly than every series of the scale.

4.5. Conditionally convergent real series

A conditionally convergent series is one that converges, but not absolutely. It was shown in C1 (93) that rearrangement of terms does not affect the sum of an absolutely convergent series, but may change the sum of a conditionally convergent one. A more precise statement is possible.

Theorem 4.51. (Riemann.) *Let u_n be real and Σu_n conditionally convergent. Then the terms of the series can be rearranged so that the sum of the first n terms tends to any given limit or has assigned upper and lower limits (finite or infinite).*

Proof. Denote the positive terms in Σu_n in the order in which they appear by $v_1, v_2, \ldots$ and the negative terms by $w_1, w_2, \ldots$. Then the series $\Sigma v_n, \Sigma w_n$ both diverge; for if both converged, $\Sigma |u_n|$ would converge and, if one converged and the other diverged, Σu_n would diverge.

We shall rearrange the series to converge to a given sum t. Take the smallest number of v's for which $v_1 + v_2 + \ldots + v_p \geqslant t$. Then take for the next terms of the rearranged series just enough w's, starting with w_1, to make the total sum $< t$. Then take more v's in order from v_{p+1} to make the sum $\geqslant t$.

Since Σv_n and Σw_n diverge, this process can be continued indefinitely. Let t_n be the sum of the first n terms of the series so constructed. To prove that $t_n \to t$ we remark that $|t - t_m| \leqslant |u_m|$, where u_m is the final term of the last complete block of positive or negative terms included in t_n; and $u_m \to 0$ as $m \to \infty$.

In view of theorem 4.21 corollary 2 it is easy to modify the process so that (t_n) has any assigned upper and lower limits. |

The tests of the last two sections are used for proving absolute convergence of series whose terms are not all positive. Theorems 4.53 and 4.54 below are often useful for establishing conditional convergence. These theorems depend on the following simple identity.

Theorem 4.52. (*Partial summation.*) *If* $a_1, \ldots, a_p, v_1, \ldots, v_p$ *are any numbers and* $s_k = \sum\limits_{i=1}^{k} a_i$ $(k = 1, \ldots, p)$, *then*

$$a_1 v_1 + \ldots + a_p v_p = s_1(v_1 - v_2) + \ldots + s_{p-1}(v_{p-1} - v_p) + s_p v_p.$$

Proof. Each side is $s_1 v_1 + (s_2 - s_1) v_2 + \ldots + (s_p - s_{p-1}) v_p$. |

Corollary. *If* $v_1 \geqslant \ldots \geqslant v_p \geqslant 0$ *and* $a_1, \ldots, a_p$ *are any real numbers such that*

$$\mu \leqslant s_k \leqslant M \quad (k = 1, \ldots, p),$$

then

$$\mu v_1 \leqslant a_1 v_1 + \ldots + a_p v_p \leqslant M v_1.$$

Proof. Since $v_1 - v_2, \ldots, v_{p-1} - v_p, v_p$ are non-negative,

$$s_1(v_1 - v_2) + \ldots + s_{p-1}(v_{p-1} - v_p) + s_p v_p$$

lies between the sums obtained by replacing each s_k by μ and M respectively. |

Theorem 4.53. (*Dirichlet.*) *Let* a_n, v_n $(n = 1, 2, \ldots)$ *be real. If*

(i) *the sequence* (s_n), *where* $s_n = \sum\limits_{r=1}^{n} a_r$, *is bounded*,

(ii) *the sequence* (v_n) *is monotonic and tends to* 0,
then $\Sigma a_n v_n$ *converges.*

Proof. We may suppose that (v_n) decreases, so that $v_n \geqslant 0$. By (i), there is an H such that $|s_n| \leqslant H$ for all n. Then

$$|a_{n+1} + \ldots + a_{n+k}| = |s_{n+k} - s_n| \leqslant 2H.$$

Therefore, by the last corollary,

$$|a_{n+1} v_{n+1} + \ldots + a_{n+p} v_{n+p}| \leqslant 2H v_{n+1}$$

for all $p \geqslant 1$. Since $v_n \to 0$, the sequence (t_n), where

$$t_n = \sum\limits_{r=1}^{n} a_r v_r,$$

is a Cauchy sequence. As R^1 is complete, $\lim t_n$ exists, i.e. $\Sigma a_r v_r$ converges. |

The alternating series theorem (C1, 91) is the special case $a_n = (-1)^n$ of theorem 4.53.

Theorem 4.54. (Abel.) *Let a_n, v_n ($n = 1, 2, \ldots$) be real. If*

(i) *Σa_n converges,*

(ii) *the sequence (v_n) is monotonic and bounded,*

then $\Sigma a_n v_n$ converges.

Proof. We can prove, as in theorem 4.53, that the partial sums of $\Sigma a_n v_n$ form a Cauchy sequence. Alternatively we can deduce 4.54 from 4.53 as follows.

From (ii), v_n tends to a limit, say v. Let $v_n = v + w_n$, so that the sequence (w_n) is monotonic and tends to 0. By (i), $\Sigma a_n v$ converges. Also (s_n), where $s_n = \sum_{r=1}^{n} a_r$, is bounded and so, by theorem 4.53, $\Sigma a_n w_n$ converges. |

Example. If (v_n) is monotonic and tends to 0, then $\Sigma v_n \sin nx$ converges for all x and $\Sigma v_n \cos nx$ converges except, possibly, for $x = 2k\pi$ ($k = 0, \pm 1, \pm 2, \ldots$).

When $x \neq 2k\pi$,

$$\left| \sum_{r=1}^{n} \sin rx \right| = \left| \frac{\cos \frac{1}{2}x - \cos(n + \frac{1}{2})x}{2 \sin \frac{1}{2}x} \right| \leqslant \frac{1}{|\sin \frac{1}{2}x|}$$

and

$$\left| \sum_{r=1}^{n} \cos rx \right| = \left| \frac{\sin(n + \frac{1}{2})x - \sin \frac{1}{2}x}{2 \sin \frac{1}{2}x} \right| \leqslant \frac{1}{|\sin \frac{1}{2}x|}$$

and the result follows from theorem 4.53.

When $x = 2k\pi$, $\sin nx = 0$ for all n and so $\Sigma v_n \sin nx$ converges; $\cos nx = 1$ for all n, so that $\Sigma v_n \cos nx$ is Σv_n, which may or may not converge.

This example is frequently used in discussing the convergence of a power series at points on its circle of convergence.

Exercises 4(e)

1. Examine for convergence the series whose nth terms are

(i) $(-1)^n a^{1/n}$ ($a > 0$),   (ii) $\dfrac{\sin nx}{\cos(\sin nx)}$,

(iii) $\dfrac{(2n)!}{n^{2n}} \cos nx$,   (iv) $\dfrac{\sin n}{\log n}$ ($n > 1$),

(v) $\dfrac{x^n}{n + x^{2n}}$,   (vi) $(n^{1/n} - 1) \cos nx$,

(vii) $(-1)^n n^p e^{q\sqrt{n}}$,   (viii) $\dfrac{(-1)^n}{n} \cos \dfrac{x}{n}$,

(ix) $\dfrac{\sin(n + n^{-1})x}{\log n}$ ($n > 1$),   (x) $\dfrac{\alpha^{1/n}}{n^{1/\alpha}} \cos n\alpha$ ($\alpha > 0$).

2. Find all the real values of x for which the series

$$\text{(i)} \quad \Sigma \left\{ \frac{1}{(n+1)^2} + \frac{1}{(n+2)^2} + \dots + \frac{1}{(2n)^2} \right\} \sin nx,$$

$$\text{(ii)} \quad \Sigma \left\{ \frac{1}{(n+1)^2} + \frac{1}{(n+2)^2} + \dots + \frac{1}{(2n)^2} \right\} \sin^2 nx$$

converge.

3. Show that the series $\quad \Sigma \left(1 + \frac{1}{2} + \dots + \frac{1}{n} \right) \frac{\sin nx}{n}$

converges absolutely for $x = k\pi$ ($k = 0, \pm 1, \pm 2, \dots$) and conditionally for all other (real) values of x.

4. The series Σa_n converges conditionally. Show that

(i) $\Sigma a_n n^k$ diverges if $k > 1$, and

(ii) if $\Sigma(a_n/n^\alpha)$ converges, so does $\Sigma(a_n/n^\beta)$ for $\beta > \alpha$.

Deduce that there is a real number $\alpha_0 \geqslant -1$ such that $\Sigma(a_n/n^\alpha)$ converges for $\alpha > \alpha_0$ and diverges for $\alpha < \alpha_0$.

5. The sequence (a_n) is monotonic and converges to the limit l. Prove that the sum s_n of the first n terms of the series

$$\sum_{n=1}^{\infty} (-1)^n a_n$$

is such that $\qquad \lim \sup s_n - \lim \inf s_n = |l|.$

6. Establish Dedekind's extension of Abel's theorem 4.54 that, if both the series Σa_n and $\Sigma |v_n - v_{n+1}|$ converge, then $\Sigma a_n v_n$ converges. State and prove a corresponding extension of Dirichlet's theorem 4.53.

The real function f on $[1, \infty)$ is such that $f''(x) > 0$ for $x \geqslant 1$ and $f(x) \to 0$ as $x \to \infty$. Prove that $f(x) > 0$, $f'(x) < 0$ for $x \geqslant 1$, and that $\sum_1^\infty f'(n)$ converges.

Use the mean value theorem to deduce the convergence of

$$\sum_1^\infty (-1)^n f(n) \sin \{\log f(n)\}.$$

7. The series Σu_n converges and the series Σv_n of positive terms diverges. Show that
$$\lim \inf (u_n/v_n) \leqslant 0 \leqslant \lim \sup (u_n/v_n).$$

Find pairs of series Σu_n, Σv_n such that

(i) $\lim \inf (u_n/v_n) = -\infty$, $\lim \sup (u_n/v_n) = \infty$;

(ii) $u_n > 0$, $\lim \inf (u_n/v_n) = 0$, $\lim \sup (u_n/v_n) = \infty$.

8. The series $\sum_0^\infty u_n$ converges absolutely and the series $\sum_0^\infty v_n$ converges. Let

$$w_n = u_0 v_n + u_1 v_{n-1} + \dots + u_n v_0 \quad (n = 0, 1, 2, \dots).$$

Prove that $\qquad \sum_{r=0}^{n} w_r = \sum_{r=0}^{n} u_r (t - \tau_{n-r}),$

where $t = \sum\limits_{0}^{\infty} v_r$, $\tau_n = \sum\limits_{n+1}^{\infty} v_r$, and deduce that

$$\sum_{n=0}^{\infty} w_n = \left(\sum_{n=0}^{\infty} u_n\right)\left(\sum_{n=0}^{\infty} v_n\right).$$

(In theorem 5.7 of C1, Σu_n and Σv_n are both assumed to be absolutely convergent and Σw_n is proved to be absolutely convergent also.)

Construct conditionally convergent series Σu_n, Σv_n such that Σw_n diverges.

9. The terms of the series $1 - \frac{1}{2} + \frac{1}{3} - \frac{1}{4} + \ldots$

are rearranged according to the following rules.

(i) The order among the positive terms and the order among the negative terms remains unchanged.

(ii) If, of the first n terms, p_n are positive and q_n are negative, $p_n/q_n \to k \ (> 0)$ as $n \to \infty$.

Prove that the sum of the series is $\frac{1}{2} \log 4k$.

10. The sequence (a_n) is decreasing and $a_n \sim l/n$. Show that, if the terms of the series

$$a_1 - a_2 + a_3 - a_4 + \ldots$$

are rearranged as in **9**, then the sum of the series is increased by $\frac{1}{2}l \log k$.

11. For what values of k is the series

$$\frac{1}{1^k} + \frac{1}{3^k} - \frac{1}{2^k} + \frac{1}{5^k} + \frac{1}{7^k} - \frac{1}{4^k} + \ldots,$$

with a negative term following two positive ones, convergent?

4.6. Power series

In C1, §5.5 and §5.6, it was shown that every power series $\Sigma a_n z^n$ has a radius of convergence R and that $R = \lim |a_n/a_{n+1}|$, *if this limit exists*. By using upper limits we can give a simpler proof of the first result and we can sharpen the second.

Theorem 4.61.

(i) *If* $\limsup |a_n|^{1/n} = 0$, *then* $\Sigma a_n z^n$ *converges absolutely for all* z.

(ii) *If* $\limsup |a_n|^{1/n} = \infty$, *then* $\Sigma a_n z^n$ *converges for* $z = 0$ *only*.

(iii) *If* $0 < \limsup |a_n|^{1/n} < \infty$ *and*

$$R = \frac{1}{\limsup |a_n|^{1/n}},$$

then $\Sigma a_n z^n$ *converges absolutely for* $|z| < R$ *and diverges for* $|z| > R$.

The radius of convergence of $\Sigma a_n z^n$ *is therefore* R, *where* R *is called* ∞ *in the case* (i) *and* 0 *in the case* (ii).

Proof.

(i) For every z, $\limsup |a_n z^n|^{1/n} = |z| \limsup |a_n|^{1/n} = 0$ and so, by theorem 4.31, $\Sigma a_n z^n$ converges absolutely for all z.

(ii) For $z \neq 0$, $\limsup |a_n z^n|^{1/n} = |z| \limsup |a_n|^{1/n} = \infty$ and so $\Sigma a_n z^n$ diverges for all $z \neq 0$.

(iii) If $|z| < R$, then $\limsup |a_n z^n|^{1/n} = |z| \limsup |a_n|^{1/n} < 1$ and so $\Sigma a_n z^n$ converges absolutely.

If $|z| > R$, then $\limsup |a_n z^n|^{1/n} > 1$ and so $\Sigma a_n z^n$ diverges. |

Theorem 4.62. *The power series $\Sigma a_n z^n$ and $\Sigma n a_n z^{n-1}$ have the same radius of convergence.*

Proof. By exercise 4(*b*), 6,

$$\limsup |n a_n|^{1/n} = \limsup |a_n|^{1/n},$$

since $\lim n^{1/n} = 1$. |

Corollary.

(i) *All the series obtained by differentiating a power series term by term any numbers of times have the same radius of convergence.*

(ii) *A similar result holds for integration.*

Exercises 4(*f*)

1. Find the (complex) values of z for which the following power series converge.

(i) $\displaystyle\sum_{2}^{\infty} \left(\frac{z}{\log n}\right)^n,$

(ii) $\displaystyle\sum_{0}^{\infty} n! \, z^n,$

(iii) $\displaystyle\sum_{2}^{\infty} \frac{z^n}{n \log^2 n},$

(iv) $\displaystyle\sum_{2}^{\infty} \frac{z^n}{\log^2 n},$

(v) $\displaystyle\sum_{0}^{\infty} 2^{-\sqrt{n}} z^n,$

(vi) $\displaystyle\sum_{0}^{\infty} \left(\frac{-nz}{n+1}\right)^n,$

(vii) $\displaystyle\sum_{1}^{\infty} \left(\sin\frac{\alpha}{n}\right) z^n$ (α real),

(viii) $\displaystyle\sum_{1}^{\infty} \frac{\log n}{n - \log n} z^n,$

(ix) $\displaystyle\sum_{0}^{\infty} \frac{z^{3n}}{8^n(1 - in)},$

(x) $\displaystyle\sum_{1}^{\infty} \frac{\sin n\alpha}{n} z^n$ (α real).

2. Show that the binomial series $\displaystyle\sum_{n=0}^{\infty} \binom{\alpha}{n} z^n$ (α real) has radius of convergence 1 unless α is a non-negative integer (in which case the series terminates). Prove also that

(i) when $\alpha \geqslant 0, \Sigma \binom{\alpha}{n} z^n$ converges absolutely for $|z| = 1$;

(ii) when $-1 < \alpha < 0, \Sigma \binom{\alpha}{n} z^n$ converges conditionally for $|z| = 1, z \neq -1$ and diverges when $z = -1$;

(iii) when $\alpha \leqslant -1, \Sigma \binom{\alpha}{n} z^n$ diverges for $|z| = 1$.

(To prove (ii) show that

$$\left| \binom{\alpha}{n} \right| \bigg/ \left| \binom{\alpha}{n+1} \right| > \frac{1}{n^k} \bigg/ \frac{1}{(n+1)^k} \tag{*}$$

if k is sufficiently small and deduce that $\binom{\alpha}{n} \to 0$ as $n \to \infty$.)

3. Investigate the convergence of the hypergeometric series

$$\sum_{n=1}^{\infty} \frac{a(a+1)\ldots(a+n-1)\, b(b+1)\ldots(b+n-1)}{n!\, c(c+1)\ldots(c+n-1)} \, z^n,$$

where a, b, c are real numbers not equal to $0, -1, -2, \ldots$.

4. The power series $\sum_0^{\infty} a_n z^n$, $\sum_0^{\infty} b_n z^n$ have radii of convergence R, S respectively.

(i) Show that the radius of convergence T_1 of $\Sigma\,(a_n+b_n)z^n$ is min (R, S) if $R \neq S$. What can be said about T_1 when $R = S$?

(ii) Show that the radius of convergence T_2 of $\Sigma\, a_n b_n z^n$ is at least RS (where $R\infty$ is interpreted as ∞ if $R > 0$). Construct an example in which $T_2 > RS$.

(iii) Show that the radius of convergence T_3 of $\Sigma\,(a_0 b_n + a_1 b_{n-1} + \ldots + a_n b_0)z^n$ is at least min (R, S). Construct an example in which $T_3 > $ min (R, S).

5. The sequence $(a_0, a_1, a_2, \ldots)$ of non-negative real numbers is decreasing and $a_0 > 0$. Show that the radius of convergence of $\Sigma\, a_n z^n$ is at least 1.

Prove that, if $|z| < 1$, then

$$\left| (1-z)\left(\sum_0^{\infty} a_n z^n \right) - a_0 \right| < a_0$$

and deduce that $\sum_0^{\infty} a_n z^n \neq 0$ for $|z| < 1$.

4.7. Double and repeated limits

Let X be any set. A *double sequence* of elements of X is a function whose range is in X and whose domain is P^2, the set of ordered pairs of positive integers. The members x_{mn} $(m = 1, 2, \ldots; n = 1, 2, \ldots)$ of a double sequence may be arranged in an array

$$
\begin{array}{cccc}
x_{11} & x_{12} & x_{13} & \ldots \\
x_{21} & x_{22} & x_{23} & \ldots \\
x_{31} & x_{32} & x_{33} & \ldots \\
\ldots & \ldots & \ldots & \ldots
\end{array}
$$

We shall only consider double sequences of real or complex numbers.

Definition. The complex double sequence (x_{mn}) is said to converge *to the number x if, given $\epsilon > 0$, there is an n_0 such that*

$$|x_{mn} - x| < \epsilon \quad \text{whenever} \quad m, n > n_0;$$

we then write

$$x_{mn} \to x \quad as \quad m, n \to \infty \quad or \quad \lim_{m,\, n \to \infty} x_{mn} = x.$$

Illustrations

(i) $\dfrac{1}{m+n} \to 0$ as $m, n \to \infty$.

(ii) $\displaystyle\lim_{m,\, n \to \infty} \dfrac{m}{m+n}$ does not exist; for $\dfrac{m}{m+m} = \dfrac{1}{2}$ for all m and $\dfrac{m}{m+2m} = \dfrac{1}{3}$.

It is natural to try to relate the double limit $\displaystyle\lim_{m,\, n \to \infty} x_{mn}$ with the two repeated limits $\displaystyle\lim_{m \to \infty} (\lim_{n \to \infty} x_{mn})$ and $\displaystyle\lim_{n \to \infty} (\lim_{m \to \infty} x_{mn})$.

Theorem 4.71. *If* $\displaystyle\lim_{m,\, n \to \infty} x_{mn}$ *exists and, for each* m, $\displaystyle\lim_{n \to \infty} x_{mn}$ *exists, then* $\displaystyle\lim_{m \to \infty} (\lim_{n \to \infty} x_{mn})$ *exists and is equal to* $\displaystyle\lim_{m,\, n \to \infty} x_{mn}$.

Proof. Let $\displaystyle\lim_{m,\, n \to \infty} x_{mn} = x$ and $\displaystyle\lim_{n \to \infty} x_{mn} = y_m$ $(m = 1, 2, \ldots)$. Given $\epsilon > 0$, there is an N such that

$$|x_{mn} - x| < \epsilon \quad for \quad m, n > N. \tag{4.71}$$

Also, given m, there is a $P(m)$ such that

$$|x_{mn} - y_m| < \epsilon \quad for \quad n > P(m). \tag{4.72}$$

Now take $m > N$. If $n > \max(N, P(m))$, (4.71) and (4.72) both hold and give

$$|y_m - x| < 2\epsilon.$$

Thus $y_m \to x$ as $m \to \infty$. |

Note that the existence of $\displaystyle\lim_{m,\, n \to \infty} x_{mn}$ does not ensure the existence of $\displaystyle\lim_{n \to \infty} x_{mn}$ for any m (cf. $x_{mn} = (-1)^n/m$).

Theorem 4.71 has an immediate corollary.

Theorem 4.72. *Suppose that* (i) *for each* m, $\displaystyle\lim_{n \to \infty} x_{mn}$ *exists and, for each* n, $\displaystyle\lim_{m \to \infty} x_{mn}$ *exists,* (ii) $\displaystyle\lim_{m,\, n \to \infty} x_{mn}$ *exists. Then* $\displaystyle\lim_{m \to \infty} (\lim_{n \to \infty} x_{mn})$ *and* $\displaystyle\lim_{n \to \infty} (\lim_{m \to \infty} x_{mn})$ *exist and are equal.*

Illustration (ii) shows that $\displaystyle\lim_{m \to \infty} (\lim_{n \to \infty} x_{mn})$ and $\displaystyle\lim_{n \to \infty} (\lim_{m \to \infty} x_{mn})$ may well exist without being equal. The inversion of successive limiting operations is generally a delicate matter; there is one simple criterion, of monotonicity, which we now give.

Theorem 4.73. *Suppose that each x_{mn} is real and that, for each m, x_{mn} increases (decreases) with n and, for each n, x_{mn} increases (decreases) with m. If one of the limits*

$$\lim_{m,\,n\to\infty} x_{mn}, \quad \lim_{m\to\infty}(\lim_{n\to\infty} x_{mn}), \quad \lim_{n\to\infty}(\lim_{m\to\infty} x_{mn})$$

exists, then the other two also exist and all three limits are equal.

Proof. We may assume that x_{mn} increases with m and n.

(i) Suppose that $x_{mn} \to x$ as $m, n \to \infty$. Then $x_{mn} \leqslant x$ for all m, n. Hence, for each m, $(x_{m1}, x_{m2}, \ldots)$ is a bounded increasing sequence and so $\lim\limits_{n\to\infty} x_{mn}$ exists. Similarly $\lim\limits_{m\to\infty} x_{mn}$ exists. It now follows from theorem 4.72 that $\lim\limits_{m\to\infty}(\lim\limits_{n\to\infty} x_{mn})$ and $\lim\limits_{n\to\infty}(\lim\limits_{m\to\infty} x_{mn})$ exist and are equal to x.

(ii) Suppose that $\lim\limits_{m\to\infty}(\lim\limits_{n\to\infty} x_{mn})$ exists. Let

$$\lim_{n\to\infty} x_{mn} = y_m \ (m = 1, 2, \ldots)$$

and let $\lim\limits_{m\to\infty} y_m = y$. Since x_{mn} increases with n, $x_{mn} \leqslant y_m$ for all n; and since x_{mn} increases with m, y_m increases with m also and so $y_m \leqslant y$ for all m. Thus $\quad x_{mn} \leqslant y_m \leqslant y$

for all m, n. It is easy to see that a bounded double sequence which increases with each index converges (exercise 4(g), 5). Therefore $\lim\limits_{m,\,n\to\infty} x_{mn}$ exists. By (i), $\lim\limits_{n\to\infty}(\lim\limits_{m\to\infty} x_{mn})$ also exists and all three limits are equal. |

The notion of a double limit is applicable to functions on R^2 (i.e. functions of two continuous real variables). For a given function f, we may consider $f(x, y)$ as x, y tend to infinity or to a finite limit. Also one variable may be continuous, the other confined to integral values. In all these cases analogues of the last three theorems hold.

Double series. Given the double sequence (u_{mn}) of complex terms, let

$$s_{mn} = \sum_{i=1}^{m} \sum_{j=1}^{n} u_{ij} \ (m, n = 1, 2, \ldots).$$

If $s_{mn} \to s$ as $m, n \to \infty$, we say that the double series

$$\sum_{m,\,n=1}^{\infty} u_{mn}$$

converges to s.

Theorem 4.71, interpreted for double series, takes the following form: *If $\sum\limits_{m,\,n=1}^{\infty} u_{mn}$ converges to s and, for each m, $\sum\limits_{n=1}^{\infty} u_{mn}$ converges, then $\sum\limits_{m=1}^{\infty} \left(\sum\limits_{n=1}^{\infty} u_{mn} \right)$ converges to s.* We use the series form of theorem 4.73 to prove a more general result.

Theorem 4.74. *Let (u_{mn}) be a double sequence of complex terms. If one of the series*

$$\sum_{m,\,n=1}^{\infty} |u_{mn}|, \quad \sum_{m=1}^{\infty} \left(\sum_{n=1}^{\infty} |u_{mn}| \right), \quad \sum_{n=1}^{\infty} \left(\sum_{m=1}^{\infty} |u_{mn}| \right) \qquad (4.73)$$

converges, then all the series

$$\sum_{m,\,n=1}^{\infty} u_{mn}, \quad \sum_{m=1}^{\infty} \left(\sum_{n=1}^{\infty} u_{mn} \right), \quad \sum_{n=1}^{\infty} \left(\sum_{m=1}^{\infty} u_{mn} \right)$$

converge and have the same sum.

Proof.

(i) First suppose that the u_{mn} are real. Let

$$p_{mn} = \tfrac{1}{2}(|u_{mn}| + u_{mn}), \quad q_{mn} = \tfrac{1}{2}(|u_{mn}| - u_{mn});$$

so that p_{mn} and q_{mn} are non-negative, $u_{mn} = p_{mn} - q_{mn}$ and

$$0 \leqslant p_{mn}, q_{mn} \leqslant |u_{mn}|.$$

If now one of the series (4.73) converges, so does the corresponding series with $|u_{mn}|$ replaced by p_{mn} (by the comparison principle or its extension to double series—see exercise 4(g), 6). Then, by theorem 4.73,

$$\sum_{m,\,n=1}^{\infty} p_{mn}, \quad \sum_{m=1}^{\infty} \left(\sum_{n=1}^{\infty} p_{mn} \right), \quad \sum_{n=1}^{\infty} \left(\sum_{m=1}^{\infty} p_{mn} \right)$$

all converge and have the same sum. This is also true for the corresponding series of q's and so subtraction gives the result for series of real terms.

(ii) When the u_{mn} are complex, let

$$u_{mn} = v_{mn} + iw_{mn},$$

where v_{mn} and w_{mn} are real. Since

$$|v_{mn}|, |w_{mn}| \leqslant |u_{mn}|,$$

if one of the series (4.73) converges, then the corresponding series with $|u_{mn}|$ replaced by $|v_{mn}|$ and $|w_{mn}|$ also converge. Hence, by (i), the real and imaginary parts of u_{mn} have the desired property and this proves the theorem. |

Exercises 4(g)

1. Find, when they exist, the double limit $\lim\limits_{m,\,n\to\infty} x_{mn}$ and the repeated limits $\lim\limits_{m\to\infty} (\lim\limits_{n\to\infty} x_{mn})$, $\lim\limits_{n\to\infty} (\lim\limits_{m\to\infty} x_{mn})$ of the sequences (x_{mn}) with the following (m, n)th terms.

(i) $\dfrac{m-n}{m+n}$,

(ii) $\dfrac{m+n}{m^2}$,

(iii) $\dfrac{m+n}{m^2+n^2}$,

(iv) $(-1)^{m+n}\left(\dfrac{1}{m}+\dfrac{1}{n}\right)$,

(v) $\dfrac{mn}{m^2+n^2}$,

(vi) $(-1)^{m+n}\dfrac{1}{n}\left(1+\dfrac{1}{m}\right)$.

2. Give an example of an unbounded, convergent double sequence.

3. Prove, without using double limits, that if x_{mn} increases (decreases) with each of m and n, then the existence of one of $\lim\limits_{m\to\infty} (\lim\limits_{n\to\infty} x_{mn})$, $\lim\limits_{n\to\infty} (\lim\limits_{m\to\infty} x_{mn})$ implies the existence of the other and also the equality of the two limits.

4. *General principle of convergence.* Show that a necessary and sufficient condition for the (complex) double sequence (x_{mn}) to converge is that, given $\epsilon > 0$, there is an n_0 such that $|x_{mn} - x_{m'n'}| < \epsilon$ if $m, m', n, n' > n_0$.

5. The double sequence (x_{mn}) of real numbers is bounded and x_{mn} increases (decreases) with each index. Prove that (x_{mn}) converges.

6. *Comparison principle.* Suppose that $\sum\limits_{m,\,n} u_{mn}$, $\sum\limits_{m,\,n} v_{mn}$ are double series of positive terms, that $u_{mn} \leqslant v_{mn}$ for all m, n and that $\sum\limits_{m,\,n} v_{mn}$ converges. Show that $\sum\limits_{m,\,n} u_{mn}$ converges.

7. The double series $\sum\limits_{m,\,n} u_{mn}$ of complex terms is said to converge absolutely if $\sum\limits_{m,\,n} |u_{mn}|$ converges. Show that an absolutely convergent double series converges.

8. Show that the double series $\sum\limits_{m,\,n=1}^{\infty} \dfrac{1}{m^k n^l}$ converges if and only if $k > 1$ and $l > 1$.

9. Show that the double series
$$\sum_{m,\,n=1}^{\infty} \frac{1}{(m+n)^k}$$
converges if and only if $k > 2$.

10. The double series $\sum\limits_{m,\,n=1}^{\infty} u_{mn}$ converges absolutely and, for $r = 1, 2, \ldots,$
$$v_r = \sum_{m+n=r} u_{mn} = u_{1,\,r-1} + u_{2,\,r-2} + \ldots + u_{r-1,\,1}.$$

Prove that
$$\sum_{r=1}^{\infty} v_r = \sum_{m,\,n=1}^{\infty} u_{mn}.$$

Deduce that, for $k > 2$,

$$\sum_{m,\,n=1}^{\infty} \frac{1}{(m+n)^k} = \sum_{n=1}^{\infty} \frac{1}{n^{k-1}}.$$

11. Let ω, ω' be non-zero complex numbers such that ω/ω' is not real. Show that the double series

$$\sum_{m,\,n=1}^{\infty} \frac{1}{(m\omega + n\omega')^k}$$

converges absolutely if and only if $k > 2$. (For the present, k must be rational; general complex powers are introduced in chapter 10.)

12. Prove that the double series $\displaystyle\sum_{m,\,n=1}^{\infty} z^{mn}$ converges absolutely for $|z| < 1$. Hence prove that, when $|z| < 1$,

$$\sum_{n=1}^{\infty} \frac{z^n}{1-z^n} = \sum_{n=1}^{\infty} \frac{z^{n^2}(1+z^n)}{1-z^n}.$$

13. The series
$$\sum_{n=1}^{\infty} a_n z^n \quad \text{and} \quad \sum_{n=1}^{\infty} b_n z^n$$

both have radius of convergence R (> 0). Their sums are denoted by $f(z)$ and $g(z)$ respectively. Establish, in a disc to be specified, the formula

$$\sum_{n=1}^{\infty} a_n g(z^n) = \sum_{n=1}^{\infty} b_n f(z^n).$$

NOTES ON CHAPTER 4

§4.1. The first mathematician to use the O, o notation systematically was E. Landau (1877–1938). P. Du Bois-Reymond (1831–89) had previously used a more elaborate notation (which included $\sim$) for comparing the rates of growth of two increasing functions which tend to infinity. (See G. H. Hardy, *Orders of Infinity* (Cambridge, 1910).)

§4.2. The upper and lower limits of a sequence of real numbers could be defined as the largest and least numbers, respectively, to which subsequences converge (with suitable provision for infinities). However this procedure would require an initial proof that such numbers exist. The method we have adopted, though less intuitive, is technically simpler.

Filters. In analysis the idea of a limit appears in many forms each of which is usually treated individually. In this book we have separately defined limits of sequences in metric spaces and of functions from one metric space into another, and other instances occur in C1. This approach, apart from being traditional, is best suited for an introductory account and we have again adopted it for defining the several different guises in which upper and lower limits appear. However a unified treatment of the limit process is possible.

Let X be a non-empty set. A non-empty set $\mathcal{F}$ of subsets of X is called a *filter* on X if the following conditions are satisfied.

F1. If $A, B \in \mathscr{F}$, then $A \cap B \in \mathscr{F}$.

F2. If $A \in \mathscr{F}$ and $B \supset A$, then $B \in \mathscr{F}$.

F3. $\varnothing \notin \mathscr{F}$.

Note that, by F2, $X \in \mathscr{F}$. It is easy to give examples of filters.

(i) In P, the set of positive integers, the sets A such that $P - A$ is finite form a filter.

(ii) In R^1, the collection of all sets A which contain an interval (x, ∞) is a filter.

(iii) Let (X_0, ρ) be a metric space, let a be a fixed limit point of X_0 and let $X = X_0 - \{a\}$. A filter on X may be defined as the collection of all subsets A of X which contain a set $B(a; r) - \{a\}$ for some $r > 0$.

(iv) Let $\mathscr{F}$ be a filter on a set X. If $A \in \mathscr{F}$, denote by $\mathscr{A}$ the subset of $\mathscr{F}$ defined by

$$\mathscr{A} = \{E \in \mathscr{F} \,|\, E \supset A\}.$$

Then
$$\mathfrak{F} = \{\mathscr{B} \subset \mathscr{F} \,|\, \mathscr{B} \supset \mathscr{A} \text{ for some } A \in \mathscr{F}\}$$

is a filter on $\mathscr{F}$.

Now let $\mathscr{F}$ be a filter on a set X and let (Y, σ) be a metric space. We then say that

$$\lim_{\mathscr{F}} f(x) = y_0$$

(where $y_0 \in Y$) if, given $\epsilon > 0$, there is an $A \in \mathscr{F}$ such that $\sigma(y_0, f(x)) < \epsilon$ for all $x \in A$.

Clearly the filter in (i) yields the limit of a sequence of points in (Y, σ); and it is easy to see that (ii) and (iii) lead to other familiar situations. For instance if, in (iii), $X = (a, \infty)$ or $(-\infty, a)$, we obtain a one-sided limit at a.

By formulating the limit concept in such generality one sacrifices, not surprisingly, some of the properties traditionally associated with limits. However these may be restored by the imposition of quite mild restrictions on Y.

Finally, let f be a bounded, real-valued function on a set X equipped with a filter $\mathscr{F}$. Then the relation

$$\phi(A) = \sup_{x \in A} f(x) \, (A \in \mathscr{F})$$

gives a bounded real-valued function ϕ on $\mathscr{F}$ and it may be shown that $\lim_{\mathfrak{F}} \phi(A)$ exists, where $\mathfrak{F}$ is the filter on $\mathscr{F}$ defined in (iv). This limit is defined as the upper limit of f with respect to $\mathscr{F}$, i.e.

$$\limsup_{\mathscr{F}} f(x) = \lim_{\mathfrak{F}} (\sup_{x \in A} f(x));$$

and similarly
$$\liminf_{\mathscr{F}} f(x) = \lim_{\mathfrak{F}} (\inf_{x \in A} f(x)).$$

In this sketch we have not considered infinite limits of real-valued functions. These present no problems of definition, but they are most simply incorporated in the theory of filters when topological rather than metric spaces form its background. A simple account of filters in this setting is given by F. Smithies in his expository article 'Abstract Analysis' (*Mathematical Gazette*, **35** (1951), 2–7). The equivalent theory of *directed sets* is also mentioned in this paper. Still another general theory of limits is described by E. J. McShane in *Studies in Modern Analysis* (edited by R. C. Buck).

§ **4.3.** In C1 and so far in this book no formal definition of the *infinite series* $\sum_{1}^{\infty} u_n$ has appeared. To fill this gap we write

$$s_n = u_1 + \ldots + u_n \ (n \geqslant 1)$$

and say that the series is the ordered pair of sequences $((u_n), (s_n))$.

There is no confusion in practice in the double usage of the symbol $\sum_{1}^{\infty} u_n$, first as designating the series formally defined as $((u_n), (s_n))$ (without any implication as to its convergence), and secondly as the number which is the sum of this series when it does converge.

Similar remarks apply to double series.

5

UNIFORM CONVERGENCE

5.1. Pointwise and uniform convergence

Consider a sequence of functions

$$f_n : X \to Y \quad (n = 1, 2, \ldots),$$

where (Y, σ) is a metric space. For the moment the space X need not carry a metric.

Suppose that there is a function $f : X \to Y$ such that, for each x in X,

$$\sigma(f_n(x), f(x)) \to 0$$

as $n \to \infty$. We then say that (f_n) *converges to f* on X, or, more specifically, (f_n) converges *pointwise* to f on X.

Illustrations

In each of these, X is a subset of R^1 and (Y, σ) is R^1 with its usual metric.

(i) $f_n(x) = \dfrac{nx}{n+x}$ for $x \geqslant 0$.

$f_n(x) \to x$.

(ii) $f_n(x) = x^n$ for $0 \leqslant x \leqslant 1$.

$f_n(x) \to f(x)$, where $f(x) = 0$ for $0 \leqslant x < 1$ and $f(1) = 1$.

(iii) $f_n(x) = \dfrac{nx}{1 + n^2 x^2}$ for $x \geqslant 0$.

$f_n(x) \to 0$.

We observe that in (ii) the sequence of functions f_n, each continuous on $[0, 1]$, has a limit function which is discontinuous at the point 1. In this example pointwise convergence is not a strong enough requirement to ensure that continuity is carried over from the functions f_n to the limit function f.

The central theme of this chapter is the transmission of properties of the individual functions of a sequence to the limit function. For this investigation we define a mode of convergence of a sequence of functions, stronger than pointwise convergence.

Definition. *Suppose that $f_n: X \to Y$ is a sequence of functions on a set X to a metric space (Y, σ). The sequence (f_n) is said to* converge uniformly *to $f: X \to Y$ if*

$$\sup_{x \in X} \sigma(f_n(x), f(x)) \to 0 \quad as \quad n \to \infty. \tag{5.11}$$

The distinction between pointwise and uniform convergence is then that, in the former $\sigma(f_n(x), f(x))$ tends to 0 as $n \to \infty$ for each separate point x and, in the latter, the supremum of all these values tends to 0. It is plain that uniform convergence implies pointwise convergence (to the same limit function). Illustration (ii) shows that convergence may hold at every point of a set, here [0, 1], but since, for every n,

$$\sup_{0 \leqslant x \leqslant 1} |f_n(x) - f(x)| = 1,$$

the convergence is not uniform.

The ϵ-N criterion for the condition (5.11) is often useful. Convergence for each separate x in X means that, given $\epsilon > 0$, there exists $N = N(\epsilon, x)$ such that $\sigma(f_n(x), f(x)) < \epsilon$ for all $n \geqslant N$. The convergence is uniform if and only if it is possible to choose $N = N(\epsilon)$, *independent of x*, such that

$$\sigma(f_n(x), f(x)) < \epsilon \quad \text{for all } n \geqslant N \text{ and all } x \in X.$$

We shall examine the illustrations (i), (ii), (iii) for uniformity of convergence. Since $X \subset R^1$ and $Y = R^1$, they lend themselves to graphical representation. If $f_n \to f$ uniformly on X, the graph of $y = f_n(x)$ for all n greater than $N(\epsilon)$ lies inside the strip of the (x, y) plane between the curves $y = f(x) + \epsilon$ and $y = f(x) - \epsilon$. The fact

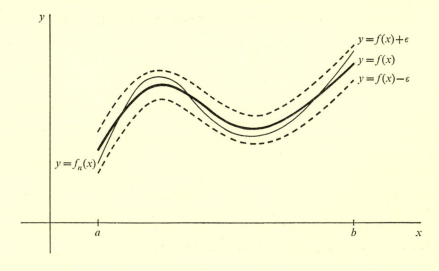

that pointwise convergence does not imply uniform convergence means that the nearness of $f_n(x)$ to $f(x)$ for each individual x does not imply the 'global' nearness of the curve $y = f_n(x)$ to the curve $y = f(x)$.

In (i) above, $f_n(x) - f(x) = -x^2/(n+x)$. In any *finite* interval, say $[0, a]$,

$$\sup_{0 \leqslant x \leqslant a} |f_n(x) - f(x)| \leqslant \frac{a^2}{n} \to 0 \quad \text{as} \quad n \to \infty,$$

and so $f_n \to f$ uniformly on $[0, a]$.

Suppose, however, that x is free to take any value in $[0, \infty)$. Since

$$\sup_{x \geqslant 0} |f_n(x) - f(x)| = \infty,$$

the convergence is not uniform in the infinite interval $[0, \infty)$.

(ii) We have already shown that the convergence of (x^n) is not uniform in $[0, 1]$. This fact is also intuitively clear. The limit function f has a jump of amount 1 at the point 1 and so, if $\epsilon < \frac{1}{2}$, none of the continuous curves $y = x^n$ can lie in the two disconnected pieces represented by $f(x) - \epsilon < y < f(x) + \epsilon$ $(0 \leqslant x \leqslant 1)$. It will be proved in theorem 5.21 that *any* sequence of continuous functions with a discontinuous limit must converge non-uniformly.

The reader should prove that in any interval $[0, 1 - \delta]$ which excludes the point 1 the convergence of x^n to its limit is uniform.

(iii) This example is more subtle than (i) or (ii). The function f_n has a maximum at the point $1/n$ and $f(1/n) = \frac{1}{2}$. We see that

$$\sup_{x \geqslant 0} |f_n(x) - f(x)| = \frac{1}{2}$$

and the convergence is not uniform.

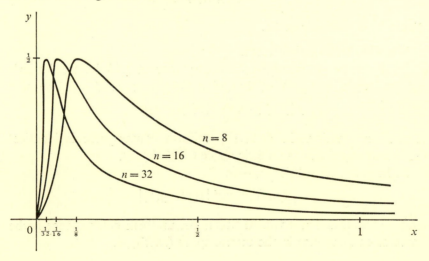

Each f_n is continuous everywhere in $[0, \infty)$ and so is f. Thus uniformity of convergence is not a necessary condition for the limit of continuous functions to be continuous.

Another example exhibits the notions of pointwise and uniform convergence in a more abstract setting.

Let C be the set of continuous and bounded real functions on R^1 and let ρ be the metric defined by

$$\rho(\phi, \psi) = \sup_{\substack{-\infty < x < \infty \\ 0 < h \leqslant 1}} \left| \int_x^{x+h} \{\phi(t) - \psi(t)\}\, dt \right| \quad (\phi, \psi \in C).$$

(It was shown in exercise 2(a), 2 that ρ *is* a metric.)

For $n \geqslant 1$, we define the functions $f_n : C \to C$ by

$$f_n(\phi)(x) = \phi\left(x + \frac{1}{n}\right) \quad (\phi \in C, x \in R^1).$$

Now take any $\phi \in C$ and let $K_\phi = \sup_{x \in R^1} |\phi(x)|$. We have

$$\rho(f_n(\phi), \phi) = \sup_{\substack{-\infty < x < \infty \\ 0 < h \leqslant 1}} \left| \int_x^{x+h} \left\{ \phi\left(t + \frac{1}{n}\right) - \phi(t) \right\} dt \right|$$

$$= \sup_{\substack{-\infty < x < \infty \\ 0 < h \leqslant 1}} \left| \int_{x+h}^{x+h+(1/n)} \phi(t)\, dt - \int_x^{x+(1/n)} \phi(t)\, dt \right|$$

$$\leqslant \frac{2}{n} K_\phi \to 0 \quad \text{as} \quad n \to \infty. \tag{5.12}$$

So, for all $\phi \in C$, $f_n(\phi) \to \phi$ as $n \to \infty$; in other words, $f_n \to I$ pointwise, where I is the identity function on C (i.e. $I(\phi) = \phi$ for all $\phi \in C$). It also follows from (5.12) that, if K is any positive number and C_K is the subset of C consisting of all ϕ such that $K_\phi \leqslant K$, then

$$\sup_{\phi \in C_K} \rho(f_n(\phi), I(\phi)) \leqslant \frac{2}{n} K.$$

Hence $f_n \to I$ uniformly on C_K.

We add three further notes.

(1) In the two paragraphs preceding theorem 3.44 the usual metric in the space $B(X, Y)$ of bounded functions $f : X \to (Y, \sigma)$ was defined by

$$\rho(f, g) = \sup_{x \in X} \sigma(f(x), g(x)).$$

Thus, for functions in $B(X, Y)$, uniform convergence is identical with convergence in the metric space $(B(X, Y), \rho)$.

If, in particular, Y is R^1, then

$$\rho(f, g) = \sup_{x \in X} |f(x) - g(x)|$$

and, for bounded real valued functions, uniform convergence is the same as convergence in the metric space $(B(X), \rho)$.

(2) We have assumed the set of functions on X to Y to be indexed by a positive integer n which tends to ∞. It is easy to adjust the definition of uniform convergence to cover, say, a double sequence indexed by (m, n) or a set of functions determined by a continuous variable t which tends to a limit t_0 or to ∞.

(3) *Convergence and uniform convergence of series.*

A statement about convergence of a series $\sum\limits_{n=1}^{\infty} u_n$ of real or complex valued functions on a set X is a translation of the corresponding statement about the sequence s_n, where $s_n = \sum\limits_{r=1}^{n} u_r$.

Illustration. For $n \geqslant 1$, let u_n be defined on the set of complex numbers z satisfying $|z| < 1$ by

$$u_n(z) = z^{n-1}.$$

Then

$$s_n(z) = \frac{1-z^n}{1-z} \to \frac{1}{1-z} = s(z), \quad \text{say,}$$

and $\sum\limits_{1}^{\infty} u_n$ converges to s at each point of the disc $|z| < 1$.

The convergence is not uniform because, for every n,

$$\sup_{|z|<1} |s_n(z) - s(z)| = \sup_{|z|<1} \left| \frac{z^n}{1-z} \right| = \infty.$$

The convergence is, however, uniform on $|z| \leqslant 1-\delta$, where δ is any fixed positive number. For we have

$$\sup_{|z| \leqslant 1-\delta} \left| \frac{z^n}{1-z} \right| \leqslant \frac{(1-\delta)^n}{\delta} \to 0 \quad \text{as} \quad n \to \infty.$$

Theorem 5.41 will show that any power series converges uniformly on a closed disc inside its circle of convergence.

Exercises 5(a)

1. The functions f_n, f on $X \cup Y$ are such that $f_n \to f$ uniformly on X and $f_n \to f$ uniformly on Y. Show that $f_n \to f$ uniformly on $X \cup Y$.

2. If complex valued functions f_n, g_n tend respectively to f, g, uniformly on X, show that, for any constants $\alpha, \beta, \alpha f_n + \beta g_n \to \alpha f + \beta g$ uniformly on X. Is it necessarily true that $f_n g_n \to fg$ uniformly on X?

3. The sequences (f_n) on $[0, 1]$, (g_n) on $(0, 1)$, (h_n) on $[0, 100]$ are defined by

$$f_n(x) = x^n(1-x), \quad g_n(x) = \frac{\sin nx}{nx}, \quad h_n(x) = \frac{nx^3}{1+nx}.$$

Prove that all three sequences converge pointwise; which of them converge uniformly?

4. Examine for uniformity of convergence on $[0, 1]$ the sequences (f_n), (g_n) defined by

$$f_n(x) = x^n(1-x^n), \quad g_n(x) = nx e^{-nx^2}.$$

5. Let (r_n) be a sequence consisting of all the rational numbers and, for $n = 1, 2, \ldots$ define the functions f_n on R^1 by

$$f_n(x) = \begin{cases} 1 & \text{if } x = r_n, \\ 0 & \text{otherwise.} \end{cases}$$

Prove that (f_n) converges pointwise, but not uniformly on every interval of R^1.

6. Let Q be the set of rational numbers in $(0, 1)$ and define the function f on Q by $f(p/q) = 1/q$ when p, q are coprime positive integers. Let (r_n) be a sequence consisting of the members of Q and, for $n = 1, 2, \ldots$, define f_n on Q by

$$f_n(x) = \begin{cases} f(x) & \text{if } x \in \{r_1, \ldots, r_n\}, \\ 0 & \text{if } x \in Q - \{r_1, \ldots, r_n\}. \end{cases}$$

Show that $f_n \to f$ uniformly on Q.

7. Examine for pointwise and uniform convergence the sequences $(f_n), (g_n), (h_n)$ of complex functions defined by

$$f_n(z) = \left(\frac{z}{\sqrt{n}}\right)^n, \quad g_n(z) = e^{nz}, \quad h_n(z) = \frac{\sin nz}{n}.$$

8. It is given that, for every continuous real function f on $[0, 1]$ for which $f(0) = 0$, the sequence (fs_n) converges uniformly on $[0, 1]$. Also $\lim_{n \to \infty} s_n(0)$ exists. Decide whether the sequence (s_n) must converge uniformly on $[0, 1]$.

9. For $n = 0, 1, 2, \ldots$, the functions u_n, v_n, w_n on $[0, 1]$ are given by

$$u_n(x) = x^n(1-x), \quad v_n(x) = x^n(1-x)^2, \quad w_n(x) = (-1)^n x^n(1-x).$$

Prove that the series $\Sigma u_n, \Sigma v_n, \Sigma w_n$ all converge pointwise; which of them converge uniformly?

10. Let $X \subset R^2$ be the rectangle $[a, b] \times [c, d]$, where $a > 0, c > 0$. The functions $f_n : X \to R^1$ are given by

$$f_n(x, y) = \frac{1 + nx}{1 + ny}.$$

Show that (f_n) converges uniformly on X.

11. The set X consists of the real functions ϕ on $[a, b]$ such that on $[a, b]$

(i) ϕ' exists and is continuous, and
(ii) $|\phi(x)|, |\phi'(x)| \leqslant M$.

For $n = 1, 2, \ldots$, the functions $f_n : X \to R^1$ are defined by

$$f_n(\phi) = \int_a^b \phi(x) \sin nx \, dx \quad (\phi \in X).$$

Prove that the sequence (f_n) converges uniformly on X.

12. Let $(X, \rho), (Y, \sigma)$ be metric spaces. Let X_0 be a subset of X and x_0 a limit point of X_0. Suppose that the functions $f_n : X_0 \to Y$ converge uniformly on X_0 and, for each n,

$$f_n(x) \to y_n \quad \text{as} \quad x \to x_0.$$

Show that, if (Y, σ) is complete, then the sequence (y_n) converges.

5.2. Properties assured by uniform convergence

Illustration (ii) of §5.1 shows that the limit function of a sequence of continuous functions may be discontinuous. We now prove that this cannot happen when the convergence is uniform.

Theorem 5.21. *Let (X, ρ) and (Y, σ) be metric spaces. Let the sequence of functions $f_n : X \to Y$ converge to f, uniformly on X. If c is a point of X at which each f_n is continuous, then f is continuous at c.*

Proof. (The argument is essentially the same as was used in theorem 3.45.)

Given $\epsilon > 0$, there is n such that

$$\sup_{x \in X} \sigma(f_n(x), f(x)) < \epsilon.$$

Since f_n is continuous at c, there is a $\delta > 0$ such that

$$\sigma(f_n(x), f_n(c)) < \epsilon \quad \text{whenever} \quad \rho(x, c) < \delta.$$

Hence, if $\rho(x, c) < \delta$,

$$\sigma(f(x), f(c)) \leqslant \sigma(f(x), f_n(x)) + \sigma(f_n(x), f_n(c)) + \sigma(f_n(c), f(c)) < 3\epsilon$$

and so f is continuous at c. |

Corollary. *If the series Σu_n of real or complex valued functions converges uniformly to s on (X, ρ) and if each u_n is continuous at c, then s is continuous at c.*

Proof. Each $s_n = u_1 + \ldots + u_n$ is continuous at c. |

The hypothesis of uniformity of convergence which has been proved in theorem 5.21 to be a sufficient condition is not a necessary condition. This is shown in illustration (iii) of §5.1. We shall prove in theorem 5.35 that, if Y is R^1 and if the sequence (f_n) is monotonic, then uniformity of convergence *is* necessary for the continuity of f.

Repeated limit problems. In theorem 5.21 continuity of f at c, when c is a limit point of X, means that

$$\lim_{x \to c} f(x) = f(c).$$

The left-hand side is $\lim_{x \to c} \lim_{n \to \infty} f_n(x)$, where the inner limit is in the σ metric and the outer in the ρ metric. The term $f(c)$ on the right-hand side is $\lim_{n \to \infty} f_n(c)$, and, if each f_n is continuous at c, this is the same as

$$\lim_{n \to \infty} \lim_{x \to c} f_n(x).$$

Hence the conclusion of theorem 5.21 is that in whichever order the two limiting operations

$$\lim_{n\to\infty} \quad \text{and} \quad \lim_{x\to c}$$

are applied to $f_n(x)$ the result is the same.

The problem of inverting the limits $m \to \infty$ and $n \to \infty$ in a double sequence has already been discussed in §4.7; the solution there lay in the sequence being monotonic in m and in n.

A common application of the notion of uniformity is to justify the inversion of the order of repeated limits. Theorems 5.22 and 5.23 will provide further instances of this. The reader will see that it is appropriate to specialize the arbitrary metric spaces of theorem 5.21 to the space R^1.

We consider sequences of real functions integrable over a finite interval. A pointwise convergent sequence of this kind need not have an integrable limit function; and, even if the limit function has an integral, the sequence of integrals need not converge to it.

To justify the first statement we take a sequence of functions f_n on [0, 1] defined as follows. If (r_n) is a sequence consisting of all the rational numbers in [0, 1], let

$$f_n(x) = \begin{cases} 1 & \text{if } x = r_1, \dots, r_n, \\ 0 & \text{otherwise.} \end{cases}$$

Then $f_n \to f$, where

$$f(x) = \begin{cases} 1 & \text{if } x \text{ is rational,} \\ 0 & \text{if } x \text{ is irrational.} \end{cases}$$

Now each function f_n is integrable over [0, 1], since every lower sum $s(\mathscr{D})$ is 0 and the upper sums $S(\mathscr{D})$ tend to 0 as the length of the largest subinterval of $\mathscr{D}$ tends to 0 (see C1, 123). However f is not integrable over [0, 1] (see C1, 122).

For the proof of the second statement take the sequence of functions g_n on [0, 1] where

$$g_n(x) = nxe^{-nx^2}.$$

$g_n \to g$ pointwise, where $g(x) = 0$ for $0 \leqslant x \leqslant 1$. Also

$$\int_0^1 g_n(x)\,dx = \tfrac{1}{2}(1 - e^{-n}) \to \tfrac{1}{2},$$

while

$$\int_0^1 g(x)\,dx = 0.$$

It is left to the reader to show that neither of the sequences (f_n), (g_n) converges uniformly (cf. exercises 5(a), 4, 5).

Theorem 5.22. *Let (f_n) be a sequence of real functions integrable over the finite interval $[a, b]$. If $f_n \to f$ uniformly on $[a, b]$, then*

   (i) *f is integrable over $[a, b]$, and*

   (ii) $\displaystyle\int_a^b f_n \to \int_a^b f.$

Proof. Given $\epsilon > 0$, let N be such that

$$\sup_{a \leqslant x \leqslant b} |f_n(x) - f(x)| < \epsilon \quad \text{for} \quad n \geqslant N.$$

(i) For $a \leqslant x \leqslant b$,

$$f_N(x) - \epsilon < f(x) < f_N(x) + \epsilon$$

and so, when $\mathscr{D}$ is any dissection of $[a, b]$,

$$s(\mathscr{D}, f_N) - \epsilon(b-a) \leqslant s(\mathscr{D}, f) \leqslant S(\mathscr{D}, f) \leqslant S(\mathscr{D}, f_N) + \epsilon(b-a).$$

Thus $S(\mathscr{D}, f) - s(\mathscr{D}, f) \leqslant S(\mathscr{D}, f_N) - s(\mathscr{D}, f_N) + 2\epsilon(b-a).$

But, since f_N is integrable over $[a, b]$,

$$S(\mathscr{D}, f_N) - s(\mathscr{D}, f_N) < \epsilon$$

if the length of the largest subinterval of $\mathscr{D}$ is sufficiently small. For such a $\mathscr{D}$,

$$S(\mathscr{D}, f) - s(\mathscr{D}, f) < \epsilon[1 + 2(b-a)]$$

and therefore f is integrable over $[a, b]$.

   (ii) For $n \geqslant N$, by C1, 127 (5),

$$\left| \int_a^b f_n - \int_a^b f \right| = \left| \int_a^b (f_n - f) \right| \leqslant \epsilon(b-a). \; \blacksquare$$

Corollary. (Term-by-term integration.) *If Σu_n converges uniformly to s on $[a, b]$ and if each u_n is integrable over $[a, b]$, then s is integrable and*

$$\int_a^b s = \Sigma \int_a^b u_n.$$

We note that uniform convergence is not necessary for the validity of term-by-term integration; this is shown by illustration (ii) of §5.1.

Theorem 5.22, proved for a finite interval, is false for an infinite interval. (See exercise 5(b), 5.)

In the last two theorems of this section we deal with sequences of functions differentiable in an open or a closed interval. When the interval is closed, derivatives at the end points are taken to be one-sided: if the interval is $[a, b]$, $f'(a)$ and $f'(b)$ are defined by

$$f'(a) = \lim_{h \to 0+} \frac{f(a+h)-f(a)}{h}, \quad f'(b) = \lim_{h \to 0+} \frac{f(b)-f(b-h)}{h}.$$

Theorem 5.23 A. *Let (f_n) be a sequence of real functions differentiable on a finite closed interval $[a, b]$. Suppose that $(f_n(x))$ converges for at least one $x \in [a, b]$ and that (f'_n) converges uniformly on $[a, b]$. Then*
  (i) *(f_n) converges uniformly on $[a, b]$,*
  (ii) *$f = \lim\limits_{n \to \infty} f_n$ is differentiable in $[a, b]$ and, for all $x \in [a, b]$,*

$$f'(x) = \lim_{n \to \infty} f'_n(x).$$

Proof. Let $\lim\limits_{n \to \infty} f'_n = g$ and take any $\epsilon > 0$. Since (f'_n) converges uniformly on $[a, b]$, there is an N such that

$$\sup_{a \leqslant x \leqslant b} |f'_n(x)-g(x)| \leqslant \epsilon \quad \text{for} \quad n \geqslant N \tag{5.21}$$

and therefore

$$\sup_{a \leqslant x \leqslant b} |f'_n(x)-f'_m(x)| \leqslant 2\epsilon \quad \text{for} \quad m, n \geqslant N.$$

If then ξ_1, ξ_2 are any points of $[a, b]$, we have by the mean value theorem for $f_n - f_m$,

$$\{f_n(\xi_1)-f_m(\xi_1)\}-\{f_n(\xi_2)-f_m(\xi_2)\} = (\xi_1-\xi_2)\{f'_n(\xi)-f'_m(\xi)\},$$

where ξ lies between ξ_1 and ξ_2; and so, for all $m, n \geqslant N$,

$$|\{f_n(\xi_1)-f_m(\xi_1)\}-\{f_n(\xi_2)-f_m(\xi_2)\}| \leqslant 2|\xi_1-\xi_2|\epsilon. \tag{5.22}$$

(i) If x_0 is a point for which $(f_n(x_0))$ converges, there is an N' such that

$$|f_n(x_0)-f_m(x_0)| \leqslant \epsilon \quad \text{for} \quad m, n \geqslant N'. \tag{5.23}$$

Now let x be any point of $[a, b]$. By (5.22) with $\xi_1 = x$ and $\xi_2 = x_0$ and by (5.23) we have, for $m, n \geqslant \max(N, N')$,

$$|f_n(x)-f_m(x)| \leqslant (1+2|x-x_0|)\epsilon \leqslant \{1+2(b-a)\}\epsilon \tag{5.24}$$

and so $\lim f_n(x)$ exists for all $x \in [a, b]$. Letting $m \to \infty$ in (5.24) we see that (f_n) converges uniformly on $[a, b]$.

(ii) Let c be an interior point of $[a, b]$ and take any $h \neq 0$ such that $c+h \in [a, b]$. By (5.22), with $\xi_1 = c+h$ and $\xi_2 = c$ we have, for $m, n \geqslant N$,

$$\left| \frac{f_n(c+h)-f_n(c)}{h} - \frac{f_m(c+h)-f_m(c)}{h} \right| \leqslant 2\epsilon. \tag{5.25}$$

Since $\qquad f_m(c+h) - f_m(c) \rightarrow f(c+h) - f(c)$

as $m \rightarrow \infty$, (5.25) shows that, for all $h \neq 0$ such that $c+h \in [a, b]$,

$$\left| \frac{f_n(c+h) - f_n(c)}{h} - \frac{f(c+h) - f(c)}{h} \right| \leqslant 2\epsilon \quad \text{for} \quad n \geqslant N. \quad (5.26)$$

By the definition of the derivative $f_N'(c)$, there is an $\eta > 0$ such that

$$\left| \frac{f_N(c+h) - f_N(c)}{h} - f_N'(c) \right| \leqslant \epsilon \quad \text{when} \quad 0 < |h| \leqslant \eta. \quad (5.27)$$

Then, by (5.26), (5.27) and (5.21),

$$\left| \frac{f(c+h) - f(c)}{h} - g(c) \right|$$

$$\leqslant \left| \frac{f(c+h) - f(c)}{h} - \frac{f_N(c+h) - f_N(c)}{h} \right| + \left| \frac{f_N(c+h) - f_N(c)}{h} - f_N'(c) \right|$$

$$+ |f_N'(c) - g(c)|$$

$$\leqslant 4\epsilon$$

when $0 < |h| \leqslant \eta$. It follows that $f'(c)$ exists and is equal to

$$g(c) = \lim f_n'(c).$$

If c is one of a, b, the above proof applies except that h must be restricted to either positive or negative values respectively. |

In dealing with open intervals we can assume rather less than uniform convergence on the whole interval.

Theorem 5.23B. *Let (f_n) be a sequence of real functions differentiable in a finite or infinite open interval (a, b). Suppose that $(f_n(x))$ converges for at least one $x \in (a, b)$ and that (f_n') converges uniformly on every finite closed subinterval of (a, b). Then*

(i) *(f_n) converges uniformly on every finite closed subinterval of (a, b),*

(ii) *$f = \lim_{n \to \infty} f_n$ is differentiable in (a, b) and, for all $x \in (a, b)$,*

$$f'(x) = \lim_{n \to \infty} f_n'(x).$$

Proof. Let x_0 be a point for which $(f_n(x))$ converges, let c be any point of (a, b) and let $[\alpha, \beta]$ be any finite, closed subinterval of (a, b). There is now a finite, closed subinterval $[p, q]$ of (a, b) such that

$$x_0, c \in (p, q) \quad \text{and} \quad [\alpha, \beta] \subset [p, q].$$

Then, by theorem 5.23 A, (f_n) converges uniformly on $[p, q]$ and hence on $[\alpha, \beta]$. Also, if $f = \lim f_n$, then $f'(c)$ exists and is equal to $\lim f_n'(c)$. |

Notes. (1) We can rewrite 5.23 A and B as theorems on the term-by-term differentiation of infinite series.

(2) The uniformity of convergence of (f_n') is not a necessary condition that $f'(x) = \lim f_n'(x)$. A counter-example is provided by exercise 5(*b*), 7.

(3) The proof of theorem 5.23 A is hard. If, however, we add to the hypothesis the requirement that each f_n' is continuous, the theorem is easily deduced from theorem 5.22 as follows:

By theorem 5.21, f_n' tends to a continuous limit function, g say.

By theorem 5.22,

$$\int_a^x g = \lim_{n \to \infty} \int_a^x f_n'$$

$$= \lim_{n \to \infty} \{f_n(x) - f_n(a)\}$$

$$= f(x) - f(a).$$

Since g is continuous, the left-hand side has derivative $g(x)$. Hence $g(x) = f'(x)$, as proved in theorem 5.23 A.

(4) Each of theorems 5.22 and 5.23 exhibits uniformity of convergence as a sufficient condition for inverting two passages to a limit. Differentiation and integration are both limiting operations (the integral $\int_a^b f$ being the limit of approximative sums—C1, 123). The conclusions of the theorems are respectively

$$\int_a^b (\lim_{n \to \infty} f_n) = \lim_{n \to \infty} \int_a^b f_n$$

and

$$\frac{d}{dx} \lim_{n \to \infty} f_n(x) = \lim_{n \to \infty} \frac{d}{dx} f_n(x)$$

or, expressed as term-by-term differentiation of series,

$$\frac{d}{dx} \sum_1^\infty u_n(x) = \sum_1^\infty u_n'(x).$$

Exercises 5(b)

1. Show that the series $\sum\limits_{0}^{\infty} x^n(1-x^n)$ converges pointwise, but not uniformly on the interval $[0, 1]$.

2. The function f_n on R^1 is given by

$$f_n(x) = \frac{x^n}{1+x^{2n}}.$$

Show that (f_n) converges uniformly on the interval $[a, b]$ if and only if $[a, b]$ does not contain either of the points $1, -1$.

3. The real function ϕ on $[0, 1]$ is continuous and f_n on $[0, 1]$ is defined by

$$f_n(x) = x^n\phi(x).$$

Prove that (f_n) converges uniformly on $[0, 1]$ if and only if $\phi(1) = 0$.

4. Give examples to show that a sequence of integrable functions with an integrable limit function may be such that the sequence of integrals (a) diverges to ∞, (b) oscillates finitely.

5. (i) Construct a sequence of functions f_n converging uniformly to a function f on an interval $[a, \infty]$ such that $\int_a^\infty f_n$ exists for each n, but $\int_a^\infty f$ does not.

   (ii) Construct a sequence of functions g_n converging uniformly to a function g on an interval $[a, \infty)$ such that $\int_a^\infty g_n$ $(n = 1, 2, ...)$ and $\int_a^\infty g$ exist, but

$$\int_a^\infty g_n \not\rightarrow \int_a^\infty g.$$

6. The functions f_n on $[0, 1]$ are given by

$$f_n(x) = \frac{nx}{1+n^2x^p} \quad (p > 0).$$

Find for what values of p the sequence (f_n) converges uniformly to its limit f. Examine whether $\int_0^1 f_n \rightarrow \int_0^1 f$ for $p = 2$ and for $p = 4$.

7. The functions f_n on $[0, 1]$ are given by

$$f_n(x) = \frac{n^2x^2}{1+n^3x^3}.$$

Show that the sequence (f_n) does not satisfy all the conditions of theorem 5.23 A, but that the derivative of the limit function exists in $[0, 1]$ and is equal to the limit of the derivatives.

8. The functions f_n on $[-1, 1]$ are defined by

$$f_n(x) = \frac{x}{1+n^2x^2}.$$

Show that (f_n) converges uniformly and that the limit function f is differentiable, but that the relation $f'(x) = \lim f_n'(x)$ does not hold for all x in $[-1, 1]$.

9. The functions f_n on $[0, 2]$ are given by

$$f_n(x) = (1+x^n)^{1/n}.$$

Prove that the sequence of functions f_n, differentiable on $[0, 2]$, converges uniformly to a limit function which is not differentiable at the point 1.

5.3. Criteria for uniform convergence

Uniform convergence of a sequence or series of functions can often be established by an extension of a criterion for the convergence of a sequence or series of numbers.

Theorem 5.31. (General principle of uniform convergence.) *A necessary and sufficient condition for the sequence (f_n) of real or complex valued functions on the set X to be uniformly convergent is that, given $\epsilon > 0$, there is an n_0 such that*

$$\sup_{x \in X} |f_m(x) - f_n(x)| < \epsilon \quad \text{whenever} \quad m, n \geqslant n_0. \qquad (5.31)$$

Proof.

(i) If (f_n) converges uniformly to f on X, then, given $\epsilon > 0$, there is an n_0 such that

$$\sup_{x \in X} |f_n(x) - f(x)| < \tfrac{1}{2}\epsilon \quad \text{when} \quad n \geqslant n_0$$

and (5.31) is an immediate consequence of this.

(ii) Suppose that, for every $\epsilon > 0$, there is an n_0 such that (5.31) holds. Then, for every x, the sequence $(f_n(x))$ of real or complex numbers is a Cauchy sequence and so, since R^1 and Z are complete, $\lim_{n \to \infty} f_n(x)$ exists. Thus there is a function f on X such that

$$f_n(x) \to f(x) \quad \text{for every } x \in X.$$

We shall show that $f_n \to f$ uniformly.

Take any $\epsilon > 0$ and let n_0 be such that (5.31) holds. Now let x be any point of X. Then, when $m, n \geqslant n_0$,

$$|f_m(x) - f_n(x)| < \epsilon$$

and so

$$|f(x) - f_n(x)| = \lim_{m \to \infty} |f_m(x) - f_n(x)| \leqslant \epsilon \quad \text{for} \quad n \geqslant n_0.$$

Since this holds for all $x \in X$,

$$\sup_{x \in X} |f(x) - f_n(x)| \leqslant \epsilon \quad \text{for} \quad n \geqslant n_0.$$

Therefore $f_n \to f$ uniformly on X. ▮

It will be noticed that the proof of the sufficiency is merely part of the proof of theorem 3.44 on the completeness of the space $B(X, Y)$.

We shall now use theorem 5.31 to establish a number of conditions ensuring the uniform convergence of series of functions.

Theorem 5.32. (Weierstrass's M-test.) *Let the functions $u_1, u_2, \ldots$ on X be real or complex valued. If, for each n, there is a constant M_n such that*
$$|u_n(x)| \leqslant M_n \quad (x \in X),$$
where ΣM_n converges, then Σu_n converges (absolutely and) uniformly on X.

Proof. Let $s_n = u_1 + \ldots + u_n$ $(n = 1, 2, \ldots)$. Since ΣM_n converges, given $\epsilon > 0$, there is an n_0 such that
$$M_{n+1} + \ldots + M_m < \epsilon \quad \text{whenever} \quad m > n \geqslant n_0.$$
Hence, for $m > n \geqslant n_0$,
$$\sup_{x \in X} |s_m(x) - s_n(x)| \leqslant M_{n+1} + \ldots + M_m < \epsilon$$
and so, by theorem 5.31, Σu_n converges uniformly on X. |

The M-test can serve only for absolutely convergent series. Theorems 4.53 and 4.54, however, have useful analogues for series which are uniformly but not absolutely convergent.

Theorem 5.33. (Dirichlet.) *The functions a_n, v_n on X are real valued and satisfy the following conditions:*

(i) *the sequence $s_n(x) = \sum_{r=1}^{n} a_r(x)$ is uniformly bounded on X, (i.e. there is a constant H such that $|s_n(x)| \leqslant H$ for all n and all $x \in X$);*

(ii) *for each x, $(v_n(x))$ is a monotonic sequence;*

(iii) *$v_n(x) \to 0$ as $n \to \infty$, uniformly on X.*

Then $\Sigma a_n v_n$ converges uniformly on X.

Proof. By (i), for $k > 0$ and all $x \in X$,
$$|a_{n+1}(x) + \ldots + a_{n+k}(x)| = |s_{n+k}(x) - s_n(x)| \leqslant 2H.$$
Let $X = X_1 \cup X_2$, where, as n increases, $v_n(x)$ decreases for $x \in X_1$ and increases for $x \in X_2$. Then, by the corollary to theorem 4.52, whenever $p \geqslant 1$,
$$|a_{n+1}(x) v_{n+1}(x) + \ldots + a_{n+p}(x) v_{n+p}(x)| \leqslant 2H v_{n+1}(x)$$
for all $x \in X_1$. Since $v_n(x) \to 0$ uniformly on X_1, it follows from theorem 5.31 that the series $\Sigma a_n v_n$ converges uniformly on X_1.

Similarly the series converges uniformly on X_2 and therefore on $X = X_1 \cup X_2$ (exercise 5(a), 1). |

Theorem 5.34. (Abel.) *The functions a_n, v_n on X are real valued and satisfy the following conditions:*
  (i) *the series $\Sigma a_n(x)$ converges uniformly on X;*
  (ii) *for each x, $(v_n(x))$ is a monotonic sequence;*
  (iii) *$v_n(x)$ is uniformly bounded on X.*
Then $\Sigma a_n v_n$ converges uniformly on X.

Proof. Let M be such that $|v_n(x)| \leqslant M$ for all $x \in X$ and all n. Again denote by X_1 and X_2 the subsets of X on which $v_n(x)$ decreases and increases respectively.

By (i), given $\epsilon > 0$, there is an n_0 such that

$$|a_{n+1}(x) + \ldots + a_{n+k}(x)| < \epsilon$$

for $n > n_0$, $k > 0$ and all $x \in X$. Then, again by the corollary to theorem 4.52,

$$|a_{n+1}(x) v_{n+1}(x) + \ldots + a_{n+p}(x) v_{n+p}(x)| < \epsilon |v_{n+1}(x)| \leqslant \epsilon M$$

for $n > n_0$, $p > 0$ and all $x \in X_1$. Hence by theorem 5.31, $\Sigma a_n v_n$ converges uniformly on X_1. Uniform convergence on X_2 is obtained similarly. |

The above proof is independent of theorem 5.33, for the type of argument used to deduce theorem 4.54 from theorem 4.53 is now inapplicable. (Why?)

A particularly simple and important case of theorem 5.34 occurs when $a_n(x) = \alpha_n$ (independent of x) and $\Sigma \alpha_n$ converges.

Example. The series $\displaystyle\sum_{n=1}^{\infty} \frac{\sin nx}{n^p}$.

When $p > 1$, by the M-test, the series converges uniformly on $(-\infty, \infty)$.

When $0 < p \leqslant 1$, the series converges uniformly in every interval $[\alpha, \beta]$ where $2k\pi < \alpha < \beta < 2(k+1)\pi$, but not in any interval containing a point $2k\pi$ ($k = 0, \pm 1, \pm 2, \ldots$). We prove this.

$\displaystyle\sum_{r=1}^{n} \sin rx$ is uniformly bounded on $[\alpha, \beta]$ for, when $\alpha \leqslant x \leqslant \beta$,

$$\left| \sum_{r=1}^{n} \sin rx \right| = \left| \frac{\cos (n+\tfrac{1}{2})x - \cos \tfrac{1}{2}x}{2 \sin \tfrac{1}{2}x} \right| \leqslant \frac{1}{|\sin \tfrac{1}{2}x|}$$

$$\leqslant \max \left(\frac{1}{|\sin \tfrac{1}{2}\alpha|}, \frac{1}{|\sin \tfrac{1}{2}\beta|} \right).$$

Also $1/n^p$ decreases and tends to 0. Therefore, by theorem 5.33, $\Sigma(\sin nx)/n^p$ converges uniformly on $[\alpha, \beta]$.

The convergence is not uniform in any interval including a point $2k\pi$. For, when $x = 2k\pi \pm \pi/(4n)$,

$$\left| \sum_{r=n+1}^{2n} \frac{\sin rx}{r^p} \right| \geqslant \sum_{r=n+1}^{2n} \frac{2}{\pi} \frac{r\pi}{4n} \frac{1}{r^p} \geqslant \frac{1}{2n} \sum_{r=n+1}^{2n} r^{1-p} \geqslant \frac{1}{2n}.n = \frac{1}{2}$$

(since $\sin \theta \geqslant 2\theta/\pi$ when $0 \leqslant \theta \leqslant \frac{1}{2}\pi$) and the necessary condition of theorem 5.31 is not satisfied. This is a recognized technique for establishing non-uniformity.

The last theorem of the section is a partial converse of theorem 5.21.

Theorem 5.35. (Dini.) *Let (f_n) be a convergent sequence of real valued functions on a compact metric space (X, ρ). If, for every $x \in X$, the sequence $(f_n(x))$ is monotonic, and if all the functions f_n $(n = 1, 2, ...)$ and $f = \lim f_n$ are continuous on X, then $f_n \to f$ uniformly on X.*

Proof. Take any $\epsilon > 0$, Given any point $c \in X$, there is an integer n_c such that

$$|f_{n_c}(c) - f(c)| < \epsilon.$$

Since $f_{n_c} - f$ is continuous, there is an open ball B_c with centre c such that

$$|f_{n_c}(x) - f(x)| < 2\epsilon$$

for all $x \in B_c$. Also $|f_n(x) - f(x)|$ decreases and so

$$|f_n(x) - f(x)| < 2\epsilon$$

for $n \geqslant n_c$ and $x \in B_c$.

As X is compact, by the Heine–Borel theorem 3.7, there is a finite number of points $c_1, ..., c_k$ such that the open sets $B_{c_1}, ..., B_{c_k}$ cover X. Hence, if

$$N = \max (n_{c_1}, ..., n_{c_k}),$$

then

$$|f_n(x) - f(x)| < 2\epsilon$$

for all $n \geqslant N$ and $x \in X$. |

By taking $X = [0, 1)$ and $f_n(x) = x^n$ we see that without the condition of compactness the conclusion of the theorem may be false.

Exercises 5(c)

1. Examine for convergence and uniformity of convergence the (real) series whose nth terms are

(i) $\dfrac{x}{n(1+nx^2)}$,   (ii) cosech nx,   (iii) $\dfrac{x}{(1+x)^n}$,

(iv) $e^{nx} \sin nx$,   (v) $(\sqrt{(n+1)} - \sqrt{n}) \cos nx$,   (vi) $\dfrac{(-1)^n}{n^x \log (n+1)}$.

2. Show that the complex series

$$\sum_{n=0}^{\infty} \frac{1}{z^2+n^2}$$

converges uniformly on any bounded subset of

$$Z^* = Z - \{\ldots, -2i, -i, 0, i, 2i, \ldots\},$$

but not on Z^*.

3. Show that the series $\displaystyle\sum_{n=1}^{\infty} \frac{x^n \sin nx}{n^p}$ $(p > 0)$

converges uniformly on the interval $[-1, 1]$.

4. The sequence (a_n) of real numbers is monotonic and $a_n \to 0$ as $n \to \infty$. Prove that the power series $\Sigma\, a_n z^n$ converges uniformly on $\overline{B(0;1)} - B(1;\delta)$ $(\delta > 0)$.

5. The functions f on $(-\infty, \infty)$ and g on $(1, \infty)$ are defined by

$$f(x) = \sum_{n=1}^{\infty} \frac{1}{n^2+n^3x^2}, \quad g(x) = \sum_{n=1}^{\infty} \frac{1}{n^x}.$$

Prove that f' and g' exist on $(-\infty, \infty)$ and $(1, \infty)$ respectively, and are obtained by term-by-term differentiation.

6. Let $\alpha \geqslant 0$ and define u_n on $[0, \infty)$ by

$$u_n(x) = \frac{x^\alpha}{1+n^3x^2}.$$

(i) Show that Σu_n converges uniformly on $[0, 1]$ if and only if $\alpha > 1$. [To prove non-uniform convergence for $\alpha \leqslant 1$ consider $\displaystyle\sum_{n+1}^{2n} u_r(1/n)$.]

(ii) Show that Σu_n converges uniformly on $[0, \infty)$ if and only if $1 < \alpha \leqslant 2$.

7. Let $\displaystyle M_n = \sup_{0 \leqslant x \leqslant 1} \frac{x^n(1-x)}{\log(n+1)}.$

Prove that $\Sigma x^n(1-x)/\log(n+1)$ converges uniformly on $[0, 1]$, but that ΣM_n diverges.

8. Find a series Σu_n which converges uniformly and absolutely on an interval $[a, b]$ and is such that $\Sigma|u_n|$ does not converge uniformly on $[a, b]$.

9. (i) Prove that the series $\displaystyle\sum_{n=1}^{\infty} \frac{(-1)^{n+1}}{n+x}$

converges uniformly on $[0, \infty)$, but that it does not converge absolutely for any x.

(ii) Let (r_n) be a sequence consisting of all the rational numbers and define the functions f_n on $(-\infty, \infty)$ by

$$f_n(x) = \begin{cases} 1 & \text{for } x = r_n, \\ 0 & \text{otherwise.} \end{cases}$$

Prove that the series $\displaystyle\sum_{n=1}^{\infty} \frac{1}{1+n^2(1-f_n(x))}$

converges (absolutely) for every x, but does not converge uniformly in any interval.

10. Denote by $[a]$ the integral part of the real number a.
Show that the series

$$\sum_{n=1}^{\infty} \frac{[nx]}{n^3}$$

converges uniformly on any finite interval. Prove that the sum function s is continuous at any irrational point; and that, at a rational point p/q (where p, q are coprime integers and $q > 0$),

$$s\left(\frac{p}{q}+\right) - s\left(\frac{p}{q}-\right) = \frac{1}{q^3} \sum_{r=1}^{\infty} \frac{1}{r^3}.$$

11. Prove that, for $0 \leqslant x \leqslant 1$, the series

$$\sum_{n=1}^{\infty} \frac{1}{n} \sin(\pi x^n)$$

converges to a function f and that the convergence is uniform in every interval $[0, a]$ $(0 < a < 1)$.
Show that the series

$$\sum_{n=1}^{\infty} \int_{0}^{1} \frac{1}{n} \sin(\pi x^n)\, dx$$

converges and deduce that this sum is equal to $\int_{0}^{1} f(x)\, dx$ $\left(= \lim_{a \to 1-} \int_{0}^{a} f(x)\, dx\right)$.

12. *Tannery's theorem.* The complex valued functions $v_0, v_1, v_2, \ldots$ defined on the set $\{1, 2, 3, \ldots\}$ satisfy the following conditions:

(i) for fixed r, $\lim_{n \to \infty} v_r(n) = w_r$,

(ii) $\sup_{n \geqslant 1} |v_r(n)| = M_r$ and ΣM_r converges.

For $n = 1, 2, \ldots$, let

$$F(n) = v_0(n) + v_1(n) + \ldots + v_{p(n)}(n),$$

where p is an increasing function of n and $p(n) \to \infty$ as $n \to \infty$. Prove that

$$\lim_{n \to \infty} F(n) = \sum_{r=0}^{\infty} w_r.$$

[Define the functions $u_0, u_1, u_2, \ldots$ on the set $\{0, 1, \frac{1}{2}, \frac{1}{3}, \ldots\}$ endowed with the usual metric on R^1 by

$$u_r\left(\frac{1}{n}\right) = \begin{cases} v_r(n) & \text{if } p(n) \geqslant r, \\ 0 & \text{if } p(n) < r; \end{cases} \quad u_r(0) = w_r$$

and use the corollary to theorem 5.21.]

13. Deduce from **12** that, for every complex number z,

$$\lim_{n \to \infty} \left(1 + \frac{z}{n}\right)^n = \exp z,$$

where $\exp z$ is defined as $\sum_{n=0}^{\infty} z^n/n!$.

14. *Sequences of monotonic functions.* Each function f_n is monotonic on the finite, closed interval $[a, b]$ and the sequence (f_n) converges pointwise on $[a, b]$ to a continuous function f. Prove that the convergence is uniform.

5.4. Further properties of power series

With the concept of uniformity at our command we can prove theorems deeper than those of §4.6.

Theorem 5.41. *If the power series $\Sigma a_n z^n$ has radius of convergence R and $0 < r < R$, then $\Sigma a_n z^n$ converges uniformly for $|z| \leqslant r$.*

Proof. When $|z| \leqslant r$, $|a_n z^n| \leqslant |a_n| r^n$ and $\Sigma |a_n| r^n$ converges by theorem 4.61. The result now follows, from the M-test. |

Theorem 5.42. *The sum of a power series is continuous within its circle of convergence.*

Proof. Let R be the radius of convergence of $\Sigma a_n z^n$ and let z_0 be any point such that $|z_0| < R$. If $|z_0| < r < R$, $\Sigma a_n z^n$ converges uniformly for $|z| \leqslant r$ and so, by the corollary to theorem 5.21, $\Sigma a_n z^n$ is continuous at z_0. |

Theorem 5.43. (Principle of equating coefficients.) *If $\Sigma a_n z^n = \Sigma b_n z^n$ for a sequence of non-zero values of z tending to 0 as limit, then $a_n = b_n$ for all n.*

Proof. Let (z_k) be a sequence of non-zero complex numbers such that $\Sigma a_n z^n = \Sigma b_n z^n$ for $z = z_k$, where $\lim z_k = 0$.

Suppose that the theorem is false and let a_m, b_m be the first pair of unequal coefficients. Then

$$z^m \sum_{n=0}^{\infty} (a_{m+n} - b_{m+n}) z^n = 0$$

for all z_k and so

$$\sum_{n=0}^{\infty} (a_{m+n} - b_{m+n}) z^n = 0$$

for all z_k. The power series clearly has a positive radius of convergence. It is therefore continuous at $z = 0$ and so its sum for $z = 0$ is 0. Thus $a_m - b_m = 0$ and we have a contradiction. |

It was proved in C1 (p. 101) that

$$\left(\sum_{n=0}^{\infty} a_n z^n \right) \left(\sum_{n=0}^{\infty} b_n z^n \right) = \sum_{n=0}^{\infty} (a_0 b_n + \ldots + a_n b_0) z^n$$

when z lies in the discs of convergence of both $\Sigma a_n z^n$ and $\Sigma b_n z^n$; in other words, the product of two power series is obtained by formal multiplication. The next theorem shows that formal manipulation also leads to the power series of composite functions.

Theorem 5.44. Let

$$f(z) = \sum_{n=0}^{\infty} a_n z^n \quad (|z| < R),$$

$$g(z) = \sum_{n=0}^{\infty} b_n z^n \quad (|z| < S).$$

If $|b_0| < R$, there is a $\rho > 0$ such that

$$f(g(z)) = \sum_{n=0}^{\infty} c_n z^n \quad (|z| < \rho),$$

where the coefficients c_n are obtained by formal substitution: defining p_{kn} $(k, n = 0, 1, 2, \ldots)$ by

$$g^k(z) = \sum_{n=0}^{\infty} p_{kn} z^n,$$

we have

$$c_n = \sum_{k=0}^{\infty} a_k p_{kn}.$$

Proof.

(i) Suppose that ζ is such that $\sum_{n=0}^{\infty} |b_n \zeta^n| < R$. Then $|g(\zeta)| < R$ and therefore

$$f(g(\zeta)) = \sum_{k=0}^{\infty} a_k g^k(\zeta) = \sum_{k=0}^{\infty} a_k \left(\sum_{n=0}^{\infty} p_{kn} \zeta^n \right) = \sum_{k=0}^{\infty} \left(\sum_{n=0}^{\infty} a_k p_{kn} \zeta^n \right).$$

If the order of the two summations on the right may be reversed, we have

$$f(g(\zeta)) = \sum_{n=0}^{\infty} \left(\sum_{k=0}^{\infty} a_k p_{kn} \zeta^n \right) = \sum_{n=0}^{\infty} c_n \zeta^n.$$

By theorem 4.74, this reversal is legitimate if

$$\sum_{k=0}^{\infty} \left(\sum_{n=0}^{\infty} |a_k p_{kn} \zeta^n| \right)$$

converges.

The power series $\Sigma |b_n| z^n$ has the same radius of convergence S as $\Sigma b_n z^n$. Let

$$h(z) = \sum_{n=0}^{\infty} |b_n| z^n \quad (|z| < S).$$

Then

$$h^k(z) = \sum_{n=0}^{\infty} q_{kn} z^n,$$

where

$$q_{kn} = \sum_{n_1 + \ldots + n_k = n} |b_{n_1}| \ldots |b_{n_k}| \geqslant \left| \sum_{n_1 + \ldots + n_k = n} b_{n_1} \ldots b_{n_k} \right| = |p_{kn}|.$$

Now $\Sigma \, |a_k| z^k$ has radius of convergence R,

$$\sum_{n=0}^{\infty} q_{kn} |\zeta^n| = h^k(|\zeta|) \quad \text{and} \quad h(|\zeta|) = \sum_{n=0}^{\infty} |b_n \zeta^n| < R.$$

Therefore

$$\sum_{k=0}^{\infty} |a_k| \left(\sum_{n=0}^{\infty} q_{kn} |\zeta^n| \right)$$

converges. Since $|p_{kn} \zeta^n| \leqslant q_{kn} |\zeta^n|$, it follows that

$$\sum_{k=0}^{\infty} |a_k| \left(\sum_{n=0}^{\infty} |p_{kn} \zeta^n| \right) = \sum_{k=0}^{\infty} \left(\sum_{n=0}^{\infty} |a_k p_{kn} \zeta^n| \right)$$

converges.

(ii) By theorem 5.42, h is continuous in the disc $|z| < S$. Since $h(0) = |b_0| < R$, there is therefore a $\rho > 0$ such that

$$\sum_{n=0}^{\infty} |b_n z^n| = h(|z|) < R$$

for all z in the disc $|z| < \rho$. |

Notes. (1) If $b_0 = 0$, then

$$p_{kn} = \sum_{n_1 + \ldots + n_k = n} b_{n_1} \ldots b_{n_k} = 0$$

for $k > n$ and so

$$c_n = \sum_{k=0}^{\infty} a_k p_{kn}$$

is the sum of a finite series.

(2) If $R = \infty$, then the inequality $\Sigma \, |b_n z^n| < R$ is satisfied by all z such that $|z| < S$; thus $\rho = S$.

We now use theorem 5.44 to obtain the reciprocal of a power series.

Theorem 5.45. *Let*

$$\phi(z) = \sum_{n=0}^{\infty} \alpha_n z^n \quad (|z| < S).$$

If $\alpha_0 \neq 0$, there is a $\rho > 0$ such that

$$\frac{1}{\phi(z)} = \sum_{n=0}^{\infty} \beta_n z^n \quad (|z| < \rho),$$

where the coefficients β_n are obtained from the equations

$$\alpha_0 \beta_0 = 1,$$

$$\alpha_0 \beta_1 + \alpha_1 \beta_0 = 0,$$

$$\alpha_0 \beta_2 + \alpha_1 \beta_1 + \alpha_2 \beta_0 = 0,$$

$$\cdots \cdots \cdots \cdots \cdots \cdots \cdots$$

Proof. In theorem 5.44 take

$$f(z) = \frac{1}{\alpha_0 + z} = \sum_{n=0}^{\infty} \frac{(-1)^n}{\alpha_0^{n+1}} z^n \quad (|z| < |\alpha_0|)$$

and

$$g(z) = \sum_{n=1}^{\infty} \alpha_n z^n \quad (|z| < S).$$

Then

$$f(g(z)) = 1/\phi(z)$$

and $1/\phi(z)$ may be expanded in a power series $\Sigma \beta_n z^n$ for all z such that

$$\sum_{n=1}^{\infty} |\alpha_n z^n| < |\alpha_0|.$$

By continuity, this inequality holds in a disc $|z| < \rho$.

The formulae of theorem 5.44 give explicit but cumbersome expressions for the β_n. It is simpler to evaluate $\beta_0, \beta_1, \beta_2 \ldots$, successively. When $|z| < \rho$,

$$\left(\sum_{n=0}^{\infty} \alpha_n z^n \right) \left(\sum_{n=0}^{\infty} \beta_n z^n \right) = \phi(z) . \frac{1}{\phi(z)} = 1$$

and so, by theorem 5.43,

$$\alpha_0 \beta_0 = 1, \quad \sum_{k=0}^{n} \alpha_k \beta_{n-k} = 0 \quad (n = 1, 2, \ldots). \mid$$

In the rest of this section we deal with real power series only.

Theorem 5.46. *Suppose that the real power series $\Sigma a_n x^n$ converges to $f(x)$ in its interval of convergence $(-R, R)$. Then*

(i) *f is differentiable in $(-R, R)$ and*

$$f'(x) = \sum_{n=1}^{\infty} n a_n x^{n-1};$$

(ii) *for $|x| < R$,*

$$\int_0^x f(t) \, dt = \sum_{n=0}^{\infty} \frac{a_n}{n+1} x^{n+1}.$$

Proof.

(i) By theorem 4.62, $\Sigma n a_n x^{n-1}$ has radius of convergence R and, by theorem 5.41, the series converges uniformly on any (finite) closed subinterval of $(-R, R)$. Theorem 5.23 B now gives the result.

(ii) If x is any point of $(-R, R)$, $\Sigma a_n x^n$ converges uniformly on $[0, x]$ (or $[x, 0]$ if $x < 0$). Now apply the corollary to theorem 5.22. $\mid$

Theorem 5.47. (Abel's continuity theorem.) *If Σa_n converges, then*

$$\sum_{n=0}^{\infty} a_n x^n \to \sum_{n=0}^{\infty} a_n \quad as \quad x \to 1-.$$

Proof. By theorem 5.34, $\Sigma a_n x^n$ converges uniformly, say to $f(x)$, for $0 \leqslant x \leqslant 1$. Then by the corollary to theorem 5.21, f is continuous on $[0, 1]$. In particular, $f(x) \to f(1)$ as $x \to 1-$. |

Example. $1 - \frac{1}{3} + \frac{1}{5} - \frac{1}{7} + \ldots = \frac{1}{4}\pi$.

If $|x| < 1$,
$$\frac{1}{1+x^2} = \sum_{n=0}^{\infty} (-1)^n x^{2n}.$$

Hence, by theorem 5.46 (ii),

$$\text{arc} \tan x = \sum_{n=0}^{\infty} \frac{(-1)^n}{2n+1} x^{2n+1}$$

for $|x| < 1$. Since $\Sigma (-1)^n/(2n+1)$ converges,

$$\lim_{x \to 1-} \text{arc} \tan x = \sum_{n=0}^{\infty} \frac{(-1)^n}{2n+1}.$$

The left-hand side is arc $\tan 1 = \frac{1}{4}\pi$.

Exercises 5(*d*)

1. Evaluate the following complex limits:

(i) $\lim\limits_{z \to 0} \dfrac{\sin z}{z}$,   (ii) $\lim\limits_{z \to a} \dfrac{\cos z - \cos a}{z - a}$,   (iii) $\lim\limits_{z \to 0} \dfrac{e^{z^2} - 1}{(e^z - 1)^2}$.

2. Find the power series expansions up to the term in z^4 of the following functions:

(i) $\cos(\sin z)$,   (ii) $e^{1/(1+z)}$,   (iii) $\sin(\cos z)$.

3. The function f on R^1 is defined by

$$f(x) = \begin{cases} (x+1)^{1/x} & \text{for } x \neq 0, \\ e & \text{for } x = 0. \end{cases}$$

Show that, at 0, f has derivatives of all orders. Evaluate $f'''(0)$.

4. Given the series $\sum\limits_{0}^{\infty} a_n$ and $\sum\limits_{0}^{\infty} b_n$, let

$$c_n = a_0 b_n + \ldots + a_n b_0 \quad (n \geqslant 0).$$

(i) Prove that, if $\Sigma a_n, \Sigma b_n, \Sigma c_n$ converge to sums A, B, C respectively, then $AB = C$.

(ii) Construct convergent series $\Sigma a_n, \Sigma b_n$ for which Σc_n diverges.

5. Let $a_0, a_1, a_2, \ldots$ be real.

(i) Show that, if Σa_n converges, then $\Sigma a_n/(n+1)$ converges and

$$\int_0^1 \left(\sum_{n=0}^{\infty} a_n x^n \right) dx = \sum_{n=0}^{\infty} \frac{a_n}{n+1}.$$

(ii) Show that, if $\Sigma a_n/(n+1)$ converges, then the last equation holds, though the integrand may be unbounded.

6. Prove that

$$1 - \frac{1}{5} + \frac{1}{9} - \frac{1}{13} + \frac{1}{17} - \ldots = \frac{1}{4\sqrt{2}} \{\pi + 2 \log (1 + \sqrt{2})\}.$$

7. Prove that, if $-1 < x < 1$,

$$\{\log (1+x)\}^2 = 2 \sum_{n=2}^{\infty} \frac{(-1)^n s_{n-1}}{n} x^n,$$

where

$$s_m = 1 + \frac{1}{2} + \frac{1}{3} + \ldots + \frac{1}{m}.$$

Is the result valid for $x = 1$?

8. Assume that, when α is any real number, the binomial expansion

$$(1+x)^\alpha = \sum_{n=0}^{\infty} \binom{\alpha}{n} x^n$$

holds for $-1 < x < 1$ (see C1, 102).

Show that, when $\alpha \geqslant 0$, $\Sigma \binom{\alpha}{n} x^n$ converges uniformly to $(1+x)^\alpha$ for $-1 \leqslant x \leqslant 1$; and, when $\alpha > -1$, the series converges uniformly to $(1+x)^\alpha$ for $-1+\delta \leqslant x \leqslant 1$ ($\delta > 0$). (See exercise 4(f), 2.)

9. Prove that, for $|x| \leqslant 1$,

$$\log \{x + \sqrt{(1+x^2)}\} = x - \frac{1}{2} \frac{x^3}{3} + \frac{1.3}{2.4} \frac{x^5}{5} - \frac{1.3.5}{2.4.6} \frac{x^7}{7} + \ldots.$$

10. Prove that, for all real α and β,

$$\sum_{r=0}^{n} \binom{\alpha}{r} \binom{\beta}{n-r} = \binom{\alpha+\beta}{n}.$$

5.5. Two constructions of continuous functions

Each of the two theorems in this section carries out the construction of a continuous function which is to have an assigned property. The method is to build up a uniformly convergent series of approximating functions designed to induce the required property in the limit function.

A continuous, nowhere differentiable function. It is easy to construct a continuous function which fails to have a derivative at the points of a finite or a countably infinite set. The extension to an uncountable set is more difficult.

Lemma. *Let f be a real function bounded on* $(-\infty, \infty)$. *If*

$$\frac{f(c)-f(x)}{c-x}$$

is unbounded, then $f'(c)$ *does not exist.*

Proof. Let $|f(x)| \leqslant M$ for all x. Suppose that $f'(c)$ does exist. Then there is a $\delta > 0$ such that

$$\left|\frac{f(c)-f(x)}{c-x} - f'(c)\right| < 1$$

for $0 < |c-x| < \delta$, so that

$$\left|\frac{f(c)-f(x)}{c-x}\right| < |f'(c)| + 1 \quad \text{for} \quad 0 < |c-x| < \delta;$$

and

$$\left|\frac{f(c)-f(x)}{c-x}\right| \leqslant \frac{2M}{\delta} \quad \text{for} \quad |c-x| \geqslant \delta.$$

The last two inequalities contradict the unboundedness of

$$\{f(c)-f(x)\}/(c-x). \;|$$

Theorem 5.51. *There exists a real function which is continuous on* $(-\infty, \infty)$, *but nowhere differentiable.*

Proof. Let g be the function on $(-\infty, \infty)$ defined by

$$g(x) = |x| \quad \text{for} \quad |x| \leqslant 2,$$

$$g(x+4n) = g(x) \quad (n = \pm 1, \pm 2, \ldots).$$

Then (i) g is continuous, (ii) $0 \leqslant g(x) \leqslant 2$ for all x, (iii) for all x, y,

$$\left|\frac{g(x)-g(y)}{x-y}\right| \leqslant 1,$$

(iv) given c, there is an x_0 such that

$$|c-x_0| = 1 \quad \text{and} \quad \left|\frac{g(c)-g(x_0)}{c-x_0}\right| = 1.$$

When $n = 0, 1, 2, \ldots$, define the functions g_n by

$$g_n(x) = a^n g(b^n x),$$

where a, b are positive constants. Then (i)' g_n is continuous, (ii)' $0 \leqslant g_n(x) \leqslant 2a^n$ for all x, (iii)' for all x, y,

$$\left|\frac{g_n(x)-g_n(y)}{x-y}\right| \leqslant a^n b^n,$$

(iv)' given c, there is an x_n such that

$$|c - x_n| = b^{-n} \quad \text{and} \quad \left| \frac{g_n(c) - g_n(x_n)}{c - x_n} \right| = a^n b^n.$$

If $0 < a < 1$, $\Sigma g_n(x)$ converges uniformly on $(-\infty, \infty)$ (by the M-test) and so

$$f = \sum_{n=0}^{\infty} g_n$$

is continuous everywhere (by theorem 5.21). However we shall show that, if a is sufficiently small and b is sufficiently large, f is nowhere differentiable.

Taking $ab > 1$ we have

$$\left| \frac{f(c) - f(x_n)}{c - x_n} \right| = \left| \sum_{i=0}^{\infty} \frac{g_i(c) - g_i(x_n)}{c - x_n} \right|$$

$$\geqslant \left| \frac{g_n(c) - g_n(x_n)}{c - x_n} \right| - \left(\sum_{i=0}^{n-1} + \sum_{i=n+1}^{\infty} \right) \left| \frac{g_i(c) - g_i(x_n)}{c - x_n} \right|$$

$$\geqslant a^n b^n - \sum_{i=0}^{n-1} a^i b^i - \sum_{i=n+1}^{\infty} \frac{2a^i}{b^{-n}}$$

$$= a^n b^n - \frac{a^n b^n - 1}{ab - 1} - \frac{2b^n a^{n+1}}{1 - a}$$

$$> \left(1 - \frac{1}{ab - 1} - \frac{2a}{1 - a} \right) a^n b^n.$$

Hence, if also

$$1 - \frac{1}{ab - 1} - \frac{2a}{1 - a} > 0,$$

the lemma shows that $f'(c)$ does not exist. This inequality holds if a is small and b large enough, for instance if $a = \frac{1}{5}$, $b = 20$. Thus the function given by

$$f(x) = \sum_{n=0}^{\infty} \frac{1}{5^n} g(20^n x)$$

is everywhere continuous and nowhere differentiable. |

The non-differentiability of f is not dependent on the lack of a derivative of g_n at the points $2k/b^n$. It is possible to start with a differentiable function g; in fact, the first continuous, nowhere differentiable function, constructed by Weierstrass, was

$$\sum_{n=0}^{\infty} a^n \cos b^n \pi x \quad (0 < a < 1, \, ab > 1 + \tfrac{3}{2}\pi).$$

Tietze's extension theorem. If a real function f is continuous on $[a, b]$, it is easy to define a function f^* continuous on R^1 such that $f^*(x) = f(x)$ for $a \leqslant x \leqslant b$. Tietze proved a theorem extending the domain of definition of a given function continuous on a closed set in an arbitrary metric space.

Lemma. *Suppose that (X, ρ) is a metric space, g is a real valued function continuous and bounded on a closed set F in X and*

$$\sup_{x \in F} |g(x)| = M.$$

Then there is a real valued continuous function h on X such that
    (i) $|h(x)| \leqslant \frac{1}{3}M$ *for $x \in X$, and*
    (ii) $|g(x) - h(x)| \leqslant \frac{2}{3}M$ *for $x \in F$.*

Proof. Let $A = \{x \in F \,|\, -M \leqslant g(x) \leqslant -\frac{1}{3}M\},$

$$B = \{x \in F \,|\, -\tfrac{1}{3}M < g(x) < \tfrac{1}{3}M\},$$

$$C = \{x \in F \,|\, \tfrac{1}{3}M \leqslant g(x) \leqslant M\}.$$

By theorem 3.23, A and C are closed and so, by exercise $3(f)$, 12, the function h on X given by

$$h(x) = \tfrac{1}{3}M \frac{\rho(x, A) - \rho(x, C)}{\rho(x, A) + \rho(x, C)} \quad (x \in X)$$

is continuous. Clearly (i) holds. To establish (ii) we compare the values of g and h on the sets A, B, C.

On A, $-M \leqslant g(x) \leqslant -\frac{1}{3}M$ and $h(x) = -\frac{1}{3}M.$

On B, $|g(x)| < \frac{1}{3}M$ and $|h(x)| < \frac{1}{3}M.$

On C, $\frac{1}{3}M \leqslant g(x) \leqslant M$ and $h(x) = \frac{1}{3}M.$ |

Theorem 5.52. (Tietze's extension theorem.) *Let (X, ρ) be a metric space and let f be a real valued function continuous and bounded on a closed set F in X. Then there is a real valued continuous function f^* on X such that*
    (i) $f^*(x) = f(x)$ *for $x \in F$, and*
    (ii) *the bounds of f^* on X are the same as the bounds of f on F.*

Proof. Since we can add a constant function to f, we may suppose that the range of f is symmetrical about 0; say

$$\inf_{x \in F} f(x) = -K, \quad \sup_{x \in F} f(x) = K.$$

Define the sequence (f_n) of real valued, continuous functions on X as follows. Let f_1 be the function h of the lemma corresponding to $g = f$. When $f_1, ..., f_{n-1}$ have been defined, let f_n be the function h of the lemma corresponding to $g = f - (f_1 + ... + f_{n-1})$. Induction shows that, for all n,

$$|f_n(x)| \leqslant \tfrac{1}{3}(\tfrac{2}{3})^{n-1}K \quad (x \in X), \tag{5.51}$$

$$|f(x) - [f_1(x) + ... + f_n(x)]| \leqslant (\tfrac{2}{3})^n K \quad (x \in F). \tag{5.52}$$

By (5.51) and the M-test, Σf_n converges uniformly on X to a function f^*. Theorem 5.21 shows that f^* is continuous on X and (5.52) shows that $f^*(x) = f(x)$ for $x \in F$. Again, by (5.51),

$$|f^*(x)| \leqslant \tfrac{1}{3}K \sum_{n=0}^{\infty} (\tfrac{2}{3})^n = K$$

for all $x \in X$. Hence

$$K = \sup_{x \in F} f^*(x) \leqslant \sup_{x \in X} f^*(x) \leqslant K$$

and so $\sup_{x \in X} f^*(x) = K$. Similarly $\inf_{x \in X} f^*(x) = -K$. |

Corollary. (Urysohn's lemma.) *Let* (X, ρ) *be a metric space and let* A, B *be disjoint, closed subsets of* X. *Then there is a continuous function* $f: X \to R^1$ *such that* $0 \leqslant f(x) \leqslant 1$ *on* X, $f(x) = 0$ *for* $x \in A$ *and* $f(x) = 1$ *for* $x \in B$.

This result shows that, given an arbitrary metric space, there is a plentiful supply of real valued functions continuous on it.

Exercises 5(e)

1. Construct a function on an interval $[a, b]$ which is everywhere n times differentiable, but nowhere has an $(n+1)$th derivative.

2. Let (r_n) be a sequence consisting of all the rational numbers in $(0, 1)$ and define the function f on $[0, 1]$ by

$$f(x) = \sum_{n=1}^{\infty} \frac{|x - r_n|}{n^2} \quad (0 \leqslant x \leqslant 1).$$

Show that f is continuous on $[0, 1]$ and not differentiable at the points r_n.
   [Define u_n $(n = 1, 2, ...)$ on $[0, 1]$ by

$$u_n(x) = \begin{cases} -1 & \text{for } 0 \leqslant x \leqslant r_n, \\ 1 & \text{for } r_n < x \leqslant 1 \end{cases}$$

and consider $\int_0^x \left(\sum_1^{\infty} u_n(x)/n^2\right) dx$.]

3. Give an example to show that the conclusion of theorem 5.52 need not hold if the condition that the domain F of f is closed is omitted.

4. Let F be a compact subset of R^k. Prove that, if f is a real valued, continuous function on F, then there is a continuous extension f^* of f which is zero outside a sufficiently large open ball $B(0; R)$ and is such that

$$\sup_{x \in R^k} |f^*(x)| = \sup_{x \in F} |f(x)|.$$

5. Prove the following extension theorem. Let (X, ρ), (Y, σ) be metric spaces and let E be a subset of X. If (Y, σ) is complete, then every uniformly continuous function $f: E \to Y$ has a unique, uniformly continuous extension to $\bar{E}$.

5.6. Weierstrass's approximation theorem and its generalization

An important problem of analysis is the approximation of functions by simpler functions. Weierstrass's theorem deals with the approximation of continuous real functions by polynomials.

Lemma 1. *There is a sequence of polynomials converging uniformly to $|x|$ for $-1 \leqslant x \leqslant 1$.*

Proof. If $u = 1 - x^2$, then $|x| = \sqrt{(1-u)}$ and the interval $0 \leqslant u \leqslant 1$ corresponds to the interval $-1 \leqslant x \leqslant 1$. By exercise $5(d), 8$, the binomial series

$$\sum_{n=0}^{\infty} \binom{\frac{1}{2}}{n} (-u)^n$$

converges uniformly to $\sqrt{(1-u)}$ for $0 \leqslant u \leqslant 1$. Hence the series

$$\sum_{n=0}^{\infty} \binom{\frac{1}{2}}{n} (x^2-1)^n$$

converges uniformly to $|x|$ for $-1 \leqslant x \leqslant 1$. |

Corollary. *Let the function g be given by*

$$g(x) = \begin{cases} 0 & \text{for } x < 0, \\ x & \text{for } x \geqslant 0. \end{cases}$$

Then, for any interval $[\alpha, \beta]$, there is a sequence of polynomials converging uniformly to $g(x)$ for $\alpha \leqslant x \leqslant \beta$.

Proof. Let k be such that $[\alpha, \beta] \subset [-k, k]$. Since

$$g(x) = \tfrac{1}{2}k \left(\frac{x}{k} + \left| \frac{x}{k} \right| \right),$$

the result immediately follows from the lemma. |

Theorem 5.61. (Weierstrass's approximation theorem.) *If the real function f is continuous on the finite closed interval* [a, b], *there is a sequence of polynomials converging uniformly to f on* [a, b].

Proof. It is sufficient to prove that, given $\epsilon > 0$, there is a polynomial p such that

$$|f(x) - p(x)| < \epsilon \quad \text{for} \quad a \leqslant x \leqslant b.$$

There is (see exercise 3(f), 9) a piecewise linear function h such that

$$|f(x) - h(x)| < \tfrac{1}{2}\epsilon \quad \text{for} \quad a \leqslant x \leqslant b.$$

Let the vertices of the graph of h be (x_0, y_0), (x_1, y_1), ..., (x_n, y_n), where $x_0 = a$ and $x_n = b$. Then, for $a \leqslant x \leqslant b$,

$$h(x) = y_0 + \sum_{i=0}^{n-1} c_i g(x - x_i),$$

where the c's are defined by the equations

$$y_1 = y_0 + c_0(x_1 - x_0),$$

$$y_2 = y_0 + c_0(x_2 - x_0) + c_1(x_2 - x_1),$$

$$\cdots\cdots\cdots\cdots\cdots\cdots\cdots$$

$$y_n = y_0 + c_0(x_n - x_0) + c_1(x_n - x_1) + \ldots + c_{n-1}(x_n - x_{n-1}).$$

By the corollary, there is a polynomial p such that

$$|h(x) - p(x)| < \tfrac{1}{2}\epsilon \quad \text{for} \quad a \leqslant x \leqslant b$$

and this gives

$$|f(x) - p(x)| < \epsilon \quad \text{for} \quad a \leqslant x \leqslant b. \; |$$

This proof, due to Lebesgue, is one of the simplest of the many proofs of Weierstrass's theorem. Some of these are easily adapted to show that the result holds in R^k ($k > 1$). M. H. Stone has put the theorem in a more general algebraic setting which exposes the essential features of the original Weierstrass theorem. To complete this chapter we shall prove Stone's theorem (5.62), but neither the theorem, nor the algebraic setting will be appealed to in the rest of the book.

Let (X, τ) be a metric space and, as usual, let $C(X)$ be the space of real valued functions bounded and continuous on X. Algebraic operations in $C(X)$ are defined in the obvious way. If $f, g \in C(X)$

and α is a real number, the (bounded and continuous) functions $f+g, fg$ and αf are such that, for all $x \in X$

$$(f+g)(x) = f(x) + g(x),$$
$$(fg)(x) = f(x)g(x),$$
$$(\alpha f)(x) = \alpha f(x).$$

Definition. *The subset A of C(X) is called an* algebra *if, whenever $f, g \in A$ and α is any real number, $f+g, fg, \alpha f \in A$.*

The set of polynomials and the set of differentiable functions on $[a, b]$ are clearly algebras in $C[a, b]$.

Using the usual metric ρ on $C(X)$ we now consider the closure $\bar{A}$ of an algebra A. $\bar{A}$ is, in fact, the set of those functions in $C(X)$ which are the limits of uniformly convergent sequences of members of A.

Lemma 2. *If (X, τ) is any metric space and A is an algebra in $C(X)$, then $\bar{A}$ is also an algebra.*

Proof. Let $f, g \in \bar{A}$. Then there are sequences $(f_n), (g_n)$ of members of A which converge uniformly on X to f, g respectively.

Clearly $f_n + g_n \to f + g$ uniformly on X and, if α is any real number, $\alpha f_n \to \alpha f$ uniformly on X. Also, since all functions in $C(X)$ are bounded, every uniformly convergent sequence is uniformly bounded. Therefore
$$f_n g_n - fg = f_n(g_n - g) + g(f_n - f) \to 0$$
uniformly on X. Thus, since $f_n + g_n, \alpha f_n, f_n g_n \in A$, it follows that $f+g, \alpha f, fg \in \bar{A}$. |

Weierstrass's approximation theorem asserts that the algebra A_p of polynomials on $[a, b]$ is dense in $C[a, b]$, i.e. that $\bar{A}_p = C[a, b]$. We shall see that, more generally, the relation $\bar{A} = C(X)$ holds if (X, τ) is compact and if the algebra A satisfies the following two conditions:

(i) A contains some non-zero constant function and so all constant functions.

(ii) A *separates points* on X, i.e. to every pair x_1, x_2 of distinct points of X there corresponds a function $f \in A$ such that $f(x_1) \neq f(x_2)$.

The algebra of polynomials on $[a, b]$ satisfies both (i) and (ii); the algebra of even polynomials on $[-1, 1]$ satisfies (i), but not (ii); and the algebra of polynomials on $[0, 1]$ with zero constant term satisfies (ii), but not (i).

Theorem 5.62. (The Stone–Weierstrass theorem.) *Let* (X, τ) *be a compact metric space. If A is an algebra in $C(X)$ which contains the constant functions and which separates points on X, then $\overline{A} = C(X)$.*

Proof.

I. We first show that, if $f \in \overline{A}$, then $|f| \in \overline{A}$ (where $|f|$ is defined by $|f|(x) = |f(x)|$). Let $\sup_{x \in X} |f(x)| = \mu$ and take any $\epsilon > 0$. By the corollary to lemma 1, there is a polynomial p given by

$$p(\xi) = \alpha_0 + \alpha_1 \xi + \ldots + \alpha_n \xi^n$$

such that $||\xi| - p(\xi)| < \epsilon$ for $-\mu \leqslant \xi \leqslant \mu$.

As $\overline{A}$ is an algebra containing the constant functions,

$$p \circ f = \alpha_0 + \alpha_1 f + \ldots + \alpha_n f^n$$

belongs to $\overline{A}$. Also

$$||f(x)| - (p \circ f)(x)| = ||f(x)| - p\{f(x)\}| < \epsilon$$

for every $x \in X$. Since $\overline{A}$ is closed in $C(X)$ (by theorem 2.36), it follows that $|f| \in \overline{A}$.

II. Defining min (f, g) and max (f, g) by

$$(\min (f, g))(x) = \min (f(x), g(x)),$$

$$(\max (f, g))(x) = \max (f(x), g(x))$$

we have

$$\min (f, g) = \tfrac{1}{2}(f + g - |f - g|),$$

$$\max (f, g) = \tfrac{1}{2}(f + g + |f - g|).$$

Hence, by I, if $f, g \in \overline{A}$, then min (f, g), max $(f, g) \in \overline{A}$.

III. Let f be any function in $C(X)$. We need only show that given $\epsilon > 0$, there is a function $g \in \overline{A}$ such that

$$|f(x) - g(x)| < \epsilon$$

for all $x \in X$.

Let u, v be two arbitrary distinct points of X. There is a function $\phi \in A$ such that $\phi(u) \neq \phi(v)$ and, since A contains the constant functions, we may assume that $\phi(u) = 0$ so that $\phi(v) \neq 0$. Also let ψ be the constant function given by $\psi(x) = f(u)$ for all $x \in X$. If α is the real number such that

$$\alpha \phi(v) + f(u) = f(v),$$

the function $g_{uv} = \alpha \phi + \psi$

is such that $g_{uv}(u) = f(u)$ and $g_{uv}(v) = f(v)$. We note that $g_{uv} \in A$.

Since g_{uv} and f are continuous, the set

$$K_v = \{x \mid x \in X; g_{uv}(x) < f(x) + \epsilon\}$$

is open; moreover it contains both u and v. Hence the collection of sets K_v, where v ranges over $X - \{u\}$ is an open covering of X. Since X is compact, by the Heine–Borel theorem there is a finite number of sets $K_{v_1}, ..., K_{v_n}$ covering X. Let

$$g_u = \min(g_{uv_1}, ..., g_{uv_n}).$$

By II, $g_u \in \bar{A}$; also $g_u(u) = f(u)$ and $g_u(x) < f(x) + \epsilon$ for all $x \in X$. The set

$$J_u = \{x \mid x \in X; g_u(x) > f(x) - \epsilon\}$$

is open. We note that

$$f(x) - \epsilon < g_u(x) < f(x) + \epsilon \quad \text{for} \quad x \in J_u.$$

The collection $\{J_u\}_{u \in X}$ is an open covering of the compact space X and so a finite number of sets $J_{u_1}, ..., J_{u_m}$ covers X. Let

$$g = \max(g_{u_1}, ..., g_{u_m}).$$

Then $g \in \bar{A}$ and $f(x) - \epsilon < g(x) < f(x) + \epsilon$ for all $x \in X$. |

Clearly the multi-dimensional form of Weierstrass's approximation theorem is a particular case of theorem 5.62. A trigonometric analogue of Weierstrass's theorem is given in exercise $5(f)$, 2.

There are also complex versions of theorems 5.61 and 5.62. For instance every complex function continuous on a compact subset of Z is the limit of a uniformly convergent sequence of polynomials in z and $\bar{z}$. That polynomials in z alone will not do is a consequence of another theorem of Weierstrass. (See exercise $13(a)$, 6.)

Exercises 5(f)

1. The real function f is continuous on $[a, b]$ and, for $n = 0, 1, 2, ...,$

$$\int_a^b f(x) x^n \, dx = 0.$$

Show that

$$\int_a^b f^2(x) \, dx = 0$$

and deduce that $f(x) = 0$ for $a \leqslant x \leqslant b$.

2. A trigonometric polynomial is a real function of the form

$$a_0 + \sum_{r=1}^{n} (a_r \cos rx + b_r \sin rx).$$

Deduce from theorem 5.62 that, if f is continuous on $[\alpha, \beta]$, where $\beta - \alpha < 2\pi$, then f is the limit of a uniformly convergent sequence of trigonometric polynomials. Show that this result is false if $\beta - \alpha \geqslant 2\pi$. Which condition of theorem 5.62 is then violated?

3. Prove that every real function continuous on the interval $[a, b]$ is the limit of a uniformly convergent sequence of even polynomials if and only if (a, b) does not include the origin.

4. *Alternative proof of Lemma* 1. Show that the only solution of the equation

$$y = \tfrac{1}{2}(y^2 + 1 - x^2)$$

which satisfies $y \leqslant 1$ is $y = 1 - |x|$.

For $-1 \leqslant x \leqslant 1$, define the sequence (y_n) of functions by

$$y_0(x) = 1, \quad y_n(x) = \tfrac{1}{2}\{y_{n-1}^2(x) + 1 - x^2\} \quad (n = 1, 2, \ldots).$$

Prove that, for $-1 \leqslant x \leqslant 1$ and $n = 0, 1, 2, \ldots$,

$$y_n(x) \geqslant 1 - |x| \quad \text{and} \quad y_{n+1}(x) \leqslant y_n(x).$$

Deduce that $y_n(x) \to 1 - |x|$ as $n \to \infty$, uniformly for $-1 \leqslant x \leqslant 1$; and hence that there is a sequence of polynomials converging uniformly to $|x|$ for $-1 \leqslant x \leqslant 1$.

NOTES ON CHAPTER 5

§5.1. The notion of uniformity of convergence was made explicit by three mathematicians independently, Weierstrass, Stokes and Seidel in the late 1840s. The idea was rediscovered a few years later by Cauchy. A critical examination of the exposition of the early writers was made by Hardy, 'Sir George Stokes and the concept of uniform convergence', *Proc. Cambridge Phil. Soc.* **19** (1918), 148–56. Hardy's note includes developments of Stokes's work and answers questions which the reader may have asked himself on §5.2. It is noticeable in theorem 5.21 that the full strength of the hypothesis of uniformity is not used in the proof. The inequality

$$\sup_{x \in X} \sigma(f_n(x), f(x)) < \epsilon$$

is quoted only for a single n, not for all $n \geqslant N$. This indicates that a condition less restrictive than uniformity should suffice for theorem 5.21. Such a wider condition, due in substance to Dini and called quasi-uniform convergence, is necessary as well as sufficient for the continuity of the limit function of a sequence of continuous functions. However it lacks the simplicity possessed by the condition of uniformity of convergence and, instead of reproducing it here, we refer the interested reader to Hardy's paper.

§5.4. N. H. Abel (1802–29) established his continuity theorem with the immediate aim of investigating the validity of the binomial expansion at the points ± 1. In the original proof, partial summation (theorem 4.52, corollary) is used to show that on the right-hand side of the identity

$$\sum_{n=0}^{\infty} a_n - \sum_{n=0}^{\infty} a_n x^n = \sum_{n=0}^{N} a_n(1 - x^n) + \sum_{n=N+1}^{\infty} a_n(1 - x^n)$$

the second term can be made arbitrarily small for all x in $(0, 1)$ by choice of a large N; the first term may then be made small by choice of x near 1.

The relation $1 - \frac{1}{3} + \frac{1}{5} - \ldots = \frac{1}{4}\pi$ was first noted by J. Gregory (1638–75) and was later obtained independently by G. W. Leibniz (1646–1716).

§5.5. Weierstrass's construction of a continuous, non-differentiable function was published in 1861. Until then arguments based on geometrical intuition were still acceptable to some analysts. The fallibility of such reasoning was finally demonstrated by this function which so strikingly contradicts the promptings of the visual imagination.

The proof of theorem 5.51 is based on a method developed by F. A. Behrend ('Crinkly Curves and Choppy Surfaces', *Amer. Math. Monthly*, **67** (1960), 971–73).

§5.6. Stone gave an account of 'The generalized Weierstrass approximation theorem' in *Math. Magazine* **21** (1948), 167–84 and 237–54. This paper is not readily accessible, but it has been reprinted, with only minor changes, in *Studies in Modern Analysis* (ed. R. C. Buck), 20–87.

6

INTEGRATION

6.1. The Riemann–Stieltjes integral

Suppose that f is a bounded real function on the finite interval $[a, b]$. If $\mathscr{D}$ is a dissection of $[a, b]$ given by

$$a = x_0 < x_1 < \dots < x_{n-1} < x_n = b,$$

let
$$\inf_{x_{i-1} \leqslant x \leqslant x_i} f(x) = m_i, \quad \sup_{x_{i-1} \leqslant x \leqslant x_i} f(x) = M_i.$$

Then the sums
$$\sum_{i=1}^{n} m_i(x_i - x_{i-1}), \quad \sum_{i=1}^{n} M_i(x_i - x_{i-1}) \tag{6.11}$$

approximate to the Riemann integral $\int_a^b f$, when it exists (C1, 123–4).

A far reaching extension of this process is due to Stieltjes who introduced a second real function g, assumed to be increasing on $[a, b]$ (in the wide sense), and replaced the increments $x_i - x_{i-1}$ in (6.11) by $g(x_i) - g(x_{i-1})$. This new procedure leads to an integral of f with respect to g. The sums corresponding to (6.11) are

$$s(\mathscr{D}) = s(\mathscr{D}, f, g) = \sum_{i=1}^{n} m_i\{g(x_i) - g(x_{i-1})\}, \tag{6.12}$$

$$S(\mathscr{D}) = S(\mathscr{D}, f, g) = \sum_{i=1}^{n} M_i\{g(x_i) - g(x_{i-1})\}; \tag{6.13}$$

they reduce to (6.11) when $g(x) = x$ $(a \leqslant x \leqslant b)$. Denoting the infimum and supremum of $f(x)$ on $[a, b]$ by m and M respectively we have
$$m\{g(b) - g(a)\} \leqslant s(\mathscr{D}, f, g) \leqslant S(\mathscr{D}, f, g) \leqslant M\{g(b) - g(a)\}.$$

Thus, for all dissections $\mathscr{D}$, the lower sums (6.12) and the upper sums (6.13) are bounded.

It is easy to see that the introduction of new points of division increases the lower sum and diminishes the upper sum. (See exercise 6(a), 1.) From this it follows that any lower sum is less than or equal to any upper sum. For let $\mathscr{D}_1$ and $\mathscr{D}_2$ be any dissections of $[a, b]$. If, now, $\mathscr{D}$ is the dissection containing all the points of division of $\mathscr{D}_1$ and $\mathscr{D}_2$, then

$$s(\mathscr{D}_1) \leqslant s(\mathscr{D}), \quad s(\mathscr{D}) \leqslant S(\mathscr{D}), \quad S(\mathscr{D}) \leqslant S(\mathscr{D}_2)$$
and so
$$s(\mathscr{D}_1) \leqslant S(\mathscr{D}_2). \tag{6.14}$$

Definition. Write

$$\sup_{\mathscr{D}} s(\mathscr{D}, f, g) = \underline{\int_a^b} f \, dg \quad \left(or \ \underline{\int_a^b} f(x) \, dg(x) \right),$$

$$\inf_{\mathscr{D}} S(\mathscr{D}, f, g) = \overline{\int_a^b} f \, dg \quad \left(or \ \overline{\int_a^b} f(x) \, dg(x) \right),$$

where the supremum and infimum are taken over all dissections $\mathscr{D}$ of $[a, b]$. The first expression is called the lower integral *of f with respect to g over $[a, b]$, the second the* upper integral.

Note that $\underline{\int} f \, dg$ and $\overline{\int} f \, dg$ exist whenever f is bounded on $[a, b]$ and g increases; also, by (6.14),

$$\underline{\int_a^b} f \, dg \leqslant \overline{\int_a^b} f \, dg.$$

Definition. If

$$\underline{\int_a^b} f \, dg = \overline{\int_a^b} f \, dg,$$

f is said to be integrable *with respect to g over $[a, b]$ and the common value of the upper and lower integrals, denoted by*

$$\int_a^b f \, dg \quad \left(or \ \int_a^b f(x) \, dg(x) \right)$$

is called the Riemann–Stieltjes (*or* RS) integral *of f with respect to g. The function g is called the* integrator, *the function f the* integrand. *The class of functions integrable with respect to g over $[a, b]$ is denoted by $R\,(g; a, b)$.*

It is convenient to complete the definition of the RS integral by putting

$$\int_b^a f \, dg = -\int_a^b f \, dg$$

(when the right-hand side exists) and

$$\int_a^a f \, dg = 0$$

(for all functions f, g).

When $g(x) = x \,(a \leqslant x \leqslant b)$, the Riemann–Stieltjes integral reduces to the Riemann integral. The definition of the Riemann integral given in C1 (123) is not obviously the same as the one given here in terms of upper and lower integrals, but the equivalence of the two definitions will be proved in theorem 6.72. Both definitions are useful. The class of functions Riemann integrable over the interval $[a, b]$ will be denoted by $R(a, b)$; and upper and lower Riemann sums by $S(\mathscr{D}, f)$, $s(\mathscr{D}, f)$.

Illustrations

(i) Every constant function k has an RS integral with respect to any increasing function g and

$$\int_a^b k \, dg = k\{g(b) - g(a)\}.$$

This follows from the fact that, for all $\mathscr{D}$, $s(\mathscr{D}) = S(\mathscr{D}) = k\{g(b) - g(a)\}$.

(ii) Let f be the function defined by

$$f(x) = \begin{cases} 1 & \text{if } x \text{ is rational,} \\ 0 & \text{if } x \text{ is irrational.} \end{cases}$$

Then $\inf f(x) = 0$ and $\sup f(x) = 1$ in every interval. Hence, when g is any increasing function,

$$\underline{\int_a^b} f \, dg = 0 \quad \text{and} \quad \overline{\int_a^b} f \, dg = g(b) - g(a).$$

Thus, if g is not constant, $f \notin R(g; a, b)$.

In the rest of this section we establish the elementary properties of the Riemann–Stieltjes integral. When defining it we made the assumption that *the integrator is increasing and the integrand is bounded.* We continue with this pre-supposition without always stating it explicity.

Theorem 6.11. *A necessary and sufficient condition that $f \in R(g; a, b)$ is that, given $\epsilon > 0$, there is a dissection $\mathscr{D}$ of $[a, b]$ such that*

$$S(\mathscr{D}, f, g) - s(\mathscr{D}, f, g) < \epsilon. \tag{6.15}$$

Proof.

(i) If $f \in R(g; a, b)$, i.e. if

$$\underline{\int_a^b} f \, dg = \overline{\int_a^b} f \, dg,$$

then, given $\epsilon > 0$, there are dissections $\mathscr{D}_1$ and $\mathscr{D}_2$ such that

$$S(\mathscr{D}_1) - \int_a^b f \, dg < \tfrac{1}{2}\epsilon \quad \text{and} \quad \int_a^b f \, dg - s(\mathscr{D}_2) < \tfrac{1}{2}\epsilon.$$

Thus $$S(\mathscr{D}_1) - s(\mathscr{D}_2) < \epsilon.$$

Now let $\mathscr{D}$ be the dissection containing all the points of division of $\mathscr{D}_1$ and of $\mathscr{D}_2$. Then

$$S(\mathscr{D}_1) \geqslant S(\mathscr{D}) \geqslant s(\mathscr{D}) \geqslant s(\mathscr{D}_2)$$

and so (6.15) holds.

(ii) Suppose that, for every $\epsilon > 0$, there is a $\mathscr{D}$ such that (6.15) holds. Then, since

$$S(\mathscr{D}) \geqslant \overline{\int_a^b} f\,dg \geqslant \underline{\int_a^b} f\,dg \geqslant s(\mathscr{D}),$$

it follows that

$$0 \leqslant \overline{\int_a^b} f\,dg - \underline{\int_a^b} f\,dg < \epsilon.$$

This holds for all $\epsilon > 0$ and so $\overline{\int_a^b} f\,dg = \underline{\int_a^b} f\,dg$. ∎

The next theorem gives the existence of the Riemann–Stieltjes integral in two simple but important cases. In its proof, and later, we call the length of the largest subinterval of a dissection $\mathscr{D}$ the *mesh* of $\mathscr{D}$ and we denote it by $\mu(\mathscr{D})$.

Theorem 6.12.
   (i) *If f is continuous on $[a, b]$, then $f \in R\,(g;\,a,\,b)$.*
   (ii) *If f is monotonic on $[a, b]$ and g is continuous (as well as increasing), then $f \in R\,(g;\,a,\,b)$.*

   Proof.
   (i) For any dissection $\mathscr{D}$ of $[a, b]$ we have, with the usual notation,

$$\begin{aligned}
S(\mathscr{D}) - s(\mathscr{D}) &= \sum_{i=1}^{n} (M_i - m_i)\{g(x_i) - g(x_{i-1})\} \\
&\leqslant \max_{1 \leqslant i \leqslant n} (M_i - m_i) \sum_{i=1}^{n} \{g(x_i) - g(x_{i-1})\} \\
&= \max_{1 \leqslant i \leqslant n} (M_i - m_i)\{g(b) - g(a)\}.
\end{aligned}$$

Since f is uniformly continuous on $[a, b]$, $\max\,(M_i - m_i)$ may be made arbitrarily small by taking $\mu(\mathscr{D})$ sufficiently small. Theorem 6.11 then shows that $f \in R\,(g;\,a,\,b)$.

   (ii) Suppose that f increases. Then

$$\begin{aligned}
S(\mathscr{D}) - s(\mathscr{D}) &= \sum_{i=1}^{n} \{f(x_i) - f(x_{i-1})\}\{g(x_i) - g(x_{i-1})\} \\
&\leqslant \max_{1 \leqslant i \leqslant n} \{g(x_i) - g(x_{i-1})\}\{f(b) - f(a)\} \to 0
\end{aligned}$$

as $\mu(\mathscr{D}) \to 0$, by the uniform continuity of g. ∎

Part (ii) of the theorem will be much improved in theorem 6.24.
   We now prove that the RS integral is linear in both the integrand and the integrator.

Theorem 6.13.

(i) (a) *If* $f \in R(g; a, b)$, *then* $kf \in R(g; a, b)$ *for every constant* k *and*

$$\int_a^b (kf)\, dg = k \int_a^b f\, dg.$$

(b) *If* $f_1, f_2 \in R(g; a, b)$, *then* $f_1 + f_2 \in R(g; a, b)$ *and*

$$\int_a^b (f_1 + f_2)\, dg = \int_a^b f_1\, dg + \int_a^b f_2\, dg.$$

(ii) (a) *If* $f \in R(g; a, b)$, *then* $f \in R(kg; a, b)$ *for every non-negative constant* k *and*

$$\int_a^b f\, d(kg) = k \int_a^b f\, dg.$$

(b) *If* $f \in R(g_1; a, b)$ *and* $f \in R(g_2; a, b)$, *then*

$$f \in R(g_1 + g_2; a, b)$$

and

$$\int_a^b f\, d(g_1 + g_2) = \int_a^b f\, dg_1 + \int_a^b f\, dg_2.$$

Proof.

(i) (a) This follows immediately from exercise 6(a), 2.

(b) We first note that, since

$$\inf f_1 + \inf f_2 \leqslant \inf (f_1 + f_2),$$

$$s(\mathscr{D}, f_1, g) + s(\mathscr{D}, f_2, g) \leqslant s(\mathscr{D}, f_1 + f_2, g). \tag{6.16}$$

Now, given $\epsilon > 0$, there are $\mathscr{D}_1, \mathscr{D}_2$ such that

$$s(\mathscr{D}_1, f_1, g) > \underline{\int}_a^b f_1\, dg - \epsilon \quad \text{and} \quad s(\mathscr{D}_2, f_2, g) > \underline{\int}_a^b f_2\, dg - \epsilon$$

and, if $\mathscr{D}$ has all the points of division of $\mathscr{D}_1$ and of $\mathscr{D}_2$,

$$s(\mathscr{D}, f_1, g) > \underline{\int}_a^b f_1\, dg - \epsilon \quad \text{and} \quad s(\mathscr{D}, f_2, g) > \underline{\int}_a^b f_2\, dg - \epsilon.$$

Hence, by (6.16),

$$\underline{\int}_a^b f_1\, dg + \underline{\int}_a^b f_2\, dg < s(\mathscr{D}, f_1 + f_2, g) + 2\epsilon \leqslant \underline{\int}_a^b (f_1 + f_2)\, dg + 2\epsilon.$$

Since this holds for every $\epsilon > 0$,

$$\underline{\int}_a^b f_1\, dg + \underline{\int}_a^b f_2\, dg \leqslant \underline{\int}_a^b (f_1 + f_2)\, dg.$$

It is proved similarly that

$$\overline{\int_a^b} f_1 \, dg + \overline{\int_a^b} f_2 \, dg \geqslant \overline{\int_a^b} (f_1 + f_2) \, dg.$$

These two inequalities show that, if $f_1, f_2 \in R(g; a, b)$, then $f_1 + f_2 \in R(g; a, b)$ and

$$\int_a^b (f_1 + f_2) \, dg = \int_a^b f_1 \, dg + \int_a^b f_2 \, dg.$$

(ii) The proof of (a) is very easy, that of (b) follows from exercise $6(a), 3$. |

Theorem 6.14.

(i) If $f_1, f_2 \in R(g; a, b)$, then $f_1 f_2 \in R(g; a, b)$.

(ii) If $\inf\limits_{a \leqslant x \leqslant b} |f(x)| > 0$ and $f \in R(g; a, b)$, then $1/f \in R(g; a, b)$.

Proof.

(i) There is a K such that $|f_1(x)|, |f_2(x)| \leqslant K$ for $a \leqslant x \leqslant b$. Then

$$|f_1(u)f_2(u) - f_1(v)f_2(v)|$$
$$= |f_1(u)f_2(u) - f_1(u)f_2(v) + f_1(u)f_2(v) - f_1(v)f_2(v)|$$
$$\leqslant K|f_1(u) - f_1(v)| + K|f_2(u) - f_2(v)|$$

and so, in any subinterval of $[a, b]$,

$$\sup f_1(x)f_2(x) - \inf f_1(x)f_2(x)$$
$$\leqslant K\{\sup f_1(x) - \inf f_1(x)\} + K\{\sup f_2(x) - \inf f_2(x)\}. \quad (6.17)$$

Now, given $\epsilon > 0$, there is a $\mathcal{D}$ such that

$$S(\mathcal{D}, f_1, g) - s(\mathcal{D}, f_1, g) < \epsilon \quad \text{and} \quad S(\mathcal{D}, f_2, g) - s(\mathcal{D}, f_2, g) < \epsilon$$

and, by (6.17),

$$S(\mathcal{D}, f_1 f_2, g) - s(\mathcal{D}, f_1 f_2, g)$$
$$\leqslant K\{S(\mathcal{D}, f_1, g) - s(\mathcal{D}, f_1, g)\} + K\{S(\mathcal{D}, f_2, g) - s(\mathcal{D}, f_2, g)\} < 2K\epsilon.$$

(ii) If $\inf\limits_{a \leqslant x \leqslant b} |f(x)| = k \ (> 0)$, we have

$$\left| \frac{1}{f(u)} - \frac{1}{f(v)} \right| = \left| \frac{f(v) - f(u)}{f(u)f(v)} \right| \leqslant \frac{|f(v) - f(u)|}{k^2}.$$

Hence

$$S(\mathcal{D}, 1/f, g) - s(\mathcal{D}, 1/f, g) \leqslant (1/k^2)\{S(\mathcal{D}, f, g) - s(\mathcal{D}, f, g)\}$$

and the right-hand side may be made arbitrarily small. |

Theorem 6.15. *If $f \in R(g; a, b)$, then $|f| \in R(g; a, b)$ and*

$$\left| \int_a^b f \, dg \right| \leqslant \int_a^b |f| \, dg.$$

Proof. Since $\quad ||f(u)| - |f(v)|| \leqslant |f(u) - f(v)|,$

$$S(\mathcal{D}, |f|, g) - s(\mathcal{D}, |f|, g) \leqslant S(\mathcal{D}, f, g) - s(\mathcal{D}, f, g)$$

and so $|f| \in R(g; a, b)$.

Also, since $-|f| \leqslant f \leqslant |f|$, it follows from exercise 6(*a*), 5 and theorem 6.13, (i) (*a*) that

$$-\int_a^b |f| \, dg \leqslant \int_a^b f \, dg \leqslant \int_a^b |f| \, dg. \; |$$

It is easy to give an example showing that $\int_a^b |f| \, dg$ may exist without $\int_a^b f \, dg$ existing. (See exercise 6(*a*), 6.)

Lemma. *If $a < b < c$,*

(i) $\displaystyle \overline{\int_a^b} f \, dg + \overline{\int_b^c} f \, dg = \overline{\int_a^c} f \, dg,$

(ii) $\displaystyle \underline{\int_a^b} f \, dg + \underline{\int_b^c} f \, dg = \underline{\int_a^c} f \, dg.$

Proof. We prove (i); the proof of (ii) is similar.

Given $\epsilon > 0$, there are dissections $\mathcal{D}_1, \mathcal{D}_2, \mathcal{D}$ of $[a, b]$, $[b, c]$, $[a, c]$ respectively such that

$$S(\mathcal{D}_1) - \overline{\int_a^b} f \, dg < \epsilon, \quad S(\mathcal{D}_2) - \overline{\int_b^c} f \, dg < \epsilon, \quad S(\mathcal{D}) - \overline{\int_a^c} f \, dg < \epsilon.$$

Let $\mathcal{D}'$ be the dissection of $[a, c]$ containing all the points of division of $\mathcal{D}$, $\mathcal{D}_1$ and $\mathcal{D}_2$. In particular, $\mathcal{D}'$ has b as a point of division. Now let $\mathcal{D}_1'$, $\mathcal{D}_2'$ be the dissections of $[a, b]$, $[b, c]$ respectively induced by $\mathcal{D}'$. Then

$$S(\mathcal{D}_1') + S(\mathcal{D}_2') = S(\mathcal{D}') \tag{6.18}$$

and

$$-\epsilon < \overline{\int_a^b} f \, dg - S(\mathcal{D}_1') \leqslant 0, \quad -\epsilon < \overline{\int_b^c} f \, dg - S(\mathcal{D}_2') \leqslant 0,$$

$$0 \leqslant S(\mathcal{D}') - \overline{\int_a^c} f \, dg < \epsilon.$$

Adding these inequalities and using (6.18) we have

$$-2\epsilon < \overline{\int_a^b} f\,dg + \overline{\int_b^c} f\,dg - \overline{\int_a^c} f\,dg < \epsilon.$$

Since this holds for all $\epsilon > 0$, (i) follows. |

Theorem 6.16.

(i) *If* $f \in R(g; a, b)$ *and* $f \in R(g; b, c)$, *then* $f \in R(g; a, c)$ *and*

$$\int_a^c f\,dg = \int_a^b f\,dg + \int_b^c f\,dg.$$

(ii) *If* $f \in R(g; a, b)$ *and if* $[c, d]$ *is any subinterval of* $[a, b]$, *then* $f \in R(g; c, d)$.

Proof.

(i) By the lemma,

$$\underline{\int_a^c} f\,dg = \int_a^b f\,dg + \int_b^c f\,dg = \overline{\int_a^c} f\,dg.$$

(ii) Again, by the lemma,

$$\underline{\int_a^c} f\,dg + \int_c^d f\,dg + \int_d^b f\,dg = \int_a^b f\,dg = \overline{\int_a^c} f\,dg + \overline{\int_c^d} f\,dg + \overline{\int_d^b} f\,dg$$

and so

$$\left(\overline{\int_a^c} f\,dg - \underline{\int_a^c} f\,dg\right) + \left(\overline{\int_c^d} f\,dg - \underline{\int_c^d} f\,dg\right) + \left(\overline{\int_d^b} f\,dg - \underline{\int_d^b} f\,dg\right) = 0.$$

Since each of the three terms on the left is non-negative, each term is 0. |

Notes.

(1) By (i), $f \in R(g; a, b)$ if f is piecewise monotonic, i.e. if $[a, b]$ may be divided into a finite number of subintervals in each of which f is monotonic.

(2) In view of the definition of $\int_\alpha^\beta f\,dg$ when $\alpha \geqslant \beta$, we need not assume in (i) that $a < b < c$.

Theorem 6.17. (Change of variable.) *Let* $f \in R(g; a, b)$. *If* ϕ *is a continuous, strictly increasing function on the interval* $[\alpha, \beta]$ *such that* $\phi(\alpha) = a$ *and* $\phi(\beta) = b$, *then* $f \circ \phi \in R(g \circ \phi; \alpha, \beta)$ *and*

$$\int_a^b f\,dg = \int_\alpha^\beta (f \circ \phi)\,d(g \circ \phi).$$

Proof. The relation $x = \phi(t)$ $(x \in [a, b],\, t \in [\alpha, \beta])$ sets up a bijection between the dissections $\mathscr{D}$ of $[a, b]$ and the dissections $\mathscr{D}^*$ of $[\alpha, \beta]$. Also, if $\mathscr{D}$, $\mathscr{D}^*$ correspond,

$$s(\mathscr{D}, f, g) = s(\mathscr{D}^*, f \circ \phi, g \circ \phi), \quad S(\mathscr{D}, f, g) = S(\mathscr{D}^*, f \circ \phi, g \circ \phi)$$

and therefore

$$\underline{\int_a^b} f\,dg = \underline{\int_\alpha^\beta} (f \circ \phi)\,d(g \circ \phi), \quad \overline{\int_a^b} f\,dg = \overline{\int_\alpha^\beta} (f \circ \phi)\,d(g \circ \phi). \ |$$

Other simple properties of the Riemann–Stieltjes integral are given in exercises 8–10 below.

Exercises 6(a)

1. Show that, if $\mathscr{D}'$ is a dissection of $[a, b]$ which contains all the points of division of the dissection $\mathscr{D}$, then

$$S(\mathscr{D}', f, g) \leqslant S(\mathscr{D}, f, g), \quad s(\mathscr{D}', f, g) \geqslant s(\mathscr{D}, f, g).$$

2. Prove that, when k is a constant,

$$\overline{\int_a^b} (kf)\,dg = k \overline{\int_a^b} f\,dg, \quad \underline{\int_a^b} (kf)\,dg = k \underline{\int_a^b} f\,dg \quad (k \geqslant 0);$$

$$\overline{\int_a^b} (kf)\,dg = k \underline{\int_a^b} f\,dg, \quad \underline{\int_a^b} (kf)\,dg = k \overline{\int_a^b} f\,dg \quad (k < 0).$$

3. Prove that

$$\overline{\int_a^b} f\,d(g_1 + g_2) = \overline{\int_a^b} f\,dg_1 + \overline{\int_a^b} f\,dg_2, \quad \underline{\int_a^b} f\,d(g_1 + g_2) = \underline{\int_a^b} f\,dg_1 + \underline{\int_a^b} f\,dg_2.$$

4. Let $f \in R(g_1; a, b) \cap R(g_2; a, b)$. Show that, if $g_1 - g_2$ is increasing, then $f \in R(g_1 - g_2; a, b)$ and

$$\int_a^b f\,d(g_1 - g_2) = \int_a^b f\,dg_1 - \int_a^b f\,dg_2.$$

5. Show that, if $f_1(x) \leqslant f_2(x)$ for $a \leqslant x \leqslant b$, then

$$\overline{\int_a^b} f_1\,dg \leqslant \overline{\int_a^b} f_2\,dg, \quad \underline{\int_a^b} f_1\,dg \leqslant \underline{\int_a^b} f_2\,dg.$$

6. Construct a function f such that, for any non-constant g, $|f| \in R(g; a, b)$, but $f \notin R(g; a, b)$.

7. Find functions f, h, g (where g is increasing) such that

$$\overline{\int_a^b} (f + h)\,dg < \overline{\int_a^b} f\,dg + \overline{\int_a^b} h\,dg, \quad \underline{\int_a^b} (f + h)\,dg > \underline{\int_a^b} f\,dg + \underline{\int_a^b} h\,dg.$$

8. Suppose that $f, h \in R(g; a, b)$ and $h(x) \geq 0$ for $a \leq x \leq b$. Show that there is a number λ between the infimum and the supremum of f on $[a, b]$ such that

$$\int_a^b fh\,dg = \lambda \int_a^b h\,dg.$$

(This result with $h(x) \equiv 1$ is the *first mean value theorem*.)

9. Show that, if f is continuous on $[a, b]$ (and g is increasing), then

$$S(\mathscr{D}, f, g), s(\mathscr{D}, f, g) \to \int_a^b f\,dg$$

as $\mu(\mathscr{D}) \to 0$.

10. Let $f \in R(g; a, b)$ and define the function F on $[a, b]$ by

$$F(x) = \int_a^x f\,dg \quad (a \leq x \leq b).$$

Prove that
   (i) F is continuous at every point of continuity of g;
   (ii) F is differentiable at a point where f is continuous and g is differentiable and, at such a point x,
$$F'(x) = f(x)g'(x).$$

11. Let g be increasing on $[a, b]$. Show that, if f is bounded on $[a, b]$ and $\int_{a+h}^b f\,dg$ exists whenever $0 < h < b-a$, then $\int_a^b f\,dg$ exists.

12. A function ϕ defined on an interval containing the point c is said to be *left continuous* at c if $\phi(c-) = \phi(c)$, i.e.

$$\lim_{x \to c-} \phi(x) = \phi(c).$$

Right continuity is similarly defined.
   Let f be bounded and g increasing on $[a, b]$. Prove that, if f and g have a common left discontinuity or a common right discontinuity, then $f \notin R(g; a, b)$.

13. Suppose that, for $n = 1, 2, \ldots, f_n \in R(g; a, b)$ and that $f_n \to f$ uniformly on $[a, b]$. Prove that $f \in R(g; a, b)$ and

$$\int_a^b f_n\,dg \to \int_a^b f\,dg$$

as $n \to \infty$.

14. Show that, if f is non-negative and continuous on $[a, b]$ and g is strictly increasing, then

$$\left(\int_a^b f^n\,dg \right)^{1/n} \to \sup_{a \leq x \leq b} f(x)$$

as $n \to \infty$.

6.2. Further properties of the Riemann–Stieltjes integral

The material in this section is rather more sophisticated than that of §6.1. Familiarity with it is not necessary for a reader who is primarily interested in the Riemann integral. We begin by showing that, if the integrator has a Riemann integrable derivative, or is an indefinite Riemann integral, then the RS integral reduces to an ordinary Riemann integral.

Theorem 6.21A. *Let* $f \in R(a, b)$. *If the increasing function* g *is differentiable in* $[a, b]$ *and* $g' \in R(a, b)$, *then* $f \in R(g; a, b)$ *and*

$$\int_a^b f(x)\,dg(x) = \int_a^b f(x)g'(x)\,dx.$$

Proof. If the theorem has been proved for non-negative f and so, in particular, for a non-negative constant, it holds for any $f \in R(a, b)$. For we can then write $f = (f + C) - C$, where C is such that

$$f(x) + C \geqslant 0$$

for $a \leqslant x \leqslant b$. We may therefore suppose that f is non-negative in $[a, b]$. Since g' is also non-negative, in any subinterval of $[a, b]$,

$$\inf f(x) \inf g'(x) \leqslant \inf f(x)g'(x) \leqslant \inf f(x) \sup g'(x). \quad (6.21)$$

As $g', fg' \in R(a, b)$, a now familiar argument shows that, given $\epsilon > 0$, there is a dissection $\mathscr{D}$ of $[a, b]$ such that the inequalities

$$S(\mathscr{D}, g') - s(\mathscr{D}, g') < \epsilon, \quad (6.22)$$

$$S(\mathscr{D}, fg') - s(\mathscr{D}, fg') < \epsilon \quad (6.23)$$

both hold. If $\mathscr{D}$ is given by

$$a = x_0 < x_1 < \ldots < x_{n-1} < x_n = b,$$

denote the infima of $f(x)$, $g'(x)$, $f(x)g'(x)$ in $[x_{i-1}, x_i]$ by m_i, m_i', p_i and the supremum of $g'(x)$ by M_i'. By (6.21), there is a number μ_i such that $m_i' \leqslant \mu_i \leqslant M_i'$ and

$$p_i = m_i \mu_i.$$

Also, by the mean value theorem, there is a number μ_i' such that $m_i' \leqslant \mu_i' \leqslant M_i'$ and

$$g(x_i) - g(x_{i-1}) = \mu_i'(x_i - x_{i-1}). \quad (6.24)$$

Then, by (6.23),

$$\int_a^b fg' - \epsilon < \sum_{i=1}^n p_i(x_i - x_{i-1}) = \sum_{i=1}^n m_i\{\mu_i' + (\mu_i - \mu_i')\}(x_i - x_{i-1})$$

$$= \sum_{i=1}^n m_i\{g(x_i) - g(x_{i-1})\} + \sum_{i=1}^n m_i(\mu_i - \mu_i')(x_i - x_{i-1}).$$

But, if $$M = \sup_{a \leqslant x \leqslant b} f(x),$$

then, by (6.22),

$$\left| \sum_{i=1}^{n} m_i(\mu_i - \mu_i')(x_i - x_{i-1}) \right| \leqslant M \sum_{i=1}^{n} (M_i' - m_i')(x_i - x_{i-1}) < M\epsilon$$

and so $\displaystyle\int_a^b fg' < s(\mathscr{D}, f, g) + (M+1)\epsilon \leqslant \underline{\int_a^b} f dg + (M+1)\epsilon.$

This holds for all $\epsilon > 0$ and therefore

$$\int_a^b fg' \leqslant \underline{\int_a^b} f dg.$$

It is shown similarly that

$$\int_a^b fg' \geqslant \overline{\int_a^b} f dg. \ |$$

Theorem 6.21B. *Let $f, g \in R(a, b)$ and let G be an indefinite integral of g, i.e.*

$$G(x) = \int_a^x g(t) dt + K \quad (a \leqslant x \leqslant b),$$

where K is a constant. If g is non-negative on $[a, b]$ (so that G increases), then $f \in R(G; a, b)$ and

$$\int_a^b f(x) dG(x) = \int_a^b f(x) g(x) dx.$$

Proof. The previous proof with g' replaced by g, and g replaced by G applies almost word for word. The only difference is that the analogue of (6.24), which reads

$$G(x_i) - G(x_{i-1}) = \mu_i'(x_i - x_{i-1}),$$

is obtained by remarking that $G(x_i) - G(x_{i-1})$ lies between $m_i'(x_i - x_{i-1})$ and $M_i'(x_i - x_{i-1})$. (Alternatively, see exercise 6(a), 8.) $|$

It will be seen later (in theorem 6.84) that, if g' is R-integrable, then g is an indefinite integral of g'. Thus, theorem B in fact includes theorem A.

Having disposed of smooth integrators we pass to the consideration of discontinuous ones.

The real function ϕ on the interval $[a, b]$ is said to have a *jump* discontinuity at the point x in (a, b) if $\phi(x-)$ and $\phi(x+)$ exist, but are not both equal to $\phi(x)$; the definition is appropriately modified for $x = a$ or $x = b$. Also, as in exercise 6(a), 12, we call ϕ continuous on the left [right] at x if $\phi(x-) = \phi(x)$ [$\phi(x+) = \phi(x)$]. Since an integrator g on $[a, b]$ is increasing, any discontinuities of g must be of the jump variety. The simplest discontinuous integrator is a function which is constant except for a jump at a single point. If the jump is not at an end point, then the function is of the form

$$g(x) = \begin{cases} \alpha & \text{for } a \leqslant x < u, \\ \beta & \text{for } u < x \leqslant b, \end{cases}$$

with $\alpha < \beta$ and $\alpha \leqslant g(u) \leqslant \beta$. Otherwise the form is

$$g(x) = \begin{cases} \alpha & \text{for } x = a \\ \beta & \text{for } a < x \leqslant b \end{cases} \quad \text{or} \quad g(x) = \begin{cases} \alpha & \text{for } a \leqslant x < b, \\ \beta & \text{for } x = b. \end{cases}$$

Theorem 6.22. *Let g on $[a, b]$ be constant except for a jump at the point u. Then $f \in R(g; a, b)$ if and only if f is continuous on the left when g is discontinuous on the left and continuous on the right when g is discontinuous on the right. If $f \in R(g; a, b)$, then*

$$\int_a^b f\,dg = \begin{cases} f(a)\{g(a+) - g(a)\} & (u = a), \\ f(u)\{g(u+) - g(u-)\} & (a < u < b), \\ f(b)\{g(b) - g(b-)\} & (u = b). \end{cases}$$

Proof. When f, g have a common discontinuity on the left or on the right, then, by exercise 6(a), 12, $f \notin R(g; a, b)$.

Suppose now that there is no such common discontinuity.

First let $a < u \leqslant b$. We shall show that $f \in R(g; a, u)$ and that

$$\int_a^u f\,dg = f(u)\{g(u) - g(u-)\}. \tag{6.25}$$

This clearly holds when $g(u) = g(u-)$, i.e. when g is constant on $[a, u]$. When $g(u) > g(u-)$, take t such that $a < t < u$ and let $\mathscr{D}$ be the dissection of $[a, u]$ with a, t, u as the sole points of division. Then

$$\inf_{t \leqslant x \leqslant u} f(x)\{g(u) - g(t)\} = s(\mathscr{D}) \leqslant \underline{\int_a^u} f\,dg \leqslant \overline{\int_a^u} f\,dg$$

$$\leqslant S(\mathscr{D}) = \sup_{t \leqslant x \leqslant u} f(x)\{g(u) - g(t)\}.$$

Since f is now continuous on the left at u, the first and last of these expressions tend to $f(u)\{g(u)-g(u-)\}$ as $t \to u-$. Thus $f \in R(g; a, u)$ and (6.25) holds.

When $a \leqslant u < b$, we prove similarly that $f \in R(g; u, b)$ and

$$\int_u^b f\,dg = f(u)\{g(u+)-g(u)\}. \enspace |$$

Lemma 1. *Any increasing function is the sum of a continuous increasing function and of countably many functions each of which is increasing and is continuous except for a single jump.*

Proof. Let g be an increasing function on $[a, b]$. By exercise 1(c), 10, the set of discontinuities of g is countable. Denote the points of discontinuity by u_k, where k ranges over a countable set Δ which may be finite, possibly empty. If $a < u_k < b$, define the function g_k by

$$g_k(x) = \begin{cases} 0 & \text{for } a \leqslant x < u_k, \\ g(u_k)-g(u_k-) & \text{for } x = u_k, \\ g(u_k+)-g(u_k-) & \text{for } u_k < x \leqslant b. \end{cases} \qquad (6.26)$$

To define g_k when $u_k = a$ take $g(a-) = g(a)$ and omit the first line of (6.26); when $u_k = b$, omit the last line. When Δ is infinite, Σg_k converges uniformly on $[a, b]$, for $\Sigma g_k(b)$ converges (the partial sums being bounded by $g(b)-g(a)$) and

$$0 \leqslant g_k(x) \leqslant g_k(b) \quad (a \leqslant x \leqslant b).$$

Hence, whether Δ is finite or infinite, the function

$$j = \sum_{k \in \Delta} g_k,$$

(taken to be identically 0 when Δ is empty) is continuous except at the points u_k ($k \in \Delta$). Thus

$$g^* = g - j$$

is continuous at all points $x \neq u_k$ ($k \in \Delta$). But, by the definition of g_k, $g - g_k$ is continuous at u_k. Since g_i ($i \in \Delta$, $i \neq k$) is also continuous at u_k, it follows that g^* is continuous on $[a, b]$.

The function g^* is increasing. To prove this we first note that the intervals $I_k = [g(u_k-), g(u_k)]$, $J_k = [g(u_k), g(u_k+)]$ are non-overlapping. Now let $a \leqslant \alpha < \beta \leqslant b$. Then the sum of the lengths of any finite number of I_k, J_k contained in the interval $[g(\alpha), g(\beta)]$ is less than or equal to $g(\beta)-g(\alpha)$. The same is therefore true of *all* the I_k, J_k in $[g(\alpha), g(\beta)]$. Thus $\sum_{k \in \Delta} \{g_k(\beta)-g_k(\alpha)\} \leqslant g(\beta)-g(\alpha)$, i.e.

$$g^*(\alpha) \leqslant g^*(\beta). \enspace |$$

Given the increasing function g, the functions g^* and j defined in the proof of the lemma are called, respectively, the *continuous component* and the *jump component* of g. To obtain the decomposition of $\int f\,dg$ corresponding to this resolution of g we need another lemma.

Lemma 2. *Suppose that the functions $g_1, g_2, \ldots$ on $[a, b]$ are increasing and that $\sum\limits_{k=1}^{\infty} g_k$ converges to a function h on $[a, b]$. Then h increases on $[a, b]$ and, if $f \in R(g_k; a, b)$ for every k, then $f \in R(h; a, b)$ and*

$$\int_a^b f\,dh = \sum_{k=1}^{\infty} \int_a^b f\,dg_k. \tag{6.27}$$

Proof. If $a \leqslant \alpha < \beta \leqslant b$, $g_k(\beta) - g_k(\alpha) \geqslant 0$ for every k and so

$$h(\beta) - h(\alpha) = \sum_{k=1}^{\infty} \{g_k(\beta) - g_k(\alpha)\} \geqslant 0.$$

Thus h increases on $[a, b]$.

Now suppose that $f \in R(g_k; a, b)$ for every k. Let

$$K = \sup_{a \leqslant x \leqslant b} |f(x)|.$$

Then, since

$$\left| \int_a^b f\,dg_k \right| \leqslant K\{g_k(b) - g_k(a)\},$$

it follows that

$$\sum_{k=1}^{\infty} \int_a^b f\,dg_k$$

converges.

Given $\epsilon > 0$, there is a p_0 such that, for $p \geqslant p_0$,

$$\sum_{k=p+1}^{\infty} \{g_k(b) - g_k(a)\} < \epsilon.$$

Since, for every dissection $\mathscr{D}$ of $[a, b]$,

$$|s(\mathscr{D}, f, g_k)|, \ |S(\mathscr{D}, f, g_k)| \leqslant K\{g_k(b) - g_k(a)\},$$

when $p \geqslant p_0$,

$$\left| \sum_{k=p+1}^{\infty} s(\mathscr{D}, f, g_k) \right|, \ \left| \sum_{k=p+1}^{\infty} S(\mathscr{D}, f, g_k) \right| < K\epsilon. \tag{6.28}$$

Take any $p \geqslant p_0$. Then there is a dissection $\mathscr{D}^*$ such that

$$\sum_{k=1}^{p} \{S(\mathscr{D}^*, f, g_k) - s(\mathscr{D}^*, f, g_k)\} < \epsilon;$$

and, using (6.28), we have

$$\sum_{k=1}^{p} \int_{a}^{b} f \, dg_k - (K+1)\epsilon < \sum_{k=1}^{p} s(\mathcal{D}^*, f, g_k) - K\epsilon$$

$$< s(\mathcal{D}^*, f, h) \leqslant \int_{\underline{a}}^{b} f \, dh \leqslant \overline{\int_{a}^{b}} f \, dh \leqslant S(\mathcal{D}^*, f, h)$$

$$< \sum_{k=1}^{p} S(\mathcal{D}^*, f, g_k) + K\epsilon < \sum_{k=1}^{p} \int_{a}^{b} f \, dg_k + (K+1)\epsilon.$$

Thus, for any $p \geqslant p_0$,

$$\sum_{k=1}^{p} \int_{a}^{b} f \, dg_k - (K+1)\epsilon < \int_{\underline{a}}^{b} f \, dh \leqslant \overline{\int_{a}^{b}} f \, dh < \sum_{k=1}^{p} \int_{a}^{b} f \, dg_k + (K+1)\epsilon$$

and so

$$\sum_{k=1}^{\infty} \int_{a}^{b} f \, dg_k - (K+1)\epsilon \leqslant \int_{\underline{a}}^{b} f \, dh \leqslant \overline{\int_{a}^{b}} f \, dh \leqslant \sum_{k=1}^{\infty} \int_{a}^{b} f \, dg_k + (K+1)\epsilon.$$

Since this is true for every $\epsilon > 0$, $f \in R(g; a, b)$ and (6.27) holds. |

Theorem 6.23. *Suppose that g is an increasing function on $[a, b]$ with discontinuities at the points u_k ($k \in \Delta$). Let $g = g^* + j$, where g^* is the continuous component and j is the jump component of g. If, now, $f \in R(g; a, b)$, then $f \in R(g^*; a, b)$ and*

$$\int_{a}^{b} f \, dg = \int_{a}^{b} f \, dg^* + \sum_{k \in \Delta} f(u_k)\{g(u_k+) - g(u_k-)\},$$

with the convention that $g(a-) = g(a)$, $g(b+) = g(b)$.

Proof. Let the g_k ($k \in \Delta$) be the functions defined by (6.26) so that $j = \Sigma \, g_k$. Since $f \in R(g; a, b)$, f and g can have no common left or common right discontinuities. Hence, by theorem 6.22, for each $k \in \Delta$, $f \in R(g_k; a, b)$ and

$$\int_{a}^{b} f \, dg_k = f(u_k)\{g_k(u_k+) - g_k(u_k-)\} = f(u_k)\{g(u_k+) - g(u_k-)\}.$$

It now follows from lemma 2 that $f \in R(j; a, b)$ and

$$\int_{a}^{b} f \, dj = \sum_{k \in \Delta} f(u_k)\{g(u_k+) - g(u_k-)\}.$$

Since $g^* = g - j$ is increasing, by exercise 6(a), 4, $f \in R(g^*; a, b)$ and

$$\int_{a}^{b} f \, dg = \int_{a}^{b} f \, dg^* + \int_{a}^{b} f \, dj. \ |$$

It is the last theorem that makes the Riemann–Stieltjes integral particularly valuable in applications of analysis. It shows, for instance, that in mathematical physics this integral can cope with point masses as well as continuous mass distributions.

We can now also improve theorem 6.12(ii).

Theorem 6.24. *If f is monotonic and g is increasing on* $[a, b]$ *and if f, g have no common left or common right discontinuities, then* $f \in R(g; a, b)$.

Proof. Let $g = g^* + j$, where $j = \Sigma g_k$. In view of the restriction on the discontinuities of f and g, theorem 6.22 shows that $f \in R(g_k; a, b)$ for all k. Therefore, by lemma 2, $f \in R(j; a, b)$. Also, by theorem 6.12(ii), $f \in R(g^*; a, b)$. |

If, in theorem 6.24, f (as well as g) is taken to be increasing, then $f \in R(g; a, b)$ and $g \in R(f; a, b)$. The symmetry of the situation is further exhibited by the formula for integration by parts.

Theorem 6.25. (Integration by parts.) *If the functions f and g are increasing on* $[a, b]$ *and have no common left and no common right discontinuities, then*

$$\int_a^b f \, dg + \int_a^b g \, df = f(b)g(b) - f(a)g(a).$$

Proof. Since f and g are increasing functions, we have, for any dissection $\mathscr{D}$ given by $a = x_0 < x_1 < \ldots < x_{n-1} < x_n = b$,

$$S(\mathscr{D}, f, g) + s(\mathscr{D}, g, f)$$

$$= \sum_{i=1}^n f(x_i)\{g(x_i) - g(x_{i-1})\} + \sum_{i=1}^n g(x_{i-1})\{f(x_i) - f(x_{i-1})\}$$

$$= f(b)g(b) - f(a)g(a). \tag{6.29}$$

But, given $\epsilon > 0$, $\mathscr{D}$ may be chosen so that

$$S(\mathscr{D}, f, g) - \epsilon < \int_a^b f \, dg \leqslant S(\mathscr{D}, f, g),$$

$$s(\mathscr{D}, g, f) \leqslant \int_a^b g \, df < s(\mathscr{D}, g, f) + \epsilon.$$

Adding these two inequalities and using (6.29) we have, for all $\epsilon > 0$,

$$f(b)g(b) - f(a)g(a) - \epsilon < \int_a^b f \, dg + \int_a^b g \, df < f(b)g(b) - f(a)g(a) + \epsilon. \; |$$

1. Show that, if f is continuous on $[0, \infty)$, then for $x \geqslant 0$,

$$\sum_{1 \leqslant n \leqslant x} f(n) = \int_0^x f(t)\,d[t],$$

where $[t]$ is the greatest integer less than or equal to t. (When $0 \leqslant x < 1$, the sum on the left is taken to be 0.)

2. The series $\sum_{n=1}^{\infty} u_n$ converges absolutely. Find an increasing function g on $[0, 1]$ and a function $f \in R(g; 0, 1)$ such that

$$\sum_{n=1}^{\infty} u_n = \int_0^1 f\,dg.$$

3. Construct functions f, g_n ($n = 1, 2, \ldots$) on an interval $[a, b]$ with the following properties:
(i) For $n = 1, 2, \ldots, g_n$ increases on $[a, b]$ and $f \in R(g_n; a, b)$,
(ii) $g_n \to g$ on $[a, b]$, but $f \notin R(g; a, b)$.

4. The functions g_n ($n = 1, 2, \ldots$) are increasing on $[a, b]$ and $g_n \to g$ uniformly on $[a, b]$. Show that, if f increases on $[a, b]$ and $f \in R(g_n; a, b)$ for every n, then $f \in R(g; a, b)$ and

$$\int_a^b f\,dg_n \to \int_a^b f\,dg.$$

6.3. Improper Riemann–Stieltjes integrals

Improper Riemann integrals were defined in C1 (138). The definitions for Riemann–Stieltjes integrals are entirely analogous.

Definition. *Let g be an increasing function on $[a, \infty)$. If $f \in R(g; a, X)$ for every $X > a$ and $\lim_{X \to \infty} \int_a^X f\,dg$ exists, then this limit is called an improper Riemann–Stieltjes integral of the first kind and is denoted by*

$$\int_a^{\infty} f\,dg.$$

Improper integrals have properties like those of infinite series and the following theorems illustrate this. The proofs are left to the reader.

Theorem 6.31. (General principle of convergence.) *If $f \in R(g; a, X)$ for every $X > a$, a necessary and sufficient condition for $\int_a^{\infty} f\,dg$ to exist is that, given $\epsilon > 0$, there is an $X_0 > a$ such that*

$$\left| \int_{X_1}^{X_2} f\,dg \right| < \epsilon \quad \text{whenever} \quad X_2 > X_1 > X_0.$$

For non-negative integrands the first comparison principle is the mainstay of proofs of convergence or divergence of integrals.

Theorem 6.32. *Suppose that $f, \phi, \psi \in R(g; a, X)$ for every $X > a$.*

(i) *If $\phi(x) \geqslant f(x) \geqslant 0$ for $x \geqslant a$ and $\int_a^\infty \phi \, dg$ exists, then $\int_a^\infty f \, dg$ exists.*

(ii) *If $f(x) \geqslant \psi(x) \geqslant 0$ for $x \geqslant a$ and $\int_a^\infty \psi \, dg$ does not exist, then $\int_a^\infty f \, dg$ does not exist.*

When f is of variable sign (and $f \in R(g; a, X)$ for every $X > a$), the simplest criterion for the convergence of $\int_a^\infty f \, dg$ is the convergence of $\int_a^\infty |f| \, dg$ (see exercise 6(c), 2). This result motivates the definition of absolute convergence.

If $f \in R(g; a, X)$ for every $X > a$ and $\int_a^\infty |f| \, dg$ exists, then $\int_a^\infty f \, dg$ is said to *converge absolutely*. If $\int_a^\infty f \, dg$ exists, but $\int_a^\infty |f| \, dg$ does not, then $\int_a^\infty f \, dg$ is said to be *non-absolutely* (or *conditionally*) *convergent*.

Illustrations. These refer to the simplest and most important case, that of the ordinary Riemann integral.

(i) $\int_1^\infty \dfrac{\sin x}{x^2} \, dx$ converges absolutely, since

$$\left| \frac{\sin x}{x^2} \right| \leqslant \frac{1}{x^2} \quad (x \geqslant 1)$$

and $\int_1^\infty \dfrac{1}{x^2} \, dx$ exists.

(ii) $\int_1^\infty \dfrac{\sin x}{x} \, dx$ converges non-absolutely.

First
$$\int_1^X \frac{\sin x}{x} \, dx = \left[-\frac{\cos x}{x} \right]_1^X - \int_1^X \frac{\cos x}{x^2} \, dx.$$

As $X \to \infty$, the first term on the right tends to $\cos 1$. The second term also converges, since $|\cos x|/x^2 \leqslant 1/x^2$. Hence $\int_1^\infty \dfrac{\sin x}{x} \, dx$ exists.

To prove that $\int_1^\infty \left| \dfrac{\sin x}{x} \right| \, dx$ diverges, note that

$$\int_{n\pi}^{(n+1)\pi} \frac{|\sin x|}{x} \, dx \geqslant \frac{1}{(n+1)\pi} \int_{n\pi}^{(n+1)\pi} |\sin x| \, dx = \frac{2}{(n+1)\pi}.$$

The other two kinds of integrals with infinite range need not detain us. The definition of $\int_{-\infty}^{a} f\,dg$ is obvious. $\int_{-\infty}^{\infty} f\,dg$ is defined as

$$\int_{-\infty}^{a} f\,dg + \int_{a}^{\infty} f\,dg;$$

the sum of these two integrals is clearly independent of a.

The second kind of improper integral is treated in exactly the same way as the first kind and we do little more than give the definitions.

Definition. *Let g be an increasing function on* $(a, b]$. *If*

$$f \in R\,(g;\, a+h,\, b)$$

whenever $0 < h < b-a$ *(so that f is bounded on every interval* $[a+h, b]$*) and if* $\lim\limits_{h\to 0+} \int_{a+h}^{b} f\,dg$ *exists, then this limit is called an improper Riemann–Stieltjes integral of the second kind and we denote it by*

$$\int_{a+}^{b} f\,dg, \quad \text{or simply} \quad \int_{a}^{b} f\,dg.$$

Notes.

(1) If f is bounded on $[a, b]$ and $f \in R\,(g;\, a+h,\, b)$ whenever $0 < h < b-a$, then $f \in R\,(g;\, a, b)$. (See exercise $6(a)$, 11.) Hence the definition of $\int_{a+}^{b} f\,dg$ is likely to be invoked only when f is unbounded near a.

(2) Suppose that $f \in R\,(g;\, a, b)$. Then $\int_{a+}^{b} f\,dg$ exists, but the improper integral need not be equal to the ordinary integral. (See exercise $6(c)$, 6.) The two integrals do coincide when g is continuous at a and so, in particular, in the case of the Riemann integral. (See exercise $6(a)$, 10(i).)

The definition of $\int_{a}^{b-} f\,dg$ (or simply $\int_{a}^{b} f\,dg$) is analogous to that of $\int_{a+}^{b} f\,dg$. Also, if $a = u_0 < u_1 < \ldots < u_{n-1} < u_n = b$ and, for $i = 1, \ldots, n$, $\int_{u_{i-1}}^{u_i} f\,dg$ exists as one of the improper integrals just defined, we call (unambiguously) the sum of these integrals the improper integral $\int_{a}^{b} f\,dg$.

The two kinds of improper integral may occur together. For we may define $\int_a^\infty f\,dg$ as $\lim\limits_{X\to\infty}\int_a^X f\,dg$ when $\int_a^X f\,dg$ exists either as an ordinary RS integral or as an improper integral of the second kind.

Series of improper integrals. We are once again concerned with the reversal of two limiting processes.

Theorem 6.33. *Suppose that g is an increasing function on $[a, \infty)$, that the functions f_n ($n = 1, 2, \ldots$) are all non-negative on $[a, \infty)$ and that*

$$\sum_{n=1}^{\infty}\left(\int_a^X f_n\,dg\right), \quad \int_a^X\left(\sum_{n=1}^{\infty} f_n\right) dg \tag{6.31}$$

exist and are equal for every $X > a$, where the integrals are ordinary or improper. If, now, one of the expressions

$$\sum_{n=1}^{\infty}\left(\int_a^\infty f_n\,dg\right), \quad \int_a^\infty\left(\sum_{n=1}^{\infty} f_n\right) dg \tag{6.32}$$

exists, then so does the other and the two are equal.

Proof. Suppose that the first expression in (6.32) exists. Since the version of theorem 4.73 with one continuous and one integral variable is applicable to $\sum\limits_{n=1}^{N}\int_a^X f_n\,dg$, and since the two expressions in (6.31) are equal, we obtain the following chain of equalities in which the existence of any term implies that of the next.

$$\sum_{n=1}^{\infty}\left(\int_a^\infty f_n\,dg\right) = \sum_{n=1}^{\infty}\left(\lim_{X\to\infty}\int_a^X f_n\,dg\right) = \lim_{X\to\infty}\sum_{n=1}^{\infty}\left(\int_a^X f_n\,dg\right)$$

$$= \lim_{X\to\infty}\int_a^X\left(\sum_{n=1}^{\infty} f_n\right) dg = \int_a^\infty\left(\sum_{n=1}^{\infty} f_n\right) dg. \quad |$$

In most applications of theorem 6.33 the interchangeability of summation and integration in (6.31) is ensured by the uniform convergence of Σf_n.

It is evident that theorem 6.33 may be restated in terms of improper integrals of the second kind.

Uniform convergence. The notion of uniform convergence may be applied to improper integrals of both kinds. It is sufficient to consider the first kind.

Definition. Suppose that the function g on $[a, \infty)$ increases, that E is a set (not necessarily equipped with a metric), and that f is a real-valued function on $[a, \infty) \times E$. If

$$\int_a^\infty f(x, y)\,dg(x) \tag{6.33}$$

exists for every $y \in E$ and if

$$\sup_{y \in E} \left| \int_X^\infty f(x, y)\,dg(x) \right| \to 0 \quad as \quad X \to \infty,$$

then the integral (6.33) *is said to* converge uniformly *on E.*

The theorems of §5.3 may be recast so as to apply to improper integrals. The analogues of the most important of these (5.31 and 5.32) are easily obtained; the rest have to await further developments. (See §6.9 and exercise 8(g), 1.)

Theorem 6.34. (General principle of uniform convergence.) *Suppose that $\int_a^b f(x, y)\,dg(x)$ exists whenever $b > a$ and $y \in E$. A necessary and sufficient condition for $\int_a^\infty f(x, y)\,dg(x)$ to converge uniformly for $y \in E$ is that, given $\epsilon > 0$, there is an X such that*

$$\sup_{y \in E} \left| \int_{X_1}^{X_2} f(x, y)\,dg(x) \right| < \epsilon \quad whenever \quad X_2 > X_1 \geqslant X.$$

Proof. The necessity of the condition is obvious. To prove the sufficiency we first note that, by theorem 6.31, $\int_a^\infty f(x, y)\,dg(x)$ exists for every $y \in E$.

Now consider a particular $y \in E$. Since

$$\left| \int_{X_1}^{X_2} f(x, y)\,dg(x) \right| < \epsilon \quad whenever \quad X_2 > X_1 \geqslant X$$

and since $\int_{X_1}^\infty f(x, y)\,dg(x)$ exists, it follows that

$$\left| \int_{X_1}^\infty f(x, y)\,dg(x) \right| \leqslant \epsilon.$$

But this holds for every $y \in E$ and therefore

$$\sup_{y \in E} \left| \int_{X_1}^\infty f(x, y)\,dg(x) \right| \leqslant \epsilon \quad \text{for} \quad X_1 \geqslant X. \; |$$

Theorem 6.35. (Weierstrass's M-test.) *Suppose that* $\int_a^b f(x, y)\, dg(x)$ *exists whenever* $b > a$ *and* $y \in E$. *If there is a function* M *on* $[a, \infty)$ *such that*

$$|f(x, y)| \leqslant M(x) \quad for \quad a \leqslant x < \infty, y \in E$$

and $\int_a^\infty M(x)\, dg(x)$ *converges, then* $\int_a^\infty f(x, y)\, dg(x)$ *converges (absolutely and) uniformly on* E.

The proof follows almost immediately from theorems 6.31 and 6.34.

Illustration. If c is any positive number,

$$\int_0^\infty e^{-xy} \sin x\, dx$$

converges uniformly for $y \geqslant c$. To prove this we can use the M-test, for, when $x \geqslant 0$ and $y \geqslant c$,

$$|e^{-xy} \sin x| \leqslant e^{-xc}.$$

Theorem 6.33 shows that the convergence is not uniform for $y > 0$, since, if n is an integer,

$$\sup_{y>0} \left| \int_{n\pi}^{(n+1)\pi} e^{-xy} \sin x\, dx \right| \geqslant \sup_{y>0} (2e^{-(n+1)\pi y}) = 2.$$

In §8.7 we shall discuss functions ϕ defined, by means of uniformly convergent integrals, in the form

$$\phi(y) = \int_a^\infty f(x, y)\, dx.$$

Exercises 6(c)

1. Prove theorem 6.31.

2. Show that, if $f \in R(g; a, X)$ for every $X > a$ and $\int_a^\infty |f|\, dg$ exists, then $\int_a^\infty f\, dg$ exists.

3. Show that

$$\int_2^\infty \frac{\sin x}{\log x}\, dx$$

exists.

4. Prove that

$$\int_1^\infty \frac{\cos x}{x + \sin x}\, dx$$

converges non-absolutely.

5. Prove that, for any constant α,

$$\frac{(n+1)^{\alpha+1} - n^{\alpha+1}}{n^\alpha} \to \alpha + 1$$

as $n \to \infty$. Deduce that, if f is any positive decreasing function on $[1, \infty)$, then

$$\sum_{n=1}^{\infty} n^{\alpha} f(n) \quad \text{and} \quad \int_{1}^{\infty} x^{\alpha} f(x) dx$$

both converge or both diverge. (The case $\alpha = 0$ is the Maclaurin–Cauchy integral theorem proved in C1, 141.)

6. Let $f \in R(g; a, b)$. Show that $\lim\limits_{h \to 0+} \int_{a+h}^{b} f dg$ exists, but that, if g is discontinuous at a and $f(a) \neq 0$, then

$$\lim_{h \to 0+} \int_{a+h}^{b} f dg \neq \int_{a}^{b} f dg.$$

7. Evaluate

$$\int_{0+}^{\infty} \frac{\log x}{(x+1)^2} dx.$$

8. Find the values of α, β for which

$$\int_{0}^{\infty} \frac{x^{\alpha}}{1+x^{\beta}} dx$$

exists.

9. Let $c > 0$. Show that $\qquad \sum\limits_{n=0}^{\infty} x^{n+c-1} \log (1/x)$

converges uniformly in every interval $[\alpha, \beta]$ such that $0 < \alpha < \beta < 1$. By applying twice the analogue of theorem 6.33 for improper integrals of the second kind prove that

$$\sum_{n=0}^{\infty} \frac{1}{(n+c)^2} = \int_{0}^{1} \frac{x^{c-1} \log (1/x)}{1-x} dx.$$

10. Show that

$$\int_{0}^{\infty} e^{-x/y} dx$$

converges uniformly on any interval $(0, Y]$, but not on $(0, \infty)$.

11. Prove that

$$\int_{2}^{\infty} \frac{\cos x}{x+\sin y} dx$$

converges uniformly on $(-\infty, \infty)$.

12. Show that

$$\int_{0}^{\infty} \frac{y}{1+x^2y^2} dx$$

(i) converges for $-\infty < y < \infty$;
(ii) converges uniformly for $|y| \geq c > 0$;
(iii) does not converge uniformly for $-\infty < y < \infty$.

6.4. Functions of bounded variation

Our interest in these functions is twofold. First, they may be used as integrators in Riemann–Stieltjes integration. Secondly, there is a natural bond between them and rectifiable curves, i.e. curves of finite length. (See §8.1.)

Let f be a real function on the finite interval $[a, b]$. If $\mathscr{D}$ is the dissection of $[a, b]$ given by

$$a = x_0 < x_1 < \ldots < x_{n-1} < x_n = b,$$

put
$$V(\mathscr{D}, f) = \sum_{i=1}^{n} |f(x_i) - f(x_{i-1})|.$$

Definition. *If $V(\mathscr{D}, f)$ is bounded for all dissections $\mathscr{D}$ of $[a, b]$, the function f is said to be of* bounded variation *on $[a, b]$ and*

$$V_a^b(f) = \sup_{\mathscr{D}} V(\mathscr{D}, f)$$

is called the total variation *of f on $[a, b]$.*

Notes.

(i) If f is monotonic on $[a, b]$, then f is of bounded variation on $[a, b]$ and
$$V_a^b(f) = |f(b) - f(a)|.$$

(ii) If f is continuous on $[a, b]$ and has a bounded derivative in (a, b), then f is of bounded variation on $[a, b]$. (See exercise $6(d)$, 1.)

(iii) A function of bounded variation is bounded.

(iv) (a) A function of bounded variation need not be continuous (since, for instance, a monotonic function need not be continuous).

(b) A continuous function need not be of bounded variation. For let f be given by

$$f(x) = \begin{cases} x \cos \dfrac{\pi}{2x} & (x \neq 0), \\ 0 & (x = 0). \end{cases}$$

Then f is continuous everywhere, but is not of bounded variation on any interval including the origin. Consider, for instance, the interval $[0, 1]$. If $\mathscr{D}_n$ is given by the points

$$0, \frac{1}{2n}, \frac{1}{2n-1}, \frac{1}{2n-2}, \frac{1}{2n-3}, \ldots, \frac{1}{3}, \frac{1}{2}, 1,$$

then
$$V(\mathscr{D}_n, f) = \frac{1}{2n} + \frac{1}{2n} + \frac{1}{2n-2} + \frac{1}{2n-2} + \ldots + \frac{1}{2} + \frac{1}{2},$$

$$= 1 + \frac{1}{2} + \ldots + \frac{1}{n-1} + \frac{1}{n}.$$

Therefore $V(\mathscr{D}_n, f) \to \infty$ as $n \to \infty$ and so f is not of bounded variation on $[0, 1]$.

It is easily shown that elementary operations on functions of bounded variation yield functions of bounded variation.

Theorem 6.41. *If f, g are of bounded variation on $[a, b]$, then so are $|f|, f+g, fg$. If, in addition, $\inf\limits_{a \leqslant x \leqslant b} |f(x)| > 0$, then $1/f$ is also of bounded variation.*

Theorem 6.42.

(i) *If f is of bounded variation on $[a, c]$ and if $a < b < c$, then f is of bounded variation on $[a, b]$ and on $[b, c]$.*

(ii) *If f is of bounded variation on $[a, b]$ and on $[b, c]$, then f is of bounded variation on $[a, c]$ and*

$$V_a^c(f) = V_a^b(f) + V_b^c(f). \tag{6.41}$$

Proof.

(i) Let $\mathscr{D}_1$, $\mathscr{D}_2$ be any dissections of $[a, b]$, $[b, c]$ respectively and let $\mathscr{D}_0$ be the corresponding dissection of $[a, c]$. Then

$$V(\mathscr{D}_1, f) + V(\mathscr{D}_2, f) = V(\mathscr{D}_0, f) \leqslant V_a^c(f).$$

Therefore f is of bounded variation on $[a, b]$ and on $[b, c]$ and also

$$V_a^b(f) + V_b^c(f) \leqslant V_a^c(f). \tag{6.42}$$

(ii) Let $\mathscr{D}$ be any dissection of $[a, c]$ and let $\mathscr{D}^*$ be $\mathscr{D}$ with b as an additional point of division if it is not already a point of $\mathscr{D}$. Let $\mathscr{D}'$, $\mathscr{D}''$ be the dissections of $[a, b]$, $[b, c]$ induced by $\mathscr{D}^*$. Then

$$V(\mathscr{D}, f) \leqslant V(\mathscr{D}^*, f) = V(\mathscr{D}', f) + V(\mathscr{D}'', f) \leqslant V_a^b(f) + V_b^c(f).$$

Hence f is of bounded variation on $[a, c]$ and also

$$V_a^c(f) \leqslant V_a^b(f) + V_b^c(f). \tag{6.43}$$

(6.42) and (6.43) give (6.41). |

We can now establish a simple alternative characterization of functions of bounded variation.

Theorem 6.43. *A function is of bounded variation if and only if it is the difference of two increasing functions.*

Proof. An increasing function is of bounded variation and, by theorem 6.41, the difference of two such functions is also of bounded variation.

Now suppose that f is of bounded variation on $[a, b]$. In view of theorem 6.42, the *variation function* v_f defined on $[a, b]$ by the equations

$$v_f(x) = \begin{cases} 0 & \text{for } x = a, \\ V_a^x(f) & \text{for } a < x \leqslant b, \end{cases}$$

exists and is increasing. Next put

$$w_f = v_f - f.$$

When $a \leqslant x < y \leqslant b$,

$$w_f(y) - w_f(x) = \{v_f(y) - v_f(x)\} - \{f(y) - f(x)\}$$
$$= V_x^y(f) - \{f(y) - f(x)\} \geqslant 0.$$

Therefore w_f increases on $[a, b]$. Also $f = v_f - w_f$. |

Corollary. *Any discontinuities of a function of bounded variation are jumps. Consequently the set of such discontinuities is countable.*

If u is any increasing function, then (in the notation of theorem 6.43) $v_f + u$ and $w_f + u$ are increasing and $f = (v_f + u) - (w_f + u)$. So the representation of a function of bounded variation as the difference of two increasing functions is not unique.

Theorem 6.43 shows why functions of bounded variation may be made to serve as integrators. The next theorem is also relevant to Riemann–Stieltjes integration.

Theorem 6.44. *Let f be of bounded variation on $[a, b]$. Then v_f is continuous on the left (right) at the point ξ in $[a, b]$ if and only if f is continuous on the left (right) at ξ.*

Proof. We consider left continuity at a point ξ such that $a < \xi \leqslant b$.

First suppose that v_f is continuous on the left at ξ, i.e. that $v_f(x) \to v_f(\xi)$ as $x \to \xi-$. Since, for $a \leqslant x < \xi$,

$$|f(\xi) - f(x)| \leqslant V_x^\xi(f) = v_f(\xi) - v_f(x),$$

it follows that $f(x) \to f(\xi)$ as $x \to \xi-$.

Now suppose that f is continuous on the left at ξ. If ϕ, ψ are increasing functions such that $f = \phi - \psi$, then

$$\{\phi(\xi) - \phi(\xi-)\} - \{\psi(\xi) - \psi(\xi-)\} = f(\xi) - f(\xi-) = 0,$$

and so

$$\phi(\xi) - \phi(\xi-) = \psi(\xi) - \psi(\xi-) = k \geqslant 0.$$

Define the functions ϕ_1, ψ_1 on $[a, b]$ by

$$\phi_1(x) = \begin{cases} \phi(x)+k & (a \leqslant x < \xi), \\ \phi(x) & (\xi \leqslant x \leqslant b), \end{cases}$$

$$\psi_1(x) = \begin{cases} \psi(x)+k & (a \leqslant x < \xi), \\ \psi(x) & (\xi \leqslant x \leqslant b). \end{cases}$$

Then ϕ_1, ψ_1 are increasing functions on $[a, b]$, $f = \phi_1 - \psi_1$, and $\phi_1(\xi-) = \phi_1(\xi)$, $\psi_1(\xi-) = \psi_1(\xi)$. Also for $a \leqslant x < \xi$, we have

$$0 \leqslant v_f(\xi) - v_f(x) = V_x^\xi(f) \leqslant V_x^\xi(\phi_1) + V_x^\xi(\psi_1)$$
$$= \{\phi_1(\xi) - \phi_1(x)\} + \{\psi_1(\xi) - \psi_1(x)\}.$$

Since the right-hand side tends to 0 as $x \to \xi-$, $v_f(\xi-) = v_f(\xi)$.

Right continuity at a point ξ ($a \leqslant \xi < b$) is dealt with similarly. |

Corollary. *Let f be of bounded variation on $[a, b]$. Then the two increasing functions v_f, w_f (which are such that $f = v_f - w_f$) are continuous on the left (right) wherever f is continuous on the left (right).*

If f is of bounded variation on $[a, b]$, define the functions p_f and q_f on $[a, b]$ by

$$p_f = \tfrac{1}{2}\{v_f + f - f(a)\}, \quad q_f = \tfrac{1}{2}\{v_f - f + f(a)\},$$

so that $\qquad p_f(a) = q_f(a) = 0 \quad \text{and} \quad p_f - q_f = f - f(a).$ $\qquad$ (6.44)

It is easily seen that p_f and q_f are increasing functions. Also, by theorem 6.44, p_f and q_f are continuous on the left (right) wherever f is continuous on the left (right). The functions p_f and q_f are called, respectively, the *positive* and *negative variation functions* of f. (Exercise 6(d), 3 supplies the reason for this nomenclature.) These functions enable us to characterize all representations of a function of bounded variation as the difference of two increasing functions. We first show that p_f and q_f are the smallest increasing functions satisfying (6.44).

Lemma. *Let f be of bounded variation on $[a, b]$. If r, s are increasing functions on $[a, b]$ such that*

$$r(a) = s(a) = 0 \quad \text{and} \quad r - s = f - f(a),$$ $\qquad$ (6.45)

then $r(x) \geqslant p_f(x)$ and $s(x) \geqslant q_f(x)$ for $a \leqslant x \leqslant b$.

Proof. Take x such that $a < x \leqslant b$ and let $\mathscr{D}$ be any dissection of $[a, x]$. Then

$$
\begin{aligned}
V(\mathscr{D},f) &= \Sigma \, |f(x_i) - f(x_{i-1})| \\
&= \Sigma \, |\{r(x_i) - r(x_{i-1})\} - \{s(x_i) - s(x_{i-1})\}| \\
&\leqslant \Sigma \, \{r(x_i) - r(x_{i-1})\} + \Sigma \, \{s(x_i) - s(x_{i-1})\} \\
&= r(x) + s(x).
\end{aligned}
$$

Since this holds for all dissections of $[a, x]$,

$$
p_f(x) + q_f(x) = v_f(x) = V_a^x(f) \leqslant r(x) + s(x).
$$

It now follows from the second equations in (6.44) and (6.45) that $p_f(x) \leqslant r(x)$ and $q_f(x) \leqslant s(x)$. $\;|$

Theorem 6.45. *Let f be of bounded variation on $[a, b]$. If u is an increasing function on $[a, b]$ and $u(a) = 0$, then $r = p_f + u$ and $s = q_f + u$ are increasing functions on $[a, b]$ such that*

$$
r(a) = s(a) = 0 \quad \text{and} \quad r - s = f - f(a).
$$

Also every pair of increasing functions r, s satisfying these equations is of the form $r = p_f + u$, $s = q_f + u$, where u increases on $[a, b]$ and $u(a) = 0$.

Proof. The first statement is obvious. To prove the second, take any points α, β such that $a \leqslant \alpha < \beta \leqslant b$ and put

$$
p^* = p_f - p_f(\alpha), \quad q^* = q_f - q_f(\alpha), \quad r^* = r - r(\alpha), \quad s^* = s - s(\alpha).
$$

Then the restrictions of p^*, q^* to $[\alpha, \beta]$ are clearly the positive and negative variation functions of f on $[\alpha, \beta]$ and

$$
r^*(\alpha) = s^*(\alpha) = 0 \quad \text{and} \quad r^* - s^* = f - f(\alpha).
$$

Therefore, by the lemma, $p^*(\beta) \leqslant r^*(\beta)$, i.e.

$$
r(\beta) - p_f(\beta) \geqslant r(\alpha) - p_f(\alpha).
$$

But α, β are arbitrary and so $u = r - p_f = s - q_f$ increases on $[a, b]$. $\;|$

Exercises 6(*d*)

1. Prove Note (ii).

2. Prove theorem 6.41.

3. Let f be of bounded variation on $[a, b]$. For the dissection $\mathscr{D}$ of $[a, b]$ given by $a = x_0 < x_1 < \ldots < x_{n-1} < x_n = b$, put

$$P(\mathscr{D}, f) = \frac{1}{2} \sum_{i=1}^{n} \{|f(x_i) - f(x_{i-1})| + [f(x_i) - f(x_{i-1})]\},$$

$$Q(\mathscr{D}, f) = \frac{1}{2} \sum_{i=1}^{n} \{|f(x_i) - f(x_{i-1})| - [f(x_i) - f(x_{i-1})]\}$$

(so that $P(\mathscr{D}, f)$, $Q(\mathscr{D}, f)$ are the sums of the terms $|f(x_i) - f(x_{i-1})|$ for which $f(x_i) - f(x_{i-1}) \geq 0$ and ≤ 0 respectively). Let

$$P_a^b(f) = \sup P(\mathscr{D}, f), \quad Q_a^b(f) = \sup Q(\mathscr{D}, f),$$

where each supremum is taken over all dissections of $[a, b]$.
   Prove that

$$P_a^b(f) + Q_a^b(f) = V_a^b(f) \quad \text{and} \quad P_a^b(f) - Q_a^b(f) = f(b) - f(a)$$

and deduce that, for $a < x \leq b$,

$$P_a^x(f) = p_f(x) \quad \text{and} \quad Q_a^x(f) = q_f(x).$$

4. Let $f \in R(g; a, b)$ and define the function F on $[a, b]$ by

$$F(x) = \int_a^x f \, dg.$$

Prove that f is of bounded variation on $[a, b]$ and that

$$V_a^b(f) = \int_a^b |f| \, dg.$$

Show also that, if $f^+ = \frac{1}{2}(|f| + f)$, $f^- = \frac{1}{2}(|f| - f)$, then

$$P_a^b(f) = \int_a^b f^+ \, dg, \quad Q_a^b(f) = \int_a^b f^- \, dg.$$

5. Construct functions f_n $(n = 1, 2, \ldots)$ on an interval $[a, b]$ such that
   (i) (f_n) converges uniformly on $[a, b]$,
   (ii) each f_n is of bounded variation on $[a, b]$,
   (iii) $\lim f_n$ is not of bounded variation on $[a, b]$.

6.5. Integrators of bounded variation

A function g of bounded variation can be expressed in the form $g = r - s$, where r and s are increasing functions. It is therefore natural to define

$$\int_a^b f \, dg = \int_a^b f \, dr - \int_a^b f \, ds$$

when both the integrals on the right exist. But the expression of g as a difference of increasing functions is not unique and so we have to show that our definition of $\int_a^b f \, dg$ is independent of the particular pair r, s. Let r_1, s_1 be another pair of increasing functions such that $g = r_1 - s_1$ and

$$\int_a^b f\,dr_1, \quad \int_a^b f\,ds_1$$

both exist. Then, since $r + s_1 = r_1 + s$, theorem 6.13 gives

$$\int_a^b f\,dr + \int_a^b f\,ds_1 = \int_a^b f\,dr_1 + \int_a^b f\,ds,$$

which is what was needed.

When f is continuous, $\int_a^b f\,dr$ and $\int_a^b f\,ds$ exist for every pair r, s. But when f has a discontinuity, then, even if f is integrable with respect to some pair r, s, there exist others with respect to which it is not integrable. For, given r and s, the increasing function u may be so chosen that $r + u$ and $s + u$ have a discontinuity at the same point as f. We therefore call f integrable with respect to g if there is *some* pair r, s such that $\int_a^b f\,dr$ and $\int_a^b f\,ds$ both exist.

The functions p_g, q_g are such that $p_g - q_g = g - g(a)$ (not g). Nevertheless it is clear that, if f is integrable with respect to p_g, q_g, then f is integrable with respect to g and

$$\int_a^b f\,dg = \int_a^b f\,dp_g - \int_a^b f\,dq_g, \tag{6.51}$$

for $p_g + g(a), q_g$ is a decomposition of g into increasing functions. The next theorem shows, conversely, that, if f is integrable with respect to g, then $\int_a^b f\,dp_g$ and $\int_a^b f\,dq_g$ exist so that $\int_a^b f\,dg$ may, in fact, be defined by (6.51).

Theorem 6.51. *Let g be of bounded variation on the interval $[a, b]$. If f is integrable over $[a, b]$ with respect to g, then f is also integrable with respect to each of the variation functions v_g, p_g, q_g of g.*

Proof. Let r, s be two increasing functions on $[a, b]$ such that $g = r - s$ and f is integrable with respect to each of r, s. Given $\epsilon > 0$, there is a dissection $\mathscr{D}$ of $[a, b]$ such that

$$S(\mathscr{D}, f, r) - s(\mathscr{D}, f, r) < \epsilon, \quad S(\mathscr{D}, f, s) - s(\mathscr{D}, f, s) < \epsilon \tag{6.52}$$

and

$$V(\mathscr{D}, g) > V_a^b(g) - \epsilon = v_g(b) - \epsilon. \tag{6.53}$$

Since

$$|g(\beta) - g(\alpha)| \leqslant [r(\beta) - r(\alpha)] + [s(\beta) - s(\alpha)],$$

the two inequalities of (6.52) give (in the usual notation)

$$\sum_{i=1}^{n} (M_i - m_i)|g(x_i) - g(x_{i-1})| < 2\epsilon. \qquad (6.54)$$

Also, if $K = \sup_{a \leqslant x \leqslant b} |f(x)|$, then by (6.53),

$$0 \leqslant \sum_{i=1}^{n} (M_i - m_i)\{[v_g(x_i) - v_g(x_{i-1})] - |g(x_i) - g(x_{i-1})|\}$$

$$\leqslant 2K \sum_{i=1}^{n} \{[v_g(x_i) - v_g(x_{i-1})] - |g(x_i) - g(x_{i-1})|\}$$

$$= 2K\{v_g(b) - V(\mathscr{D}, g)\} < 2K\epsilon. \qquad (6.55)$$

From (6.54) and (6.55) we now obtain

$$\sum_{i=1}^{n} (M_i - m_i)\{v_g(x_i) - v_g(x_{i-1})\} < 2(K+1)\epsilon$$

and so $f \in R(v_g; a, b)$.

As $p_g = \frac{1}{2}\{v_g + g - g(a)\} = \frac{1}{2}\{v_g + r - s - g(a)\}$ is an increasing function, exercise $6(a)$, 4 shows that $f \in R(p_g; a, b)$. Similarly $f \in R(q_g; a, b)$. |

Most of the theorems on Riemann–Stieltjes integration with an increasing integrator are easily extended to integration with respect to a function of bounded variation. Where there are differences, integrators of bounded variation usually lead to a more symmetrical result.

The integral is linear in the integrator as well as the integrand: *If k_1, k_2 are any constants,*

$$\int_a^b (k_1 f_1 + k_2 f_2)\,dg = k_1 \int_a^b f_1\,dg + k_2 \int_a^b f_2\,dg \qquad (6.56)$$

and $$\int_a^b f\,d(k_1 g_1 + k_2 g_2) = k_1 \int_a^b f\,dg_1 + k_2 \int_a^b f\,dg_2, \qquad (6.57)$$

where in each case the existence of both integrals on the right implies the existence of the integral on the left. This is the analogue of theorem 6.13. We have now removed the restriction $k \geqslant 0$ which was necessary in part (ii)(a) of that theorem since integrators had to be increasing. Theorem 6.14 takes the same form as before, but the analogue of theorem 6.15 does not have quite the appearance of the original.

Theorem 6.52. *If g is of bounded variation on $[a, b]$ and $f \in R(g; a, b)$, then $|f| \in R(v_g; a, b)$ and*

$$\left| \int_a^b f \, dg \right| \leqslant \int_a^b |f| \, dv_g.$$

Proof. By theorems 6.51 and 6.15, $|f|$ is integrable with respect to v_g, p_g, q_g. Using also the relation $v_g = p_g + q_g$ and theorem 6.13 we have

$$\left| \int_a^b f \, dg \right| \leqslant \left| \int_a^b f \, dp_g \right| + \left| \int_a^b f \, dq_g \right|$$

$$\leqslant \int_a^b |f| \, dp_g + \int_a^b |f| \, dq_g \ = \int_a^b |f| \, dv_g. \ |$$

Corollary. $\qquad \left| \int_a^b f \, dg \right| \leqslant \sup_{a \leqslant x \leqslant b} |f(x)| \cdot V_a^b(g).$

Sometimes $\int_a^b f \, dv_g$ is written $\int_a^b f |dg|$. In this notation the conclusion of theorem 6.52 has the form

$$\left| \int_a^b f \, dg \right| \leqslant \int_a^b |f| \, |dg|.$$

Theorem 6.16 holds also for integrators of bounded variation. The new version of theorem 6.17 appears with a slightly relaxed hypothesis.

Theorem 6.53. (Change of variable.) *Let $f \in R(g; a, b)$, where g is of bounded variation. If ϕ is a continuous, strictly monotonic function such that $\phi(\alpha) = a$ and $\phi(\beta) = b$, then $f \circ \phi$ is integrable with respect to $g \circ \phi$ and*

$$\int_a^b f \, dg = \int_\alpha^\beta (f \circ \phi) \, d(g \circ \phi).$$

Proof. Let $g = r - s$, where r, s are increasing functions and $f \in R(r; a, b) \cap R(s; a, b)$. When ϕ increases the result follows immediately from theorem 6.17. When ϕ decreases (so that $\beta < \alpha$), then $-(r \circ \phi)$ and $-(s \circ \phi)$ are increasing functions and the argument of theorem 6.17 shows that

$$\int_a^b f \, dr = \int_\beta^\alpha (f \circ \phi) \, d(-(r \circ \phi)), \quad \int_a^b f \, ds = \int_\beta^\alpha (f \circ \phi) \, d(-(s \circ \phi)).$$

Hence, using also (6.57) we have

$$\int_a^b f \, dg = \int_\beta^\alpha (f \circ \phi) \, d(-(g \circ \phi)) = \int_\alpha^\beta (f \circ \phi) \, d(g \circ \phi). \ |$$

Corollary. *The function ϕ of the theorem may be taken to be* piecewise *strictly monotonic when $f \in R\,(g;\,c,\,d)$, where*

$$c = \inf_{\alpha \leqslant t \leqslant \beta} \phi(t), \quad d = \sup_{\alpha \leqslant t \leqslant \beta} \phi(t).$$

Lastly we combine the analogues of theorems 6.24 and 6.25.

Theorem 6.54. (Integration by parts.) *If the functions f and g are of bounded variation on $[a,\,b]$ and have no common left and no common right discontinuities, then $f \in R\,(g;\,a,\,b)$, $g \in R\,(f;\,a,\,b)$ and*

$$\int_a^b f\,dg + \int_a^b g\,df = f(b)g(b) - f(a)g(a). \tag{6.58}$$

Proof. By theorem 6.44, the pairs of functions p_f and p_g, p_f and q_g, q_f and p_g, q_f and q_g have no common left and no common right discontinuities. It follows from theorem 6.24 that $f \in R\,(g;\,a,\,b)$ and $g \in R\,(f;\,a,\,b)$. Theorem 6.25 and simple algebra now yield (6.58). |

Improper integrals with respect to integrators of bounded variation can be defined in the now familiar way, but the manipulation of these integrals may be awkward. The comparison principle does not hold for them (see exercise $6(e)$, 6); nor does absolute convergence imply convergence (see exercise $6(e)$, 7(ii).)

Exercises 6(e)

1. Prove theorem 6.21 B with the restriction $g(x) \geqslant 0$ removed. (For the corresponding extension of theorem 6.21 A see theorem 6.85.)

2. The function f is continuous on $[a,\,b]$ and g is of bounded variation. Denoting by $\mathscr{D}$ a dissection
$$a = x_0 < x_1 < \ldots < x_{n-1} < x_n = b$$
and by ξ_i any point in $[x_{i-1},\,x_i]$, prove that
$$\sum_{i=1}^{n} f(\xi_i)\{g(x_i) - g(x_{i-1})\} \to \int_a^b f\,dg$$
as $\mu(\mathscr{D}) \to 0$. (See exercise $6(a)$, 9.)

3. Let $f_n \to f$ uniformly on $[a,\,b]$ and let g be of bounded variation. Prove that if, for $n = 1, 2, \ldots, f_n \in R\,(g;\,a,\,b)$, then $f \in R\,(g;\,a,\,b)$ and
$$\int_a^b f_n\,dg \to \int_a^b f\,dg.$$
(See exercise $6(a)$, 13.)

4. Show that the first mean value theorem (see exercise $6(a)$, 8) does not hold for the RS integral with an integrator of bounded variation.

5. Let f be a positive decreasing function on $[a, \infty)$ with $\lim_{x \to \infty} f(x) = 0$. Also let g be bounded on $[a, \infty)$ and of bounded variation on every interval $[a, X]$ $(X > a)$. Show that, if $f \in R(g; a, X)$ for every $X > a$, then $\int_a^{\infty} f\,dg$ exists.

6. Let f on $[\pi, \infty)$ be defined by

$$f(x) = \begin{cases} 0 & \text{for } (2n-1)\pi < x < 2n\pi, \\ 1/x & \text{for } 2n\pi \leqslant x \leqslant (2n+1)\pi. \end{cases}$$

Show that $\int_\pi^{\infty} \frac{1}{x}\,d(\cos x)$ exists, but that $\int_\pi^{\infty} f(x)\,d(\cos x)$ does not.

7. (i) Let the functions f, g on $[a, \infty)$ be such that, for every $X > a$, g is of bounded variation on $[a, X]$ and $f \in R(g; a, X)$. Show that, if $\int_a^{\infty} |f|\,dv_g$ exists, then $\int_a^{\infty} f\,dg$ exists.

(ii) Prove that $\int_\pi^{\infty} \left| \frac{\sin x}{x} \right|\,d(\cos x)$ exists, but that $\int_\pi^{\infty} \frac{\sin x}{x}\,d(\cos x)$ does not.

6.6. The Riesz representation theorem

Every linear function from R^m to R^n may be represented by an $n \times m$ matrix (§3.2). This is one of a group of theorems on the general form of linear functions associated with various vector spaces. The most famous of these results is the Riesz representation theorem which characterizes certain linear functions on the space $C[a, b]$ (of real functions continuous on the interval $[a, b]$). It fittingly illustrates the power and importance of the Riemann–Stieltjes integral. (It is not, however, subsequently used in this book and the proof, which is difficult, may be omitted without fear of repercussions.)

Theorem 6.61.

(i) *To every continuous linear function* $\lambda \colon C[a, b] \to R^1$ *there corresponds a function g of bounded variation on $[a, b]$ such that, for every $f \in C[a, b]$,*

$$\lambda(f) = \int_a^b f\,dg. \tag{6.61}$$

(ii) *Conversely, given a function g of bounded variation on $[a, b]$,* (6.61) *determines a continuous linear function* $\lambda \colon C[a, b] \to R^1$.

The second part follows immediately from (6.56) and theorem 6.52, corollary. The proof of the first part is easy if one can use the fact that a bounded linear function on $C[a, b]$ can be extended, without change of norm, to a linear function on $B[a, b]$ (the space of bounded real functions on $[a, b]$). Unfortunately we cannot prove this with the tools at our command and we have to content ourselves with a less interesting result which, nevertheless, is sufficient to establish the theorem.

The sequence (f_n) of real functions on $[a, b]$ is *increasing* if, for every x in $[a, b]$, the sequence $(f_n(x))$ increases; and *uniformly bounded* if there is a constant K such that $|f_n(x)| < K$ for $a \leqslant x \leqslant b$ and all n. We denote by $C_+[a, b]$ the vector space of all (bounded) real functions on $[a, b]$ which are of the form

$$\lim f_n - \lim h_n,$$

where (f_n) and (h_n) are uniformly bounded, increasing sequences from $C[a, b]$.

Lemma. *A real valued, continuous linear function on $C[a, b]$ may be extended, without change of norm, to the space $C_+[a, b]$.*

Proof. Let $\lambda \colon C[a, b] \to R^1$ be a bounded linear function.

(i) Consider a uniformly bounded, increasing sequence (f_n) from $C[a, b]$. The sequence converges and its limit function f is bounded. We begin by showing that $(\lambda(f_n))$ converges. For $k \geqslant 2$, let

$$\epsilon_k = \begin{cases} 1 & \text{if } \lambda(f_k) - \lambda(f_{k-1}) \geqslant 0, \\ -1 & \text{if } \lambda(f_k) - \lambda(f_{k-1}) < 0. \end{cases}$$

Then

$$\sum_{k=2}^{n} |\lambda(f_k) - \lambda(f_{k-1})| = \sum_{k=2}^{n} \epsilon_k \{\lambda(f_k) - \lambda(f_{k-1})\}$$

$$= \lambda \left\{ \sum_{k=2}^{n} \epsilon_k (f_k - f_{k-1}) \right\}$$

$$\leqslant \|\lambda\| \sup_{a \leqslant x \leqslant b} \left| \sum_{k=2}^{n} \epsilon_k \{f_k(x) - f_{k-1}(x)\} \right|$$

$$\leqslant \|\lambda\| \sup_{a \leqslant x \leqslant b} \sum_{k=2}^{n} \{f_k(x) - f_{k-1}(x)\}$$

$$= \|\lambda\| \sup_{a \leqslant x \leqslant b} \{f_n(x) - f_1(x)\}$$

$$\leqslant \|\lambda\| \sup_{a \leqslant x \leqslant b} \{f(x) - f_1(x)\},$$

i.e. the partial sums of $\sum_{k=2}^{\infty} |\lambda(f_k) - \lambda(f_{k-1})|$ are bounded. It follows that the series

$$\lambda(f_1) + \sum_{k=2}^{\infty} \{\lambda(f_k) - \lambda(f_{k-1})\}$$

converges absolutely and so its sequence of partial sums, i.e. $(\lambda(f_n))$ converges.

(ii) Next, let (f_n^*) be any other uniformly bounded, increasing sequence from $C[a, b]$ which also has the limit f. For $n = 1, 2, \ldots,$ put

$$\phi_n = f_n - \frac{1}{n}, \quad \phi_n^* = f_n^* - \frac{1}{n}.$$

Then (ϕ_n) and (ϕ_n^*) are both *strictly* increasing sequences and clearly

$$\lim \lambda(f_n) = \lim \lambda(\phi_n), \quad \lim \lambda(f_n^*) = \lim \lambda(\phi_n^*).$$

Take an integer m_1. There is an integer n_1 such that $\phi_{m_1}(x) < \phi_{n_1}^*(x)$ for all x in $[a, b]$. To prove this let F_k be the closed set of points x for which $\phi_{m_1}(x) \geqslant \phi_k^*(x)$. Since the F_k form a contracting sequence, if none of them were empty, there would be a point ξ common to them all (exercise 3(f), 6) and we should have

$$\phi_{m_1}(\xi) \geqslant \lim \phi_k^*(\xi) = f(\xi).$$

But this is impossible since (ϕ_n) strictly increases, and so there is an n_1 such that $F_{n_1} = \varnothing$. Continuing in this way we obtain two sequences of integers (m_i), (n_j) such that

$$\phi_{m_1} < \phi_{n_1}^* < \phi_{m_2} < \phi_{n_2}^* < \ldots$$

By (i), the sequence

$$\lambda(\phi_{m_1}), \quad \lambda(\phi_{n_1}), \quad \lambda(\phi_{m_2}), \quad \lambda(\phi_{n_2}), \ldots$$

converges and its limit is equal to both $\lim \lambda(\phi_n)$ and $\lim \lambda(\phi_n^*)$. Hence

$$\lim \lambda(f_n) = \lim \lambda(f_n^*).$$

We can now, without ambiguity, define $\lambda(f)$ to be $\lim \lambda(f_n)$.

(iii) It is clear that, if f, h are limits of uniformly bounded, increasing sequences from $C[a, b]$, then so are cf, where c is a non-negative constant, and $f+h$. Moreover

$$\lambda(cf) = c\lambda(f), \quad \lambda(f+h) = \lambda(f)+\lambda(h).$$

We still have to define $\lambda(f-h)$. To do so we suppose that f^*, h^* are of the same type as f, h and that

$$f-h = f^*-h^*.$$

Since $f+h^* = f^*+h$, we have

$$\lambda(f)+\lambda(h^*) = \lambda(f^*)+\lambda(h),$$

i.e.

$$\lambda(f)-\lambda(h) = \lambda(f^*)-\lambda(h^*).$$

Hence we may define $\lambda(f-h)$ unambiguously as $\lambda(f)-\lambda(h)$. It follows immediately that λ is linear on the vector space $C_+[a, b]$.

(iv) Any member of $C_+[a, b]$ is of the form $f-h$, where f, h are the limits of uniformly bounded, increasing sequences (f_n), (h_n) from $C[a, b]$. To prove that the norm of the extension of λ to $C_+[a, b]$ is still $\|\lambda\|$ we have to show that

$$|\lambda(f-h)| \leqslant \|\lambda\|K,$$

where $K = \sup|f(x)-h(x)|$. We define the functions

$$f_n^* \ (n = 1, 2, ...)$$

by the equations

$$f_n^*(x) = \begin{cases} f_n(x) & \text{if } |f_n(x)-h_n(x)| \leqslant K, \\ h_n(x)+K & \text{if } f_n(x)-h_n(x) > K, \\ h_n(x)-K & \text{if } f_n(x)-h_n(x) < -K. \end{cases}$$

Clearly

$$|f_n^*(x)-h_n(x)| \leqslant K \quad (a \leqslant x \leqslant b; n = 1, 2, ...).$$

It is also easy to see that the functions f_n^* are continuous and that $f_n^* \to f$. Finally, by considering the various forms that $f_{n+1}^*(x)-f_n^*(x)$ may take we can show that $f_n^*(x)$ increases with n. (Exercise 6(f), 1.) Hence

$$|\lambda(f-h)| = |\lim \lambda(f_n^*) - \lim \lambda(h_n)| = \lim |\lambda(f_n^*-h_n)| \leqslant \|\lambda\|K. \ |$$

Proof of theorem 6.61 (i). The functions $h_a, h_x \ (a < x < b), h_b$ given by

$$h_a(t) = 0 \quad \text{for } a \leqslant t \leqslant b,$$

$$h_x(t) = \begin{cases} 1 & \text{for } a \leqslant t \leqslant x, \\ 0 & \text{for } x < t \leqslant b, \end{cases}$$

$$h_b(t) = 1 \quad \text{for } a \leqslant t \leqslant b$$

all belong to $C_+[a, b]$. This is obvious for h_a and h_b; in the case of h_x we need only remark that $-h_x = \lim \phi_n$, where, for $n > 1/(b-x)$, $\phi_n(t) = -1$ in $[a, x]$, $\phi_n(t) = 0$ in $[x+(1/n), b]$, and ϕ_n is linear in $[x, x+(1/n)]$ (in the elementary sense—see note at foot of p. 66).

Let g on $[a, b]$ be the function such that

$$g(x) = \lambda(h_x) \quad (a \leqslant x \leqslant b);$$

in particular, $g(a) = 0$.

If a dissection $\mathscr{D}$ of $[a, b]$ has the points of division

$$a = t_0 < t_1 < \ldots < t_{n-1} < t_n = b,$$

put $\epsilon_i = 1$ or -1 according as $g(t_i) - g(t_{i-1}) \geqslant 0$ or < 0. Then

$$V(\mathscr{D}, g) = \sum_{i=1}^{n} \epsilon_i \{g(t_i) - g(t_{i-1})\} = \lambda \left\{ \sum_{i=1}^{n} \epsilon_i (h_{t_i} - h_{t_{i-1}}) \right\} \leqslant \|\lambda\|,$$

since

$$\left| \sum_{i=1}^{n} \epsilon_i [h_{t_i}(x) - h_{t_{i-1}}(x)] \right| = 1$$

for $a \leqslant x \leqslant b$. Thus g is of bounded variation on $[a, b]$.

Now let f be any member of $C[a, b]$ and take $\epsilon > 0$. Since f is integrable with respect to the variation functions p_g, q_g of g and is also uniformly continuous on $[a, b]$, there is a dissection $\mathscr{D}$ of $[a, b]$ given by

$$a = x_0 < x_1 < \ldots < x_{n-1} < x_n = b$$

which has the following properties:

$$S(\mathscr{D}, f, p_g) - \int_a^b f \, dp_g < \epsilon, \quad S(\mathscr{D}, f, q_g) - \int_a^b f \, dq_g < \epsilon; \quad (6.62)$$

and

$$M_i - m_i < \epsilon \quad (i = 1, \ldots, n), \quad (6.63)$$

where M_i, m_i denote the supremum and infimum, respectively, of f in $[x_{i-1}, x_i]$. By (6.62),

$$\left| \sum_{i=1}^{n} M_i \{g(x_i) - g(x_{i-1})\} - \int_a^b f \, dg \right| < 2\epsilon; \quad (6.64)$$

while, by (6.63), the step function

$$\phi = \sum_{i=1}^{n} M_i (h_{x_i} - h_{x_{i-1}})$$

is such that

$$\sup_{a \leqslant x \leqslant b} |f(x) - \phi(x)| < \epsilon$$

and therefore

$$|\lambda(f) - \lambda(\phi)| = |\lambda(f - \phi)| < \|\lambda\| \epsilon. \quad (6.65)$$

But

$$\lambda(\phi) = \sum_{i=1}^{n} M_i \{\lambda(h_{x_i}) - \lambda(h_{x_{i-1}})\} = \sum_{i=1}^{n} M_i \{g(x_i) - g(x_{i-1})\}$$

and so, by (6.64),

$$\left| \lambda(\phi) - \int_a^b f \, dg \right| < 2\epsilon.$$

Using also (6.65) we finally have

$$\left| \lambda(f) - \int_a^b f \, dg \right| < (\|\lambda\| + 2)\epsilon,$$

which proves (6.61). $\blacksquare$

Exercises 6(f)

1. Prove that the sequence (f_n^*), defined in part (iv) of the proof of the lemma, is increasing.

2. A necessary and sufficient condition for the function ϕ of bounded variation on $[a, b]$ to be such that

$$\int_a^b f \, d\phi = 0$$

for all $f \in C[a, b]$ is that $\phi(b) = \phi(a)$ and $\phi(x) = \phi(a)$ when x is a point of continuity of ϕ in (a, b).

Prove the necessity of this condition by considering the functions f, $f_{x, \delta}$ defined as follows:

   (i) $f(t) = 1$ for $a \leqslant t \leqslant b$;

   (ii) when x is a point of continuity of ϕ in (a, b) and $0 < \delta < b - x$, $f_{x, \delta}(t) = 1$ for $a \leqslant t \leqslant x$, $f_{x, \delta}(t) = 0$ for $x + \delta \leqslant t \leqslant b$ and $f_{x, \delta}$ is linear in $[x, x + \delta]$.

Prove the sufficiency of the condition by using exercise 6(e), 2.

3. Let c be a fixed point in the interval (a, b) and define the linear function $\lambda : C[a, b] \to R^1$ by
$$\lambda(f) = f(c).$$

Enumerate all the functions g which have the property that

$$\lambda(f) = \int_a^b f \, dg \quad (f \in C[a, b]).$$

6.7. The Riemann integral

In the rest of this chapter we develop the theory of the Riemann integral. Our first task is to reconcile the definition of the integral given in C1 and the definition in terms of upper and lower integrals given in §6.1. The instrument for this is Darboux's theorem, a proof of which was indicated in C1 (125). The reader is reminded that, given a dissection $\mathscr{D}$ of an interval $[a, b]$, $\mu(\mathscr{D})$, the mesh of $\mathscr{D}$, is the length of the largest subinterval; and $S(\mathscr{D}, f)$, $s(\mathscr{D}, f)$ denote the upper and lower Riemann sums of a bounded function f on $[a, b]$.

Theorem 6.71. (Darboux.) *If the function f on $[a, b]$ is bounded, then, as $\mu(\mathscr{D}) \to 0$,*

$$\text{(i)} \quad S(\mathscr{D}, f) \to \overline{\int_a^b} f, \qquad \text{(ii)} \quad s(\mathscr{D}, f) \to \underline{\int_a^b} f.$$

Proof. Let $\mathscr{D}$ be the dissection of $[a, b]$ given by

$$a = x_0 < x_1 < \ldots < x_{n-1} < x_n = b$$

and let $\mathscr{D}'$ be the dissection with additional points of division $x^{(1)}, x^{(2)}, \ldots, x^{(k)}$, where

$$x_{i-1} < x^{(1)} < x^{(2)} < \ldots < x^{(k)} < x_i.$$

Denote, as usual, by m, M, m_i, M_i the infima and suprema of f on $[a, b]$ and on $[x_{i-1}, x_i]$ respectively. Also let $M^{(1)}$, $M^{(2)}$, ..., $M^{(k+1)}$ be the suprema of f on the $k+1$ additional subintervals of $\mathscr{D}'$. We then have

$$0 \leqslant S(\mathscr{D}) - S(\mathscr{D}')$$

$$= M_i(x_i - x_{i-1})$$

$$- \{M^{(1)}(x^{(1)} - x_{i-1}) + M^{(2)}(x^{(2)} - x^{(1)}) + \ldots + M^{(k+1)}(x_i - x^{(k)})\}$$

$$\leqslant M_i(x_i - x_{i-1}) - m_i(x_i - x_{i-1}) \leqslant (M - m)(x_i - x_{i-1}).$$

Hence, if $\mathscr{D}^*$ is any dissection obtained from $\mathscr{D}$ by adding new points of division,

$$0 \leqslant S(\mathscr{D}) - S(\mathscr{D}^*) \leqslant (M - m) \Sigma^* (x_i - x_{i-1}), \qquad (6.71)$$

where Σ^* ranges over the intervals containing new points of division.

Now, given $\epsilon > 0$, there is a $\mathscr{D}_0$ such that

$$S(\mathscr{D}_0) < \overline{\int_a^b} f + \epsilon.$$

Let n_0 be the number of subintervals of $\mathscr{D}_0$ and take δ so that

$$0 < \delta < \epsilon/n_0.$$

Let $\mathscr{D}_\delta$ be a dissection with $\mu(\mathscr{D}_\delta) < \delta$ and let $\mathscr{D}^*$ be the dissection with all the points of division of $\mathscr{D}_0$ and of $\mathscr{D}_\delta$. Then

$$S(\mathscr{D}^*) \leqslant S(\mathscr{D}_0)$$

and, by (6.71),

$$0 \leqslant S(\mathscr{D}_\delta) - S(\mathscr{D}^*) \leqslant (M - m)n_0\delta < \epsilon(M - m),$$

since at most n_0 points are added to $\mathscr{D}_\delta$ to form $\mathscr{D}^*$. Therefore

$$\overline{\int_a^b} f \leqslant S(\mathscr{D}_\delta) \leqslant S(\mathscr{D}^*) + \epsilon(M - m)$$

$$\leqslant S(\mathscr{D}_0) + \epsilon(M - m) < \overline{\int_a^b} f + \epsilon(1 + M - m),$$

i.e. $$\overline{\int_a^b} f \leqslant S(\mathscr{D}_\delta) < \overline{\int_a^b} f + \epsilon(1 + M - m).$$

This proves (i). The proof of (ii) is similar. |

Theorem 6.71 does not extend unrestrictedly to the Riemann–Stieltjes integral; but the analogue *is* valid if the integrand or the integrator is continuous. (See exercises 6(*a*), 9; 6(*g*), 1 and 2.)

The next theorem establishes the equivalence of the two definitions of the Riemann integral. In it we use the standard notation

$$a = x_0 < x_1 < \ldots < x_{n-1} < x_n = b$$

for a dissection $\mathscr{D}$ of $[a, b]$ and, for $i = 1, \ldots, n$, we allow ξ_i to range over all points of $[x_{i-1}, x_i]$.

Theorem 6.72.

(i) *If f is integrable over $[a, b]$, then*

$$\sum_{i=1}^{n} f(\xi_i)(x_i - x_{i-1}) \to \int_a^b f \quad as \quad \mu(\mathscr{D}) \to 0.$$

(ii) *If* $\quad \sum_{i=1}^{n} f(\xi_i)(x_i - x_{i-1}) \to \sigma \quad as \quad \mu(\mathscr{D}) \to 0,$

then f is integrable over $[a, b]$ and $\int_a^b f = \sigma$.

Proof.

(i) For every set of ξ_i,

$$s(\mathscr{D}) \leqslant \sum_{i=1}^{n} f(\xi_i)(x_i - x_{i-1}) \leqslant S(\mathscr{D})$$

and, by theorem 6.71, $s(\mathscr{D}) \to \int_a^b f$, $S(\mathscr{D}) \to \int_a^b f$ as $\mu(\mathscr{D}) \to 0$.

(ii) Take any $\epsilon > 0$. Given $\mathscr{D}$, we choose, for each i, the points ξ_i, ξ_i' so that

$$f(\xi_i) > M_i - \epsilon \quad and \quad f(\xi_i') < m_i + \epsilon.$$

Then $\qquad \sum_{i=1}^{n} f(\xi_i)(x_i - x_{i-1}) > S(\mathscr{D}) - \epsilon(b-a)$

and $\qquad \sum_{i=1}^{n} f(\xi_i')(x_i - x_{i-1}) < s(\mathscr{D}) + \epsilon(b-a),$

so that

$$\sum_{i=1}^{n} f(\xi_i')(x_i - x_{i-1}) - \epsilon(b-a) < s(\mathscr{D}) \leqslant \underline{\int_a^b} f$$

$$\leqslant \overline{\int_a^b} f \leqslant S(\mathscr{D}) < \sum_{i=1}^{n} f(\xi_i)(x_i - x_{i-1}) + \epsilon(b-a).$$

Letting $\mu(\mathscr{D}) \to 0$, we obtain

$$\sigma - \epsilon(b-a) \leqslant \underline{\int_a^b} f \leqslant \overline{\int_a^b} f \leqslant \sigma + \epsilon(b-a). \;|$$

Exercises 6(g)

1. Show that the analogue of Darboux's theorem holds for the RS integral with a *continuous* increasing integrator.

2. Give an example of an increasing function g on an interval $[a, b]$ and a function $f \in R(g; a, b)$ such that $\lim_{\mu(\mathscr{D}) \to 0} S(\mathscr{D}, f, g)$ does not exist.

3. Let f be continuous on $[a, b]$. Show that f is of bounded variation on $[a, b]$ if and only if $\lim_{\mu(\mathscr{D}) \to 0} V(\mathscr{D}, f)$ exists; and that, if the limit exists, then it is $V_a^b(f)$.

4. Give an example of a function f which is of bounded variation on $[a, b]$, but such that $\lim_{\mu(\mathscr{D}) \to 0} V(\mathscr{D}, f)$ does not exist.

5. Deduce from theorem 6.71 that, if $a < b < c$,

$$\overline{\int_a^b} f + \overline{\int_b^c} f = \overline{\int_a^c} f \quad \text{and} \quad \underline{\int_a^b} f + \underline{\int_b^c} f = \underline{\int_a^c} f.$$

6. The function f is continuous and strictly increasing on $[a, b]$. Show that, if $f(a) = \alpha, f(b) = \beta$, then

$$\int_a^b f + \int_\alpha^\beta f^{-1} = b\beta - a\alpha.$$

6.8. Content

It is easy to see that a bounded function with only a finite number of discontinuities is Riemann integrable (see exercise 6(h), 1). We show in the present section that this remains true for suitably restricted infinite sets of discontinuities. The extension of our results to multiple integrals will play an essential part in chapter 8.

Definition. Let X be a set. If E is any subset of X, the function $\chi_E: X \to R^1$ given by

$$\chi_E(x) = \begin{cases} 1 & \text{for } x \in E, \\ 0 & \text{for } x \in X - E \end{cases}$$

is called the indicator *(or* characteristic*) function of E.*

In this section we always take X to be R^1. If, for instance, $E = [\alpha, \beta]$,

$$\chi_{[\alpha, \beta]}(x) = \begin{cases} 1 & \text{for } \alpha \leqslant x \leqslant \beta, \\ 0 & \text{for } x < \alpha \text{ or } x > \beta. \end{cases}$$

Now let $[a, b] \supset [\alpha, \beta]$. Then

$$\int_a^b \chi_{[\alpha, \beta]}(x) \, dx = \beta - \alpha. \tag{6.81}$$

(In particular, $\int_a^b \chi_E = 0$ when E has just one point in $[a, b]$.) For, if $a < \alpha < \beta < b$ and $\mathscr{D}$ is a dissection of $[a, b]$ containing α, β as poi nts of division, then

$$\beta - \alpha = s(\mathscr{D}) \leqslant \underline{\int_a^b} \chi_{[\alpha,\beta]} \leqslant \overline{\int_a^b} \chi_{[\alpha,\beta]} \leqslant S(\mathscr{D}) = \beta - \alpha + \delta + \delta', \quad (6.82)$$

where δ, δ' are the lengths of the two subintervals of $\mathscr{D}$ adjoining $[\alpha, \beta]$; and, if $a = \alpha$ or $\beta = b$, then (6.82) holds with $\delta = 0$ or $\delta' = 0$.

It is proved similarly, or from (6.81), that

$$\int_a^b \chi_{(\alpha,\beta)} = \int_a^b \chi_{(\alpha,\beta]} = \int_a^b \chi_{[\alpha,\beta)} = \beta - \alpha. \quad (6.83)$$

Also, if E is an arbitrary bounded set in R^1 and $[a, b], [a', b']$ are any intervals containing E, then, by the lemma preceding theorem 6.16,

$$\overline{\int_a^b} \chi_E = \overline{\int_{a'}^{b'}} \chi_E, \quad \underline{\int_a^b} \chi_E = \underline{\int_{a'}^{b'}} \chi_E.$$

This consideration and the identities (6.81) and (6.83) lead to an extension of the notion of the length of an interval.

Definition. *Let E be any bounded set in R^1. The* outer Jordan content *$\bar{c}(E)$ of E is the common value of $\overline{\int_a^b} \chi_E$ for all intervals $[a, b]$ containing E; the* inner Jordan content *$\underline{c}(E)$ is the common value of $\overline{\int_a^b} \chi_E$. If $\underline{c}(E) = \bar{c}(E)$, this number is called the* Jordan content *of E and is denoted by $c(E)$.*

If the set E has content, then χ_E is integrable over any interval $[a, b]$ containing E and $\int_a^b \chi_E$ has the same value for all such intervals.

We have shown that the content of a finite interval is its length. An example of a bounded set that does not have content is the set of rational points in a finite interval.

Although we have defined content in general, in this section we are mainly interested in sets of zero content.

Theorem 6.81. *The bounded set E has zero content if and only if, given $\epsilon > 0$, there is a finite number of open intervals, or a finite number of closed intervals, which cover E and whose total length is less than ϵ.*

Proof. Let $E \subset [a, b]$.

(i) Suppose that $c(E) = \int_a^b \chi_E = 0$. Then, given $\epsilon > 0$, there is a $\mathscr{D}$

such that $S(\mathscr{D}, \chi_E) < \epsilon$; and $S(\mathscr{D}, \chi_E)$ is clearly the sum of the lengths of the closed intervals $[x_{i-1}, x_i]$ which contain points of E.

Each $[x_{i-1}, x_i]$ may be replaced by a slightly larger open interval so that the total length of the open intervals is still less than ϵ.

(ii) Suppose that, given $\epsilon > 0$, there are open intervals $I_1, ..., I_m$ covering E and of total length less than ϵ. Since a pair of overlapping intervals may be replaced by a single open interval, we may assume that $I_1, ..., I_m$ are pairwise disjoint. Their end points (between a and b) determine a dissection $\mathscr{D}$ of $[a, b]$. Since $[a, b] - (I_1 \cup ... \cup I_m)$ contains no points of E, $S(\mathscr{D}, \chi_E) < \epsilon$. Hence $c(E) = 0$.

If we had started with closed intervals, we could have enlarged them slightly into open intervals. |

That every finite set has zero content followed already from (6.81). It is now easy to see that a bounded infinite set with no more than a finite number of limit points has zero content. (See exercise 6(h), 3.) For instance, if $E = \{1, \frac{1}{2}, \frac{1}{3}, ...\}$, $c(E) = 0$.

Theorem 6.82. *If f is bounded on $[a, b]$ and the set E of points of discontinuity of f has zero content, then f is integrable over $[a, b]$.*

Proof. Given $\epsilon > 0$, there is a finite number of pairwise disjoint open intervals $I_1, ..., I_m$, of total length less than ϵ, such that

$$G = I_1 \cup ... \cup I_m \supset E.$$

The function f is continuous on

$$H = [a, b] - G = [a, b] \cap G'$$

and, since H is bounded and closed, f is uniformly continuous on H. Thus there is a $\delta > 0$ such that $|f(x) - f(x')| < \epsilon$ whenever $x, x' \in H$ and $|x - x'| < \delta$.

Now let $\mathscr{D}$ be a dissection of $[a, b]$ which contains all the end points (in $[a, b]$) of $I_1, ..., I_m$ and all of whose subintervals in H are of length less than δ. If the subintervals of $\mathscr{D}$ are $[x_{i-1}, x_i]$ $(i = 1, ..., n)$ and M_i, m_i, M, m have their usual meaning,

$$S(\mathscr{D}) - s(\mathscr{D}) = \sum_{i=1}^{n} (M_i - m_i)(x_i - x_{i-1})$$
$$= (\Sigma_G + \Sigma_H)(M_i - m_i)(x_i - x_{i-1}),$$

where Σ_G and Σ_H run over the subintervals in $\overline{G}$ and H, respectively. Clearly

$$\Sigma_G(M_i - m_i)(x_i - x_{i-1}) \leqslant \Sigma_G(M - m)(x_i - x_{i-1}) < (M - m)\epsilon$$

and
$$\Sigma_H (M_i - m_i)(x_i - x_{i-1}) \leqslant \Sigma_H \epsilon(x_i - x_{i-1}) \leqslant \epsilon(b-a).$$

Thus
$$S(\mathscr{D}) - s(\mathscr{D}) < (M - m + b - a)\epsilon. \;|$$

Theorem 6.83. *Suppose that f and g are bounded on $[a, b]$ and that $f(x) = g(x)$ except when x belongs to a set of zero content. Then either both f and g are integrable over $[a, b]$ or neither is integrable; and, if f, g are integrable,*

$$\int_a^b f = \int_a^b g.$$

Proof. All that is necessary is to show that $h = f - g$ is integrable and $\int_a^b h = 0$.

Let E be the set of points ξ such that $h(\xi) \neq 0$ and let M, m be the supremum and infimum of h on $[a, b]$. Then

$$m\chi_E(x) \leqslant h(x) \leqslant M\chi_E(x)$$

for $a \leqslant x \leqslant b$ and so, since $c(E) = 0$,

$$0 = m \int_a^b \chi_E = \int_a^b m\chi_E \leqslant \underline{\int_a^b} h \leqslant \overline{\int_a^b} h \leqslant \int_a^b M\chi_E = M \int_a^b \chi_E = 0. \;|$$

The theorem shows that, as far as integration is concerned, sets of zero content are immaterial. It is therefore legitimate, and sometimes convenient, to leave a function undefined in a set of zero content. For instance

$$\int_0^1 \sin \frac{1}{x} \, dx$$

has a definite meaning without the integrand being defined at $x = 0$: whatever value is assigned to the integrand at 0, the integral exists and its value remains unaltered. The next theorem similarly illustrates the utility of this feature of the Riemann integral.

Theorem 6.84. *Let the function f be continuous on the closed interval $[a, b]$ and differentiable in the open interval (a, b). If f' (which need not be defined at a or b) is integrable over $[a, b]$, then*

$$\int_a^b f' = f(b) - f(a).$$

(This is a stronger result than theorem 7.63 of C1.)

Proof. Let $\mathscr{D}$ be the dissection of $[a, b]$ given by

$$a = x_0 < x_1 < \ldots < x_{n-1} < x_n < b.$$

By the mean value theorem,

$$f(b) - f(a) = \sum_{i=1}^{n} \{f(x_i) - f(x_{i-1})\} = \sum_{i=1}^{n} f'(\xi_i)(x_i - x_{i-1}),$$

where $x_{i-1} < \xi_i < x_i$ $(i = 1, \ldots, n)$. Since f' is integrable, theorem 6.72(i) shows that the right-hand side tends to $\int_a^b f'$ as $\mu(\mathscr{D}) \to 0$. |

In the above proof the use of theorem 6.72(i) (and so of Darboux's theorem) is convenient, but not essential. (See exercise 6(h), 10.)

Theorem 6.84 means that, if f' is integrable, then f is an indefinite integral of f'. It therefore shows that theorem 6.21A is included in 6.21B. Equally, the following convenient extension of theorem 6.21A is only a special case of exercise 6(e), 1.

Theorem 6.85. *Let $f \in R\,(a, b)$ and let g be continuous and of bounded variation on $[a, b]$. If g is differentiable in (a, b) and $g' \in R\,(a, b)$, then $f \in R\,(g; a, b)$ and*

$$\int_a^b f\,dg = \int_a^b fg'.$$

Exercises 6(h)

1. The function f on $[a, b]$ is bounded and is continuous except at a finite number of points. Prove, without using the notion of content, that f is integrable over $[a, b]$.

2. Show that, if $[\alpha, \beta] \subset [a, b]$, then

$$\int_a^b \chi_{(\alpha, \beta)} = \beta - \alpha.$$

3. Show that a bounded infinite set with a finite number of limit points has zero content.

4. Given any set E, let $E^{(1)}$ be the set of limit points of E and, for $n = 2, 3, \ldots$, define $E^{(n)}$ to be the set of limit points of $E^{(n-1)}$. Show that, if E is bounded and $E^{(n)}$ is finite, then $c(E) = 0$.

5. Let E_1, E_2 be subsets of R^1. Prove that

$$\chi_{E_1 \cup E_2}(x) = \max\,(\chi_{E_1}(x), \chi_{E_2}(x)), \quad \chi_{E_1 \cap E_2}(x) = \min\,(\chi_{E_1}(x), \chi_{E_2}(x)).$$

Assuming the identities

$$\max\,(a, b) = \tfrac{1}{2}\{(a+b) + |a-b|\}, \quad \min\,(a, b) = \tfrac{1}{2}\{(a+b) - |a-b|\},$$

show that, if $c(E_1)$, $c(E_2)$ exist, then so do $c(E_1 \cup E_2)$, $c(E_1 \cap E_2)$ and

$$c(E_1 \cup E_2) + c(E_1 \cap E_2) = c(E_1) + c(E_2).$$

6. Prove that, if $c(E_1) = \ldots = c(E_n) = 0$, then $c(E_1 \cup \ldots \cup E_n) = 0$; and show that the relation $c(E_n) = 0$ $(n = 1, 2, \ldots)$ does not imply that $c(E_1 \cup E_2 \cup \ldots) = 0$.

7. The functions f, g are bounded in $[a, b]$ and $f(x) = g(x)$ in $[a, b]$ except for x in a set of zero content. Prove that

$$\overline{\int_a^b} f = \overline{\int_a^b} g, \quad \underline{\int_a^b} f = \underline{\int_a^b} g.$$

8. For $x \neq 1, \tfrac{1}{2}, \tfrac{1}{3}, \ldots,$

$$f(x) = \sum_{n=1}^{\infty} \frac{1}{n^2} \sin \frac{n}{1 - nx}.$$

Show that f is integrable over $[0, 1]$.

9. The function f on $[a, b]$ is given by

$$f(x) = \begin{cases} 0 & \text{if } x \text{ is irrational,} \\ 1/q & \text{if } x = p/q, \end{cases}$$

where p, q are coprime integers. By finding a sequence of integrable functions converging uniformly to f, or otherwise, prove that f is integrable over $[a, b]$. (It was shown in §3.1 that f is discontinuous at all rational points of $[a, b]$. The set of discontinuities is therefore not of zero content, which shows that the converse of theorem 6.82 is false.)

10. Prove theorem 6.84 without using any results from §6.7.

11. The function f on $[0, 1]$ is given by

$$f(x) = \begin{cases} 0 & \text{for } x = 0, \\ x \sin (\log x) & \text{for } 0 < x \leqslant 1. \end{cases}$$

Prove that f is differentiable in $(0, 1]$, but not at 0, that f' is integrable over $[0, 1]$ and that

$$\int_0^1 f' = 0.$$

12. Show that the function f on $[0, 1]$ given by

$$f(x) = \begin{cases} 0 & \text{for } x = 0, \\ x^2 \sin \dfrac{\pi}{2x^2} & \text{for } 0 < x \leqslant 1, \end{cases}$$

is differentiable in $[0, 1]$. Show also that f' is not integrable over $[0, 1]$, but that $\displaystyle\int_{0+}^1 f'$ exists and is 1.

13. The function f has a continuous derivative in the interval $[a, b]$. Show that f is the difference of two increasing functions with continuous derivatives in $[a, b]$.

6.9. Some manipulative theorems

The first theorem of this section is the most useful version of the formula for integration by parts.

Theorem 6.91. Let f, g be integrable over $[a, b]$ and let F, G be indefinite integrals of f, g respectively, i.e. functions given by

$$F(x) = \int_a^x f(t)\,dt + K, \quad G(x) = \int_a^x g(t)\,dt + L \quad (a \leqslant x \leqslant b),$$

where K, L are arbitrary constants. Then

$$\int_a^b fG + \int_a^b Fg = F(b)\,G(b) - F(a)\,G(a). \tag{6.91}$$

Proof. Since F, G are continuous and of bounded variation (exercises $6(a)$, 10; $6(d)$, 4), the result easily follows from the theorem on integration by parts for the RS integral (theorem 6.54) and from theorem 6.21 B as extended in exercise $6(e)$, 1. However a direct proof is also desirable.

We first note that the continuity of F and G ensures the existence of the two integrals in (6.91).

For any dissection $\mathscr{D}$ of $[a, b]$ given by

$$a = x_0 < x_1 < \ldots < x_{n-1} < x_n = b,$$

we have

$$\sum_{i=1}^n G(x_i) \int_{x_{i-1}}^{x_i} f(x)\,dx + \sum_{i=1}^n F(x_{i-1}) \int_{x_{i-1}}^{x_i} g(x)\,dx$$

$$= \sum_{i=1}^n G(x_i)\{F(x_i) - F(x_{i-1})\} + \sum_{i=1}^n F(x_{i-1})\{G(x_i) - G(x_{i-1})\}$$

$$= \sum_{i=1}^n \{F(x_i)\,G(x_i) - F(x_{i-1})\,G(x_{i-1})\}$$

$$= F(b)\,G(b) - F(a)\,G(a). \tag{6.92}$$

We now show that

$$\sum_{i=1}^n G(x_i) \int_{x_{i-1}}^{x_i} f(x)\,dx \to \int_a^b G(x)f(x)\,dx \quad \text{as} \quad \mu(\mathscr{D}) \to 0. \tag{6.93}$$

The function G is uniformly continuous on $[a, b]$, i.e. given $\epsilon > 0$, there is a $\delta > 0$ such that

$$|G(x) - G(x')| < \epsilon \quad \text{whenever} \quad x, x' \in [a, b] \quad \text{and} \quad |x - x'| < \delta.$$

Thus, when $\mu(\mathscr{D}) < \delta$,

$$\left| \int_a^b G(x) f(x)\, dx - \sum_{i=1}^n G(x_i) \int_{x_{i-1}}^{x_i} f(x)\, dx \right|$$

$$= \left| \sum_{i=1}^n \int_{x_{i-1}}^{x_i} \{G(x) - G(x_i)\} f(x)\, dx \right|$$

$$\leqslant \sum_{i=1}^n \int_{x_{i-1}}^{x_i} |G(x) - G(x_i)|\, |f(x)|\, dx$$

$$\leqslant \sum_{i=1}^n \int_{x_{i-1}}^{x_i} \epsilon |f(x)|\, dx = \epsilon \int_a^b |f(x)|\, dx.$$

This proves (6.93). Similarly

$$\sum_{i=1}^n F(x_{i-1}) \int_{x_{i-1}}^{x_i} g(x)\, dx \to \int_a^b F(x) g(x)\, dx \quad \text{as} \quad \mu(\mathscr{D}) \to 0. \quad (6.94)$$

Letting $\mu(\mathscr{D}) \to 0$ in (6.92) and using (6.93) and (6.94) we obtain (6.91). |

The formula (6.91) with continuous f and g was proved in C1 (130) and has been used by us on previous occasions. The proof under these assumptions is very simple, but the strength of theorem 6.91 lies in the hypothesis of mere integrability.

Theorem 6.92. (Change of variable.) *Let f be integrable over $[a, b]$ and suppose that ϕ on $[\alpha, \beta]$ satisfies the following conditions:*
(i) *ϕ increases on $[\alpha, \beta]$ and*

$$\phi(\alpha) = a, \quad \phi(\beta) = b;$$

(ii) *ϕ' exists in and is integrable over $[\alpha, \beta]$.*
Then

$$\int_a^b f = \int_\alpha^\beta (f \circ \phi) \phi'$$

i.e.
$$\int_a^b f(x)\, dx = \int_\alpha^\beta f\{\phi(t)\} \phi'(t)\, dt.$$

Proof. When ϕ is *strictly* increasing this is easily deduced from theorems 6.17 and 6.21 A. A proof of the theorem as stated, and independent of the Stieltjes integral, proceeds as follows.

Let $\mathscr{D}$ be the dissection of $[a, b]$ given by

$$a = x_0 < x_1 < \dots < x_{n-1} < x_n = b.$$

For each i, choose a t_i such that $\phi(t_i) = x_i$. The t_i are not necessarily unique (unless ϕ is strictly increasing), but we always have $t_{i-1} < t_i$. For simplicity, take $t_0 = \alpha$, $t_n = \beta$. Note that, by theorem 6.84,

$$x_i - x_{i-1} = \phi(t_i) - \phi(t_{i-1}) = \int_{t_{i-1}}^{t_i} \phi'(t)\,dt.$$

As usual, let

$$m_i = \inf_{x_{i-1} \leqslant x \leqslant x_i} f(x) = \inf_{t_{i-1} \leqslant t \leqslant t_i} f\{\phi(t)\},$$

$$M_i = \sup_{x_{i-1} \leqslant x \leqslant x_i} f(x) = \sup_{t_{i-1} \leqslant t \leqslant t_i} f\{\phi(t)\}.$$

Then, since $\phi'(t) \geqslant 0$ for $\alpha \leqslant t \leqslant \beta$,

$$s(\mathcal{D}, f) = \sum_{i=1}^{n} m_i(x_i - x_{i-1}) = \sum_{i=1}^{n} \int_{t_{i-1}}^{t_i} m_i \phi'(t)\,dt$$

$$\leqslant \sum_{i=1}^{n} \underline{\int}_{t_{i-1}}^{t_i} f\{\phi(t)\}\phi'(t)\,dt = \underline{\int}_{\alpha}^{\beta} f\{\phi(t)\}\phi'(t)\,dt$$

$$\leqslant \overline{\int}_{\alpha}^{\beta} f\{\phi(t)\}\phi'(t)\,dt = \sum_{i=1}^{n} \overline{\int}_{t_{i-1}}^{t_i} f\{\phi(t)\}\phi'(t)\,dt$$

$$\leqslant \sum_{i=1}^{n} \int_{t_{i-1}}^{t_i} M_i \phi'(t)\,dt = \sum_{i=1}^{n} M_i(x_i - x_{i-1}) = S(\mathcal{D}, f).$$

Now $s(\mathcal{D}, f)$ and $S(\mathcal{D}, f)$ can be taken arbitrarily close to $\int_a^b f$. Hence $\int_{\alpha}^{\beta} (f \circ \phi)\phi'$ exists and is equal to $\int_a^b f$. |

A similar theorem holds for a decreasing function ϕ and therefore also for a piecewise monotonic ϕ, provided that f is integrable over $[c, d]$, where

$$c = \inf_{\alpha \leqslant t \leqslant \beta} \phi(t), \quad d = \sup_{\alpha \leqslant t \leqslant \beta} \phi(t).$$

There are other sets of conditions for changing the variable. For instance, when f is continuous, ϕ need not be piecewise monotonic (see exercise 6(i), 1).

If f is integrable over $[a, b]$, then there is a number λ between the infimum and the supremum of f on $[a, b]$ such that

$$\int_a^b f = \lambda(b - a).$$

This rather obvious statement is usually called the *first mean value theorem*. However the *second mean value theorem* is more interesting.

Theorem 6.93. (Second mean value theorem.) *Suppose that g is integrable over* $[a, b]$.

(i) *If* f *is non-negative and decreases in* $[a, b]$, *then there is a* ξ *in* $[a, b]$ *such that*

$$\int_a^b fg = f(a) \int_a^\xi g.$$

(ii) *If* f *is non-negative and increases in* $[a, b]$, *then there is an* η *in* $[a, b]$ *such that*

$$\int_a^b fg = f(b) \int_\eta^b g.$$

Proof.

(i) Let
$$K = \sup_{a \leqslant x \leqslant b} |g(x)|,$$

so that $g(x) + K \geqslant 0$ for $a \leqslant x \leqslant b$. Then, for any dissection $\mathscr{D}$ given by
$$a = x_0 < x_1 < \ldots < x_{n-1} < x_n = b,$$

$$\int_a^b f(g+K) = \sum_{i=1}^n \int_{x_{i-1}}^{x_i} f(g+K) \leqslant \sum_{i=1}^n f(x_{i-1}) \int_{x_{i-1}}^{x_i} (g+K)$$

$$= \sum_{i=1}^n f(x_{i-1}) \int_{x_{i-1}}^{x_i} g + K \sum_{i=1}^n f(x_{i-1})(x_i - x_{i-1}). \quad (6.95)$$

If
$$G(x) = \int_a^x g,$$

then, since $G(a) = 0$, f decreases, and $f(b) \geqslant 0$,

$$\sum_{i=1}^n f(x_{i-1}) \int_{x_{i-1}}^{x_i} g = \sum_{i=1}^n f(x_{i-1})\{G(x_i) - G(x_{i-1})\}$$

$$= \sum_{i=1}^n f(x_{i-1}) G(x_i) - \sum_{i=0}^{n-1} f(x_i) G(x_i)$$

$$= \sum_{i=1}^n \{f(x_{i-1}) - f(x_i)\} G(x_i) + f(b) G(b)$$

$$\leqslant \sum_{i=1}^n \{f(x_{i-1}) - f(x_i)\} \sup_{a \leqslant x \leqslant b} G(x) + f(b) G(b)$$

$$= \{f(a) - f(b)\} \sup_{a \leqslant x \leqslant b} G(x) + f(b) G(b)$$

$$\leqslant f(a) \sup_{a \leqslant x \leqslant b} G(x).$$

Also
$$\sum_{i=1}^n f(x_{i-1})(x_i - x_{i-1}) \to \int_a^b f \quad \text{as} \quad \mu(\mathscr{D}) \to 0.$$

Hence, letting $\mu(\mathscr{D}) \to 0$ in (6.95) we obtain

$$\int_a^b f(g+K) \leqslant f(a) \sup_{a \leqslant x \leqslant b} G(x) + K \int_a^b f,$$

i.e.

$$\int_a^b fg \leqslant f(a) \sup_{a \leqslant x \leqslant b} G(x). \qquad (6.96)$$

Applying (6.96) with g replaced by $-g$ we have

$$\int_a^b f(-g) \leqslant f(a) \sup_{a \leqslant x \leqslant b} \{-G(x)\} = -f(a) \inf_{a \leqslant x \leqslant b} G(x),$$

so that

$$\int_a^b fg \geqslant f(a) \inf_{a \leqslant x \leqslant b} G(x). \qquad (6.97)$$

Since G is continuous, (6.96) and (6.97) show that there is a ξ in $[a, b]$ such that

$$\int_a^b fg = f(a) G(\xi) = f(a) \int_a^\xi g.$$

(ii) This can be proved by an analogous argument or by applying (i) to $f(b-x)$ and $g(b-x)$. For then there is a ξ in $[0, b-a]$ such that

$$\int_0^{b-a} f(b-x)g(b-x)dx = f(b) \int_0^\xi g(b-x)dx$$

and theorem 6.92 (with decreasing ϕ) gives

$$\int_a^b f(t)g(t)dt = f(b) \int_\eta^b g(t)dt,$$

where $\eta = b-\xi$. |

Corollary. *If g is integrable over $[a, b]$ and f is monotonic in $[a, b]$, then there is a ξ in $[a, b]$ such that*

$$\int_a^b fg = f(a) \int_a^\xi g + f(b) \int_\xi^b g.$$

Proof. Suppose that f decreases. Then $f-f(b)$ is non-negative and decreasing so that, by the theorem, there is a ξ in $[a, b]$ such that

$$\int_a^b \{f-f(b)\}g = \{f(a)-f(b)\} \int_a^\xi g,$$

i.e.

$$\int_a^b fg = f(a) \int_a^\xi g + f(b) \left\{ \int_a^b g - \int_a^\xi g \right\} = f(a) \int_a^\xi g + f(b) \int_\xi^b g. |$$

Note. Theorem 6.93 may be extended to RS integrals with continuous integrators. (See exercise 6(i), 4.)

The second mean value theorem is often used to establish the existence or uniform convergence of improper integrals. The analogues for integrals of Dirichlet's and Abel's tests may be deduced from it (see exercises $6(i)$, 5, 6), but the following example shows that a direct attack is simpler.

Example. The integral

$$\int_0^\infty \frac{\sin xy}{x}\, dx. \tag{6.98}$$

Convergence of the integral. The integral converges for all values of y. To prove this we first remark that the integral is not improper at 0, since the integrand is continuous and bounded on $(0, \infty)$. Also we need only consider the case $y \neq 0$. Then, when $X_2 > X_1 > 0$,

$$\left| \int_{X_1}^{X_2} \frac{\sin xy}{x}\, dx \right| = \left| \frac{1}{X_1} \int_{X_1}^{\xi} \sin xy\, dx \right| \leqslant \frac{2}{X_1|y|} \tag{6.99}$$

and $2/(X_1|y|)$ may be made arbitrarily small for all sufficiently large X_1. The general principle of convergence (theorem 6.31) therefore shows that (6.98) exists.

We had previously proved the existence of the integral (with $y = 1$) by using integration by parts. In fact both methods are frequently applicable.

Uniform convergence of the integral. The integral converges uniformly in any set of the form $|y| \geqslant c > 0$. For, by (6.99),

$$\sup_{|y| \geqslant c} \left| \int_{X_1}^{X_2} \frac{\sin xy}{x}\, dx \right| \leqslant \frac{2}{X_1 c}$$

and we can now use the general principle of uniform convergence (theorem 6.34).

The convergence of (6.98) is not uniform in any interval containing the point $y = 0$. Consider, for instance, the interval $[0, 1]$. Take any $X \geqslant 1$ and let $u = \pi/(4X)$. Then

$$\int_X^{2X} \frac{\sin xu}{x}\, dx \geqslant \int_X^{2X} \frac{\sin \frac{1}{4}\pi}{x}\, dx = \frac{1}{\sqrt{2}} \log 2.$$

Therefore, for every $X \geqslant 1$,

$$\sup_{0 \leqslant y \leqslant 1} \left| \int_X^{2X} \frac{\sin xy}{x}\, dx \right| \geqslant \frac{1}{\sqrt{2}} \log 2$$

and theorem 6.34 shows that the integral does not converge uniformly on $[0, 1]$.

Value of the integral. We prove that

$$\int_0^\infty \frac{\sin xy}{x}\, dx = \begin{cases} -\tfrac{1}{2}\pi & \text{for } y < 0, \\ 0 & \text{for } y = 0, \\ \tfrac{1}{2}\pi & \text{for } y > 0. \end{cases}$$

Given y, denote the value of the integral by $\phi(y)$. It is clear that $\phi(0) = 0$ and $\phi(-y) = -\phi(y)$. Moreover, if $y > 0$,

$$\int_0^X \frac{\sin xy}{x}\, dx = \int_0^{Xy} \frac{\sin t}{t}\, dt \to \phi(1) \quad \text{as} \quad X \to \infty,$$

and so $\phi(y) = \phi(1)$. Hence we need only show that $\phi(1) = \tfrac{1}{2}\pi$.

For $n = 1, 2, \ldots$, put

$$u_n = \int_0^{\frac{1}{2}\pi} \sin 2nx \cot x\, dx, \quad v_n = \int_0^{\frac{1}{2}\pi} \frac{\sin 2nx}{x}\, dx.$$

The following three steps establish the required identity:

(i) $u_n = \tfrac{1}{2}\pi$;   (ii) $\lim_{n\to\infty} (u_n - v_n) = 0$;   (iii) $\lim_{n\to\infty} v_n = \phi(1)$.

(i) is proved by induction. We have

$$u_1 = \int_0^{\frac{1}{2}\pi} 2\cos^2 x\, dx = \tfrac{1}{2}\pi;$$

and

$$u_{n+1} - u_n = \int_0^{\frac{1}{2}\pi} 2\cos(2n+1)x \cos x\, dx$$

$$= \int_0^{\frac{1}{2}\pi} \{\cos(2n+2)x + \cos 2nx\}\, dx = 0.$$

(ii) Since $x^{-1} - \cot x$ and $-(x^{-2} - \operatorname{cosec}^2 x)$ converge as $x \to 0+$, by theorem 6.84 the former is an indefinite integral of the latter. Hence, by theorem 6.91,

$$v_n - u_n = \int_0^{\frac{1}{2}\pi} \sin 2nx \left(\frac{1}{x} - \cot x\right) dx$$

$$= \left[-\frac{\cos 2nx}{2n}\left(\frac{1}{x} - \cot x\right)\right]_0^{\frac{1}{2}\pi} - \int_0^{\frac{1}{2}\pi} \frac{\cos 2nx}{2n}\left(\frac{1}{x^2} - \operatorname{cosec}^2 x\right) dx$$

$$\to 0 \text{ as } n \to \infty.$$

(iii) follows from the identity

$$v_n = \int_0^{n\pi} \frac{\sin t}{t}\, dt.$$

Exercises 6(i)

1. Suppose that the functions f, ϕ satisfy the following conditions:

(i) ϕ' exists in and is integrable over an interval $[\alpha, \beta]$,

(ii) f is continuous on $[c, d]$ where c, d are, respectively, the infimum and supremum of ϕ in $[\alpha, \beta]$.

Show that, if $\phi(\alpha) = a$ and $\phi(\beta) = b$, then

$$\int_a^b f = \int_\alpha^\beta (f \circ \phi)\,\phi'.$$

(In C1, 130, ϕ' was taken to be continuous.)

2. *Frullani's integral.* The function f on $(0, \infty)$ is integrable over every finite interval $[0, X)$ and $f(x) \to l$ as $x \to 0+$, $f(x) \to L$ as $x \to \infty$. Show that, if $a, b > 0$, then

$$\int_{0+}^\infty \frac{f(ax) - f(bx)}{x}\,dx = (L - l)\log\frac{a}{b}.$$

3. Prove the following versions of the second mean value theorem and its corollary:

Under the conditions of theorem 6.93 (i) there is a ξ in $[a, b]$ such that

$$\int_a^b fg = f(a+)\int_a^\xi g.$$

Under the conditions of theorem 6.93 (ii) there is an η in $[a, b]$ such that

$$\int_a^b fg = f(b-)\int_\eta^b g.$$

Under the conditions of theorem 6.93, corollary there is a ξ in $[a, b]$ such that

$$\int_a^b fg = f(a+)\int_a^\xi g + f(b-)\int_\xi^b g.$$

4. Show that the second mean value theorem holds for RS integrals with continuous integrators of bounded variation: if g is continuous and of bounded variation on $[a, b]$ and if f is non-negative and decreasing on $[a, b]$, then there is a number ξ in $[a, b]$ such that

$$\int_a^b f\,dg = f(a)\int_a^\xi dg = f(a)\{g(\xi) - g(a)\}.$$

Give an example to show that the result becomes false if the condition that g is continuous is omitted.

5. The functions f, g on $[a, \infty)$ are such that

(i) $\displaystyle\int_a^x f$ exists and is bounded for $x \geqslant a$,

(ii) g is monotonic on $[a, \infty)$ and $g(x) \to 0$ as $x \to \infty$.

Prove that $\displaystyle\int_a^\infty fg$ exists.

6. The functions f, g on $[a, \infty)$ are such that

(i) $\displaystyle\int_a^\infty f$ exists,

(ii) g is monotonic and bounded on $[a, \infty)$.

Prove that $\displaystyle\int_a^\infty fg$ exists.

7. Show that

$$\int_1^\infty \sin(x^\alpha)\,dx$$

converges when $|\alpha| > 1$ and diverges when $|\alpha| \leqslant 1$.

8. Show that all the following integrals converge:

$$(a)\ \int_0^\infty \frac{\sin x}{x + \sin x}\,dx, \qquad (b)\ \int_0^\infty \frac{x \sin x}{(x+1)(x+\sin x)}\,dx,$$

$$(c)\ \int_2^\infty \frac{x \cos(x^2)}{\log x}\,dx, \qquad (d)\ \int_0^\infty \cos x \sin\frac{1}{x}\,dx.$$

9. Show that

$$\int_0^\infty \frac{\sin x}{x+y}\,dx$$

converges uniformly for $y \geqslant 0$.

10. Show that

$$\int_0^\infty e^{-xy} \sin\frac{x}{y}\,dx$$

converges uniformly for $y > 0$.

NOTES ON CHAPTER 6

§6.1. Riemann's theory of integration was published posthumously in 1867. He defined the integral as $\lim \Sigma f(\xi_i)(x_i - x_{i-1})$ (cf. theorem 6.72); upper and lower sums and integrals were introduced subsequently by Darboux. Stieltjes's idea of integrating one function with respect to another was initially hidden in a long and now famous paper of 1894 on continued fractions. The importance of the integral was recognized only fifteen years later when F. Riesz used it in his representation theorem (6.61).

§6.3. It is easy to see that, if $\displaystyle\int_{-\infty}^\infty f\,dg$ exists, then

$$\lim_{X\to\infty} \int_{-X}^{X} f\,dg \qquad\qquad (*)$$

exists and is equal to $\displaystyle\int_{-\infty}^\infty f\,dg$. On the other hand, $(*)$ may exist without $\displaystyle\int_{-\infty}^\infty f\,dg$ existing (e.g. when $f(x) = g(x) = x$). In this case $(*)$ is called the *principal value* of $\displaystyle\int_{-\infty}^\infty f\,dg$ and is denoted by $(P) \displaystyle\int_{-\infty}^\infty f\,dg$.

Similarly, if f is unbounded near the point $u \in (a, b)$, $(P) \int_a^b f \, dg$ is defined as

$$\lim_{h \to 0+} \left(\int_a^{u-h} + \int_{u+h}^b \right) f \, dg$$

when this limit exists, but the improper integral $\int_a^b f \, dg$ does not. For instance the improper integral $\int_{-1}^1 \frac{1}{x} \, dx$ does not exist, but $(P) \int_{-1}^1 \frac{1}{x} \, dx$ does.

§6.4. The concept of a function of bounded variation is due to Jordan (1893) who used it for the characterization of curves of finite length. (See §8.1.)

§6.6. Riesz's theorem 6.61 has been followed by many similar theorems involving other types of spaces. For instance let l be the space of complex sequences $x = (x_n)$ such that $\Sigma |x_n|$ converges. Then every complex valued linear function f on l is of the form

$$f(x) = \Sigma c_n x_n,$$

where the sequence (c_n) is bounded. This representation theorem and others are given by Banach in his book *Théorie des Opérations Linéaires* (61–68).

§6.8. It is pointed out in exercise 6(h), 9 that the converse of theorem 6.82 is false. However a modified form of the condition in the theorem is both necessary and sufficient for Riemann integrability. A set E in R^1 is said to have zero *measure* if, given $\epsilon > 0$, there is a countable set of intervals (a_n, b_n) such that

$$\cup(a_n, b_n) \supset E \quad \text{and} \quad \Sigma(b_n - a_n) < \epsilon.$$

Theorem 6.81 shows that every set of zero content also has zero measure. On the other hand, a set of zero measure need not have zero content; the set of rational numbers in an interval is of this kind. The improved version of theorem 6.82 is that *a bounded function is Riemann integrable if and only if its points of discontinuity form a set of zero measure.*

It is curious that such a satisfying result on the Riemann integral should use a concept from *Lebesgue* integration. For measure, of which we have given the simplest example, is a generalization of content and it is related to Lebesgue integration in the same way as content is to Riemann integration. The essential advance which the idea of measure marks over that of content is that content is only *finitely additive*, whereas measure is *completely additive*. By this is meant that, if $E_1, \ldots, E_n$ are pairwise disjoint sets with content, then

$$c(E_1 \cup \ldots \cup E_n) = c(E_1) + \ldots + c(E_n)$$

(see exercises 6(h), 5, 6); but, if (E_n) is a *sequence* of pairwise disjoint sets possessing measure, then

$$m(E_1 \cup E_2 \cup \ldots) = m(E_1) + m(E_2) + \ldots$$

(where $m(E)$ denotes the measure of E).

The Riemann theory of integration is attractive, simple, and adequate for many purposes. However the Lebesgue theory, though more elaborate, provides not only a more powerful integral, but one which in many ways is also simpler to manipulate. Since 1902, when Lebesgue published his first results, a good many other integrals have been developed to solve various specific problems in analysis but none has that unique combination of properties which makes the Lebesgue integral pre-eminent. For a short account of the Lebesgue integral see J. C. Burkill, *The Lebesgue Integral* or R. G. Bartle, *The Elements of Integration*.

7

FUNCTIONS FROM R^m TO R^n

7.1. Differentiation

It is desirable to express a number of results in this chapter by means of matrices. We shall therefore conform to the practice in matrix algebra of regarding points in R^m (vectors) as column matrices. If $x \in R^m$, we denote the components of x by $x_1, ..., x_m$; and, to save space, we write $x = (x_1, ..., x_m)^T$, where the superscript T indicates transposition. Similarly, a function f with values in R^n is written as $(f_1, ..., f_n)^T$. The origin of any space R^m will be denoted by θ. (It will always be clear from the context to which space the symbol θ refers.) Since subscripts are now used to identify coordinates, we shall often use superscripts to designate different vectors. For instance a sequence of vectors may be denoted by $(x^1, x^2, ...)$ or (x^k).

Our aim is to define the process of differentiation for a function from R^m to R^n. Before doing so we consider the particular cases $m = 1$ and $n = 1$.

Vector valued functions on (subsets of) R^1 present no difficulty. If X is a subset of R^1 and ξ is an interior point of X, the function $f = (f_1, ..., f_n)^T : X \to R^n$ is differentiable at ξ if

$$\lim_{h \to 0} \frac{f(\xi + h) - f(\xi)}{h}$$

exists, i.e. if there is a vector $f'(\xi)$ such that

$$\left\| \frac{f(\xi + h) - f(\xi) - f'(\xi)h}{h} \right\| \to 0 \quad \text{as} \quad h \to 0.$$

Evidently $f'(\xi)$ exists if and only if $f_1'(\xi), ..., f_n'(\xi)$ all exist, and

$$f'(\xi) = (f_1'(\xi), ..., f_n'(\xi))^T.$$

Note that f' is again a function from R^1 to R^n.

For functions from R^m to R^1 the notion of a partial derivative is familiar. If we regard partial differentiation as differentiation along a line parallel to a coordinate axis, we are immediately led to the idea of differentiation in an arbitrary direction.

Definition. *Let ξ be an interior point of a subset X of R^m and let u be a unit vector in R^m (i.e. $u = (u_1, ..., u_m)^T$ is such that*

$$\|u\| = \sqrt{(u_1^2 + ... + u_m^2)} = 1).$$

Given the function $f: X \to R^1$, if

$$\lim_{t \to 0} \frac{f(\xi + tu) - f(\xi)}{t}$$

exists, then this limit is called the derivative *of f in the direction u and is denoted by $D_u f(\xi)$.*

Clearly the partial derivatives of f are the derivatives in the directions

$$e^1 = (1, 0, \ldots, 0)^T, \quad e^2 = (0, 1, \ldots, 0)^T, \ldots, \quad e^m = (0, 0, \ldots, 1)^T.$$

We shall generally use the expressions $D_1 f, D_2 f, \ldots, D_m f$ for them, since the classical notation for partial derivatives can lead to confusion.

In spite of possible first appearances, the directional derivative is not the counterpart of the derivative of real functions. For instance the existence of all directional derivatives at a point does not ensure continuity at that point. For let the function $f: R^2 \to R^1$ be given by

$$f(x, y) = \begin{cases} \dfrac{x^2 y}{x^4 + y^2} & \text{when } (x, y) \neq (0, 0), \\[2mm] 0 & \text{when } (x, y) = (0, 0). \end{cases}$$

Then, if $u^2 + v^2 = 1$ and $t \neq 0$,

$$\frac{f(tu, tv) - f(0, 0)}{t} = \frac{u^2 v}{u^4 t^2 + v^2},$$

so that

$$D_{(u, v)} f(0, 0) = \begin{cases} u^2/v & \text{when } v \neq 0, \\ 0 & \text{when } v = 0. \end{cases}$$

However $f(x, x^2) = \frac{1}{2}$ for all $x \neq 0$ and so f is not continuous at $(0, 0)^T$.

In C1 (156) the property of differentiability of a function from R^2 to R^1 was defined. The definition is easily extended to functions from R^m to R^1. If X is a subset of R^m and $\xi = (\xi_1, \ldots, \xi_m)^T$ is an interior point of X, the function $f: X \to R^1$ is said to be differentiable at ξ if there are constants $a_1, \ldots, a_m$ such that

$$\frac{f(\xi_1 + h_1, \ldots, \xi_m + h_m) - f(\xi_1, \ldots, \xi_m) - (a_1 h_1 + \ldots + a_m h_m)}{\sqrt{(h_1^2 + \ldots + h_m^2)}} \to 0 \quad (7.11)$$

as $\sqrt{(h_1^2 + \ldots + h_m^2)} \to 0$. The motive for this definition is that, if f is differentiable at $\xi = (\xi_1, \ldots, \xi_m)^T$ and if $f(\xi_1, \ldots, \xi_m) = \xi_{m+1}$, then the surface $x_{m+1} = f(x_1, \ldots, x_m)$ in R^{m+1} has the tangent plane

$$x_{m+1} - f(\xi_1, \ldots, \xi_m) - a_1(x_1 - \xi_1) - \ldots - a_m(x_m - \xi_m) = 0$$

at the point $(\xi_1, \ldots, \xi_m, \xi_{m+1})^T$. (Here we have used the terms surface and tangent plane in an intuitive sense; surfaces are not, in fact, discussed in this book.)

We have previously noted (see p. 40) that every linear function $\lambda : R^m \to R^1$ is given by a formula

$$\lambda(x) = c_1 x_1 + \ldots + c_m x_m \quad (x = (x_1, \ldots, x_m)^T \in R^m)$$

where $c_1, \ldots, c_m$ are constants. Hence the condition (7.11) for differentiability can be put in the form:

$$\frac{f(\xi + h) - f(\xi) - \lambda(h)}{\|h\|} \to 0 \quad \text{as} \quad \|h\| \to 0$$

for a suitable linear function λ on R^m to R^1. The definition of differentiability of a function from R^m to R^n now follows quite naturally.

Definition. *Let X be a subset of R^m and ξ an interior point of X. The function $f : X \to R^n$ is said to be* differentiable *at ξ if there exists a linear function $\lambda : R^m \to R^n$ such that*

$$\frac{\|f(\xi + h) - f(\xi) - \lambda(h)\|}{\|h\|} \to 0 \quad \textit{as} \quad \|h\| \to 0 \tag{7.12}$$

(where the norms in numerator and denominator are those in R^n and R^m respectively); the linear function λ is called the linear derivative *of f at ξ and is denoted by $Df(\xi)$.*

To justify calling $Df(x)$ *the* linear derivative we need the following uniqueness theorem.

Theorem 7.11. *If $f : X \to R^n$ (where $X \subset R^m$) is differentiable at ξ, then the linear derivative of f at ξ is unique.*

Proof. Let λ^1 and λ^2 both be linear derivatives of f at ξ and put $\lambda^1 - \lambda^2 = \mu$. Then, for $i = 1, 2$,

$$\frac{\|f(\xi + h) - f(\xi) - \lambda^i(h)\|}{\|h\|} \to 0 \quad \text{as} \quad \|h\| \to 0$$

and therefore, by the triangle inequality,

$$\frac{\|\mu(h)\|}{\|h\|} \to 0 \quad \text{as} \quad \|h\| \to 0. \tag{7.13}$$

Now take any $x \in R^m$. Since λ^1 and λ^2 are linear, so is μ and, by (7.13),

$$\|\mu(x)\| = \|x\| \lim_{t \to 0} \frac{\|\mu(tx)\|}{\|tx\|} = 0.$$

Thus $\lambda^1 = \lambda^2$. $\blacksquare$

We now add a number of comments on the definition of differentiability.

(*a*) It is sometimes convenient to write (7.12) in the form

$$\|f(\xi+h)-f(\xi)-\lambda(h)\| = o(\|h\|) \quad \text{as} \quad \|h\| \to 0.$$

(*b*) A linear function $\lambda: R^1 \to R^n$ is such that

$$\lambda(x) = ax \quad (x \in R^1)$$

for some vector a; and there is a bijection between the λ's and the a's. If $f = (f_1, ..., f_n)^T$ is a differentiable function from R^1 to R^n, this bijection provides the connexion between the linear derivative $Df(\xi)$ of f at ξ and the derivative $f'(\xi) = (f_1'(\xi), ..., f_n'(\xi))^T$; for $Df(\xi)$ is the linear function on R^1 to R^n determined by

$$Df(\xi)(x) = f'(\xi)x = (f_1'(\xi)x, ..., f_n'(\xi)x)^T \quad (x \in R^1).$$

We shall see later (theorem 7.13) that, when $f = (f_1, ..., f_n)^T$ is a differentiable function from R^m to R^n, there is a corresponding connexion between $Df(\xi)$ and the partial derivatives of $f_1, ..., f_n$ at ξ.

(*c*) We have stressed that, when $f: X \to R^n$ (where $X \subset R^m$) is differentiable at ξ, then $Df(\xi)$ is a linear function on R^m to R^n. On the other hand Df is a function from R^m to $L(R^m, R^n)$, the space of linear functions on R^m to R^n (see §3.2). The domain of Df is the subset of X on which f is differentiable.

(*d*) If the function $f: R^m \to R^n$ is constant, then clearly for every $\xi \in R^m$, $Df(\xi)$ is the zero linear function $\Theta: R^m \to R^n$ (defined by $\Theta(x) = \theta$ for all $x \in R^m$).

(*e*) Let $f: R^m \to R^n$ be a linear function. The identity

$$f(\xi+h)-f(\xi)-f(h) = \theta$$

shows that, at every point ξ of R^m, f is differentiable and $Df(\xi) = f$. Note that this is an entirely different relation from

$$(d/dx) \exp x = \exp x \quad (x \in R^1)$$

which is described by the equation $g' = g$.

(*f*) Terms other than linear derivative which are used for $Df(\xi)$ are derivative (or total derivative) and differential, but these suffer from the disadvantage of having other well established meanings.

(*g*) The relation (7.12) can be used to define the property of differentiability and the linear derivative of a function whose domain and range lie in arbitrary normed vector spaces.

It is easy to prove that differentiability implies continuity and the existence of partial derivatives.

Theorem 7.12. *If $X \subset R^m$ and the function $f: X \to R^n$ is differentiable at the (interior) point ξ of X, then f is continuous at ξ.*

Proof. Denote the norm of the linear function $Df(\xi)$ by k (see p. 41), so that

$$\|Df(\xi)(x)\| \leqslant k\|x\|$$

for all $x \in R^m$. By the definition of differentiability, there is a $\delta > 0$ such that

$$\|f(\xi+h) - f(\xi) - Df(\xi)(h)\| < \|h\|$$

when $\|h\| < \delta$. Hence

$$\|f(\xi+h) - f(\xi)\| < \|Df(\xi)(h)\| + \|h\| \leqslant (k+1)\|h\|$$

for $\|h\| < \delta$. |

Theorem 7.13. *Let the function $f = (f_1, ..., f_n)^T : X \to R^n$ $(X \subset R^m)$ be differentiable at the point ξ of X. Then all the partial derivatives $D_i f_j(\xi)$ $(i = 1, ..., m; j = 1, ..., n)$ exist and the linear derivative $Df(\xi)$ is given by*

$$Df(\xi)(x) = \begin{pmatrix} D_1 f_1(\xi) \dots D_m f_1(\xi) \\ \cdots\cdots\cdots\cdots \\ D_1 f_n(\xi) \dots D_m f_n(\xi) \end{pmatrix} \begin{pmatrix} x_1 \\ \vdots \\ x_m \end{pmatrix} \quad (x = (x_1, ..., x_m)^T \in R^m).$$

Proof. The linear function $Df(\xi): R^m \to R^n$ is determined (see p. 40) by an equation of the form

$$Df(\xi)(x) = \begin{pmatrix} a_{11} \dots a_{1m} \\ \cdots\cdots\cdots \\ a_{n1} \dots a_{nm} \end{pmatrix} \begin{pmatrix} x_1 \\ \vdots \\ x_m \end{pmatrix} \quad (x = (x_1, ..., x_m)^T \in R^m).$$

Now $\|f(\xi+h) - f(\xi) - Df(\xi)(h)\| / \|h\| \to 0$

as $\|h\| \to 0$. Considering the jth component in the numerator and taking $h = te^i$, we have

$$|f_j(\xi+te^i) - f_j(\xi) - a_{ji}t| / |t| \to 0$$

as $t \to 0$. Thus $D_i f_j(\xi)$ exists and is equal to a_{ji}. |

Note. If $m = 1$, i.e. if $f = (f_1, ..., f_n)^T$ is a function from R^1 to R^n, the matrix of $Df(\xi)$ is the column matrix $(f_1'(\xi), ..., f_n'(\xi))^T$ which has previously been defined as $f'(\xi)$.

If $n = 1$, i.e. if f is a function from R^m to R^1, the matrix of $Df(\xi)$ is the row matrix $(D_1 f(\xi), ..., D_m f(\xi))$.

When $m = n$, the determinant

$$\begin{vmatrix} D_1 f_1(\xi) \dots D_n f_1(\xi) \\ \dots\dots\dots\dots\dots \\ D_1 f_n(\xi) \dots D_n f_n(\xi) \end{vmatrix}$$

associated with the linear derivative $Df(\xi)$ is called the *Jacobian* of f at ξ and is denoted by $Jf(\xi)$. The classical notation is

$$\frac{\partial(f_1, ..., f_n)}{\partial(\xi_1, ..., \xi_n)}.$$

The fact that the linear function $Df(\xi): R^n \to R^n$ is bijective if and only if $Jf(\xi) \neq 0$ (see p. 40) is crucial for the implicit function theorem (7.43) and related results. The geometrical significance of the Jacobian is that it measures local magnification: if f is a function from R^n to R^n and E is a set none of whose points is far from ξ, then the volume of $f(E)$ is approximately $|Jf(\xi)|$ times the volume of E. This statement and the ideas it involves will be made precise in § 8.6.

We have previously given an example of a discontinuous function whose partial derivatives exist. The converse of theorem 7.13 is therefore false. Differentiability, however, *is* ensured if the partial derivatives not only exist, but are also continuous.

Lemma. *The function* $f = (f_1, ..., f_n)^T : X \to R^n$ $(X \subset R^m)$ *is differentiable at the (interior) point* ξ *if and only if each component function* $f_j : X \to R^1$ *is differentiable at* ξ.

Proof. Let λ be the linear function on R^m to R^n determined by the matrix

$$\begin{pmatrix} a_{11} \dots a_{1m} \\ \dots\dots\dots \\ a_{n1} \dots a_{nm} \end{pmatrix}$$

and let λ_j $(j = 1, ..., n)$ be the linear function on R^m to R^1 given by the row matrix

$$(a_{j1}, ..., a_{jm}).$$

To prove the lemma we need only remark that

$$\|f(\xi+h) - f(\xi) - \lambda(h)\| = o(\|h\|) \quad \text{as} \quad \|h\| \to 0$$

if and only if, for $j = 1, ..., n$,

$$|f_j(\xi+h) - f_j(\xi) - \lambda_j(h)| = o(\|h\|) \quad \text{as} \quad \|h\| \to 0. \;|$$

Theorem 7.14. *Let X be a subset of R^m and let ξ be an interior point of X. If the function $f = (f_1, ..., f_n)^T : X \to R^n$ is such that each partial derivative $D_i f_j$ $(i = 1, ..., m; j = 1, ..., n)$ exists in an open ball $B(\xi; \delta)$ and is continuous at ξ, then f is differentiable at ξ.*

Proof. In view of the lemma it is sufficient to prove that each function $f_j : X \to R^1$ is differentiable at ξ.

If $h = (h_1, ..., h_m)^T$, let $k^0 = \theta$ and

$$k^i = (h_1, ..., h_i, 0, ..., 0)^T = h_1 e^1 + ... + h_i e^i \quad (i = 1, ..., m),$$

so that $\qquad f_j(\xi+h) - f_j(\xi) = \sum_{i=1}^{m} \{f_j(\xi+k^i) - f_j(\xi+k^{i-1})\}.$ (7.14)

For each i, $k^i - k^{i-1} = h_i e^i$ and so, by the mean value theorem, there is a number α_i between 0 and 1 such that

$$f_j(\xi+k^i) - f_j(\xi+k^{i-1}) = h_i D_i f_j(\xi+k^{i-1}+\alpha_i h_i e^i). \quad (7.15)$$

Since the functions $D_i f_j$ are continuous at ξ and

$$\|k^{i-1}+\alpha_i h_i e^i\| \leqslant \|k^i\| \leqslant \|h\|,$$

it follows that

$$h_i D_i f_j(\xi+k^{i-1}+\alpha_i h_i e^i) = h_i\{D_i f_j(\xi)+o(1)\} = h_i D_i f_j(\xi)+o(\|h\|)$$

as $\|h\| \to 0$. Hence, by (7.14) and (7.15),

$$f_j(\xi+h) - f_j(\xi) - \sum_{i=1}^{m} h_i D_i f_j(\xi) = o(\|h\|). \;|$$

Exercises 7(a)

1. Let X be a subset of R^m and let ξ be an interior point of X. Show that, if the function $f = (f_1, ..., f_n)^T : X \to R^n$ is such that all partial derivatives $D_i f_j$ exist and are bounded in an open ball $B(\xi; \delta)$, then f is continuous at ξ.

2. Let ξ be an interior point of the subset X of R^m. Show that, if the function $f : X \to R^1$ possesses the directional derivative $D_u f(\xi)$, then $D_{-u} f(\xi)$ exists and is equal to $-D_u f(\xi)$.

3. The function $f : X \to R^1 (X \subset R^m)$ is differentiable at the (interior) point ξ of X. Prove that every directional derivative $D_u f(\xi)$ exists and that

$$D_u f(\xi) = Df(\xi)(u).$$

4. If the function $f: X \to R^1 (X \subset R^m)$ is differentiable at the interior point ξ of X, the *gradient* $\nabla f(\xi)$ of f at ξ is defined as the matrix of $Df(\xi)$, i.e.

$$\nabla f(\xi) = (D_1 f(\xi), \ldots, D_m f(\xi)).$$

Show that, if $\nabla f(\xi) \neq (0, \ldots, 0)$, then

$$u_0 = \frac{(\nabla f(\xi))^T}{\|\nabla f(\xi)\|}$$

is the unique vector such that $\|u_0\| = 1$ and

$$\|Df(\xi)\| = \sup_{\|u\|=1} |D_u f(\xi)| = |D_{u_0} f(\xi)|.$$

5. The function $f: R^2 \to R^1$ is given by

$$f(x,y) = \begin{cases} \dfrac{x^2 y}{x^2 + y^2} & \text{when } (x, y) \neq (0, 0), \\ 0 & \text{when } (x, y) = (0, 0). \end{cases}$$

Prove that at $(0, 0)^T$ f is continuous and possesses all directional derivatives, but is not differentiable.

6. The function $f: R^2 \to R^1$ is defined by $f(x, y) = |xy| ((x, y)^T \in R^2)$. Prove that f is differentiable at $\theta = (0, 0)^T$, but that there is no disc $B(\theta; \delta)$ at all points of which $D_1 f$ and $D_2 f$ exist.

7. The functions $\phi = (\phi_1, \ldots, \phi_n)^T$ and $\psi = (\psi_1, \ldots, \psi_p)^T$ have a common domain $X \subset R^m$ and ranges in R^n, R^p respectively. The function $f: X \to R^{n+p}$ is defined by

$$f(x) = (\phi_1(x), \ldots, \phi_n(x), \psi_1(x), \ldots, \psi_p(x))^T \quad (x \in X)$$

and we abbreviate the right-hand side to $(\phi(x), \psi(x))^T$. Prove that f is differentiable at the point $\xi \in X$ if and only if ϕ, ψ are differentiable at ξ and that, when there is differentiability, $Df(\xi)$ is given by

$$Df(\xi)(h) = (D\phi(\xi)(h), D\psi(\xi)(h))^T \quad (h \in R^m).$$

8. The real functions ϕ, ψ on R^1 are differentiable at the points a, b respectively. Prove that the function $f: R^2 \to R^1$ given by $f(x, y) = \phi(x)\psi(y) ((x, y)^T \in R^2)$ is differentiable at $(a, b)^T$.

9. Find a function $f: R^2 \to R^1$ which is everywhere continuous and nowhere differentiable.

10. The function $f: R^3 \to R^2$ is defined by

$$f(x, y) = (\sin x \cos y, \sin x \sin y, \cos x)^T \quad ((x, y)^T \in R^2).$$

Show that f is everywhere differentiable and that, for all $(x, y)^T \in R^2$,

$$\|Df(x, y)\| = 1.$$

7.2. Operations on differentiable functions

Given two functions f, g from R^m to R^n, we may form their sum $f+g$ (a function from R^m to R^n) and their inner product $f.g$ (a function from R^m to R^1). We also consider scalar multiplication of a function from R^m to R^n by a function from R^m to R^1.

Theorem 7.21.

(i) *Suppose that the functions* $f: X \to R^n$ *and* $g: X \to R^n$ $(X \subset R^m)$ *are differentiable at* ξ. *Then* $f+g$ *and* $f.g$ *are differentiable at* ξ *and*

$$D(f+g)(\xi) = Df(\xi) + Dg(\xi),$$

$$D(f.g)(\xi) = f(\xi).Dg(\xi) + g(\xi).Df(\xi).$$

(ii) *If the functions* $\phi: X \to R^1$ *and* $f: X \to R^n$ $(X \subset R^m)$ *are differentiable at* ξ, *then* ϕf *is differentiable at* ξ *and the linear function* $D(\phi f)(\xi)$ *from* R^m *to* R^n *is given by*

$$D(\phi f)(\xi)(x) = \phi(\xi)[Df(\xi)(x)] + [D\phi(\xi)(x)]f(\xi) \quad (x \in R^m).$$

Proof.

(i) The proof of the first statement may be left to the reader. To prove the second we write

$$(f.g)(\xi+h) - (f.g)(\xi) - \{f(\xi).[Dg(\xi)(h)] + g(\xi).[Df(\xi)(h)]\}$$

$$= \{f(\xi+h) - f(\xi) - Df(\xi)(h)\}.g(\xi+h)$$

$$+ \{g(\xi+h) - g(\xi) - Dg(\xi)(h)\}.f(\xi)$$

$$+ [Df(\xi)(h)].\{g(\xi+h) - g(\xi)\}$$

$$= p+q+r,$$

say. By Cauchy's inequality (theorem 2.21), $|q| = o(\|h\|)$ as $\|h\| \to 0$. Using also the continuity of g at ξ we have $|p| = o(\|h\|)$ and

$$|r| \leqslant \|Df(\xi)\| \, \|h\| \, \|g(\xi+h) - g(\xi)\| = o(\|h\|).$$

(ii) The proof is similar to the one we have just given. |

We now turn to the composition of two differentiable functions.

Theorem 7.22. *Let* X, Y *be subsets of* R^m *and* R^n *respectively. If the function* $f = (f_1, \ldots, f_n)^T: X \to Y$ *is differentiable at the point* ξ *and* $g = (g_1, \ldots, g_p)^T: Y \to R^p$ *is differentiable at the point* $f(\xi) = \eta$, *then the composite function* $g \circ f: X \to R^p$ *is differentiable at* ξ *and*

$$D(g \circ f)(\xi) = Dg(\eta) \circ Df(\xi),$$

so that the matrix of $D(g \circ f)(\xi)$ *is equal to the matrix product*

$$\begin{pmatrix} D_1 g_1(\eta) \ldots D_n g_1(\eta) \\ \cdots\cdots\cdots\cdots \\ D_1 g_p(\eta) \ldots D_n g_p(\eta) \end{pmatrix} \begin{pmatrix} D_1 f_1(\xi) \ldots D_m f_1(\xi) \\ \cdots\cdots\cdots\cdots \\ D_1 f_n(\xi) \ldots D_m f_n(\xi) \end{pmatrix}.$$

Proof. Put $Df(\xi) = \lambda$, $Dg(\eta) = \mu$, so that, if

$$f(\xi+h)-f(\xi)-\lambda(h) = q(h)$$

and

$$g(\eta+k)-g(\eta)-\mu(k) = r(k),$$

then

$$\|q(h)\| = o(\|h\|) \quad \text{and} \quad \|r(k)\| = o(\|k\|).$$

Let

$$(g\circ f)(\xi+h)-(g\circ f)(\xi)-(\mu\circ\lambda)(h) = s(h).$$

We have to show that $\|s(h)\| = o(\|h\|)$.

Since μ is linear,

$$s(h) = g(f(\xi+h))-g(f(\xi))-\mu(f(\xi+h)-f(\xi))$$

$$+\mu(f(\xi+h)-f(\xi)-\lambda(h))$$

$$= r(f(\xi+h)-f(\xi))+\mu(q(h)).$$

We saw in the proof of theorem 7.12 that, when $\|h\|$ is small,

$$\|f(\xi+h)-f(\xi)\| \leqslant (\|\lambda\|+1)\|h\|.$$

Since also $\|r(k)\| = o(\|k\|)$ and $\|r(\theta)\| = 0$, it follows that, given $\epsilon > 0$,

$$\|r(f(\xi+h)-f(\xi))\| \leqslant \epsilon\|f(\xi+h)-f(\xi)\| \leqslant \epsilon(\|\lambda\|+1)\|h\| \quad (7.21)$$

for all sufficiently small $\|h\|$. Thus

$$\|r(f(\xi+h)-f(\xi))\| = o(\|h\|).$$

(Notice that in (7.21) we cannot write

$$\|r(f(\xi+h)-f(\xi))\| = o(\|f(\xi+h)-f(\xi)\|),$$

since $\|f(\xi+h)-f(\xi)\|$ may be 0.) Moreover

$$\|\mu(q(h))\| \leqslant \|\mu\|\,\|q(h)\| = o(\|h\|).$$

This proves that $\|s(h)\| = o(\|h\|)$.

The last clause of the theorem now follows from the fact that the matrix of the composition of two linear functions is the product of their matrices. ▮

Corollary. *Let X be a subset of R^n and suppose that the function $f: X \to R^n$ is injective. If f is differentiable at the point ξ of X and f^{-1} is differentiable at the point $\eta = f(\xi)$, then $Df(\xi): R^n \to R^n$ is bijective and*

$$Df^{-1}(\eta) = [Df(\xi)]^{-1},$$

so that, if f_i^{-1} denotes the i-th component of f^{-1},

$$\begin{pmatrix} D_1 f_1^{-1}(\eta) \ldots D_n f_1^{-1}(\eta) \\ \cdots\cdots\cdots\cdots \\ D_1 f_n^{-1}(\eta) \ldots D_n f_n^{-1}(\eta) \end{pmatrix} = \begin{pmatrix} D_1 f_1(\xi) \ldots D_n f_1(\xi) \\ \cdots\cdots\cdots\cdots \\ D_1 f_n(\xi) \ldots D_n f_n(\xi) \end{pmatrix}^{-1}.$$

Using the functions f and g of theorem 7.22 put

$$g \circ f = \phi = (\phi_1, \ldots, \phi_p)^T.$$

Then the (j, i)th element $D_i \phi_j(\xi)$ in the matrix of $D\phi(\xi)$ is given by

$$D_i \phi_j(\xi) = \sum_{k=1}^{n} D_k g_j(\eta) \, D_i f_k(\xi). \tag{7.22}$$

The formula (7.22) is the *chain rule*. Now denote points in R^m and R^n by $(x_1, \ldots, x_m)^T$ and $(y_1, \ldots, y_n)^T$ respectively. Then the classical form for (7.22) is

$$\frac{\partial \phi_j}{\partial x_i} = \sum_{k=1}^{n} \frac{\partial g_j}{\partial y_k} \frac{\partial y_k}{\partial x_i},$$

where $\partial y_k / \partial x_i$ and $\partial \phi_j / \partial x_i$ are evaluated at ξ and $\partial g_j / \partial y_k$ is evaluated at η. The chain rule was given in this form in C1 (158). Theorem 7.22 expresses the rule concisely and explains its origin.

Note. When $m = 1$, so that f and $\phi = g \circ f$ are vector valued functions on R^1, then

$$\begin{pmatrix} \phi_1'(\xi) \\ \vdots \\ \phi_p'(\xi) \end{pmatrix} = \begin{pmatrix} D_1 g_1(\eta) \ldots D_n g_1(\eta) \\ \cdots\cdots\cdots\cdots \\ D_1 g_p(\eta) \ldots D_n g_p(\eta) \end{pmatrix} \begin{pmatrix} f_1'(\xi) \\ \vdots \\ f_n'(\xi) \end{pmatrix},$$

i.e.
$$\phi'(\xi) = Dg(\eta)(f'(\xi)).$$

When $m = n = 1$, so that f is a real function, and g is a vector valued function on R^1, then

$$\begin{pmatrix} \phi_1'(\xi) \\ \vdots \\ \phi_p'(\xi) \end{pmatrix} = \begin{pmatrix} g_1'(\eta) \\ \vdots \\ g_p'(\eta) \end{pmatrix} (f'(\xi)) = \begin{pmatrix} g_1'(\eta) f'(\xi) \\ \vdots \\ g_p'(\eta) f'(\xi) \end{pmatrix},$$

i.e.
$$\phi'(\xi) = f'(\xi) g'(f(\xi)).$$

If also $p = 1$, we have the formula, proved in C1 (68), for differentiating the composition of two real functions.

Exercises 7(b)

1. Given the function $f: X \to R^n$ ($X \subset R^m$), the function $\|f\| : X \to R^1$ is defined by

$$\|f\|(x) = \|f(x)\| (x \in R^m).$$

Prove that, if f is differentiable at the point $\xi \in X$, then so is $\|f\|$, provided that $\|f(\xi)\| \neq 0$.

2. Show that, if the function $f: X \to R^n$ ($X \subset R^m$) is differentiable at the point $\xi \in X$, then so is the function $\|f\|^2 : X \to R^1$. Prove also that, if f is such that $\|f\|$ is constant, then

$$f(\xi) . Df(\xi)(x) = 0$$

for all $x \in R^m$.

3. Let X be a subset of R^n such that $tx \in X$ whenever $x \in X$ and $t > 0$. The function $g: X \to R^p$ is said to be *homogeneous* of degree α (where α is real) if

$$g(tx) = t^\alpha g(x)$$

for all $x \in X$ and $t > 0$.

  Let X be open and let g be differentiable at all points of X.

  (i) Show that, if g is homogeneous of degree α, then each function $D_i g = (D_i g_1, ..., D_i g_p)^T : X \to R^p$ is homogeneous of degree $\alpha - 1$.

  (ii) Prove that a necessary and sufficient condition for g to be homogeneous of degree α is that

$$Dg(x)(x) = \alpha g(x)$$

for all $x \in X$.

4. The function $f: R^2 - \{\theta\} \to R^2$ is given by

$$f(x, y) = \frac{(x, y)^T}{\|(x, y)\|^\alpha}$$

(α real). Use the polar transformation $p: R^2 \to R^2$ defined by

$$(x, y)^T = p(r, \phi) = (r \cos \phi, r \sin \phi)^T   ((r, \phi)^T \in R^2)$$

to prove that $Jf(x, y) = (1 - \alpha) \|(x, y)\|^{-2\alpha}$.

5. The function $f = (f_1, f_2, f_3)^T : R^3 \to R^3$ is given by

$$f_1(x) = x_1 - x_1 x_2, \quad f_2(x) = x_1 x_2 - x_1 x_2 x_3, \quad f_3(x) = x_1 x_2 x_3.$$

Prove that f is injective in $X = R^3 - \{x \mid x_1 x_2 = 0\}$ and find $Y = f(X)$. Show also that $f^{-1} : Y \to R^3$ is differentiable and evaluate $Jf^{-1}(y)$ for all $y \in Y$.

7.3. Some properties of differentiable functions

In the study of real functions the mean value theorem is a powerful tool. Its analogue

$$f(v) - f(u) = Df(\xi)(v - u)$$

(for suitable ξ) holds for functions f from R^m to R^1 (see exercise 7(c), 1). The corresponding result for vector valued functions is false (see exercise 7(c), 2), but the theorem below is a partial substitute. In it we use the notion of convexity. A set S in R^m is *convex* if, whenever two points belong to S, then so does the segment joining them. An interval, an open ball, or a line segment provide examples of convex sets.

Theorem 7.31. *Let X be a subset of R^m and let S be a convex subset of X. If the function $f: X \to R^n$ is differentiable at all points of S and $\|Df(x)\|$ is bounded for $x \in S$, then, for all $u, v \in S$,*

$$\|f(u) - f(v)\| \leqslant (\sup_{x \in S} \|Df(x)\|) \|u - v\|. \tag{7.31}$$

Proof. Take any points u, v in S and put $v - u = h$. If $f(u) = f(v)$, (7.31) clearly holds. Otherwise let ω be the unit vector in R^n such that

$$\|f(v) - f(u)\| = \omega.\{f(v) - f(u)\}. \tag{7.32}$$

Now define the function $\phi: [0, 1] \to R^1$ by the relation

$$\phi(t) = \omega.f(u + th).$$

Then $\phi(0) = \omega.f(u)$, $\phi(1) = \omega.f(v)$ and ϕ is clearly continuous on $[0, 1]$. Also ϕ is differentiable in $(0, 1)$ and (see note following theorem 7.22)

$$\phi'(t) = \omega.Df(u + th)(h).$$

By the mean value theorem, there is a τ in $(0, 1)$ such that

$$\phi(1) - \phi(0) = \phi'(\tau),$$

i.e. $\qquad\qquad \omega.\{f(v) - f(u)\} = \omega.Df(u + \tau h)(h).$

Then, by (7.32),

$$\|f(v) - f(u)\| = \omega.Df(u + \tau h)(h) \leqslant \|Df(u + \tau h)\| \|h\|$$

and, since $u + \tau h \in S$, this implies (7.31). $\;\blacksquare$

Under certain conditions there may also be a lower estimate for $\|f(u) - f(v)\|$. We need only a simple result of this kind.

Theorem 7.32. *Let X be a subset of R^n and let the function $f: X \to R^n$ be differentiable at the point ξ. If $Jf(\xi) \neq 0$, then there is a $\delta > 0$ such that*

$$\|f(\xi + h) - f(\xi)\| > \frac{1}{2\|Df(\xi)^{-1}\|} \|h\| \tag{7.33}$$

whenever $\|h\| < \delta$.

Proof. We have

$$f(\xi + h) - f(\xi) = Df(\xi)(h) + r(h),$$

where $\|r(h)\|/\|h\| \to 0$ as $\|h\| \to 0$. Then

$$\|f(\xi+h)-f(\xi)\| \geqslant \|Df(\xi)(h)\| - \|r(h)\|$$

$$\geqslant \frac{1}{\|Df(\xi)^{-1}\|}\|h\| - \|r(h)\|,$$

since

$$\|h\| = \|Df(\xi)^{-1}\{Df(\xi)(h)\}\| \leqslant \|Df(\xi)^{-1}\| \, \|Df(\xi)(h)\|;$$

and we choose δ so that

$$\|r(h)\| < \frac{1}{2\|Df(\xi)^{-1}\|}\|h\|$$

for $\|h\| < \delta$. $\blacksquare$

An immediate consequence of the mean value theorem is the fact that a real function whose derivative is zero in an interval is constant in that interval. Theorem 7.31 enables us to prove a similar theorem for functions from R^m to R^n.

Theorem 7.33. *Suppose that the set G in R^m is open and connected. If the differentiable function $f: G \to R^n$ is such that, for every $\xi \in G$, $Df(\xi)$ is the zero linear function Θ (see p. 200), then f is a constant function.*

Proof. Let u be a fixed point of G and let x be any other point of G. By theorem 3.32, u and x may be connected by a polygonal arc lying entirely in G. Denote the vertices of this arc by

$$u = x^0, x^1, \ldots, x^{p-1}, x^p = x.$$

Since the straight line segment joining x^{i-1} to x^i is a convex subset of G, it follows from theorem 7.31 that $f(x^i) = f(x^{i-1})$. This holds for $i = 1, \ldots, p$ and therefore $f(x) = f(u)$. $\blacksquare$

In the theorems of the next section (and in others) the functions under consideration will not only have to be differentiable; their linear derivatives will also be required to be continuous. The sense in which continuity is to be understood is entirely natural. Let the function $f: X \to R^n$ ($X \subset R^m$) be differentiable in the subset X_1 of X. If $\xi \in X_1$, $Df(\xi)$ is a linear function on R^m to R^n, i.e. $Df(\xi) \in L(R^m, R^n)$; and, as we have previously remarked, the linear derivative Df is a function with domain X_1 and range in $L(R^m, R^n)$. The set X_1 is equipped with the metric induced by R^m and the set $L(R^m, R^n)$ has a natural norm (see (3.22), p. 41) which produces a metric.

Therefore the usual notion of continuity applies to the functions on X_1 to $L(R^m, R^n)$, of which Df is one. If all points of X are interior points, i.e. if X is open, then X_1 may coincide with X and we are led to the following definition.

Definition. *A function $f: G \to R^n$, where G is an open subset of R^m, is said to be* continuously differentiable *if $Df(\xi)$ exists for all $\xi \in G$ and if the function $Df: G \to L(R^m, R^n)$ is continuous on G.*

Let (X, ρ) be an arbitrary metric space. A function

$$\lambda: X \to L(R^m, R^n)$$

may be represented by a matrix

$$\begin{pmatrix} a_{11} \dots a_{1m} \\ \cdots \cdots \\ a_{n1} \dots a_{nm} \end{pmatrix}$$

each element a_{ji} of which is a function on X to R^1. The inequalities (theorem 3.25)

$$\max_{j, i} |a_{ji}(x) - a_{ji}(\xi)| \leqslant \|\lambda(x) - \lambda(\xi)\| \leqslant (\sum_{j, i} [a_{ji}(x) - a_{ji}(\xi)]^2)^{\frac{1}{2}}$$

show that λ is continuous at the point $\xi \in X$ if and only if each of the functions a_{ji} is continuous at ξ. Combining this result with theorems 7.13 and 7.14 we obtain a simple criterion for a function to be continuously differentiable.

Theorem 7.34. *Let G be an open subset of R^m. The function*

$$f = (f_1, \dots, f_n)^T : G \to R^n$$

is continuously differentiable if and only if each of the partial derivatives $D_i f_j (i = 1, \dots, m; j = 1, \dots, n)$ exists and is continuous on G.

It is evident from the definition of continuous differentiability (or from theorem 7.34) that, for real functions, the property is equivalent to the possession of a continuous derivative.

Exercises 7(c)

1. Let X be a subset of R^m which contains the points u, v and the segment

$$\sigma = \{x \,|\, x = u + t(v - u); 0 < t < 1\}.$$

The function $f: X \to R^1$ is continuous at u, v and differentiable at all points of σ. Show that there is a point ξ of σ such that

$$f(v) - f(u) = Df(\xi)(v - u).$$

2. The function $f: R^1 \to R^2$ is given by

$$f(x) = (\cos x, \sin x)^T.$$

Prove that there are points $u, v \in R^1$ such that

$$f(v) - f(u) \neq Df(\xi)(v-u)$$

for all $\xi \in R^1$.

3. Prove that, under the conditions of theorem 7.31,

$$\|f(v) - f(u) - Df(u)(v-u)\| \leqslant (\sup_{x \in S} \| Df(x) - Df(u)\|)\|v - u\|$$

for all $u, v \in S$.

4. The set G in R^m is open and the function $f: G \to R^n$ is continuously differentiable. Show that, if K is a compact subset of G, then
   (i) $\|f(x+h) - f(x)\| \to 0$,
   (ii) $\|f(x+h) - f(x) - Df(x)(h)\|/\|h\| \to 0$
as $\|h\| \to 0$, uniformly for $x \in K$.

5. Let G be an open, connected subset of R^m and let (f^ν) be a sequence of continuously differentiable functions on G to R^n. Suppose that $(f^\nu(x))$ converges for at least one $x \in G$ and that (Df^ν) converges uniformly on every compact subset of G. Show that
   (i) (f^ν) converges uniformly on every compact subset of G;
   (ii) $f = \lim f^\nu$ is continuously differentiable in G and, for all $x \in G$,

$$Df(x) = \lim_{\nu \to \infty} Df^\nu(x).$$

7.4. The implicit function theorem

As an application of Banach's fixed point principle we obtained in chapter 3 a theorem on the existence of implicitly defined functions. We begin by proving a variant of that theorem in which the hypothesis is a little more stringent and the conclusion is correspondingly strengthened.

Theorem 7.41. *Let* $(\xi, \eta)^T$ *be an interior point of a set E in R^2 and suppose that the function $f: E \to R^1$ satisfies the following conditions:*
   (i) $f(\xi, \eta) = 0$;
   (ii) *f is continuously differentiable in an open set G containing $(\xi, \eta)^T$;*
   (iii) $D_2 f(\xi, \eta) \neq 0$.
Then there exist a rectangle

$$M \times N = [\xi - \alpha, \xi + \alpha] \times [\eta - \beta, \eta + \beta]$$

and a continuous function $\phi: M \to N$ *such that $y = \phi(x)$ is the only solution lying in $M \times N$ of the equation $f(x, y) = 0$; moreover ϕ is continuously differentiable in M° (the interior of M) and*

$$\phi'(x) = -D_1 f(x, \phi(x))/D_2 f(x, \phi(x))$$

for $\xi - \alpha < x < \xi + \alpha$.

Proof. Theorem 3.47 gives a unique solution ϕ, but only guarantees its continuity. To prove continuous differentiability we first remark that M and N may be taken so small that $M \times N \subset G$ and, for $(x, y)^T \in M \times N$, $D_2 f(x, y) \neq 0$. Next, take a point x in the open interval $(\xi - \alpha, \xi + \alpha)$ and let h be any non-zero number such that $x + h \in [\xi - \alpha, \xi + \alpha]$. Then

$$\{f(x+h, \phi(x+h)) - f(x, \phi(x+h))\} + \{f(x, \phi(x+h)) - f(x, \phi(x))\} = 0$$

and so, by the mean value theorem,

$$h D_1 f(u, \phi(x+h)) + \{\phi(x+h) - \phi(x)\} D_2 f(x, v) = 0,$$

where u lies between x and $x + h$, while v lies between $\phi(x)$ and $\phi(x+h)$. Thus, since $D_2 f(x, v) \neq 0$,

$$\frac{\phi(x+h) - \phi(x)}{h} = -\frac{D_1 f(u, \phi(x+h))}{D_2 f(x, v)}.$$

It now follows from the continuity of the functions ϕ, $D_1 f$ and $D_2 f$ that, first, $\phi'(x)$ exists and is equal to

$$-D_1 f(x, \phi(x))/D_2 f(x, \phi(x))$$

and, secondly, that ϕ' is continuous on $(\xi - \alpha, \xi + \alpha)$. |

A simple example will serve to emphasize the *local* character of the solution of the equation considered in theorem 7.41. If f is given by

$$f(x, y) = x + \sin y,$$

then in the rectangle $[-1, 1] \times [-\frac{1}{2}\pi, \frac{1}{2}\pi]$ there is a unique solution passing through the point $(0, 0)^T$. There is no solution through any point $(x, y)^T$ with $|x| > 1$; and, if $\beta > \frac{1}{2}\pi$, solutions in

$$[-1, 1] \times [-\beta, \beta]$$

are not unique, i.e. there are values of x in $[-1, 1]$ to which there correspond two or more values of y in $[-\beta, \beta]$ such that $f(x, y) = 0$.

The problem of theorem 7.41 has an obvious generalization in keeping with the spirit of the present chapter. In the equation $f(x, y) = 0$ we replace the real numbers x, y by points in R^m and R^n respectively. The function f will then have its domain in $R^m \times R^n$ and it is easy to see that the range of f should be in R^n. For, writing

$x = (x_1, ..., x_m)^T$, $y = (y_1, ..., y_n)^T$ and $f = (f_1, ..., f_n)^T$ we then have the system of equations

$$\left.\begin{aligned} f_1(x_1, ..., x_m, y_1, ..., y_n) &= 0, \\ \cdots\cdots\cdots\cdots\cdots\cdots\cdots \\ f_n(x_1, ..., x_m, y_1, ..., y_n) &= 0 \end{aligned}\right\} \qquad (7.41)$$

which is to lead to unique expressions for $y_1, ..., y_n$ in terms of $x_1, ..., x_m$; and our experience shows that the number of equations should, in fact, be equal to the number of unknowns.

The equations (7.41) suggest that the domain of f is in R^{m+n}. On the other hand the vector equation

$$f(x, y) = \theta$$

indicates a domain in $R^m \times R^n$, as was originally stipulated. This dual notation is legitimate in view of the natural identification of points in $R^m \times R^n$ and R^{m+n}. By equipping $R^m \times R^n$ with the norm induced by R^{m+n} we make the two spaces entirely interchangeable; and we shall use either or both as convenient. For instance we denote by

$$D_i f_j(x_1, ..., x_m, y_1, ..., y_n)$$

the partial derivative of f_j with respect to the ith co-ordinate in R^{m+n} (an x if $1 \leqslant i \leqslant m$, a y if $m+1 \leqslant i \leqslant m+n$) and we also use the abbreviation $D_i f_j(x, y)$.

The indicated generalization of theorem 7.41 can be formulated and proved with the use only of partial derivatives, but it is more illuminating to use also a concept of partial differentiation which is akin to linear differentiation.

Definition. *Let f be a function from $R^m \times R^n$ to R^p and suppose that $(\xi, \eta)^T$ ($\xi \in R^m$, $\eta \in R^n$) is an interior point of the domain E of f. If there is a linear function $\lambda: R^m \to R^p$ such that*

$$\|f(\xi+h, \eta) - f(\xi, \eta) - \lambda(h)\| = o(\|h\|) \quad as \quad \|h\| \to 0, \quad (7.42)$$

then λ is the linear partial derivative $D_{(1)} f(\xi, \eta)$ *of f at $(\xi, \eta)^T$. The* linear partial derivative $D_{(2)} f(\xi, \eta)$, *a linear function on R^n to R^p, is similarly defined.*

The argument of theorem 7.11 shows that linear partial derivatives are unique.

Theorem 7.42. *Suppose that the function $f: E \to R^p$ ($E \subset R^m \times R^n$) is differentiable at the point $(\xi, \eta)^T$. Then $D_{(1)} f(\xi, \eta)$ and $D_{(2)} f(\xi, \eta)$*

exist; moreover, for all $x \in R^m$ and all $y \in R^n$,

$$D_{(1)} f(\xi, \eta)(x) = Df(\xi, \eta)(x, \theta)$$

and $$D_{(2)} f(\xi, \eta)(y) = Df(\xi, \eta)(\theta, y).$$

Proof. By the definition of $Df(\xi, \eta)$,

$$\|f(\xi+h, \eta+k) - f(\xi, \eta) - Df(\xi, \eta)(h, k)\| = o(\|(h, k)\|)$$

as $\|(h, k)\| \to 0$. Hence, taking $k = \theta$, we have (7.42) with $\lambda(h)$ replaced by $Df(\xi, \eta)(h, \theta)$. The uniqueness of linear partial derivatives now gives the result for $D_{(1)} f(\xi, \eta)$. |

Corollary. *Under the conditions of the theorem*

$$\|D_{(1)} f(\xi, \eta)\|, \|D_{(2)} f(\xi, \eta)\| \leqslant \|Df(\xi, \eta)\|.$$

If the function $f = (f_1, ..., f_p)^T$ from $R^m \times R^n$ to R^p possesses the linear partial derivative $D_{(1)} f(\xi, \eta)$, then the partial derivatives $D_i f_j(\xi, \eta)$ $(i = 1, ..., m; j = 1, ..., p)$ exist and $D_{(1)} f(\xi, \eta)$ is represented by the matrix

$$\begin{pmatrix} D_1 f_1(\xi, \eta) \dots D_m f_1(\xi, \eta) \\ \dots \dots \dots \dots \dots \dots \\ D_1 f_p(\xi, \eta) \dots D_m f_p(\xi, \eta) \end{pmatrix}. \tag{7.43}$$

This is proved in the same way as theorem 7.13. Similarly, when $D_{(2)} f(\xi, \eta)$ exists, then the partial derivatives $D_i f_j(\xi, \eta)$ $(i = m+1, ..., m+n; j = 1, ..., p)$ exist and the matrix of $D_{(2)} f(\xi, \eta)$ is

$$\begin{pmatrix} D_{m+1} f_1(\xi, \eta) \dots D_{m+n} f_1(\xi, \eta) \\ \dots \dots \dots \dots \dots \dots \dots \\ D_{m+1} f_p(\xi, \eta) \dots D_{m+n} f_p(\xi, \eta) \end{pmatrix}. \tag{7.44}$$

Thus, if f is differentiable at $(\xi, \eta)^T$, the matrices of $D_{(1)} f(\xi, \eta)$ and of $D_{(2)} f(\xi, \eta)$ consist, respectively, of the first m and last n columns of the matrix of $Df(\xi, \eta)$.

When $p = m$, the matrix (7.43) of $D_{(1)} f(\xi, \eta)$ is square and we denote its determinant by $J_{(1)} f(\xi, \eta)$; when $p = n$, the determinant of (7.44) is denoted by $J_{(2)} f(\xi, \eta)$.

We have now laid the groundwork for the principal theorem of this chapter.

Theorem 7.43. *(Implicit function theorem.) Let $(\xi, \eta)^T (\xi \in R^m, \eta \in R^n)$ be an interior point of a set E in $R^m \times R^n$ and suppose that the function $f: E \to R^n$ satisfies the following conditions:*

(i) $f(\xi, \eta) = \theta$;

(ii) *f is continuously differentiable in an open set G containing* $(\xi, \eta)^T$;

(iii) $J_{(2)} f(\xi, \eta) \neq 0$.

Then there exist intervals

$$M = [\xi_1 - \alpha, \xi_1 + \alpha] \times \ldots \times [\xi_m - \alpha, \xi_m + \alpha],$$

$$N = [\eta_1 - \beta, \eta_1 + \beta] \times \ldots \times [\eta_n - \beta, \eta_n + \beta]$$

and a continuous function $\phi: M \to N$ such that $y = \phi(x)$ is the only solution lying in $M \times N$ of the equation $f(x, y) = \theta$; moreover ϕ is continuously differentiable in $M°$ and

$$D\phi(x) = -[D_{(2)} f(x, \phi(x))]^{-1} \circ D_{(1)} f(x, \phi(x)). \qquad (7.45)$$

for $x \in M°$.

Proof. The proof follows the lines laid down in the proofs of theorems 3.47 and 7.41, except for additional complications which arise from the absence of a mean value theorem.

By (ii), the function $J_{(2)} f$ is continuous on G. Hence, by (iii), we may suppose that, for $(x, y)^T \in G$, $J_{(2)} f(x, y) \neq 0$, so that the linear function $D_{(2)} f(x, y): R^n \to R^n$ is bijective. Put $D_{(2)} f(\xi, \eta) = Q$.

Since the function Df is continuous on G, there are intervals

$$K = [\xi_1 - \delta, \xi_1 + \delta] \times \ldots \times [\xi_m - \delta, \xi_m + \delta],$$

$$N = [\eta_1 - \beta, \eta_1 + \beta] \times \ldots \times [\eta_n - \beta, \eta_n + \beta]$$

such that, for $(x, y)^T \in K \times N$,

$$\| Df(\xi, \eta) - Df(x, y) \| < 1/(2 \| Q^{-1} \|).$$

Then, denoting by I the identity function on R^n and using the existence and linearity of Q^{-1}, we have

$$\| I - Q^{-1} \circ D_{(2)} f(x, y) \| = \| Q^{-1} \circ \{ D_{(2)} f(\xi, \eta) - D_{(2)} f(x, y) \} \|$$

$$\leqslant \| Q^{-1} \| \, \| D_{(2)} f(\xi, \eta) - D_{(2)} f(x, y) \|$$

$$\leqslant \| Q^{-1} \| \, \| Df(\xi, \eta) - Df(x, y) \|$$

$$< \tfrac{1}{2}. \qquad (7.46)$$

By (i) and (ii) we can also find a positive number $\alpha \leqslant \delta$ such that, for $x \in M = [\xi_1 - \alpha, \xi_1 + \alpha] \times \ldots \times [\xi_m - \alpha, \xi_m + \alpha]$,

$$\|f(x, \eta)\| < \beta/(2\|Q^{-1}\|)$$

and therefore

$$\|Q^{-1}[f(x, \eta)]\| < \tfrac{1}{2}\beta. \tag{7.47}$$

Denote, as usual, by $B(M, N)$ the set of all (bounded) functions $\psi : M \to N$, and by $C(M, N)$ the set consisting of all the continuous functions in $B(M, N)$. By theorems 3.44 and 3.45, the two spaces, equipped with their customary metrics, are both complete.

Let Ω be the mapping on $B(M, N)$ defined by $\Omega(\psi) = \chi$, where

$$\chi(x) = \psi(x) - Q^{-1}[f(x, \psi(x))] \quad (x \in M).$$

Note that, if ψ is continuous on M, then so is χ.

We first show that Ω maps $B(M, N)$ into itself. Let $\psi \in B(M, N)$. Then, for $x \in M$,

$$\Omega(\psi)(x) - \eta = \psi(x) - Q^{-1}[f(x, \psi(x))] - \eta$$

$$= \{\psi(x) - Q^{-1}[f(x, \psi(x))]\} - \{\eta - Q^{-1}[f(x, \eta)]\} + Q^{-1}[f(x, \eta)].$$

Now, for fixed x, the function

$$I - Q^{-1} \circ f(x, .) : N \to R^n$$

(where $f(x, .)$ is the function $g : N \to R^n$ defined by $g(y) = f(x, y)$ for $y \in N$) has at y the linear derivative

$$I - Q^{-1} \circ D_{(2)} f(x, y);$$

for, at any point, the linear derivative of a linear function is the function itself. Hence, by theorem 7.31 and by (7.46) and (7.47),

$$\|\Omega(\psi)(x) - \eta\| < \tfrac{1}{2}\|\psi(x) - \eta\| + \tfrac{1}{2}\beta \leqslant \beta.$$

We next use a similar argument to prove that Ω is a contraction mapping. If $\psi_1, \psi_2 \in B(M, N)$ and $x \in M$,

$$\Omega(\psi_1)(x) - \Omega(\psi_2)(x)$$

$$= \{\psi_1(x) - Q^{-1}[f(x, \psi_1(x))]\} - \{\psi_2(x) - Q^{-1}[f(x, \psi_2(x))]\}$$

and so, by theorem 7.31 and (7.46),

$$\|\Omega(\psi_1)(x) - \Omega(\psi_2)(x)\| \leqslant \tfrac{1}{2}\|\psi_1(x) - \psi_2(x)\|.$$

Thus

$$\rho(\Omega(\psi_1), \Omega(\psi_2)) \leqslant \tfrac{1}{2}\rho(\psi_1, \psi_2).$$

It now follows from theorem 3.46 that Ω has a unique fixed point. In other words there is a unique function $\phi: M \to N$ such that $\Omega(\phi) = \phi$, i.e. $f(x, \phi(x)) = \theta$ for $x \in M$. (The uniqueness of ϕ implies, in particular, that $\phi(\xi) = \eta$.)

To prove that ϕ is continuously differentiable we proceed in stages. First we note that ϕ is continuous. For we have remarked that, if ψ is continuous, then so is $\Omega(\psi)$. Hence, if Ω^* is the restriction of Ω to $C(M, N)$, then Ω^* is a contraction mapping on $C(M, N)$. Therefore Ω^* has a fixed point and, since a fixed point of Ω^* is also a fixed point of Ω, it follows that ϕ is the fixed point of Ω^*. Thus $\phi \in C(M, N)$.

The next step is to show that, for $x \in M^\circ$,

$$\|\phi(x+h) - \phi(x)\| = O(\|h\|) \quad \text{as} \quad \|h\| \to 0.$$

Put $D_{(2)} f(x, \phi(x)) = S$. Since S is bijective and ϕ is continuous, it follows from theorem 7.32 that, when $\|h\|$ is sufficiently small,

$$\|f(x, \phi(x+h))\| = \|f(x, \phi(x+h)) - f(x, \phi(x))\|$$

$$\geq \frac{1}{2\|S^{-1}\|} \|\phi(x+h) - \phi(x)\|.$$

Also, since Df is continuous on $M \times N$, for $(u, v)^T \in M \times N$,

$$\|D_{(1)} f(u, v)\| \leq \|Df(u, v)\| \leq A,$$

say. Hence, by theorem 7.31,

$$\|f(x, \phi(x+h))\| = \|f(x+h, \phi(x+h)) - f(x, \phi(x+h))\| \leq A\|h\|.$$

Therefore, when $\|h\|$ is small enough,

$$\frac{1}{2\|S^{-1}\|} \|\phi(x+h) - \phi(x)\| \leq \|f(x, \phi(x+h))\| \leq A\|h\|,$$

or $$\|\phi(x+h) - \phi(x)\| \leq B\|h\|.$$

We can now prove that $D\phi(x)$ exists and is given by (7.45). Putting $\phi(x) = y$ we have

$$f(x+h, y+k) - f(x, y) - Df(x, y)(h, k) = r(h, k),$$

where $$\frac{\|r(h, k)\|}{\|(h, k)\|} \to 0 \quad \text{as} \quad \|(h, k)\| \to 0.$$

Now take $k = \phi(x+h) - \phi(x)$ so that $f(x, y) = f(x+h, y+k) = \theta$.

Then $$Df(x, y)(h, k) = -r(h, k),$$

i.e., by theorem 7.42,

$$D_{(1)}f(x, y)(h) + D_{(2)}f(x, y)(k) = -r(h, k),$$

and so

$$\{[D_{(2)}f(x, y)]^{-1} \circ D_{(1)}f(x, y)\}(h) + k = -[D_{(2)}f(x, y)]^{-1}(r(h, k)).$$

Thus

$$\phi(x+h) - \phi(x) + \{[D_{(2)}f(x, \phi(x))]^{-1} \circ D_{(1)}f(x, \phi(x))\}(h) = R(h),$$

where

$$\frac{\|R(h)\|}{\|h\|} \leqslant \frac{\|S^{-1}\| \|r(h, \phi(x+h) - \phi(x))\|}{\|h\|}$$

$$\leqslant \|S^{-1}\| \frac{\|r(h, \phi(x+h) - \phi(x))\|}{\|(h, \phi(x+h) - \phi(x))\|} \frac{\|h\| + B\|h\|}{\|h\|}$$

$$= C \frac{\|r(h, \phi(x+h) - \phi(x))\|}{\|(h, \phi(x+h) - \phi(x))\|}, \tag{7.48}$$

say. Since $\|\phi(x+h) - \phi(x)\| \to 0$ as $\|h\| \to 0$, the right-hand side of (7.48) tends to 0 as $\|h\| \to 0$.

Finally we show that $D\phi$ is continuous or, equivalently, that the partial derivatives of $\phi_1, \ldots, \phi_n$ are continuous. This follows from the continuity of ϕ and from the matrix equivalent of (7.45) which gives each $D_i\phi_j(x)$ as a rational expression in the partial derivatives of f at the point $(x, \phi(x))^T$. |

Notes.

(i) The formula (7.45) agrees with the result obtained on differentiating the equation

$$f(x, \phi(x)) = \theta \quad (x \in M^\circ)$$

when ϕ is known to be differentiable. In the case $m = n = 1$ the chain rule immediately leads to the identity

$$D_1 f(x, \phi(x)) + D_2 f(x, \phi(x)) \phi'(x) = 0$$

(cf. theorem 7.41). In general define the function $g: M^\circ \to R^m \times R^n$ by

$$g(x) = (x, \phi(x))^T.$$

Then (see exercise 7(*a*), 7) $Dg(x)$ is given by

$$Dg(x)(h) = (h, D\phi(x)(h))^T.$$

Hence, putting $g(x) = z$ we have, by theorems 7.22 and 7.42,

$$\theta = D(f \circ g)(x)(h) = [Df(z) \circ Dg(x)](h) = Df(z)[Dg(x)(h)]$$

$$= Df(z)[(h, D\phi(x)(h))] = D_{(1)}f(z)(h) + D_{(2)}f(z)[D\phi(x)(h)].$$

Thus $D_{(2)} f(x, \phi(x)) \circ D\phi(x) = -D_{(1)} f(x, \phi(x))$.

(ii) If, in the theorem, $m = n$, so that the matrices of $D_{(1)} f(x, y)$, $D_{(2)} f(x, y)$ and $D\phi(x)$ all have n rows and n columns, then

$$J\phi(x) = (-1)^n \frac{J_{(1)} f(x, \phi(x))}{J_{(2)} f(x, \phi(x))}.$$

(For if A is an $n \times n$ matrix, then det $(-A) = (-1)^n$ det A.)

(iii) The condition $J_{(2)} f(\xi, \eta) \neq 0$ is not necessary to ensure the existence, in the neighbourhood of the point $(\xi, \eta)^T$, of a unique solution $y = \phi(x)$ of the equation $f(x, y) = \theta$. For instance the equation

$$f(x, y) = x^3 - y^3 = 0 \quad (x, y \in R^1)$$

has the unique solution $y = x$, but $J_{(2)} f(0, 0) = 0$.

Illustration. Here, as so often in particular examples, we discard the suffix notation for vectors and use x, y, u, etc. as co-ordinates.

The equations

$$\left. \begin{array}{l} f_1(x, y, u, v, w) = x(u+v)+yw-3 = 0, \\ f_2(x, y, u, v, w) = x(u^2+v^2)+yw^2-5 = 0, \\ f_3(x, y, u, v, w) = x(u^3+v^3)+yw^3-9 = 0 \end{array} \right\} \tag{7.49}$$

are satisfied at the point $(1, 1, 0, 1, 2)^T$.

If $p = (x, y)^T, q = (u, v, w)^T$ and $f = (f_1, f_2, f_3)^T$, the equations (7.49) may be written
$$f(p, q) = \theta.$$
Then

$$J_{(2)} f(p, q) = \begin{vmatrix} x & x & y \\ 2xu & 2xv & 2yw \\ 3xu^2 & 3xv^2 & 3yw^2 \end{vmatrix} = 6x^2 y(v-w)(w-u)(u-v)$$

and so, if $p_0 = (1, 1)^T, q_0 = (0, 1, 2)^T$,

$$J_{(2)} f(p_0, q_0) = 12.$$

Therefore there is a solution $q = \phi(p)$, i.e.

$$u = \phi_1(x, y), \quad v = \phi_2(x, y) \quad w = \phi_3(x, y)$$

which passes through the point $(p_0, q_0)^T$ or $(1, 1, 0, 1, 2)^T$, and the solution is unique in the neighbourhood of this point.

The matrices of $D_{(1)} f(p, q)$ and $D_{(2)} f(p, q)$ are

$$\begin{pmatrix} u+v & w \\ u^2+v^2 & w^2 \\ u^3+v^3 & w^3 \end{pmatrix}, \quad \begin{pmatrix} x & x & y \\ 2xu & 2xv & 2yw \\ 3xu^2 & 3xv^2 & 3yw^2 \end{pmatrix},$$

respectively. Therefore, by (7.45), the matrix of $-D\phi(p_0)$ is

$$\begin{pmatrix} 1 & 1 & 1 \\ 0 & 2 & 4 \\ 0 & 3 & 12 \end{pmatrix}^{-1} \begin{pmatrix} 1 & 2 \\ 1 & 4 \\ 1 & 8 \end{pmatrix} = \frac{1}{12} \begin{pmatrix} 12 & -9 & 2 \\ 0 & 12 & -4 \\ 0 & -3 & 2 \end{pmatrix} \begin{pmatrix} 1 & 2 \\ 1 & 4 \\ 1 & 8 \end{pmatrix} = \frac{1}{12} \begin{pmatrix} 5 & 4 \\ 8 & 16 \\ -1 & 4 \end{pmatrix}.$$

Thus
$$D_1\phi_1(1, 1) = -\tfrac{5}{12}, \qquad D_2\phi_1(1, 1) = -\tfrac{1}{3},$$
$$D_1\phi_2(1, 1) = -\tfrac{2}{3}, \qquad D_2\phi_2(1, 1) = -\tfrac{4}{3},$$
$$D_1\phi_3(1, 1) = \tfrac{1}{12}, \qquad D_2\phi_3(1, 1) = -\tfrac{1}{3}.$$

Exercises 7(d)

1. Let f be a function with domain $E \subset R^m \times R^n$ and range in R^p and let $(\xi, \eta)^T$ be an interior point of E.

(i) Show that both $D_{(1)}f(\xi,\eta)$ and $D_{(2)}f(\xi, \eta)$ may exist without f being differentiable at $(\xi, \eta)^T$.

(ii) Show that, if both linear partial derivatives of f exist in an open ball with centre $(\xi, \eta)^T$ and $D_{(1)}f$, $D_{(2)}f$ are continuous at $(\xi, \eta)^T$, then f is differentiable at $(\xi, \eta)^T$.

2. Prove that, if $(\xi, \eta)^T$ satisfies the equation

$$xy - \log x + \log y = 0 \quad (x, y > 0),$$

then there is a continuously differentiable solution $y = \phi(x)$ which passes through this point and is unique in its neighbourhood. Prove also that there is one point $(\xi, \eta)^T$ such that the corresponding solution $y = \phi(x)$ has a maximum at ξ.

3. The function $f: R^1 \to R^1$ is continuously differentiable, f' is strictly increasing and $\lim_{x \to \infty} f'(x) = \infty$, $\lim_{x \to -\infty} f'(x) = -\infty$. Prove that, to any $\xi \neq 0$, there corresponds an η such that

$$f(\xi + \eta) = f(\xi) + f(\eta)$$

and that through such a point $(\xi, \eta)^T$ there passes a solution $y = \phi(x)$ of the equation

$$f(x + y) = f(x) + f(y)$$

which is unique in the neighbourhood of $(\xi, \eta)^T$.

Construct an example to show that, when $\xi = 0$, there may be no corresponding η.

4. Find a point $(\xi, \eta, \zeta)^T$ in the neighbourhood of which the equation

$$\sin yz + \sin zx + \sin xy = 0$$

has a unique solution $z = \phi(x, y)$.

5. Show that in the neighbourhood of any point which satisfies the equations

$$x^4 + (x + z)y^3 - 3 = 0,$$
$$x^4 + (2x + 3z)y^3 - 6 = 0$$

there is a unique solution $y = \phi_1(x)$, $z = \phi_2(x)$ of these equations.

6. Show that the equations

$$xy^5 + yu^5 + zv^5 = 1,$$
$$x^5y + y^5u + z^5v = 1$$

have a unique solution $u = \phi_1(x, y, z)$, $v = \phi_2(x, y, z)$ in the neighbourhood of the point

$$(x, y, z, u, v)^T = (0, 1, 1, 1, 0)^T$$

and find the matrix of $D\phi(0, 1, 1)$ (where $\phi = (\phi_1, \phi_2)^T$).

7. Verify that there are points which satisfy the equations

$$x^2 + u + e^v = 0,$$

$$y^2 + v + e^w = 0,$$

$$z^2 + w + e^u = 0.$$

Prove that, if $(x_0, y_0, z_0, u_0, v_0, w_0)^T$ is such a point, then in its neighbourhood there exists a unique solution $u = \phi_1(x, y, z)$, $v = \phi_2(x, y, z)$, $w = \phi_3(x, y, z)$. Express the matrix of $D\phi(x, y, z)$ (where $\phi = (\phi_1, \phi_2, \phi_3)^T$) in terms of x, y, z, u, v, w.

8. Show that there are points which satisfy the equations

$$x - e^u \cos v = 0,$$

$$v - e^y \sin x = 0.$$

Prove that, if $(x_0, y_0, u_0, v_0)^T$ is such a point, then in its neighbourhood there exists a unique solution $(u, v)^T = \phi(x, y)$. Show that

$$J\phi(x, y) = v/x.$$

9. The functions $f, g : R^1 \to R^1$ are continuously differentiable and f', g' are strictly positive. Suppose that the equations

$$f(xu) + g(yv) = 0,$$

$$f(yv) + g(zw) = 0,$$

$$f(zw) + g(xu) = 0$$

are satisfied at the point $(x_0, y_0, z_0, u_0, v_0, w_0)^T$, where $x_0, y_0, z_0 \neq 0$. Show that, in a neighbourhood of this point, there is a unique solution $(u, v, w)^T = \phi(x, y, z)$ with Jacobian

$$J\phi(x, y, z) = -\frac{uvw}{xyz}.$$

7.5. The inverse function theorem

Let f be a function with domain and range in R^n. The question whether f has an inverse can be put in the form: given $y = (y_1, ..., y_n)^T$ in the range of f, does the vector equation

$$f(x) - y = \theta,$$

i.e. the system $f_1(x_1, ..., x_n) - y_1 = 0,$

$$\cdots\cdots\cdots\cdots\cdots\cdots\cdots$$

$$f_n(x_1, ..., x_n) - y_n = 0,$$

have a unique solution for the vector $x = (x_1, ..., x_n)^T$ in terms of $y = (y_1, ..., y_n)^T$? The problem of inversion is therefore little more than part of the larger problem, solved in the previous section, of the existence of implicitly defined functions.

Theorem 7.51. (Inverse function theorem.) *Let ξ be an interior point of a set E in R^n and suppose that the function $f: E \to R^n$ satisfies the following conditions:*

(i) *f is continuously differentiable in an open set containing ξ;*

(ii) *$Jf(\xi) \neq 0$.*

Then there exists an open set U containing ξ such that the restriction of f to U is injective; the inverse function ϕ is continuously differentiable on the open set $V = f(U)$; and if $y \in V$ and $y = f(x)$, where $x \in U$, then

$$D\phi(y) = [Df(x)]^{-1}. \tag{7.51}$$

Proof. Define the function $g: E \times R^n \to R^n$ by

$$g(x, y) = f(x) - y \quad (x \in E, y \in R^n).$$

Then g is continuously differentiable at a point $(x, y)^T$ as long as f is continuously differentiable at x. This follows from the continuity of the partial derivatives $D_i g_j$ which are given by

$$D_i g_j(x, y) = \begin{cases} D_i f_j(x) & \text{if } 1 \leqslant i \leqslant n, \\ -1 & \text{if } n+1 \leqslant i \leqslant 2n. \end{cases}$$

Hence, if $f(\xi) = \eta$, then $g(\xi, \eta) = \theta$ and g is continuously differentiable in an open ball with centre $(\xi, \eta)^T$. Also clearly $D_{(1)}g(\xi, \eta) = Df(\xi)$ and so $J_{(1)}g(\xi, \eta) = Jf(\xi) \neq 0$. We may therefore apply theorem 7.43 to g (with the roles of x and y interchanged). This means that there are intervals

$$M = [\xi_1 - \alpha, \xi_1 + \alpha] \times \ldots \times [\xi_n - \alpha, \xi_n + \alpha],$$

$$N = [\eta_1 - \beta, \eta_1 + \beta] \times \ldots \times [\eta_n - \beta, \eta_n + \beta]$$

and a continuous function $\phi: N \to M$ such that $x = \phi(y)$ is the only solution lying in $M \times N$ of the equation $f(x) = y$. In other words, the restriction of f to $\phi(N) (\subset M)$ is injective and the inverse function ϕ is continuous.

Since ϕ has a continuous inverse, by theorem 3.23, $\phi(N^\circ)$ is open; and, since $\eta \in N^\circ$, $\xi = \phi(\eta) \in \phi(N^\circ)$. Hence there is an open ball $U = B(\xi; \delta)$ contained in $\phi(N^\circ)$ and, again by theorem 3.23, $V = f(U)$ is open.

Finally, ϕ is continuously differentiable in N° and so in V and

$$D\phi(y) = -[D_{(1)}g(\phi(y), y)]^{-1} \circ D_{(2)}g(\phi(y), y).$$

Thus, if $\phi(y) = x$ and I is the identity function on R^n,

$$D\phi(y) = -[Df(x)]^{-1} \circ (-I) = [Df(x)]^{-1}. \ |$$

(The formula 7.51 also follows from theorem 7.22 once the existence and differentiability of ϕ have been established.)

Corollary. *Under the conditions of the theorem*

$$J\phi(y) = 1/Jf(x).$$

Illustration. The function $f: R^2 \to R^2$ is given by

$$f(x, y) = (xe^y, xe^{-y})^T.$$

We have
$$Jf(x, y) = \begin{vmatrix} e^y & xe^y \\ e^{-y} & -xe^{-y} \end{vmatrix} = -2x.$$

Hence theorem 7.51 guarantees the existence of a local inverse $(x, y)^T = \phi(u, v)$, i.e. a unique solution of the equations

$$xe^y = u, \ xe^{-y} = v, \tag{7.52}$$

in the neighbourhood of any point $(\xi, \eta)^T$ such that $\xi \neq 0$. The matrix of $D\phi(u, v)$ is

$$\begin{pmatrix} e^y & xe^y \\ e^{-y} & -xe^{-y} \end{pmatrix}^{-1} = \frac{1}{2x}\begin{pmatrix} xe^{-y} & xe^y \\ e^{-y} & -e^y \end{pmatrix}. \tag{7.53}$$

In fact it is easy to find ϕ explicitly. The range of f is the set $\{(u, v)^T \,|\, uv \geqslant 0\}$ and the equations (7.52) lead to

$$x = \pm(uv)^{\frac{1}{2}}, \ y = \log(u/v)^{\frac{1}{2}}$$

when $u, v \neq 0$, i.e. $x \neq 0$. If $\xi > 0$, the local inverse is $x = (uv)^{\frac{1}{2}}$, $y = \log(u/v)^{\frac{1}{2}}$ and, if $\xi < 0$, it is $x = -(uv)^{\frac{1}{2}}$, $y = \log(u/v)^{\frac{1}{2}}$. When $\xi = 0$, there is no inverse for then $u = v = 0$ and f maps any point $(0, y)^T$ into $(0, 0)^T$. Finally we note that the matrix of $D\phi(u, v)$ is

$$\begin{pmatrix} \pm\frac{1}{2}u^{-\frac{1}{2}}v^{\frac{1}{2}} & \frac{1}{2}u^{\frac{1}{2}}v^{-\frac{1}{2}} \\ \frac{1}{2}u^{-1} & -\frac{1}{2}v^{-1} \end{pmatrix}$$

according as $x = (uv)^{\frac{1}{2}}$ or $x = -(uv)^{\frac{1}{2}}$, and this is the same as (7.53).

The one-dimensional form of the inverse function theorem is not new (see C1, 62 and 69) and may be extended to provide for a global instead of only a local inverse. For if a real function has a continuous, non-vanishing derivative in an interval, then the function is strictly monotonic and therefore injective. In higher dimensions there is no corresponding extension. If the Jacobian of a function f does not vanish on an open set, then f need not be injective on that set (as a whole) even if the set is connected. For instance the function $f: R^2 \to R^2$ given by

$$f_1(x, y) = e^x \cos y, \quad f_2(x, y) = e^x \sin y$$

is not injective, since all the points $(x, y + 2n\pi)^T$ $(n = 0, \pm 1, \pm 2, \dots)$ have the same image; but $Jf(x, y) = e^{2x} > 0$ for all $(x, y)^T \in R^2$. Nevertheless the next theorem shows that the non-vanishing of the Jacobian entails a property which had previously been known only for functions with continuous inverses.

Theorem 7.52. *Let E be a subset of R^n. Suppose that the function $f: E \to R^n$ is continuously differentiable on the open set $G \subset E$ and that $Jf(x) \neq 0$ for $x \in G$. Then $f(G)$ is open.*

Proof. Take any point $\eta \in f(G)$ and let $\xi \in G$ be such that $f(\xi) = \eta$. Then ξ is an interior point of G and, by theorem 7.51, there is an open set U in G which contains ξ and is such that $V = f(U)$ is open. Since $\eta \in V$ and $V \subset f(G)$, it follows that η is an interior point of $f(G)$. |

Exercises 7(e)

1. The function f from R^n to R^n is differentiable at the point ξ. Prove that the condition $Jf(\xi) \neq 0$ is not necessary to secure the existence of an inverse in the neighbourhood of the point ξ; but that it *is* necessary for ensuring the differentiability of this inverse.

2. Show that the function $f: R^2 \to R^2$ given by

$$f(x, y) = (\cos x + \cos y, \quad \sin x + \sin y)^T$$

has an inverse in the neighbourhood of all points $(\xi, \eta)^T$ such that $\xi - \eta \neq n\pi$ $(n = 0, \pm 1, \pm 2, \dots)$ and that at all other points no local inverse exists.

3. Prove that, if $\xi, \eta, \zeta \neq 0$, then the function $f: R^3 \to R^3$ given by

$$f(x, y, z) = (yz, zx, xy)^T$$

has an inverse ϕ in the neighbourhood of $(\xi, \eta, \zeta)^T$ and find the matrix of $D\phi(yz, zx, xy)$.

4. Prove that the function $f: R^2 \to R^2$ given by

$$f(x, y) = \left(\frac{x}{\sqrt{(1 + x^2 + y^2)}}, \quad \frac{y}{\sqrt{(1 + x^2 + y^2)}} \right)^T$$

has an inverse in the neighbourhood of every point of R^2. If ϕ is the local inverse at $(x, y)^T$, find $J\phi(f(x, y))$.

5. The function $f = (f_1, f_2, f_3)^T : R^3 \to R^3$ is given by

$$f_1(x, y, z) = x+y+z,$$
$$f_2(x, y, z) = yz+zx+xy,$$
$$f_3(x, y, z) = xyz;$$

and ξ, η, ζ are (pairwise) unequal real numbers. Show that f has an inverse ϕ in the neighbourhood of the point $(\xi, \eta, \zeta)^T$ and that

$$J\phi(f(x, y, z)) = -[(y-z)(z-x)(x-y)]^{-1}.$$

6. Show that the function $f: R^n \to R^n$ $(n \geqslant 3)$ given by

$$f(x_1, ..., x_n) = (x_1+x_2, \; x_1 x_2+x_3, \; ..., \; x_1 x_2...x_{n-1}+x_n, \; x_2...x_{n-1}+x_n)^T$$

has an inverse ϕ in the neighbourhood of the point $(1, 1, ..., 1)^T$ and prove that $J\phi(2, 2, ..., 2) = (-1)^n$.

7. Let G be an open subset of R^n and suppose that the function $f: G \to R^n$ is continuously differentiable and that $Jf(x) \neq 0$ for all x in G. Show that the image $f(F)$ of a closed subset F of G need not be closed.

7.6. Functional dependence

The last theorem was stated as a result on mappings. It acquires a different aspect if its conclusion is replaced by the following (weaker) statement: a function Ψ from R^n to R^1 for which $\Psi(f(x)) = 0$ when $x \in G$ is such that $\Psi(y) = 0$ in an open set. In this form the theorem is related to the notion of *functional dependence*.

Definition. *The real-valued functions $f_1, ..., f_n$ with domain E in R^m are said to be* functionally dependent *in an open subset G of E if there is a function Ψ from R^n to R^1 which does not vanish identically in any open set and which is such that $\Psi(f_1(x), ..., f_n(x)) = 0$ for $x \in G$.*

The term *functional independence* has the obvious meaning and theorem 7.52 now leads to the following result.

Theorem 7.61. *Let E be a subset of R^n. If $f = (f_1, ..., f_n)^T : E \to R^n$ is continuously differentiable in the open set $G \subset E$ and $Jf(x)$ is not identically zero for $x \in G$, then $f_1, ..., f_n$ are functionally independent in G.*

Proof. Let $\xi \in G$ be such that $Jf(\xi) \neq 0$. Then $Jf(x) \neq 0$ for x in an open subset H of G and $f(H)$ is open. Hence, if $\Phi(f(x)) = 0$ for all $x \in G$, then $\Phi(y) \equiv 0$ in the open set $f(H)$. |

The pair of functions f_1, f_2 on R^2 given by

$$f_1(x, y) = x \cos y, \quad f_2(x, y) = x \sin y$$

is independent in any open subset of R^2 since every such set contains points at which the Jacobian x does not vanish. A simple example of functional dependence is provided by

$$g_1(x, y) = \cos(x+y), \quad g_2(x, y) = \sin(x+y).$$

For the function Ψ on R^2 given by

$$\Psi(u, v) = u^2 + v^2 - 1$$

only vanishes on the circle $u^2 + v^2 = 1$ and $\Psi(g_1(x, y), g_2(x, y)) = 0$ for all $(x, y)^T \in R^2$.

The property of *linear* dependence, familiar from algebra, is, of course, a particular case of functional dependence. Another special case arises from the next definition.

Definition. *Let E be a subset of R^m and let G be an open set contained in E. If the functions*

$$f = (f_1, ..., f_n)^T : E \to R^n, \quad g = (g_1, ..., g_p)^T : E \to R^p$$

are such that $$g(x) = \Phi(f(x)) \quad (x \in G)$$

for a suitable function Φ from R^n to R^p, then g is said to be functionally dependent on f in G.

It is easy to see that, if g is functionally dependent on f, then each set of functions $f_1, ..., f_n, g_j$ $(j = 1, ..., p)$ is functionally dependent. (See exercise $7(f), 7$.)

Theorem 7.62. *Let $(\xi, \eta)^T$ ($\xi \in R^m$, $\eta \in R^n$) be an interior point of a set E in $R^m \times R^n$ and suppose that the functions $f = (f_1, ..., f_m)^T : E \to R^m$ and $g = (g_1, ..., g_p)^T : E \to R^p$ satisfy the following conditions:*

(i) *f and g are continuously differentiable in an open set G containing $(\xi, \eta)^T$;*

(ii) *$J_{(1)} f(\xi, \eta) \neq 0$;*

(iii) *the function $h = (f_1, ..., f_m, g_1, ..., g_p)^T$ is such that, for $(x, y)^T \in G$, the rank of the matrix of $Dh(x, y)$ is m.*

Then there is an open set containing $(\xi, \eta)^T$ in which g is functionally dependent on f.

Proof. Define the function $f^* : E \times R^m \to R^m$ by

$$f^*(x, y, z) = f(x, y) - z \quad ((x, y)^T \in E, z \in R^m). \tag{7.61}$$

Then f^* is clearly continuously differentiable in an open set containing $(\xi, \eta, \zeta)^T$, where $\zeta = f(\xi, \eta)$. Also $f^*(\xi, \eta, \zeta) = \theta$ and

$J_{(1)}f^*(\xi, \eta, \zeta) = J_{(1)}f(\xi, \eta) \neq 0$. Therefore, by theorem 7.43, there are closed intervals

$$M = [\xi_1 - \alpha, \xi_1 + \alpha] \times \dots \times [\xi_m - \alpha, \xi_m + \alpha],$$
$$N = [\eta_1 - \beta, \eta_1 + \beta] \times \dots \times [\eta_n - \beta, \eta_n + \beta],$$
$$P = [\zeta_1 - \beta, \zeta_1 + \beta] \times \dots \times [\zeta_m - \beta, \zeta_m + \beta]$$

and a continuous function $\phi: N \times P \to M$ such that $x = \phi(y, z)$ is the only solution lying in $M \times N \times P$ of the equation $f^*(x, y, z) = \theta$. The intervals M, N are taken to be so small that $M \times N \subset G$ and $J_{(1)}f(x, y) \neq 0$ in $M \times N$.

For $i = 1, \dots, m$,

$$f_i^*(\phi(y, z), y, z) = 0 \quad ((y, z)^T \in N \times P).$$

Since ϕ has components $\phi_1, \dots, \phi_m$, the chain rule gives

$$D_1 f_i^*(\phi(y, z), y, z) D_j \phi_1(y, z) + \dots + D_m f_i^*(\phi(y, z), y, z) D_j \phi_m(y, z)$$
$$+ D_{m+j} f_i^*(\phi(y, z), y, z) = 0$$

for $j = 1, \dots, n$. Writing $\phi(y, z) = x$ and using (7.61) we now have

$$D_1 f_i(x, y) D_j \phi_1(y, z) + \dots + D_m f_i(x, y) D_j \phi_m(y, z)$$
$$= - D_{m+j} f_i(x, y). \quad (7.62)$$

Now let $g^*: N \times P \to R^p$ be the function defined by

$$g^*(y, z) = g(\phi(y, z), y) \quad ((y, z)^T \in N \times P).$$

Considering kth components we obtain from the chain rule

$$D_1 g_k(x, y) D_j \phi_1(y, z) + \dots + D_m g_k(x, y) D_j \phi_m(y, z)$$
$$- D_j g_k^*(y, z) = - D_{m+j} g_k(x, y) \quad (7.63)$$

for $j = 1, \dots, n$, where again $x = \phi(y, z)$.

Let j and k be fixed. From the linear equations (7.62) corresponding to $i = 1, \dots, m$ and the equation (7.63) we obtain

$$D_j g_k^*(y, z) \begin{vmatrix} D_1 f_1(x, y) \dots D_m f_1(x, y) & 0 \\ \dots\dots\dots\dots\dots\dots\dots\dots & \\ D_1 f_m(x, y) \dots D_m f_m(x, y) & 0 \\ D_1 g_k(x, y) \dots D_m g_k(x, y) & -1 \end{vmatrix}$$

$$= \begin{vmatrix} D_1 f_1(x, y) \dots D_m f_1(x, y) & - D_{m+j} f_1(x, y) \\ \dots\dots\dots\dots\dots\dots\dots\dots\dots\dots & \\ D_1 f_m(x, y) \dots D_m f_m(x, y) & - D_{m+j} f_m(x, y) \\ D_1 g_k(x, y) \dots D_m g_k(x, y) & - D_{m+j} g_k(x, y) \end{vmatrix}.$$

The determinant on the left is equal to $-J_{(1)}f(x, y) \neq 0$, and the determinant on the right is zero, by condition (iii). Hence

$$D_j g_k^*(y, z) = 0$$

for $(y, z)^T \in N \times P$. This holds for $j = 1, \ldots, n$ and $k = 1, \ldots, m$. Therefore, by theorem 7.33, given any $z \in P$, $g^*(y, z)$ is constant for $y \in N$. Let $\Phi: P \to R^p$ be the function defined by

$$\Phi(z) = g^*(y, z) \quad (y \in N, z \in P).$$

There is a positive number $r < \min(\alpha, \beta)$ such that

$$\|f(x, y) - f(\xi, \eta)\| < \beta$$

if $\|x - \xi\| < r$ and $\|y - \eta\| < r$. Denote the open balls $B(\xi; r)(\subset R^m)$ and $B(\eta; r)(\subset R^n)$ by U, V respectively.

Take any point $(x, y)^T \in U \times V$ and put $z = f(x, y)$. We then have $x \in M, y \in N, z \in P$ and $f^*(x, y, z) = \theta$. But, given any point $(v, w)^T \in N \times P$, $u = \phi(v, w)$ is the only point in M satisfying

$$f^*(u, v, w) = \theta.$$

It follows that $x = \phi(y, z)$. Therefore the function Φ is such that, for $(x, y)^T \in U \times V$,

$$g(x, y) = g(\phi(y, z), y) = g^*(y, z) = \Phi(z) = \Phi(f(x, y)).$$

Incidentally, since ϕ is continuously differentiable on $N^\circ \times P^\circ$, Φ is continuously differentiable on P°. $\mathbf{|}$

The theorem is again *local*. Even if condition (ii) is strengthened to $J_{(1)}f(x, y)$ being non-zero at all points of the open set G (which may be connected), there need not be a function Φ such that $g(x) = \Phi(f(x))$ *throughout* G. Take, for instance,

$$f_1(x_1, x_2, y) = e^{x_1+y} \cos x_2, \quad f_2(x_1, x_2, y) = e^{x_1+y} \sin x_2,$$

$$g(x_1, x_2, y) = e^{x_1+y} x_2.$$

Then $J_{(1)}f(x, y) = e^{2x_1+2y} > 0$ in R^3 and the matrix of $Dh(x, y)$ has constant rank 2, but clearly different functions Φ have to be used in the strips $2k\pi \leqslant x_2 < (2k+2)\pi$ $(k = 0, \pm 1, \pm 2, \ldots)$.

The analogue of theorem 7.62 in which $n = 0$ (i.e. y is absent) takes the obvious form. The proof only requires an application of the inverse function theorem. (See exercise $7(f), 9$.) Changing our notation slightly we now combine the two cases $n > 0$ and $n = 0$.

Theorem 7.63. *Let ξ be an interior point of a set E in R^m and suppose that the function $f = (f_1, \ldots, f_n)^T : E \to R^n$ satisfies the following conditions:*

(i) *f is continuously differentiable in an open set G containing ξ;*

(ii) *for $x \in G$, the rank of the matrix of $Df(x)$ is r, where $0 < r < n$.*

Then there is an open set containing ξ in which $n - r$ of the functions $f_1, \ldots, f_n$ are functionally dependent on the remaining r functions.

(When $r = 0$ all the functions $f_1, \ldots, f_n$ are constant in all connected open subsets of G.)

Exercises 7(f)

1. Show that the functions f, g, h on R^3 to R^1 given by

$$f(x, y, z) = x \sin y \cos z, \quad g(x, y, z) = x \sin y \sin z, \quad h(x, y, z) = x \cos y$$

are functionally independent in every open subset of R^3.

2. Show that the functions f, g on R^2 to R^1 given by

$$f(x, y) = xy + e^{2xy}, \quad g(x, y) = xy - e^{2xy}$$

are functionally dependent in R^2.

3. Prove that there is an interval $[-\alpha, \alpha]^3 \times [-\beta, \beta]^3$ in R^6 in which the equations

$$x + y + z = u^3 + v,$$
$$x^2 + y^2 + z^2 = v^3 + w,$$
$$x^3 + y^3 + z^3 = w^3 + u$$

have a unique solution $u = \phi_1(x, y, z)$, $v = \phi_2(x, y, z)$, $w = \phi_3(x, y, z)$. Prove also that ϕ_1, ϕ_2, ϕ_3 are functionally independent in $(-\alpha, \alpha)^3$.

4. Prove that the functions $f_1, \ldots, f_n$ on $E(\subset R^m)$ to R^1 are functionally dependent in the open subset G of E if and only if $f(G)$ (where $f = (f_1, \ldots, f_n)^T$) has no interior point.

5. Given a real-valued function F on a metric space (X, ρ), denote the set $\{x \in X \mid F(x) \neq 0\}$ by A. Then the *support* of F is defined as the set $\bar{A}$ (i.e. as the smallest closed set outside which $F(x) \equiv 0$).

Let $f = (f_1, \ldots, f_n)^T$ be a function with domain E in R^m and range in R^n. Show that $f_1, \ldots, f_n$ are functionally dependent in an open subset G of E if and only if there exists a function Ψ from R^n to R^1 which is such that its support contains $f(G)$ and $\Psi(f(x)) = 0$ for $x \in G$.

6. The real-valued functions $f_1, \ldots, f_n$ with domain $E \subset R^m$ are functionally dependent in an open subset G of E. Show that, if g is any real-valued function on E, then $f_1, \ldots, f_n, g$ are functionally dependent in G.

7. Let E be a subset of R^m and let G be an open subset of E. Show that if, in G, $g : E \to R^1$ is functionally dependent on $f = (f_1, \ldots, f_n)^T : E \to R^n$, then the set of functions $f_1, \ldots, f_n, g$ is functionally dependent in G.

8. The (continuously differentiable) real functions f_1, f_2 are defined by

$$f_1(x) = x^2, \quad f_2(x) = \begin{cases} x^4 \sin(1/x) & (x \neq 0), \\ 0 & (x = 0). \end{cases}$$

Show (by use of **4**) that f_1, f_2 are functionally dependent in R^1, but that neither of f_1, f_2 is functionally dependent on the other in any open set containing the origin.

9. The set G in R^m is open and the function $f: G \to R^n$ is injective. Prove that any function $g: G \to R^p$ is functionally dependent on f in G.

10. The functions f, g, h on R^2 to R^1 are given by

$$f(x,y) = \tfrac{1}{2}(x^2 - y^2), \quad g(x, y) = xy, \quad h(x, y) = (x^2 + y^2) \cos x.$$

Show that, in R^2, h is functionally dependent on f, g.

11. The functions f, g, h on R^4 to R^1 are defined by

$$f(w, x, y, z) = w + x + y + z, \quad g(w, x, y, z) = w^2 + x^2 + y^2 + z^2,$$

$$h(w, x, y, z) = w^3 + x^3 + y^3 + z^3 - 3(xyz + yzw + zwx + wxy).$$

Show that, in R^4, h is functionally dependent on f, g.

7.7. Maxima and minima

A real valued function f on a set X in R^m is said to have a *(local)* *maximum* at the interior point ξ of X if there is a $\delta > 0$ such that $f(x) \leqslant f(\xi)$ whenever $x \in B(\xi; \delta)$; a *minimum* of f is similarly defined. The corresponding definitions in C1 for functions on R^1 (§4.7) and on R^2 (§8.8) differ slightly in requiring strict inequalities for x in $B(\xi; \delta) - \{\xi\}$. We have here adopted the more comprehensive definition because in this chapter it will be convenient to be sure that a function has a maximum (minimum) at a point where it assumes its supremum (infimum).

It is easy to see that, when f is differentiable at ξ, a necessary condition for f to have a maximum or minimum at ξ is that

$$Df(\xi) = \Theta,$$

the zero linear function. (See exercise 7(g), 1.) There are also sufficient conditions for a maximum or minimum; the cases $m = 1, 2$ are treated in C1 (§4.7, 8.8); for the general case see exercise 7(h), 6. However our present concern is to extend the theory in a different direction: we shall consider maxima and minima of functions relative to the values taken on subsets of their domains. The most interesting case is that in which the subsets are determined by equations of the form $g(x) = \theta$.

Definition. Let X be a subset of R^m and let g be a function on X with values in R^q, where $1 \leqslant q < m$. Denote the set $\{x \,|\, g(x) = \theta\}$ by X_0 and suppose that the point ξ in X_0 is an interior point of X. The function $f\colon X \to R^1$ is now said to have, at ξ, a maximum [minimum] subject to the constraint $g(x) = \theta$ if there is a $\delta > 0$ such that $f(x) \leqslant f(\xi)$ $[f(x) \geqslant f(\xi)]$ whenever $x \in X_0 \cap B(\xi; \delta)$.

The next theorem gives a necessary condition for a constrained maximum or minimum to occur at a given point.

Theorem 7.71. Let ξ be an interior point of a set X in R^m $(m \geqslant 2)$ and suppose that the function $g = (g_1, ..., g_q)^T \colon X \to R^q$, where $1 \leqslant q < m$, satisfies the following conditions:

(i) $g(\xi) = \theta$;

(ii) g is continuously differentiable in an open set G containing ξ;

(iii) the matrix of $Dg(\xi)$ has (maximum) rank q.

If the function $f\colon X \to R^1$ is also differentiable in G and if f, subject to the constraint $g(x) = \theta$, has a maximum or a minimum at the point ξ, then there are real numbers $\lambda_1, ..., \lambda_q$ such that

$$D_i f(\xi) + \lambda_1 D_i g_1(\xi) + ... + \lambda_q D_i g_q(\xi) = 0 \qquad (7.71)$$

for $i = 1, ..., m$.

The numbers $\lambda_1, ..., \lambda_q$ are called Lagrange's multipliers.

Proof. We may suppose that

$$\begin{vmatrix} D_{m-q+1} g_1(\xi) ... D_m g_1(\xi) \\ \cdots\cdots\cdots\cdots\cdots \\ D_{m-q+1} g_q(\xi) ... D_m g_q(\xi) \end{vmatrix} \neq 0. \qquad (7.72)$$

Now put $p = m - q$ and regard f, g as functions with domains in $R^p \times R^q$. Also write $\xi = (\eta, \zeta)^T$ $(\eta \in R^p, \zeta \in R^q)$.

By the implicit function theorem there are an open interval

$$M^\circ = (\eta_1 - \alpha, \eta_1 + \alpha) \times ... \times (\eta_p - \alpha, \eta_p + \alpha)$$

and a continuously differentiable function $\phi = (\phi_1, ..., \phi_q)^T \colon M^\circ \to R^q$ such that $\phi(\eta) = \zeta$ and

$$g(y, \phi(y)) = \theta \quad \text{for} \quad y \in M^\circ.$$

Take any $h \in R^p$ and let $D\phi(\eta)(h) = k(\in R^q)$. By equation (7.45) we then have $D_{(1)} g(\eta, \zeta)(h) + D_{(2)} g(\eta, \zeta)(k) = \theta$;

and, by theorem 7.42, this is equivalent to

$$Dg(\eta, \zeta)(h, k) = \theta. \tag{7.73}$$

Define the function $\psi = (\psi_1, ..., \psi_m)^T : (-\alpha, \alpha) \to R^m = R^{p+q}$ by

$$\psi_i(t) = \begin{cases} \eta_i + th_i & (1 \leqslant i \leqslant p), \\ \phi_{i-p}(\eta + th) & (p+1 \leqslant i \leqslant p+q). \end{cases}$$

Then, by the chain rule,

$$\psi_i'(0) = \begin{cases} h_i & (1 \leqslant i \leqslant p), \\ D\phi_{i-p}(\eta)(h) & (p+1 \leqslant i \leqslant p+q). \end{cases}$$

Since $D\phi(\eta)(h) = k$, the last line may be written

$$\psi_{p+l}'(0) = k_l \quad (1 \leqslant l \leqslant q).$$

Next, let $\chi = f \circ \psi$, so that

$$\chi(t) = f(\eta + th, \phi(\eta + th)) \quad (-\alpha < t < \alpha). \tag{7.74}$$

Then χ has a maximum or a minimum at 0 and so $\chi'(0) = 0$. Using the note after theorem 7.22 we therefore have

$$0 = \chi'(0) = Df(\eta, \phi(\eta))(\psi'(0)) = Df(\eta, \zeta)(\psi'(0))$$
$$= D_1 f(\eta, \zeta) h_1 + ... + D_p f(\eta, \zeta) h_p$$
$$+ D_{p+1} f(\eta, \zeta) k_1 + ... + D_{p+q} f(\eta, \zeta) k_q. \tag{7.75}$$

We have shown that, given $h \in R^p$, there is $k \in R^q$ such that (7.75) and (7.73) hold. Putting

$$D_i f(\eta, \zeta) = a_i, \quad D_i g_j(\eta, \zeta) = b_{ji}$$

for $i = 1, ..., m = p+q$ and $j = 1,..., q$ we express these equations as

$$a_1 h_1 + ... + a_p\, h_p + a_{p+1}\, k_1 + ... + a_{p+q}\, k_q = 0,$$
$$b_{11} h_1 + ... + b_{1p} h_p + b_{1,p+1} k_1 + ... + b_{1,p+q} k_q = 0,$$
$$\cdots\cdots\cdots\cdots\cdots\cdots\cdots\cdots\cdots\cdots\cdots$$
$$b_{q1} h_1 + ... + b_{qp} h_p + b_{q,p+1} k_1 + ... + b_{q,p+q} k_q = 0.$$

Moreover (7.72) now takes the form

$$\begin{vmatrix} b_{1,p+1} \dots b_{1,p+q} \\ \cdots\cdots\cdots\cdots \\ b_{q,p+1} \dots b_{q,p+q} \end{vmatrix} \neq 0. \tag{7.76}$$

The $q+1$ vectors

$$(a_{p+1}, ..., a_{p+q})^T, \quad (b_{1,p+1}, ..., b_{1,p+q})^T, \quad ..., \quad (b_{q,p+1}, ..., b_{q,p+q})^T$$

(each with q components) are linearly dependent, i.e. there are numbers $\lambda_0, \lambda_1, ..., \lambda_q$, not all zero, such that

$$\lambda_0(a_{p+1}, ..., a_{p+q})^T + \lambda_1(b_{1,p+1}, ..., b_{1,p+q})^T + ...$$
$$+ \lambda_q(b_{q,p+1}, ..., b_{q,p+q})^T = \theta.$$

But, by (7.76),

$$(b_{1,p+1}, ..., b_{1,p+q})^T, \quad ..., \quad (b_{q,p+1}, ..., b_{q,p+q})^T$$

are linearly independent and so we may take $\lambda_0 = 1$. Thus

$$a_{p+l} + \lambda_1 b_{1,p+l} + ... + \lambda_q b_{q,p+l} = 0 \qquad (7.77)$$

for $l = 1, ..., q$.

Now, for a fixed i between 1 and p, take $h_i = 1$ and $h_r = 0$ for $1 \leqslant r \leqslant p, r \neq i$. There are $k_1, ..., k_q$ such that

$$a_i + a_{p+1}k_1 + ... + a_{p+q}\ k_q = 0,$$
$$b_{1i} + b_{1,p+1}k_1 + ... + b_{1,p+q}k_q = 0,$$
$$\cdots\cdots\cdots\cdots\cdots\cdots\cdots\cdots\cdots$$
$$b_{qi} + b_{q,p+1}k_1 + ... + b_{q,p+q}k_q = 0.$$

Multiply these equations by $1, \lambda_1, ..., \lambda_q$ respectively and add. In view of (7.77) we have

$$a_i + \lambda_1 b_{1i} + ... + \lambda_q b_{qi} = 0 \qquad (7.78)$$

and this holds for $i = 1, ..., p$. The two sets of equations (7.78) and (7.77) combined are equivalent to (7.71). |

In a situation of the kind described in theorem 7.71, the multipliers $\lambda_1, ..., \lambda_q$ are of no particular interest and their role is mainly auxiliary. The q equations $g_1(\xi) = 0, ..., g_q(\xi) = 0$ and the m equations (7.71) make up a set of $m+q$ equations for the $m+q$ unknowns

$$\xi_1, ..., \xi_m; \ \lambda_1, ..., \lambda_q.$$

The points $(\xi_1, ..., \xi_m)^T$ obtained in this way are called *critical points*, and among them are any points of X at which constrained maxima or minima occur. There remains the final investigation whether a given critical point actually yields a maximum or minimum. This is usually best carried out on an *ad hoc* basis, for, although a general procedure exists, it is cumbersome. (References will be found in the notes at the end of the chapter.) However the related problem of finding the largest and least values of f (as distinct from *local* maxima and minima) is often quite simple. For suppose that f is differentiable and g is continuously differentiable on the open set X, that $Dg(x)$ has maximum rank throughout X and that the set X_0 defined by the equation $g(x) = \theta$ is compact. Then f assumes its bounds on X_0. Moreover, if ξ is a point of X_0 at which a bound is attained, then the corresponding function χ of (7.74) has a maximum or minimum at 0. This means that ξ is a critical point. Thus, to obtain the bounds of f on X_0 it is only necessary to compare the values of f at all critical points.

Example. An ellipse in R^3 is given by the equations

$$\begin{cases} g_1(x, y, z) = 2x^2 + y^2 - 4 = 0, \\ g_2(x, y, z) = x + y + z = 0. \end{cases}$$

Find the points on the ellipse which are (*a*) nearest to and (*b*) furthest from the *y*-axis.

We wish to find the points at which

$$f(x, y, z) = x^2 + z^2,$$

subject to the constraint $g(x, y, z) = \theta$, takes its largest and least values.

The two determinants

$$\begin{vmatrix} D_1 g_1(x, y, z) & D_3 g_1(x, y, z) \\ D_1 g_2(x, y, z) & D_3 g_2(x, y, z) \end{vmatrix} = \begin{vmatrix} 4x & 0 \\ 1 & 1 \end{vmatrix} = 4x,$$

$$\begin{vmatrix} D_2 g_1(x, y, z) & D_3 g_1(x, y, z) \\ D_2 g_2(x, y, z) & D_3 g_2(x, y, z) \end{vmatrix} = \begin{vmatrix} 2y & 0 \\ 1 & 1 \end{vmatrix} = 2y$$

cannot both vanish at a point satisfying $g_1(x, y, z) = 0$. Hence theorem 7.71 may be used and, at a point giving a constrained maximum or minimum, the equations

$$2x + \lambda_1 4x + \lambda_2 = 0,$$

$$\lambda_1 2y + \lambda_2 = 0,$$

$$2z \qquad + \lambda_2 = 0$$

hold for suitable values of λ_1 and λ_2. Therefore, at such a point,

$$\begin{vmatrix} 2x & 4x & 1 \\ 0 & 2y & 1 \\ 2z & 0 & 1 \end{vmatrix} = 0,$$

i.e.

$$xy + z(2x - y) = 0.$$

Substituting for z from $g_2(x, y, z) = 0$ we get

$$2x^2 - y^2 = 0.$$

The equation $g_1(x, y, z) = 0$ now gives $x^2 = 1$. The critical points are therefore

$$p = (1, \sqrt{2}, -1-\sqrt{2})^T, \quad q = (1, -\sqrt{2}, -1+\sqrt{2})^T,$$

$$r = (-1, \sqrt{2}, 1-\sqrt{2})^T, \quad s = (-1, -\sqrt{2}, 1+\sqrt{2})^T.$$

Since

$$f(p) = f(s) = 4+2\sqrt{2}$$

and

$$f(q) = f(r) = 4-2\sqrt{2},$$

the points q, r are nearest to and the points p, s are furthest from the y-axis.

Exercises 7(g)

1. Show that, if the real-valued function f on the set $X \subset R^m$ is differentiable at the (interior) point ξ of X and has a maximum or a minimum at ξ, then $Df(\xi) = \Theta$.

2. Prove the following generalization of Rolle's theorem. Let G be a bounded, open set in R^m and let the function $f : \bar{G} \to R^1$ be continuous on $\bar{G}$ and differentiable in G. If f is constant on $\operatorname{fr} G$, then there exists a point $\xi \in G$ such that $Df(\xi) = \Theta$.

3. Find the largest and least values of xy subject to the constraint $x^2 + xy + y^2 = 1$.

4. Find the largest and least values of $(y-z)(z-x)(x-y)$ subject to the condition $x^2 + y^2 + z^2 = 2$.

5. Obtain the maxima and minima of

$$x^2 + y^2 + z^2 - yz - zx - xy$$

subject to the condition

$$x^2 + y^2 + z^2 - 2x + 2y + 6z + 9 = 0.$$

6. Calculate the dimensions of the rectangular box of largest volume which can be made from a sheet of area a^2, if the box is (i) to have a top, (ii) not to have a top.

7. Find the largest and least values of $x^3 + y^3 + z^3$ subject to the conditions $x^2 + y^2 + z^2 = 1$ and $x + y + z = 1$.

8. Obtain the largest and least values of

$$2(x+y+z) - xyz$$

on the closed ball $\{(x, y, z)^T \,|\, x^2 + y^2 + z^2 \leqslant 9\}$.

(Consider $2(x+y+z) - xyz$ (i) without restriction, and (ii) under the constraint $x^2 + y^2 + z^2 = 9$.)

9. Find the points of the set in R^3 defined by

$$x + z = 1, \quad 4x^2 + y^2 \leqslant 16$$

which are, respectively, nearest to and furthest from the origin.

(Consider $x^2 + y^2 + z^2$ under the constraints (i) $x + z = 1$, (ii) $x + z = 1$ and $4x^2 + y^2 = 16$.)

7.8. Second and higher derivatives

The purpose of this section is to give a short account of linear derivatives of order higher than the first. The notion of partial derivative of arbitrary order is taken for granted (see C1, 154) and we now only fix the notation. If f is a function from R^m to R^1 $D_i D_j f(\xi)$ means $D_i(D_j f(\xi))$ and $D_i D_i f(\xi)$ is abbreviated to $D_i^2 f(\xi)$. For partial derivatives of higher order the notation is analogous.

Although in §7.1 linear derivatives were defined only for functions from R^m to R^n, we remarked that precisely the same definition may be adopted for functions with domain and range in any normed vector spaces. Now, when f is a function from R^m to R^n, Df is a function from R^m to the normed vector space $L(R^m, R^n)$. We can therefore speak of the linear derivative of Df and the definition below is the natural consequence.

Definition. *Let the function* $f: X \to R^n$, *where* $X \subset R^m$, *be differentiable on the subset* X_1 *of* X; *and let* ξ *be an interior point of* X_1. *Then* f *is said to be* twice differentiable *at* ξ *if* Df *is differentiable at* ξ, *i.e. if there exists a linear function* $\mu: R^m \to L(R^m, R^n)$ *such that*

$$\frac{\|Df(\xi+h) - Df(\xi) - \mu(h)\|}{\|h\|} \to 0 \quad as \quad \|h\| \to 0$$

(where the norms in numerator and denominator are those in $L(R^m, R^n)$ *and* R^m *respectively)*; *the linear function* μ *is called the* second linear derivative *of* f *at* ξ *and is denoted by* $D^2 f(\xi)$.

It is easy to prove (as in theorem 7.11) that second linear derivatives are unique.

When f is a function from R^m to R^n and is twice differentiable at ξ,

$$D^2 f(\xi) \in L(R^m, L(R^m, R^n)),$$
$$D^2 f(\xi)(u) \in L(R^m, R^n) \quad (u \in R^m),$$
$$D^2 f(\xi)(u)(v) \in R^n \quad (u, v \in R^m).$$

The function $D^2 f(\xi)$ is not easy to visualize; but $D^2 f(\xi)(.)(.)$ is simply a function from $R^m \times R^m$ to R^n. It is the function χ defined by

$$\chi(u, v) = D^2 f(\xi)(u)(v) \quad (u, v \in R^m),$$

which turns out to be *bilinear*. By this we means that $\chi(u, .)$ is a linear function (on R^m) for every $u \in R^m$ and $\chi(., v)$ is linear for every $v \in R^m$. In other words,

$$\chi(cu, v) = \chi(u, cv) = c\chi(u, v)$$

and

$$\chi(t+u, v) = \chi(t, v) + \chi(u, v), \quad \chi(u, v+w) = \chi(u, v) + \chi(u, w)$$

for every $c \in R^1$ and all $t, u, v, w \in R^m$. This is very easy to see. First $\chi(u, .) = D^2f(\xi)(u)$ is linear, as $D^2f(\xi)$ has values in $L(R^m, R^n)$. Secondly, since $D^2f(\xi)$ is linear,

$$D^2f(\xi)(cu) = cD^2f(\xi)(u)$$

and $$D^2f(\xi)(t+u) = D^2f(\xi)(t) + D^2f(\xi)(u),$$

that is, for every v,

$$D^2f(\xi)(cu)(v) = cD^2f(\xi)(u)(v)$$

and $$D^2f(\xi)(t+u)(v) = D^2f(\xi)(t)(v) + D^2f(\xi)(u)(v).$$

We shall show further that $\chi(u, v)$ is symmetric in u, v, but the proof is rather more difficult.

Theorem 7.81. *If the function $f: X \to R^n$, where $X \subset R^m$, is twice differentiable at the point ξ, then the bilinear function $D^2f(\xi)(.)(.)$ is symmetric, i.e.* $\quad D^2f(\xi)(u)(v) = D^2f(\xi)(v)(u)$
for all $u, v \in R^m$.

Proof. Let ϵ be any positive number. There is an open ball $B(\xi; 3\delta)$ (in X) within which f is differentiable and such that

$$\|Df(\xi+\eta) - Df(\xi) - D^2f(\xi)(\eta)\| \leqslant \epsilon \|\eta\| \tag{7.81}$$

whenever $\eta \in B(\xi; 3\delta)$. Take any points h, k in $B(\xi; \delta)$ and put

$$\Delta = f(\xi+h+k) - f(\xi+h) - f(\xi+k) + f(\xi).$$

Now define the function $g: B(\theta; 2\delta) \to R^n$ by

$$g(x) = f(\xi+h+x) - f(\xi+x) \quad (\|x\| < 2\delta),$$

so that $$\Delta = g(k) - g(\theta).$$

Then g is differentiable in $B(\theta; 2\delta)$ and

$$Dg(x) = Df(\xi+h+x) - Df(\xi+x). \tag{7.82}$$

Also, by exercise 7(c), 3, if S is the closed ball with centre θ and radius $\|k\| (< \delta)$,

$$\|\Delta - Dg(\theta)(k)\| = \|g(k) - g(\theta) - Dg(\theta)(k)\|$$
$$\leqslant (\sup_{x \in S} \|Dg(x) - Dg(\theta)\|) \|k\|. \tag{7.83}$$

Let $x \in S$. Then, by (7.81),

$$\|Df(\xi + h + x) - Df(\xi) - D^2f(\xi)(h + x)\| \leqslant \epsilon(\|h\| + \|x\|)$$

and

$$\|Df(\xi + x) - Df(\xi) - D^2f(\xi)(x)\| \leqslant \epsilon\|x\|,$$

so that, by (7.82),

$$\|Dg(x) - D^2f(\xi)(h)\| \leqslant \epsilon(\|h\| + 2\|x\|) \leqslant 2\epsilon(\|h\| + \|k\|). \quad (7.84)$$

It follows from (7.84) and (7.83) that, first,

$$\|Dg(x) - Dg(\theta)\| \leqslant 4\epsilon(\|h\| + \|k\|) \quad (x \in S)$$

and then

$$\|\Delta - D^2f(\xi)(h)(k)\| \leqslant 6\epsilon(\|h\| + \|k\|)\|k\|.$$

Since Δ is symmetric in h and k we also have

$$\|\Delta - D^2f(\xi)(k)(h)\| \leqslant 6\epsilon(\|h\| + \|k\|)\|h\|$$

and therefore

$$\|D^2f(\xi)(h)(k) - D^2f(\xi)(k)(h)\| \leqslant 6\epsilon(\|h\| + \|k\|)^2.$$

The last inequality was proved under the assumption that $\|h\|$, $\|k\| < \delta$. But, if u, v are any points of R^m, we may choose the positive number α so that $\|\alpha u\|$, $\|\alpha v\| < \delta$ and then

$$\|D^2f(\xi)(u)(v) - D^2f(\xi)(v)(u)\| = \|D^2f(\xi)(\alpha u)(\alpha v) - D^2f(\xi)(\alpha v)(\alpha u)\|/\alpha^2$$
$$\leqslant 6\epsilon(\|\alpha u\| + \|\alpha v\|)^2/\alpha^2 = 6\epsilon(\|u\| + \|v\|)^2.$$

As this inequality holds for all $\epsilon > 0$, the left-hand side must be 0. |

It is not difficult to show that any bilinear function χ on $R^m \times R^m$ to R^n is of the form

$$\chi(u, v) = \sum_{i=1}^{m} \sum_{j=1}^{m} a^{ij} u_i v_j \quad (u = (u_1, ..., u_m)^T, v = (v_1, ..., v_m)^T \in R^m),$$

where the a^{ij} are constant vectors in R^n. We shall show that, for $D^2f(\xi)(.)(.)$,

$$a^{ij} = D_i D_j f(\xi) = (D_i D_j f_1(\xi), ..., D_i D_j f_n(\xi))^T.$$

The proof does not require the general result just mentioned; it follows from the next theorem which has independent interest.

Theorem 7.82. *Suppose that the function* $f = (f_1, ..., f_n)^T : X \to R^n$, *where* $X \subset R^m$, *is differentiable in the set* $X_1 \subset X$ *and let* v *be any given point of* R^m. *If* f *is twice differentiable at* $\xi (\in X_1)$, *the function*

$$g = Df(.)(v) = \begin{pmatrix} D_1 f_1 v_1 + ... + D_m f_1 v_m \\ \cdots\cdots\cdots\cdots\cdots \\ D_1 f_n v_1 + ... + D_m f_n v_m \end{pmatrix} \quad (7.85)$$

from X_1 to R^n is differentiable at ξ and

$$Dg(\xi) = D^2f(\xi)(.)(v),$$

i.e. $$Dg(\xi)(u) = D^2f(\xi)(u)(v) \quad (u \in R^m).$$

Proof. Let
$$Df(\xi+h) - Df(\xi) - D^2f(\xi)(h) = r(h)$$

so that $r(h) \in L(R^m, R^n)$ and $\|r(h)\| = o(\|h\|)$ as $\|h\| \to 0$. Then

$$Df(\xi+h)(v) - Df(\xi)(v) - D^2f(\xi)(h)(v) = r(h)(v) \qquad (7.86)$$

and $\|r(h)(v)\| \leqslant \|r(h)\| \|v\| = o(\|h\|)$ as $\|h\| \to 0$. But the left-hand side of (7.86) is
$$g(\xi+h) - g(\xi) - D^2f(\xi)(h)(v)$$

and $D^2f(\xi)(.)(v)$ is a linear function on R^m to R^n. Hence g is differentiable at ξ and

$$Dg(\xi) = D^2f(\xi)(.)(v). \mid$$

Theorem 7.83. *If the function $f = (f_1, ..., f_n)^T : X \to R^n$, where $X \subset R^m$, is twice differentiable at the point ξ, then*

(i) *all the partial derivatives $D_i D_j f_k(\xi)$ $(i, j = 1, ..., m; k = 1, ..., n)$ exist;*

(ii) $D_i D_j f_k(\xi) = D_j D_i f_k(\xi)$ $(i, j = 1 ..., m; k = 1, ..., n)$;

(iii) *for all $u = (u_1, ..., u_m)^T, v = (v_1, ..., v_m)^T \in R^m$,*

$$D^2f(\xi)(u)(v) = \sum_{i=1}^m \sum_{j=1}^n D_i D_j f(\xi) u_i v_j = \begin{pmatrix} \sum_{i=1}^m \sum_{j=1}^m D_i D_j f_1(\xi) u_i v_j \\ \cdots\cdots\cdots\cdots\cdots \\ \sum_{i=1}^m \sum_{j=1}^m D_i D_j f_n(\xi) u_i v_j \end{pmatrix}.$$
$$(7.87)$$

Proof. In theorem 7.82 take $v = e^j$ (the vector with jth component 1 and other components 0). Then the function

$$g^j = D_j f = (D_j f_1, ..., D_j f_n)^T$$

is differentiable at ξ and so, by theorem 7.13, all partial derivatives of $D_j f_1, ..., D_j f_n$ exist at ξ. This is true for all j and therefore (i) follows.

Now take an arbitrary $v \in R^m$. The function g given by (7.85) is differentiable at ξ and the matrix of $Dg(\xi)$ is

$$
\begin{pmatrix}
D_1[D_1 f_1(\xi) v_1 + \ldots + D_m f_1(\xi) v_m] \ldots D_m[D_1 f_1(\xi) v_1 + \ldots + D_m f_1(\xi) v_m] \\
\cdots\cdots\cdots\cdots\cdots\cdots\cdots\cdots\cdots\cdots\cdots\cdots\cdots\cdots\cdots\cdots \\
D_1[D_1 f_n(\xi) v_1 + \ldots + D_m f_n(\xi) v_m] \ldots D_m[D_1 f_n(\xi) v_1 + \ldots + D_m f_n(\xi) v_m]
\end{pmatrix}.
$$

Since each $D_i D_j f_k(\xi)$ exists, the formula $D^2 f(\xi)(u)(v) = Dg(\xi)(u)$ now gives the expression (7.87).

To prove (ii) we use (iii) and theorem 7.81. We have

$$
D_i D_j f(\xi) = D^2 f(\xi)(e^i)(e^j) = D^2 f(\xi)(e^j)(e^i) = D_j D_i f(\xi). \quad |
$$

It may be shown that the existence and continuity of all second order partial derivatives of a function ensure that the function is twice differentiable (cf. theorem 7.14). For this reason, part (ii) of theorem 7.83 invites comparison with theorem 7.83 of C1, where it is proved that $D_i D_j f_k(\xi) = D_j D_i f_k(\xi)$ if $D_i D_j f_k$ and $D_j D_i f_k$ are continuous at ξ. Two further results of the same type are embodied in exercise 7(h) 3, and in the theorem below which will be needed in chapter 10.

Theorem 7.84. *Let* $f = (f_1, \ldots, f_n)^T$ *be a function with domain* $X \subset R^m$ *and range in* R^n; *and let* ξ *be an interior point of* X. *If* $D_i f_k$ *and* $D_j f_k$ *exist in an open set containing* ξ, *and are differentiable at* ξ, *then* $D_i D_j f_k(\xi) = D_j D_i f_k(\xi)$.

Proof. It is clearly sufficient to take $m = 2$ and $n = 1$. We therefore have to show (with a slight change of notation) that

$$
D_1 D_2 f(\xi, \eta) = D_2 D_1 f(\xi, \eta).
$$

Suppose that $D_1 f$, $D_2 f$ exist in an open ball with centre $(\xi, \eta)^T$ and radius 2δ. For $0 < |h| < \delta$, put

$$
\Delta(h) = f(\xi+h, \eta+h) - f(\xi+h, \eta) - f(\xi, \eta+h) + f(\xi, \eta).
$$

We now define the differentiable real function g on $(-\delta, \delta)$ by

$$
g(x) = f(\xi+x, \eta+h) - f(\xi+x, \eta) \quad (|x| < \delta).
$$

Then
$$
\Delta(h) = g(h) - g(0)
$$

and, by the mean value theorem, there is an α such that $0 < \alpha < 1$ and

$$
\Delta(h) = hg'(\alpha h) = h\{D_1 f(\xi+\alpha h, \eta+h) - D_1 f(\xi+\alpha h, \eta)\}.
$$

But, since $D_1 f$ is differentiable at ξ,

$$D_1 f(\xi + \alpha h, \eta + h) - D_1 f(\xi, \eta)$$
$$= D_1^2 f(\xi, \eta)\alpha h + D_2 D_1 f(\xi, \eta)h + o(\|(\alpha h, h)\|)$$

and $\quad D_1 f(\xi + \alpha h, \eta) - D_1 f(\xi, \eta) = D_1^2 f(\xi, \eta)\alpha h + o(\|(\alpha h, 0)\|).$

Hence, as $h \to 0$,

$$\Delta(h)/h^2 = D_2 D_1 f(\xi, \eta) + o(\sqrt{(\alpha^2 + 1)} + \alpha),$$

i.e. $\qquad\qquad \Delta(h)/h^2 \to D_2 D_1 f(\xi, \eta).$

By symmetry, we also have

$$\Delta(h)/h^2 \to D_1 D_2 f(\xi, \eta) \quad \text{as} \quad h \to 0. \;|$$

Linear derivatives of any order are defined inductively and give rise to *multilinear* functions. For instance $D^3 f(\xi)$ is the linear derivative at ξ of the function $D^2 f$ whose domain is in R^m and range in $L(R^m, L(R^m, R^n))$; thus

$$D^3 f(\xi) \in L(R^m, L(R^m, L(R^m, R^n))).$$

The function $D^3 f(\xi)(.)(.)(.)$ is much less complex. It is simply a *trilinear* function from $R^m \times R^m \times R^m$ to R^n and is given by

$$D^3 f(\xi)(u)(v)(w) = \sum_{i=1}^m \sum_{j=1}^m \sum_{k=1}^m D_i D_j D_k f(\xi) u_i v_j w$$

Exercises 7(h)

1. Let X be an open subset of R^m such that $tx \in X$ whenever $x \in X$ and $t > 0$. Let the function $f: X \to R^1$ be twice differentiable and homogeneous of degree α (see exercise 7(b), 3).

(i) Prove that
$$D^2 f(x)(x)(x) = \alpha(\alpha - 1)f(x)$$
for all $x \in X$.

(ii) Prove that, if $\alpha = 1$ and all second order partial derivatives of f are continuous, then

$$\begin{vmatrix} D_1 D_1 f(x) & \dots & D_m D_1 f(x) \\ \dots\dots\dots\dots\dots\dots\dots\dots\dots & & \\ D_1 D_m f(x) & \dots & D_m D_m f(x) \end{vmatrix} = 0$$

for all $x \in X$.

(The determinant is called the *Hessian* of f and is denoted by $Hf(x)$; it is equal to $Jg(x)$, where $g = (D_1 f, \dots, D_m f)^T$.)

2. Show that the function $f: R^2 \to R^1$ given by

$$f(x, y) = \begin{cases} \dfrac{xy(x^2 - y^2)}{x^2 + y^2} & ((x, y) \neq (0, 0)), \\ 0 & ((x, y) = (0, 0)) \end{cases}$$

is differentiable everywhere and that all four second order partial derivatives exist everywhere, but that $D_1 D_2 f(0, 0) \neq D_2 D_1 f(0, 0)$.

3. The function f from R^2 to R^1 is continuously differentiable in an open set G containing the point $(\xi, \eta)^T$. Prove that, if $D_2 D_1 f$ exists in G and is continuous at $(\xi, \eta)^T$, then $D_1 D_2 f(\xi, \eta)$ exists and is equal to $D_2 D_1 f(\xi, \eta)$.

(By considering

$$\Delta(h, k) = f(\xi + h, \eta + k) - f(\xi, \eta + k) - f(\xi + h, \eta) + f(\xi, \eta)$$

show that, given $h, k \neq 0$, there exist α, β such that $0 < \alpha, \beta < 1$ and

$$\{D_1 f(\xi + \alpha h, \eta + k) - D_1 f(\xi + \alpha h, \eta)\}/k = \{D_2 f(\xi + h, \eta + \beta k) - D_2 f(\xi, \eta + \beta k)\}/h;$$

and prove that the left-hand side tends to $D_2 D_1 f(\xi, \eta)$ as $(h, k)^T \to (0, 0)^T$.)

4. Three sets of conditions for the equality of mixed partial derivatives have been given in the present section. Construct a function which satisfies none of these sets of conditions but is nevertheless such that $D_1 D_2 f(\xi, \eta)$ and $D_2 D_1 f(\xi, \eta)$ exist and are equal.

5. *Taylor's theorem in R^m.* Let X be a subset of R^m and ξ an interior point of X. The function $f: X \to R^1$ is such that all partial derivatives of order p are continuous in an open ball $B(\xi; r)$. Show that, when $\|h\| < r$, there is a number α between 0 and 1 such that

$$f(\xi + h) = f(\xi) + Df(\xi)(h) + \frac{1}{2!} D^2 f(\xi)(h)^2 + \ldots + \frac{1}{(p-1)!} D^{p-1} f(\xi)(h)^{p-1}$$

$$+ \frac{1}{p!} D^p f(\xi + \alpha h)(h)^p,$$

where $D^2 f(\xi)(h)^2 = D^2 f(\xi)(h)(h)$, etc.

$(D^k f(\xi)(h)^k$ may here be regarded as shorthand for

$$\sum_{i_1 = 1}^{m} \ldots \sum_{i_k = 1}^{m} D_{i_1} \ldots D_{i_k} f(\xi) h_{i_1} \ldots h_{i_k},$$

though, in fact, the continuity of the pth order partial derivatives implies the existence of the linear derivatives up to order p.

Define the real function g by $g(t) = f(\xi + th)$ $(|t| < r/\|h\|)$ and use Taylor's theorem in R^1 (C1, 81); the case $m = 2$ is dealt with in C1, 163.)

6. Let ξ be an interior point of the set X in R^m. Suppose that the function $f: X \to R^1$ has continuous second order partial derivatives in an open set containing ξ and that $D_1 f(\xi) = \ldots = D_m f(\xi) = 0$. Using the notation

$$H_k f(\xi) = \begin{vmatrix} D_1 D_1 f(\xi) & \ldots & D_k D_1 f(\xi) \\ \cdots\cdots\cdots\cdots\cdots\cdots\cdots \\ D_1 D_k f(\xi) & \ldots & D_k D_k f(\xi) \end{vmatrix}$$

for $k = 1, \ldots, m$, show that

(i) if $H_k f(\xi) > 0$ for all k, then f has a minimum at ξ;

(ii) if $H_k f(\xi) < 0$ when k is odd and $H_k f(\xi) > 0$ when k is even, then f has a maximum at ξ;

(iii) if either $H_k f(\xi) < 0$ for an even k, or $H_k f(\xi) > 0$ for an odd k and also $H_k f(\xi) < 0$ for an odd k, then f has neither a maximum nor a minimum at ξ. (For the case $m = 2$ see C1, theorem 8.8.

The hypotheses in (i), (ii) and (iii) respectively, are equivalent to the following properties of $D^2 f(\xi)$:

(i)' $D^2 f(\xi)(u)^2 > 0$ for all $u \neq \theta$;

(ii)' $D^2 f(\xi)(u^2) < 0$ for all $u \neq \theta$;

(iii)' $D^2 f(\xi)(.)^2$ takes values > 0 and < 0.

This is proved in most books on linear algebra (e.g. L. Mirsky, *Linear Algebra*, 400–407).)

7. Find the maxima and minima of the function $f: R^3 \to R^1$ defined by

$$f(x, y, z) = x^2 + y^2 + z^2 + 2xyz.$$

What is the least value of f in the closed ball

$$\{(x, y, z)^T \,|\, x^2 + y^2 + z^2 \leqslant a^2\}?$$

8. Show that the function $f: R^2 \to R^1$ given by

$$f(x, y) = x^2 y - 4x^2 - 2y^2$$

has no minima. Deduce that, if K is any compact subset of R^2, then f attains its infimum on K at a point of the frontier of K.

NOTES ON CHAPTER 7

§7.1. In the context of the long history of the calculus a rigorous treatment of functions of several variables is surprisingly new. As late as 1909 the definition of differentiability of a function from R^2 to R^1 was the subject of a research paper (W. H. Young, 'On differentials', *Proc. London Math. Soc.* 7, 157–80). However subsequent progress was rapid. By 1925 Fréchet had produced a theory of differentiation on normed vector spaces; and in 1927 T. H. Hildebrandt and L. M. Graves published a parallel but independent account which included an implicit function theorem ('Implicit functions and their differentials in general analysis', *Trans. Amer. Math. Soc.* 29, 127–53). Recently, Dieudonné's *Foundations of Modern Analysis* has been most influential in disseminating the modern theory of differentiation.

§7.4. The limitations of the implicit function theorem 7.43 are vividly illustrated by the simple case of $f(x, y) = (x - y)^2$. Although the equation $f(x, y) = 0$ has the unique solution $y = x$, the conditions of 7.43 are nowhere satisfied since $D_2 f(x, x) = 0$ for all x. There exists a theory in which differentiability conditions are avoided, but only at the cost of considerable complications. (W. Quade and S. Schottlaender, 'Über implizite Funktionen', *Math. Zeitschrift* **89** (1965), 137–80.)

§7.6. A converse of theorem 7.61 is true: if the function $f = (f_1, ..., f_n)^T$ is continuously differentiable in an open set G and $Jf(x) = 0$ for $x \in G$, then $f_1, ..., f_n$ are functionally dependent in every bounded open subset of G. Proofs of this result are long and not easy. The first one is due to K. Knopp and R. Schmidt ('Funktionaldeterminanten und Abhängigkeit von Funktionen', *Math. Zeitschrift* **25** (1926), 373–87); a highly sophisticated account has recently been given by W. F. Newns ('Functional dependence', *Amer. Math. Monthly* **74** (1967), 911–19). However if the condition $Jf(x) = 0$ for $x \in G$ is replaced by the more restrictive one that the matrix of $Df(x)$ has constant rank $r < n$, then we have the situation of theorem 7.63 where there is, though only locally, the stronger conclusion that $n - r$ of the functions $f_1, \dots, f_n$ are functionally dependent on the remaining r functions.

The fact that in theorem 7.63 the rank must not be variable is shown by the function $f = (f_1, f_2)^T : R^2 \to R^2$ defined by

$$f_1(x, y) = x^2, \quad f_2(x, y) = \begin{cases} x^4 \sin{(1/x)} & (x \neq 0), \\ 0 & (x = 0). \end{cases}$$

The rank of the matrix of $Df(x, y)$ is 1 when $x \neq 0$ and is 0 when $x = 0$; and it follows from exercise $7(f)$, 8 that, in any open set containing a portion of the line $x = 0$, neither of the functions f_1, f_2 is functionally dependent on the other.

§7.7. For conditions sufficient to ensure the existence of a constrained maximum or minimum at a critical point, see H. Hancock, *Maxima and Minima*, chapter VI.

§7.8. Linear derivatives of arbitrary order are best studied in the general setting of normed vector spaces (as in Dieudonné's book). For even if the function f has values in R^n, so that $Df(\xi)$ has values in the same space, $D^2 f(\xi)$ and higher linear derivatives have their values in vector spaces of linear functions.

INTEGRALS IN R^n

8.1. Curves

A function f on the *closed* interval $[a, b]$ (and with range in R^n) is said to be differentiable at a or b if the appropriate one-sided derivatives exist at these points (C1, 66); and f is called *continuously differentiable* if the derivative exists at all points of $[a, b]$ and is continuous on $[a, b]$. (This supplements the definition of continuous differentiability for functions with *open* domains which was so extensively used in chapter 7.) The function f is called *piecewise continuously differentiable* on $[a, b]$ if there is a dissection

$$a = a_0 < a_1 < \ldots < a_{k-1} < a_k = b$$

such that the restriction of f to each interval $[a_{i-1}, a_i]$ is continuously differentiable. Thus, in particular, f is continuous on $[a, b]$ and has one-sided derivatives at the points a_i.

We are now ready to discuss the notion of a curve in R^n ($n > 1$) or Z.

Definition. A transit *in R^n (or Z) is a continuous function whose domain is a finite closed interval in R^1 and whose range lies in R^n (or Z).*

We distinguish between the transit, which is a function, and the range of this function, which is a set of points in R^n (or Z). Many different transits may have the same range.

To obtain a *curve* we group together certain transits. Two transits $\phi : [a, b] \to R^n$ (or Z), $\psi : [c, d] \to R^n$ (or Z) are called *equivalent*, and we write $\phi \sim \psi$, if $\phi = \psi \circ \alpha$, where α is a real function on $[a, b]$ with the following properties:

(i) $\alpha(a) = c$, $\alpha(b) = d$;

(ii) α is piecewise continuously differentiable;

(iii) for $a \leqslant t \leqslant b$, $\alpha'(t)$ and, where this does not exist, each one-sided derivative of α, is strictly positive.

It is not difficult to see that the relation $\sim$ so defined is, in fact, an equivalence relation (see exercise 8 (a), 1).

Definition. An equivalence class of transits is called a curve. *Any member of the equivalence class constituting a curve γ is called a representation of γ.*

All the representations of a curve γ have the same range. This set of points is called the *trace* of γ; it is compact and connected, being the image of a compact and connected set by a continuous function. If $\phi:[a, b] \to R^n$ (or Z) is a representation of γ, the points $\phi(a)$, $\phi(b)$ (which are the same for all representations of γ) are called the *initial point* and the *end point* of γ, respectively.

Our definition of a curve is intended to convey the idea that a curve is a method of description of a certain set of points (its trace); also that such a set of points may be described in many different ways which, however, do not differ in essentials. The definition automatically excludes the possibility of a single curve being indicated by both sets of arrows in each of the illustrations below.

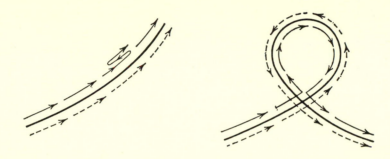

A curve γ' is said to be an *arc* of a curve γ if there exist a representation ϕ of γ with domain $[a, b]$ and a representation ψ of γ' with domain $[c, d]$ such that $[c, d] \subset [a, b]$ and ψ is the restriction of ϕ to $[c, d]$.

If a transit is piecewise continuously differentiable, then every equivalent transit has the same property. We can therefore speak of piecewise continuously differentiable curves. They will be called *paths*.

A rudimentary algebra of curves is useful. We begin with the operation of addition. Let $\gamma_1, ..., \gamma_m$ be curves such that, for $i = 1, ..., m-1$, the end point of γ_i coincides with the initial point of γ_{i+1}. For each i take a representation ϕ^i of γ_i whose domain is the interval $[i-1, i]$. Since $\phi^i(i) = \phi^{i+1}(i)$, the function ϕ on $[0, m]$ defined by

$$\phi(t) = \phi^i(t) \quad \text{for} \quad i-1 \leqslant t \leqslant i \quad (i = 1, ..., m)$$

is continuous and so generates a curve γ. It is not difficult to see that γ does not depend on the particular choice of functions ϕ^i (see exercise 8(a), 3). We therefore define γ to be the sum of $\gamma_1, ..., \gamma_m$ and write

$$\gamma = \gamma_1 + ... + \gamma_m.$$

Clearly, if $\gamma_1, ..., \gamma_m$ are paths, then γ is a path.

If ϕ on $[a, b]$ is a representation of a curve γ, then ϕ^* on $[-b, -a]$ given by

$$\phi^*(t) = \phi(-t) \quad (-b \leqslant t \leqslant -a)$$

generates a curve with the same trace as γ, but with initial point $\phi(b)$ and end point $\phi(a)$. It is natural to denote this curve by $-\gamma$. Evidently $-(-\gamma) = \gamma$. A sum $\gamma_1 + (-\gamma_2)$ is, as usual, written $\gamma_1 - \gamma_2$. Note that

$$-(\gamma_1 + ... + \gamma_m) = -\gamma_m - ... - \gamma_1.$$

It is easy to formalize the intuitive idea of the length of a curve. To avoid repetition we confine ourselves to curves in R^n; the treatment for Z is entirely analogous.

Our definition of curve length is derived from the ancient device of inscribing polygons. Let ϕ on $[a, b]$ be a transit in R^n. For any dissection $\mathscr{D}$ of $[a, b]$ given by

$$a = t_0 < t_1 < ... < t_{m-1} < t_m = b,$$

put $\qquad\qquad \Lambda(\mathscr{D}, \phi) = \sum_{i=1}^{m} \|\phi(t_i) - \phi(t_{i-1})\|.$

We note that, by the triangle inequality for norms, the insertion of new points of subdivision increases a sum $\Lambda(\mathscr{D}, \phi)$.

Definition. *If $\Lambda(\mathscr{D}, \phi)$ is bounded for all dissections $\mathscr{D}$ of $[a, b]$, the transit ϕ is said to be* rectifiable *and*

$$\Lambda_a^b(\phi) = \sup_{\mathscr{D}} \Lambda(\mathscr{D}, \phi)$$

is called the length *of ϕ.* (For an alternative definition see exercise 8(a), 10.)

Let ϕ on $[a, b]$, ψ on $[c, d]$ be equivalent transits in R^n. Then there is a continuous, strictly increasing function α on $[a, b]$ such that $\alpha(a) = c$, $\alpha(b) = d$ and

$$\phi(t) = \psi\{\alpha(t)\} \quad (a \leqslant t \leqslant b).$$

Since the inverse function α^{-1} is also strictly increasing, the relations

$$u = \alpha(t) \quad (a \leqslant t \leqslant b), \quad t = \alpha^{-1}(u) \quad (c \leqslant u \leqslant d)$$

set up a bijection between all the dissections of $[a, b]$ and all the dissections of $[c, d]$. Also, if $\mathscr{D}$, $\mathscr{D}'$ are corresponding dissections of $[a, b]$ and $[c, d]$ respectively, then

$$\Lambda(\mathscr{D}, \phi) = \Lambda(\mathscr{D}', \psi).$$

Hence ϕ and ψ are either both rectifiable or both not rectifiable and, if they are rectifiable, they have the same length. It follows that we are justified in applying the terms *rectifiability* and *length* to curves as well as to transits. We shall denote the length of a rectifiable curve γ by $\Lambda(\gamma)$.

Theorem 8.11. *The curve in R^n represented by ϕ on $[a, b]$ is rectifiable if and only if each of the components $\phi_1, ..., \phi_n$ of ϕ is of bounded variation on $[a, b]$.*

Proof. For $y = (y_1, ..., y_n)$ in R^n we have

$$|y_i| \leqslant \|y\| \leqslant |y_1| + ... + |y_n| \quad (i = 1, ..., n).$$

Therefore, for any dissection $\mathscr{D}$ of $[a, b]$ and $i = 1, ..., n$,

$$V(\mathscr{D}, \phi_i) \leqslant \Lambda(\mathscr{D}, \phi) \leqslant V(\mathscr{D}, \phi_1) + ... + V(\mathscr{D}, \phi_n) \qquad (8.11)$$

and these inequalities lead to the required result. |

Theorem 8.12. *Let $\gamma, \gamma_1, \gamma_2$ be curves such that $\gamma = \gamma_1 + \gamma_2$. Then* (i) *$\gamma$ is rectifiable if and only if γ_1 and γ_2 are rectifiable and* (ii) *when γ is rectifiable*

$$\Lambda(\gamma) = \Lambda(\gamma_1) + \Lambda(\gamma_2).$$

The proof is very similar to that of theorem 6.42 and is left to the reader. Part (i) is also an immediate consequence of theorems 8.11 and 6.42.

Theorem 8.13. *Every path in R^n is rectifiable; and, if ϕ on $[a, b]$ is a representation of a path γ in R^n, then*

$$\Lambda(\gamma) = \int_a^b \|\phi'\|. \qquad (8.12)$$

Proof. The function ϕ is piecewise continuously differentiable. Hence $[a, b]$ may be divided into a finite number of subintervals in each of which the components $\phi_1, ..., \phi_n$ have bounded derivatives. Thus $\phi_1, ..., \phi_n$ are of bounded variation in $[a, b]$ and so, by theorem 8.11, γ is rectifiable. We next note that the integral in (8.12) exists although the integrand may be undefined at a finite number of points; for, elsewhere in $[a, b]$, ϕ' is bounded and continuous. To prove the identity (8.12) we shall suppose that ϕ is continuously differentiable throughout $[a, b]$. This is permissible by theorem 8.12 and exercise 8(a), 4.

Take any $\epsilon > 0$. Since $\phi'_1, ..., \phi'_n$ are uniformly continuous, there is a $\delta > 0$ such that, for $j = 1, ..., n$,

$$|\phi'_j(u) - \phi'_j(v)| < \epsilon \text{ whenever } a \leqslant u, v \leqslant b \text{ and } |u - v| < \delta. \quad (8.13)$$

Also there is a dissection $\mathscr{D}$ of $[a, b]$ such that

$$\Lambda(\mathscr{D}, \phi) > \Lambda_a^b(\phi) - \epsilon = \Lambda(\gamma) - \epsilon.$$

We add, if necessary, points of division to $\mathscr{D}$ to form a dissection $\mathscr{D}^*$ for which $\mu(\mathscr{D}^*) < \delta$. Then

$$\Lambda(\mathscr{D}^*, \phi) \geqslant \Lambda(\mathscr{D}, \phi) > \Lambda(\gamma) - \epsilon. \quad (8.14)$$

Let the points of division of $\mathscr{D}^*$ be

$$a = t_0 < t_1 < ... < t_{m-1} < t_m = b.$$

For each j ($j = 1, ..., n$), the interval (t_{i-1}, t_i) contains a point τ_{ij} such that
$$\phi_j(t_i) - \phi_j(t_{i-1}) = \phi'_j(\tau_{ij})(t_i - t_{i-1}).$$

Therefore, by (8.13), for every t in $[t_{i-1}, t_i]$,

$$|\phi_j(t_i) - \phi_j(t_{i-1}) - \phi'_j(t)(t_i - t_{i-1})| < \epsilon(t_i - t_{i-1}) \quad (j = 1, ..., n)$$

and so
$$\big| \|\phi(t_i) - \phi(t_{i-1})\| - \|\phi'(t)\|(t_i - t_{i-1}) \big| < n\epsilon(t_i - t_{i-1}). \quad (8.15)$$

But, by exercise 6(a), 8, there is a u_i in $[t_{i-1}, t_i]$ such that

$$\int_{t_{i-1}}^{t_i} \|\phi'\| = \|\phi'(u_i)\|(t_i - t_{i-1}).$$

Hence, by (8.15),

$$\left| \|\phi(t_i) - \phi(t_{i-1})\| - \int_{t_{i-1}}^{t_i} \|\phi'\| \right| < n\epsilon(t_i - t_{i-1}).$$

This holds for $i = 1, ..., m$. Adding these inequalities we obtain

$$\left| \Lambda(\mathscr{D}^*, \phi) - \int_a^b \|\phi'\| \right| < n\epsilon(b - a)$$

and therefore, by (8.14),

$$\left| \Lambda(\gamma) - \int_a^b \|\phi'\| \right| < [1 + n(b - a)]\epsilon. \quad \blacksquare$$

Exercises 8 (a)

1. Prove that the relation $\sim$ between transits is an equivalence relation.

2. A curve in R^n is said to be *smooth* if one of its representations $\phi : [a, b] \to R^n$ is continuously differentiable and such that $\phi'(t) \neq \theta$ in $[a, b]$. Show that not all representations of a smooth curve are continuously differentiable. (Contrast *paths* all of whose representations are piecewise continuously differentiable.)

3. Show that, in the definition of the sum $\gamma_1 + \ldots + \gamma_m$ of the curves γ_i $(i = 1, \ldots, m)$ in R^n, the choice of the representations $\phi^i : [i-1, i] \to R^n$ of γ_i is immaterial.

4. Let ϕ on $[a, b]$ be a representation of a curve γ and let $u \in (a, b)$. Denote the arcs corresponding to the restrictions of ϕ to $[a, u]$ and $[u, b]$ respectively by γ_1 and γ_2. Prove that $\gamma = \gamma_1 + \gamma_2$.

5. A rectifiable curve γ has initial point p and end point q. Prove that

$$\Lambda(\gamma) \geqslant \|p - q\|.$$

6. Curves $\gamma_1, \gamma_2, \gamma_3$ in R^2 are given by the functions $\phi^i = (\phi_1^i, 0)$ $(i = 1, 2, 3)$, all on $[-1, 1]$, where

$$\phi_1^1(t) = t,$$

$$\phi_1^2(t) = -\sin \tfrac{3}{2}\pi t,$$

$$\phi_1^3(t) = \begin{cases} 0 & (t = 0), \\ -t \cos (\pi/t) & (0 < |t| \leqslant 1). \end{cases}$$

Show that $\gamma_1, \gamma_2, \gamma_3$ have all the same trace and the same initial and end points, but that γ_1, γ_2 are rectifiable whereas γ_3 is not. Show also that $\Lambda(\gamma_1) \neq \Lambda(\gamma_2)$.

7. Let p, q be coprime positive integers. Find the length of the epicycloid represented by $\phi = (\phi_1, \phi_2) : [0, 2q\pi] \to R^2$, where

$$\phi_1(t) = \left(\frac{p}{q}+1\right) a \cos t + a \cos \left(\frac{p}{q}+1\right) t, \quad \phi_2(t) = \left(\frac{p}{q}+1\right) a \sin t + a \sin \left(\frac{p}{q}+1\right) t.$$

(The curve is generated by the movement of a point fixed on a circle of radius a as this circle rolls on a fixed circle of radius $(p/q) a$.)

8. Find the length of the curve in R^3 represented by ϕ on $[0, 2]$, where

$$\phi(t) = (t \cos t, t \sin t, t) \quad (0 \leqslant t \leqslant 2).$$

9. Let ϕ on $[a, b]$ be a rectifiable transit in R^n. Prove that the function s on $[a, b]$ given by

$$s(t) = \begin{cases} 0 & \text{for } t = a, \\ \Lambda_a^t(\phi) & \text{for } a < t \leqslant b \end{cases}$$

is continuous on $[a, b]$. (Arc length is a continuous function of the 'parameter' t.)

10. Let ϕ on $[a, b]$ be a transit in R^n. Show that ϕ is rectifiable if and only if $\lim_{\mu(\mathscr{D}) \to 0} \Lambda(\mathscr{D}, \phi)$ exists and that, if the limit does exist, then it is equal to $\Lambda_a^b(\phi)$. (Cf. theorem 6.71 and exercise 6(g), 3.)

11. *A space filling curve.* The continuous function $f: R^1 \to R^1$ is defined as follows: f has period 2, is even and $f(t) = 0$ for $t \in [0, \frac{1}{3}]$, $f(t) = 1$ for $t \in [\frac{2}{3}, 1]$, f is linear in $[\frac{1}{3}, \frac{2}{3}]$. The functions $\phi, \psi : [0, 1] \to R^1$ are now defined by

$$\phi(t) = \frac{1}{2}f(t) + \frac{1}{2^2}f(3^2 t) + \ldots + \frac{1}{2^n}f(3^{2n-2}t) + \ldots,$$

$$\psi(t) = \frac{1}{2}f(3t) + \frac{1}{2^2}f(3^3 t) + \ldots + \frac{1}{2^n}f(3^{2n-1}t) + \ldots.$$

Show that ϕ and ψ are continuous and that the curve represented by (ϕ, ψ) passes through every point of the square $[0, 1] \times [0, 1]$.

(If $\xi, \eta \in [0, 1]$, let

$$\xi = 0 \cdot \alpha_1 \alpha_2 \alpha_3 \ldots, \quad \eta = 0 \cdot \beta_1 \beta_2 \beta_3 \ldots$$

in the binary scale (so that each α_i or β_i is 0 or 1). Let $\tau \in [0, 1]$ be the number defined in the ternary scale by

$$\tau = 0 \cdot (2\alpha_1)(2\beta_1)(2\alpha_2)(2\beta_2) \ldots$$

and show that $\xi = \phi(\tau)$, $\eta = \psi(\tau)$.)

8.2. Line integrals

Let G be an open subset of R^n $(n > 1)$. We wish to define an integral of a continuous function $f: G \to R^n$ along a rectifiable curve γ in G. (We say that 'a curve lies in a set' when the trace of the curve is contained in that set.)

Take a representation $\phi : [a, b] \to R^n$ of γ and a dissection $\mathcal{D}$ of $[a, b]$ given by

$$a = t_0 < t_1 < \ldots < t_{r-1} < t_r = b.$$

When $t_{j-1} \leqslant \tau_j \leqslant t_j$, $f\{\phi(\tau_j)\}$ is a value of f on the trace of the arc corresponding to $[t_{j-1}, t_j]$ and we accordingly consider a sum

$$\sum_{j=1}^{r} f\{\phi(\tau_j)\} \cdot \{\phi(t_j) - \phi(t_{j-1})\} = \sum_{j=1}^{r} f_1\{\phi(\tau_j)\}\{\phi_1(t_j) - \phi_1(t_{j-1})\}$$

$$+ \ldots + \sum_{j=1}^{r} f_n\{\phi(\tau_j)\}\{\phi_n(t_j) - \phi_n(t_{j-1})\}. \quad (8.21)$$

Each function ϕ_i is of bounded variation on $[a, b]$ and, since f and ϕ are continuous, $f \circ \phi$ is continuous. Hence $f_i \circ \phi \in R(\phi_i; a, b)$ for $i = 1, \ldots, n$, and, by exercise 6(g), 1,

$$\sum_{j=1}^{n} f_i\{\phi(\tau_j)\}\{\phi_i(t_j) - \phi_i(t_{j-1})\} \to \int_a^b (f_i \circ \phi)\, d\phi_i$$

as $\mu(\mathcal{D}) \to 0$. Thus the sum in (8.21) converges to

$$\sum_{i=1}^{n} \int_a^b (f_i \circ \phi)\, d\phi_i.$$

But we cannot define this expression to be the integral of f along γ until we have shown that it is independent of the particular representation ϕ of γ.

Theorem 8.21. *Let G be an open set in R^n and suppose that the function $f: G \to R^n$ is continuous. If γ is a rectifiable curve in G and $\phi:[a,b] \to R^n$, $\psi:[c,d] \to R^n$ are both representations of γ, then*

$$\sum_{i=1}^{n} \int_a^b (f_i \circ \phi)\, d\phi_i = \sum_{i=1}^{n} \int_c^d (f_i \circ \psi)\, d\psi_i.$$

Proof. Since $\phi \sim \psi$, there exists a continuous, strictly increasing function $\alpha:[a,b] \to [c,d]$ such that $\alpha(a) = c$, $\alpha(\beta) = d$ and $\phi = \psi \circ \alpha$. By theorem 6.53, for each i,

$$\int_c^d (f_i \circ \psi)\, d\psi_i = \int_a^b (f_i \circ (\psi \circ \alpha))\, d(\psi_i \circ \alpha) = \int_a^b (f_i \circ \phi)\, d\phi_i. \; |$$

We are now in a position to define the first of the two kinds of *line integral.*

Definition. *Let G be an open set in R^n and suppose that the function $f: G \to R^n$ is continuous. If γ is a rectifiable curve in G, then $\int_\gamma f$, the integral of f along γ, is defined by*

$$\int_\gamma f = \sum_{i=1}^{n} \int_a^b (f_i \circ \phi)\, d\phi_i, \tag{8.22}$$

where $\phi:[a,b] \to R^n$ is any representation of γ.

When γ is a *path* the sum on the right in (8.22) reduces to a single Riemann integral. For then ϕ is piecewise continuously differentiable and each ϕ_i is the difference of two increasing functions which are piecewise continuously differentiable (see exercise 6(h), 13). Therefore, by theorem 6.21 A,

$$\int_a^b (f_i \circ \phi)(t)\, d\phi_i(t) = \int_a^b (f_i \circ \phi)(t)\, \phi_i'(t)\, dt$$

and

$$\int_\gamma f = \sum_{i=1}^{n} \int_a^b (f_i \circ \phi)\, \phi_i' = \int_a^b (f \circ \phi) \cdot \phi'.$$

The form $\quad \int_a^b \{f_1(\phi(t))\, \phi_1'(t) + \ldots + f_n(\phi(t))\, \phi_n'(t)\}\, dt$

of $\int_\gamma f$ (when γ is a path) suggests the notation

$$\int_\gamma f_i(x)\, dx_i = \int_a^b f_i(\phi(t))\, \phi_i'(t)\, dt.$$

Therefore $\int_{\gamma} f$ is often denoted by

$$\int_{\gamma} (f_1(x)\,dx_1 + \ldots + f_n(x)\,dx_n).$$

Illustration. If γ is the path in R^2 given by

$$x = \cos t, \quad y = \sin t \quad (0 \leqslant t \leqslant 2\pi)$$

then

$$\int_{\gamma} (x^2 y\,dx - x^3\,dy) = \int_0^{2\pi} (-\cos^2 t \sin^2 t - \cos^4 t)\,dt = -\int_0^{2\pi} \cos^2 t\,dt = -\pi.$$

The path γ is called *the unit circle described in the positive* (*anti-clockwise*) *direction*. A similar phrase may be used for the path given by

$$x = a + r\cos t, \quad y = b + r\sin t \quad (0 \leqslant t \leqslant 2\pi).$$

Another frequently encountered path (in R^n) is one represented in the form

$$x = p + (q-p)t \quad (0 \leqslant t \leqslant 1)$$

This is what will be understood by *the* (*straight line*) *segment from p to q*.

In the next theorem we collect the elementary properties of line integrals.

Theorem 8.22. *Let* $\gamma, \gamma_1, \gamma_2$ *be rectifiable curves in an open subset G of* R^n *and let the functions* $f, g : G \to R^n$ *be continuous. Then*

(i) $\displaystyle\int_{\gamma} (f+g) = \int_{\gamma} f + \int_{\gamma} g$;

(ii) $\displaystyle\int_{\gamma} cf = c\int_{\gamma} f$ *(where c is a real number)*;

(iii) $\displaystyle\int_{-\gamma} f = -\int_{\gamma} f$;

(iv) $\displaystyle\int_{\gamma_1 + \gamma_2} f = \int_{\gamma_1} f + \int_{\gamma_2} f$ *(when* $\gamma_1 + \gamma_2$ *is defined).*

Proof. Parts (i) and (ii) are obvious and (iii) follows from the identity (see theorem 6.53)

$$\int_{-b}^{-a} f_i\{\phi(-t)\}\,d\phi_i(-t) = \int_b^a f_i\{\phi(t)\}\,d\phi_i(t) = -\int_a^b f_i\{\phi(t)\}\,d\phi_i(t).$$

To prove (iv), let ϕ on $[0, 1]$ and ψ on $[1, 2]$ be representations of γ_1 and γ_2 respectively. Then $\gamma_1 + \gamma_2$ is represented by χ on $[0, 2]$, where

$$\chi(t) = \begin{cases} \phi(t) & \text{for } 0 \leqslant t \leqslant 1, \\ \psi(t) & \text{for } 1 \leqslant t \leqslant 2 \end{cases}$$

and
$$\int_0^2 (f_i \circ \chi)\, d\chi_i = \int_0^1 (f_i \circ \phi)\, d\phi_i + \int_1^2 (f_i \circ \psi)\, d\psi_i. \ |$$

Integrals with respect to arc. The integrand is now real valued, not, as previously, vector valued. Given an open subset G of R^n, a continuous function $f: G \to R^1$ and a rectifiable curve γ in G, we define an integral of f with respect to the arc length of γ.

Let $\phi : [a, b] \to R^n$ be a representation of γ and define the increasing function $s_\phi : [a, b] \to R^1$ by

$$s_\phi(t) = \begin{cases} 0 & \text{for} \quad t = a, \\ \Lambda_a^t(\phi) & \text{for} \quad a < t \leqslant b. \end{cases}$$

For a dissection $\mathcal{D}$ of $[a, b]$ into subintervals $[t_0, t_1], \ldots, [t_{r-1}, t_r]$ we consider the sum

$$\sum_{j=1}^r f\{\phi(\tau_j)\}\{s_\phi(t_j) - s_\phi(t_{j-1})\},$$

where $t_{j-1} \leqslant \tau_j \leqslant t_j$. Since $f \circ \phi$ is continuous and s_ϕ increases, this sum tends to

$$\int_a^b (f \circ \phi)\, ds_\phi \tag{8.23}$$

as $\mu(\mathcal{D}) \to 0$.

Now, if $\psi : [c, d] \to R^n$ is another representation of γ, so that $\phi = \psi \circ \alpha$, where α satisfies the usual conditions, then

$$\Lambda_a^t(\phi) = \Lambda_c^{\alpha(t)}(\psi) \quad (a < t \leqslant b),$$

i.e.
$$s_\phi = s_{\psi \circ \alpha} = s_\psi \circ \alpha.$$

Hence, as in the proof of theorem 8.21,

$$\int_c^d (f \circ \psi)\, ds_\psi = \int_a^b (f \circ (\psi \circ \alpha))\, d(s_\psi \circ \alpha) = \int_a^b (f \circ \phi)\, ds_\phi.$$

Thus the integrals (8.23) do not depend on the particular representation ϕ of γ and we can therefore denote them all by

$$\int_\gamma f\, ds \quad \text{or} \quad \int_\gamma f(x)\, ds.$$

When γ is a *path*, then, by theorem 8.13, s_ϕ is an indefinite integral of $\|\phi'\|$ and so, by theorem 6.21 B,

$$\int_\gamma f\, ds = \int_a^b (f \circ \phi)\, ds_\phi = \int_a^b (f \circ \phi)\|\phi'\|.$$

Line integrals with respect to arc have the properties (i), (ii) and (iv) stated in theorem 8.22. However (iii) is replaced by

(iii)′ $$\int_{-\gamma} f\,ds = \int_{\gamma} f\,ds.$$

For if $\phi:[a, b] \to R^n$ is a representation of γ, let $\phi^*:[-b, -a] \to R^n$ be the representation of $-\gamma$ defined by

$$\phi^*(t) = \phi(-t) \quad (-b \leqslant t \leqslant -a).$$

Then clearly $\qquad s_{\phi^*}(t) = \Lambda(\gamma) - s_\phi(-t)$

and so

$$\int_{-b}^{-a} f\{\phi^*(t)\}\,ds_{\phi^*}(t) = \int_{-b}^{-a} f\{\phi(-t)\}\,d(\Lambda(\gamma) - s_\phi(-t))$$

$$= -\int_{-b}^{-a} f\{\phi(-t)\}\,ds_\phi(-t)$$

$$= \int_a^b f\{\phi(t)\}\,ds_\phi(t).$$

In analysis, line integrals with respect to arc are comparatively unimportant. However, in mathematical physics, they have a part to play. (See exercises 8(b), 9 and 10.)

Exercises 8(b)

1. Let G be an open subset of R^n ($n > 1$); let γ be a path in G and suppose that the functions $f: G \to R^n$, $g: G \to R^1$ are continuous. Prove, without using the Stieltjes integral, that, if $\phi : [a, b] \to R^n$ and $\psi : [c, d] \to R^n$ are both representations of γ, then

$$\int_a^b (f \circ \phi).\phi' = \int_c^d (f \circ \psi).\psi'$$

and

$$\int_a^b (g \circ \phi)\|\phi'\| = \int_c^d (g \circ \psi)\|\psi'\|.$$

2. Prove theorem 8.22 (iii) and the corresponding result for integrals with respect to arc, without making use of the Stieltjes integral, when γ is a path.

3. Let G be an open subset of R^n and suppose that the function $f: G \to R^n$ is continuous. Show that, if γ is a rectifiable curve whose trace E lies in G, then

$$\left|\int_\gamma f\right| \leqslant \int_\gamma \|f\|\,ds \leqslant M\Lambda(\gamma),$$

where $M = \sup\limits_{x \in E} \|f(x)\|$.

4. Let γ be the triangular path in R^2 consisting of the segments from $(0, 0)$ to $(1, 0)$, from $(1, 0)$ to $(1, 1)$, and from $(1, 1)$ to $(0, 0)$. Evaluate

$$\int_\gamma \frac{y^2\,dx + x^2\,dy}{1 + x^2 + y^2}.$$

5. Given that $a \neq 0$, let γ be the path in R^3 defined by

$$x = a(1 + \cos 2t), \quad y = a \sin 2t, \quad z = 2a \cos t \quad (0 \leqslant t \leqslant 2\pi).$$

Show that the trace of γ is the intersection of a (hollow) cylinder and a (hollow) sphere. Evaluate

$$\int_\gamma \{(y^2 + z^2)\,dx + (z^2 + x^2)\,dy + (x^2 + y^2)\,dz\}.$$

6. The function $f: R^3 \to R^3$ is defined by

$$f(x, y, z) = (yz, 0, xy);$$

and γ is the (helical) path given by

$$x = a \cos t, \quad y = b \sin t, \quad z = ct \quad (0 \leqslant t \leqslant \tfrac{1}{2}\pi).$$

Evaluate $\displaystyle\int_\gamma f$, and also $\displaystyle\int_\sigma f$, where σ is the segment from the initial point of γ to the end point of γ.

7. Let G be an open set in R^n. Let the function $g: G \to R^1$ be continuously differentiable and let the function $f: G \to R^n$ be defined by

$$f = \operatorname{grad} g = (D_1 g, \ldots, D_n g).$$

Prove that, if the path γ with initial point u and end point v lies in G, then

$$\int_\gamma f = g(v) - g(u).$$

(Thus $\displaystyle\int_\gamma f$ has the same value for all paths γ in G which have the same initial and end points; the integral is *independent of path* (contrast **6**).)

8. Let D be an open *connected* set (region) in R^n. Prove that, if the function $f: D \to R^n$ is continuous and the line integral of f is independent of path, then there exists a continuously differentiable function $g: D \to R^1$ such that $f = \operatorname{grad} g$.

(Thus, in view also of **7**, in a *region* D the integral of a continuous function $f: D \to R^n$ is independent of path if and only if $f = \operatorname{grad} g$ for some continuously differentiable function $g: D \to R^1$. By theorem 7.33, any two functions g differ by a constant.

For the proof, take a fixed point $a \in D$ and, for any $x \in D$, define $g(x) = \displaystyle\int_{\gamma_x} f$, where γ_x is any path with initial point a and end point x; such paths exist, by theorem 3.32.)

9. The ends of a uniform chain of length $2l$ are held at the same height and a distance $2a(<2l)$ apart. It is known that, if the co-ordinate system is suitably chosen, the form of the chain is given by

$$x = t, \quad y = c \cosh (t/c) \quad (-a \leqslant t \leqslant a),$$

where c is a constant. Show that c is uniquely determined by a and l. Show also that the centroid of the chain is a distance

$$(l\sqrt{(c^2 + l^2)} - ac)/(2l)$$

below the ends.

(The centroid $(\xi_1, \ldots, \xi_n)$ of a curve γ in R^n is given by $\xi_i \Lambda(\gamma) = \displaystyle\int_\gamma x_i\,ds$.)

10. A wire is shaped in the form of the twisted cubic given by

$$x = 3t - t^3, \quad y = 3t^2, \quad z = 3t + t^3 \quad (0 \leqslant t \leqslant \sqrt{2}).$$

Show that the total length of the wire is 10, and that the arc length between the origin and the point (x, y, z) is $\sqrt{2}z$.

Find the moment of inertia of the wire about the y-axis, if its mass per unit length varies uniformly with arc length, from 0 at the origin to 1 at the other end.

(If a wire is shaped as a curve γ in R^3, its moment of inertia about a straight line λ is $\displaystyle\int_\gamma \mu\rho^2 \, ds$, where $\mu(x, y, z)$ is the mass per unit length at (x, y, z) and $\rho(x, y, z)$ is the distance of (x, y, z) from λ.)

8.3. Integration over intervals in R^n

We first dispose of some matters of terminology. By a *closed interval* in R^n we mean a set I of the form

$$[a_1, b_1] \times \ldots \times [a_n, b_n],$$

where $a_i < b_i$ for $i = 1, \ldots, n$. The interval is *cubical* if

$$b_1 - a_1 = \ldots = b_n - a_n.$$

The *faces* of I are the $2n$ sets

$$I \cap \{(x_1, \ldots, x_n) | x_i = a_i\}, \quad I \cap \{(x_1, \ldots, x_n) | x_i = b_i\} \quad (i = 1, \ldots, n).$$

The *volume* $V(I)$ of I is defined to be

$$(b_1 - a_1) \ldots (b_n - a_n).$$

If $\mathscr{D}_i$ is the dissection of the interval $[a_i, b_i]$ given by

$$a_i = x_i^0 < x_i^1 < \ldots < x_i^{r_i - 1} < x_i^{r_i} = b_i,$$

the dissection $\mathscr{D} = \mathscr{D}_1 \times \ldots \times \mathscr{D}_n$ of I is the division of I into the $r = r_1 \ldots r_n$ subintervals

$$[x_1^{j_1 - 1}, x_1^{j_1}] \times \ldots \times [x_n^{j_n - 1}, x_n^{j_n}],$$

where $1 \leqslant j_1 \leqslant r_1, \ldots, 1 \leqslant j_n \leqslant r_n$. To define integration over I we now proceed in exactly the same way as in the one-dimensional case.

Let the function $f : I \to R^1$ be bounded. Given a dissection

$$\mathscr{D} = \mathscr{D}_1 \times \ldots \times \mathscr{D}_n$$

of I into subintervals $I_1, \ldots, I_r$, put

$$m_j = \inf_{x \in I_j} f(x), \quad M_j = \sup_{x \in I_j} f(x)$$

for $j = 1, \ldots, r$ and

$$s(\mathscr{D}) = s(\mathscr{D}, f) = \sum_{j=1}^{r} m_j V(I_j), \quad S(\mathscr{D}) = S(\mathscr{D}, f) = \sum_{j=1}^{r} M_j V(I_j).$$

These sums are, as usual, called *lower* and *upper sums* respectively. Further subdivision of I clearly increases the lower sum and diminishes the upper sum. It follows as in §6.1 that, if $\mathscr{D}$ and $\mathscr{D}'$ are any dissections of I,

$$s(\mathscr{D}) \leqslant S(\mathscr{D}'). \tag{8.31}$$

We now write

$$\sup_{\mathscr{D}} s(\mathscr{D}, f) = \underline{\int_I} f, \quad \inf_{\mathscr{D}} S(\mathscr{D}, f) = \overline{\int_I} f$$

and call the first expression the *lower integral* and the second the *upper integral* of f over I. By (8.31),

$$\underline{\int_I} f \leqslant \overline{\int_I} f. \tag{8.32}$$

If there is equality in (8.32), f is said to be *integrable* over I and the common value of the upper and lower integrals, called the *integral* of f over I, is denoted by

$$\int_I f \quad \text{or} \quad \int_I f(x)\,dx.$$

An integral over an interval in R^n, where $n > 1$, may, for emphasis, be called a *multiple* integral. In R^2 and R^3 the corresponding terms are *double* integral and *triple* integral, respectively. These integrals are often written

$$\iint_I f \quad \text{or} \quad \iint_I f(x, y)\,dx\,dy$$

and

$$\iiint_I f \quad \text{or} \quad \iiint_I f(x, y, z)\,dx\,dy\,dz.$$

Multiple integrals and integrals in R^1 have the same elementary properties. The proofs of the following statements are easy adaptations of their analogues in chapter 6.

(i) If f, g are integrable over I, then so are $|f|$, $1/f$ provided that $\inf_{x \in I} f(x) > 0$, $af + bg$ (where a, b are real numbers) and fg; also

$$\left| \int_I f \right| \leqslant \int_I |f|$$

and

$$\int_I (af + bg) = a \int_I f + b \int_I g.$$

(ii) Let I_1, I_2 be intervals which have a face in common and are *non-overlapping* (i.e. have no common interior points) so that

$I_1 \cup I_2$ is an interval (distinct from I_1 and from I_2). Then, for any bounded function f on $I_1 \cup I_2$,

$$\underline{\int}_{I_1 \cup I_2} f = \underline{\int}_{I_1} f + \underline{\int}_{I_2} f \quad \text{and} \quad \overline{\int}_{I_1 \cup I_2} f = \overline{\int}_{I_1} f + \overline{\int}_{I_2} f. \qquad (8.33)$$

Thus, if f is integrable over I_1 and over I_2, then f is integrable over $I_1 \cup I_2$ and

$$\int_{I_1 \cup I_2} f = \int_{I_1} f + \int_{I_2} f.$$

It also follows that integrability over an interval implies integrability over any subinterval.

The multi-dimensional form of Darboux's theorem (6.71) is important. For a dissection $\mathscr{D}$ of the interval I into subintervals $I_1, \ldots, I_r$, the *mesh* $\mu(\mathscr{D})$ of $\mathscr{D}$ is the largest of the diameters

$$\rho(I_1), \ldots, \rho(I_r)$$

(see exercise 3(d), 7). If now f is any bounded function on I,

$$\underline{\int}_I f = \lim_{\mu(\mathscr{D}) \to 0} s(\mathscr{D}, f) \quad \text{and} \quad \overline{\int}_I f = \lim_{\mu(\mathscr{D}) \to 0} S(\mathscr{D}, f).$$

Now let f be a real-valued function on the interval $K = I \times J$ of $R^{m+n} = R^p$, where $I \subset R^m$ and $J \subset R^n$. Suppose that, for every $y \in J$, the function $f(., y)$ is integrable over I and that the function ϕ on J given by

$$\phi(y) = \int_I f(x, y) \, dx$$

is integrable over J. We then write

$$\int_J \phi = \int_J \left(\int_I f(x, y) \, dx \right) dy$$

as

$$\int_J dy \int_I f(x, y) \, dx$$

and call this expression a *repeated* (or *iterated*) integral. The same decomposition of K leads to the integral

$$\int_I dx \int_J f(x, y) \, dy$$

(if it exists).

A repeated integral may involve any number of integrations. For instance if

$$K = [a_1, b_1] \times \ldots \times [a_p, b_p],$$

one such integral is

$$\int_{a_p}^{b_p} dx_p \ldots \int_{a_1}^{b_1} f(x_1, \ldots, x_p) \, dx_1.$$

The next theorem and its corollaries serve to connect the integral of f over K and the associated repeated integrals; exercises 8(c), 6–9 emphasize the distinctiveness of these integrals.

Theorem 8.31. *Let I, J be closed intervals in R^m and R^n respectively and suppose that the function $f : I \times J \to R^1$ is integrable over $I \times J$. Then the functions ϕ, ψ on J defined by*

$$\phi(y) = \underline{\int}_I f(x, y) \, dx, \quad \psi(y) = \overline{\int}_I f(x, y) \, dx$$

are integrable over J and

$$\int_J dy \underline{\int}_I f(x, y) \, dx = \int_{I \times J} f(x, y) \, dx \, dy = \int_J dy \overline{\int}_I f(x, y) \, dx. \quad (8.34)$$

Proof. Let $\mathscr{D}$ be any dissection of $I \times J$. Then $\mathscr{D}$ is of the form $\mathscr{D}_I \times \mathscr{D}_J$, where $\mathscr{D}_I, \mathscr{D}_J$ are dissections of I and J into subintervals $I_1, \ldots, I_r$ and $J_1, \ldots, J_s$ respectively. The subintervals of $I \times J$ are therefore $I_k \times J_l$ $(k = 1, \ldots, r; l = 1, \ldots, s)$.

Put
$$M_{kl} = \sup_{(x, y) \in I_k \times J_l} f(x, y), \quad N_l = \sup_{y \in J_l} \psi(y)$$

and, for any $y \in J$, let
$$M_k(y) = \sup_{x \in I_k} f(x, y).$$

If $y \in J_l$, we have

$$\sum_{k=1}^{r} M_{kl} V(I_k) \geqslant \sum_{k=1}^{r} M_k(y) V(I_k) \geqslant \overline{\int}_I f(x, y) \, dx = \psi(y)$$

and, since this is true for all $y \in J_l$,

$$\sum_{k=1}^{r} M_{kl} V(I_k) \geqslant N_l.$$

Also $V(I_k \times J_l) = V(I_k) V(J_l)$ and therefore

$$\sum_{k=1}^{r} \sum_{l=1}^{s} M_{kl} V(I_k \times J_l) \geqslant \sum_{l=1}^{s} N_l V(J_l) \geqslant \overline{\int}_J \psi.$$

This inequality holds for all dissections of $I \times J$. Hence

$$\int_{I \times J} f \geqslant \overline{\int}_J \psi = \overline{\int}_J dy \overline{\int}_I f(x, y) \, dx.$$

An argument similar to the one above shows that

$$\int_{I \times J} f \leqslant \int_{\underline{J}} dy \int_{\underline{I}} f(x, y) \, dx.$$

To prove the integrability of ψ and the right-hand identity in (8.34) we also use the inequalities

$$\int_{\underline{J}} dy \int_{\underline{I}} f(x, y) \, dx \leqslant \int_{\underline{J}} dy \int_{I}^{\overline{}} f(x, y) \, dx \leqslant \int_{J}^{\overline{}} dy \int_{I}^{\overline{}} f(x, y) \, dx,$$

for these then give

$$\int_{I \times J} f \leqslant \int_{\underline{J}} dy \int_{I}^{\overline{}} f(x, y) \, dx \leqslant \int_{J}^{\overline{}} dy \int_{I}^{\overline{}} f(x, y) \, dx \leqslant \int_{I \times J} f. \mid$$

In the last theorem the functions ϕ, ψ may, of course, be replaced by ϕ^*, ψ^*, where

$$\phi^*(x) = \int_{\underline{J}} f(x, y) \, dy, \quad \psi^*(x) = \int_{J}^{\overline{}} f(x, y) \, dy \quad (x \in I).$$

We are therefore led to a sufficient condition for the invertibility of the order of integration in a repeated integral.

Corollary 1. *Let I, J be closed intervals in R^m and R^n respectively. If*

   (i) *the function $f: I \times J \to R^1$ is integrable,*

   (ii) $\int_I f(x, y) \, dx$ *exists for every $y \in J$,*

   (iii) $\int_J f(x, y) \, dy$ *exists for every $x \in I$,*

then $\quad \int_J dy \int_I f(x, y) \, dx \quad and \quad \int_I dx \int_J f(x, y) \, dy$

exist and are equal.

It is easily seen that every continuous function is integrable (cf. theorem 6.12(i)). The result of corollary 1 therefore certainly holds if f is continuous on $I \times J$. (The existence of the two repeated integrals is also a consequence of the fact, to be proved in §8.7, that $\int_I f(x, .) \, dx$ and $\int_J f(., y) \, dy$ are continuous on J and I respectively.)

Repeated application of corollary 1 shows that, for continuous functions, the process of multiple integration may be replaced by a number of single integrations.

Corollary 2. *Let* $I = [a_1, b_1] \times \ldots \times [a_n, b_n]$ *be an interval in* R^n. *If the function* $f : I \to R^1$ *is continuous, then*

$$\int_I f(x_1, \ldots, x_n)\, dx_1 \ldots dx_n = \int_{a_n}^{b_n} dx_n \ldots \int_{a_1}^{b_1} f(x_1, \ldots, x_n)\, dx_1.$$

Moreover the integral on the right may be replaced by

$$\int_{a_{i_n}}^{b_{i_n}} dx_{i_n} \ldots \int_{a_{i_1}}^{b_{i_1}} f(x_1, \ldots, x_n)\, dx_{i_1},$$

where $(i_1, \ldots, i_n)$ *is any permutation of* $(1, \ldots, n)$.

Another corollary is a relation between double integrals and line integrals.

Corollary 3. (Green's theorem for intervals in R^2.) *Let* I *be the interval* $[a, b] \times [c, d]$ *in* R^2 *and denote by* ∂I *the path determined by the four straight line segments from* (a, c) *to* (b, c) *to* (b, d) *to* (a, d) *to* (a, c). *If* G *is an open set containing* I *and the function*

$$f = (f_1, f_2) : G \to R^2$$

is continuously differentiable, then

$$\int_{\partial I} f = \iint_I (D_1 f_2 - D_2 f_1).$$

Proof. In the statement of the corollary we have used our convention regarding the paths determined by straight line segments (see p. 254). Thus $\partial I = \gamma_1 + \gamma_2 - \gamma_3 - \gamma_4$, where $\gamma_1, \gamma_2, \gamma_3, \gamma_4$ may be represented by

$$x = t, y = c \quad (a \leqslant t \leqslant b), \qquad x = b, y = t \quad (c \leqslant t \leqslant d),$$
$$x = t, y = d \quad (a \leqslant t \leqslant b), \qquad x = a, y = t \quad (c \leqslant t \leqslant d),$$

respectively. We now have, by corollary 2 and exercise 6(*h*), 10,

$$\iint_I D_1 f_2 = \int_c^d dy \int_a^b D_1 f_2(x, y)\, dx$$
$$= \int_c^d f_2(b, y)\, dy - \int_c^d f_2(a, y)\, dy = \int_{\gamma_2} f - \int_{\gamma_4} f;$$

and similarly $\qquad \displaystyle \iint_I D_2 f_1 = \int_{\gamma_3} f - \int_{\gamma_1} f. \;\blacksquare$

Exercises 8(c)

1. The real-valued function f on $[a, b] \times [c, d] (\subset R^2)$ is increasing in each variable, i.e., for each $y \in [c, d]$, $f(x, y)$ increases with x and, for each $x \in [a, b]$, $f(x, y)$ increases with y. Prove that f is integrable over $[a, b] \times [c, d]$ (cf. theorem 6.12 (ii)). Extend this result to functions on intervals in R^n ($n = 1, 2, \ldots$).

2. Let I be a closed interval in R^m and suppose that the function $f : I \to R^1$ is integrable over I. Define the function $\phi : I \times R^n \to R^1$ by

$$\phi(x, y) = f(x) \quad (x \in I, y \in R^n).$$

Prove that ϕ is integrable over every interval $K = I \times J$ in R^{m+n}.

Suppose, in addition, that the function $g : J \to R^1$ is integrable and that the function $h : K \to R^1$ is defined by

$$h(x, y) = f(x)g(y) \quad (x \in I, y \in J).$$

Show that h is integrable over K and that

$$\int_K h = \left(\int_I f \right) \left(\int_J g \right).$$

3. Prove that, if $a, b > 0$, $c \neq 0$ and $I = [0, a] \times [0, b]$, then

$$\int_I \frac{dx \, dy}{(x^2 + y^2 + c^2)^{\frac{3}{2}}} = \frac{1}{c} \arctan \frac{ab}{c \sqrt{(a^2 + b^2 + c^2)}}.$$

4. If I is the interval $[-1, 2] \times [-1, 0] \times [0, 2]$ of R^3, show that

$$\int_I \left(\frac{z}{1 - |x|y} \right)^2 dx \, dy \, dz = 9 \log 6.$$

5. Evaluate

$$\int_0^1 dy \int_0^\alpha \sqrt{(1 - y \cos^2 x)} \, dx \quad (0 < \alpha < \tfrac{1}{2}\pi).$$

6. The function f on $I = [0, 1] \times [0, 1] (\subset R^2)$ is defined by

$$f(x, y) = \begin{cases} 0 & \text{if at least one of } x, y \text{ is irrational,} \\ 1/q & \text{if } y \text{ is rational and } x = p/q, \end{cases}$$

where p, q are coprime integers and $q > 0$. Prove that

$$\int_I f(x, y) \, dx \, dy \quad \text{and} \quad \int_0^1 dy \int_0^1 f(x, y) \, dx$$

exist and are equal, but that

$$\int_0^1 dx \int_0^1 f(x, y) \, dy$$

does not exist.

Find a function h on I such that $\int_I h$ exists, but neither of the associated repeated integrals exists.

7. Define the function f on $I = [0, \pi] \times [0, 1]$ by

$$f(x, y) = \begin{cases} \cos x & \text{if } y \text{ is rational,} \\ 0 & \text{if } y \text{ is irrational.} \end{cases}$$

Show that

$$\int_0^1 dy \int_0^\pi f(x, y)\, dx$$

exists, but that

$$\int_I f(x, y)\, dx\, dy \quad \text{and} \quad \int_0^\pi dx \int_0^1 f(x, y)\, dy$$

do not exist.

8. Let $(p_1, p_2, p_3, \ldots)$ be the sequence of primes $(2, 3, 5, \ldots)$; let

$$P_k = \left\{ \left(\frac{m}{p_k}, \frac{n}{p_k} \right) \,\middle|\, m, n = 1, \ldots, p_k - 1 \right\}$$

and let

$$P = \bigcup_{k=1}^{\infty} P_k.$$

Show that the set P is dense in $[0, 1] \times [0, 1]$, but that any line parallel to the axes contains no more than a finite number of points of P.

Define the function $f: I = [0, 1] \times [0, 1] \to R^1$ by

$$f(x, y) = \begin{cases} 1 & \text{for } x \in P, \\ 0 & \text{for } x \in I - P. \end{cases}$$

Show that

$$\int_0^1 dx \int_0^1 f(x, y)\, dy \quad \text{and} \quad \int_0^1 dy \int_0^1 f(x, y)\, dx$$

exist and are both 0, but that $\int_I f$ does not exist.

9. The function $f: [0, 1] \times [0, 1] \to R^1$ is defined by

$$f(x, y) = \begin{cases} 1/y^2 & \text{if } 0 < x < y < 1, \\ -1/x^2 & \text{if } 0 < y < x < 1, \\ 0 & \text{otherwise.} \end{cases}$$

Prove that

$$\int_0^1 dy \int_0^1 f(x, y)\, dx \quad \text{and} \quad \int_0^1 dx \int_0^1 f(x, y)\, dy$$

both exist, but are unequal. (By theorem 8.31, f cannot be integrable over $[0, 1] \times [0, 1]$; in fact, f is unbounded near the axes.)

8.4. Integration over arbitrary bounded sets in R^n

We often wish to integrate functions over sets which are not intervals. There is a simple device for doing this (which applies to R^1 as well as to multi-dimensional Euclidean spaces).

Let E be a bounded subset of R^n and suppose that the function $f: E \to R^1$ is bounded. Define the function $f_E: R^n \to R^1$ by

$$f_E(x) = \begin{cases} f(x) & \text{if } x \in E, \\ 0 & \text{if } x \in E'. \end{cases}$$

It then follows easily from (8.33) (see exercise 8 (d), 1) that

$$\underline{\int}_I f_E \quad \text{and} \quad \overline{\int}_I f_E$$

have the same values for all closed intervals I containing E. We call these two numbers the lower integral and the upper integral of f over E, denoting them by

$$\underline{\int}_E f \quad \text{and} \quad \overline{\int}_E f,$$

respectively. If the upper and lower integrals are equal, their common value is called the integral of f over E; it is written

$$\int_E f$$

and f is said to be integrable over E.

When $f \equiv 1$, f_E is the indicator (or characteristic) function χ_E (first mentioned in §6.8). The inner content $\underline{c}(E)$ of E and the outer content $\bar{c}(E)$ are defined by

$$\underline{c}(E) = \underline{\int}_E 1, \quad \bar{c}(E) = \overline{\int}_E 1.$$

If $\underline{c}(E) = \bar{c}(E)$, the common value is called the content of E and is denoted by $c(E)$. (Cf. §6.8.) Clearly, the content of a closed interval is its volume. It is also easy to verify that the content of an open interval

$$(a_1, b_1) \times \ldots \times (a_n, b_n)$$

is $(b_1 - a_1) \ldots (b_n - a_n)$, the content (or volume) of its closure (exercise 8(d), 2). For other elementary results on integration and content see exercises 8(d), 3–7.

The theorem below generalizes theorem 6.81.

Theorem 8.41.

(i) *The bounded set E in R^n has zero content if and only if, given $\epsilon > 0$, there is a finite number of closed intervals (or a finite number of open intervals) which cover E and whose total content is less than ϵ.*

(ii) *The statement (i) is true when all intervals are restricted to be cubical.*

Proof. We prove (i) and (ii) at the same time.

First suppose that $c(E) = 0$. Let K be any cubical interval containing E. Since $\int_K \chi_E = 0$, Darboux's theorem shows that there is a dissection $\mathcal{D}$ of K into cubical subintervals such that $S(\mathcal{D}, \chi_E) < \epsilon$; and $S(\mathcal{D}, \chi_E)$ is the sum of the contents of the closed cubical subintervals $K_1, \ldots, K_s$ which contain points of E.

Each K_l may be replaced by a larger open cubical interval so that the content of the open intervals is still less than ϵ.

Now suppose that, given $\epsilon > 0$, there is a finite number of open intervals J_k covering E and of total content less than ϵ. Let I be a closed interval containing all the J_k and form a dissection $\mathscr{D}$ of I into subintervals $I_1, \ldots, I_r$ by producing the faces of the J_k. (The process for $n = 2$ is illustrated in the figure; the intervals J_k are shaded.)

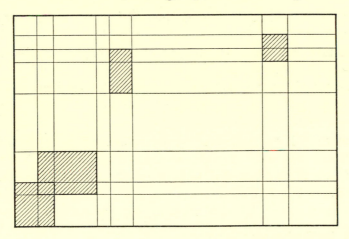

Since the I_j in $I - \cup J_k$ do not contain points of E and the total content of the rest is less than ϵ, $S(\mathscr{D}, \chi_E) < \epsilon$. This holds for all $\epsilon > 0$ and so $\int_I \chi_E = 0$.

If we start with closed intervals, we may enlarge them slightly into open ones and then proceed as before. |

We have previously noted that continuous functions are integrable. This result can be significantly improved.

Theorem 8.42. *If f is bounded on the closed interval I and the set of points of discontinuity of f has zero content, then f is integrable over I.*

The proof, in which the points of discontinuity are covered by open intervals of small content, is easily adapted from the proof of the analogous theorem 6.82.

An informal interpretation of the next theorem is that the sets with content are those whose edges are not too blurred.

Theorem 8.43. *A bounded set in R^n has content if and only if its frontier has zero content.*

Proof. Let E be a bounded set in R^n and denote its frontier by F. The function χ_E is continuous on the open set F'. Hence, by theorem 8.42, if $c(F) = 0$, then χ_E is integrable over every interval containing E, i.e. $c(E)$ exists.

Now suppose that F does not have zero content, so that

$$\overline{\int_F} 1 = a > 0.$$

If I is an interval containing F, then $S(\mathscr{D}, \chi_F) \geqslant a$ for every dissection $\mathscr{D}$ of I. Take I to be a cubical interval whose frontier lies in E' and take $\mathscr{D}$ to be a dissection of I into congruent cubical subintervals $I_1, \ldots, I_r$.

Let $\xi \in F$. If ξ is in I_k and in no other subinterval, then

$$\sup_{x \in I_k} \chi_E(x) - \inf_{x \in I_k} \chi_E(x) = 1. \tag{8.41}$$

If, however, ξ is on the frontier of two or more of the I_k, then (8.41) need hold for no more than one of these. Since at most 2^n of the I_k contain any given point of F, (8.41) holds for at least a fraction $1/2^n$ of the I_k meeting F. Hence

$$S(\mathscr{D}, \chi_E) - s(\mathscr{D}, \chi_E) \geqslant a/2^n$$

and so, by Darboux's theorem,

$$\overline{\int_I} \chi_E - \underline{\int_I} \chi_E \geqslant a/2^n > 0. \;|$$

A combination of theorems 8.42 and 8.43 gives a particularly useful condition for integrability.

Theorem 8.44. *Let E be a subset of R^n which has content. If the bounded function $f : E \to R^1$ is continuous except on a set of zero content, then f is integrable over E.*

Proof. Let X be the subset of E on which f is discontinuous. Then f_E is continuous on $(\bar{E})'$ and on $E^\circ - X$. Hence, if F is the frontier of E, all discontinuities of f_E lie in the set $X \cup F$ which has zero content. It follows from theorem 8.42 that f_E is integrable over every interval containing E. $\;|$

Theorem 8.45. *The trace of any rectifiable curve (in R^n) has zero content.*

Proof. Let the rectifiable curve γ have trace Γ and length λ. If $\psi : [a, b] \to R^n$ is a representation of γ, the function s on $[a, b]$ given by $s(t) = \Lambda_a^t(\psi)$ is continuous (exercise $8(a), 9$). Hence there are points

$$a = u^0 < u^1 < \ldots < u^{r-1} < u^r = b$$

such that $\qquad s(u^j) - s(u^{j-1}) = \lambda/r \quad (j = 1, \ldots, r).$

Now, for each j, let K_j be the cubical interval with centre $\psi(u^j)$ and sides of length $2\lambda/r$. Then, by exercise $8(a), 5$, the K_j cover Γ and their total content is

$$r\left(\frac{2\lambda}{r}\right)^n = \frac{(2\lambda)^n}{r^{n-1}} \leqslant \frac{(2\lambda)^n}{r},$$

which may be made arbitrarily small by taking r sufficiently large. $\mid$

The previous theorem is mainly useful in R^2; in higher dimensional spaces the sets of points with which it deals are not 'large' enough.

Let m, n be integers such that $1 \leqslant m < n$. A *surface* in R^n is defined to be a continuously differentiable function $\psi: G \to R^n$, where G is an open subset of R^m. Also, if H is any compact subset of G, we call the restriction of ψ to H a *surface section* and the set $\psi(H)$ its *trace*. A continuously differentiable transit is a surface section for which $m = 1$ and H is a finite closed interval. (The notes at the end of this chapter include some remarks about surfaces.)

Lemma. *Let G be an open subset of R^n and suppose that the function $\phi: G \to R^n$ is continuously differentiable. If the set E has zero content and $\bar{E} \subset G$, then the set $\phi(E)$ also has zero content.*

Proof. By theorem 8.43, $\bar{E}$ has zero content. It is enough to prove that $\phi(\bar{E})$ has zero content. Given $\epsilon > 0$, there are closed cubical intervals $I_1, \ldots, I_r$ which cover $\bar{E}$ and the total content of which is less than ϵ. Since the distance between the compact set $\bar{E}$ and the closed set G' is positive (exercise $3(f), 14$), we may suppose that each I_j is contained in G. Put

$$\bigcup_{j=1}^r I_j = H.$$

The function $D\phi: G \to L(R^n, R^n)$ is continuous and therefore bounded on the compact set H. Let

$$\sup_{x \in H} \|D\phi(x)\| = M.$$

Then, by theorem 7.31,

$$\|\phi(x) - \phi(y)\| \leqslant M\|x - y\|$$

whenever $x, y \in I_j \ (j = 1, \ldots, r)$. Hence, if the sides of I_j are of length s_j, $\phi(I_j)$ can be enclosed in a cubical interval whose sides are of length $(M\sqrt{n})s_j$. Thus $\phi(\bar{E})$ can be covered by a finite number of cubical intervals whose total content is less than $(M\sqrt{n})^n \epsilon$. |

Theorem 8.46. *The trace of a surface section in R^n ($n > 1$) has zero content.*

Proof. Let the trace in question be $\psi(H)(\subset R^n)$, where ψ is a continuously differentiable function on the open set G in R^m, where $1 \leqslant m < n$ and H is a compact subset of G. The set

$$H^* = \{(x_1, ..., x_m, 0, ..., 0)|(x_1, ..., x_m) \in H\}$$

in R^n is clearly compact and has zero content. Also the function ψ^* on the open subset $G \times R^{n-m}$ of R^n given by

$$\psi^*(x_1, ..., x_n) = \psi(x_1, ..., x_m)$$

is, by consideration of its partial derivatives, seen to be continuously differentiable. Since $\psi(H) = \psi^*(H^*)$, it therefore follows from the lemma that $\psi(H)$ has zero content. |

We now introduce a convenient phrase. Let E be a subset of R^n and suppose that every point of E either does or does not possess a certain property P. If P holds for all points of E except for those in a set of zero content, we shall say that P holds *nearly everywhere* in E (or, for *nearly all* points of E). The n-dimensional form of theorem 6.83 may then be stated as follows.

Theorem 8.47. *Suppose that f and g are bounded on a bounded subset E of R^n and that $f(x) = g(x)$ nearly everywhere in E. Then either both f and g are integrable over E or neither is integrable; and, if f, g are integrable,*

$$\int_E f = \int_E g.$$

The proof is similar to that of theorem 6.83. The theorem shows, as was pointed out in similar circumstances in §6.8, that the process of integration over a given set does not require the integrand to be defined everywhere on that set—nearly everywhere is enough. This remark leads to a useful corollary of theorem 8.31.

Theorem 8.48. *Let I, J be closed intervals in R^m and R^n respectively. If*

(i) *the function $f: I \times J \to R^1$ is integrable, and*

(ii) $\phi(y) = \int_I f(x, y)\, dx$ *exists for nearly all $y \in J$,*

then ϕ is integrable over J and

$$\int_J dy \int_I f(x, y)\, dx = \int_{I \times J} f(x, y)\, dx\, dy.$$

This theorem is the basis for evaluating multiple integrals over sets which need not be intervals. We illustrate the method with a simple example.

Let the functions $u, v: [a, b] \to R^1$ be such that $u(x) \leqslant v(x)$ for $a \leqslant x \leqslant b$ and let E be the set of points (x, y) in R^2 such that $a \leqslant x \leqslant b$ and $u(x) \leqslant y \leqslant v(x)$. If, now, $f: E \to R^1$ is integrable and

$$\int_{u(x)}^{v(x)} f(x, y)\, dy$$

exists for nearly every x in $[a, b]$, then

$$\int_E f(x, y)\, dx\, dy = \int_a^b dx \int_{u(x)}^{v(x)} f(x, y)\, dy.$$

The identity follows from theorem 8.48 and the observation that, if $[a, b] \times [c, d] \supset E$, then

$$\int_c^d f_E(x, y)\, dy = \int_{u(x)}^{v(x)} f(x, y)\, dy.$$

In particular, take E to be the triangle with vertices $(0, 0)$, $(0, 1)$, $(1, 1)$ and let $f: E \to R^1$ be given by

$$f(x, y) = 4(x^2 + 2y + 1)^{-2}.$$

Then E has content, since its frontier is the trace of a rectifiable curve (theorems 8.43, 8.45). Also f is continuous on E. Therefore f is integrable over E (theorem 8.44) and $\int_x^1 f(x, y)\, dy$ exists for every $x \in [0, 1]$. Thus

$$\int_E \frac{4}{(x^2 + 2y + 1)^2}\, dx\, dy = \int_0^1 dx \int_x^1 \frac{4}{(x^2 + 2y + 1)^2}\, dy$$

$$= \int_0^1 \left\{ \frac{2}{(x + 1)^2} - \frac{2}{x^2 + 3} \right\} dx = 1 - \frac{\pi}{3\sqrt{3}}.$$

We also have

$$\int_E \frac{4}{(x^2+2y+1)^2}\, dx\, dy = \int_0^1 dy \int_0^y \frac{4}{(x^2+2y+1)^2}\, dx,$$

but the evaluation of the repeated integral on the right is tedious.

Exercises 8(d)

1. The set E in R^n is bounded. Prove that, if the function $f: E \to R^1$ is bounded, then $\underline{\int}_I f_E$ and $\overline{\int}_I f_E$ have the same values for all closed intervals I containing E.

2. Prove that $\quad c\{(a_1, b_1) \times \ldots \times (a_n, b_n)\} = (b_1 - a_1)\ldots(b_n - a_n)$.

3. Show that, if E is a subset of R^n such that $c(E) = 0$, then, for any bounded function $f: E \to R^1$, $\int_E f$ exists and is 0.

4. The function f is integrable over a set E in R^n. (i) Show that f need not be integrable over every subset of E, but that f *is* integrable over every subset with content. (ii) Show that, if f is integrable over the subset X of E, then f is also integrable over $E - X$.

5. Let E_1, E_2 be bounded sets in R^n and suppose that the function $f: E_1 \cup E_2 \to R^1$ is non-negative and integrable over E_1 and over E_2. Use the method of exercise 6(*h*), 5 to prove that f is integrable over $E_1 \cup E_2$ and over $E_1 \cap E_2$ and that

$$\int_{E_1 \cup E_2} f + \int_{E_1 \cap E_2} f = \int_{E_1} f + \int_{E_2} f.$$

Prove also that f is integrable over $E_1 - E_2$ and that

$$\int_{E_1 - E_2} f = \int_{E_1} f - \int_{E_1 \cap E_2} f.$$

Deduce that both results hold without the restriction that f is non-negative.

6. Let the function f be bounded on the (bounded) set E in R^n; and suppose that there exists a sequence (X_k) of subsets of E such that f is integrable over each set $E - X_k$ and $\overline{c}(X_k) \to 0$ as $k \to \infty$. Prove that f is integrable over E and that

$$\int_{E - X_k} f \to \int_E f \quad \text{as} \quad k \to \infty.$$

7. *The mean value theorem.* The set E in R^n has content and the function $f: E \to R^1$ is integrable over E. Prove that there is a number μ between the upper and lower bounds of f on E such that

$$\int_E f = \mu c(E).$$

If E is also closed and connected and f is continuous, prove that there is a point ξ in E such that $\mu = f(\xi)$.

8. Let I be the closed interval

$$[a_1, b_1] \times \ldots \times [a_n, b_n]$$

in R^n and let the function $f: I \to R^1$ be integrable. For $x = (x_1, \ldots, x_n) \in I$, put

$$J_x = [a_1, x_1] \times \ldots \times [a_n, x_n].$$

Prove that the function $F: I \to R^1$ defined by

$$F(x) = \int_{J_x} f \quad (x \in I)$$

is continuous. ($F(x)$ is defined for all $x \in I$, since each J_x has content; if $x_i = a_i$ for some i, then $c(J_x) = 0$.)

9. Prove that the content of an open or closed disc of radius r is πr^2. (The content of an n-dimensional ball will be found in exercise $8(f)$, 12.)

10. (i) Let I be a closed interval in R^n and suppose that the function $f: I \to R^1$ is integrable over I. Show that the set

$$S = \{(x_1, \ldots, x_n, x_{n+1}) | (x_1, \ldots, x_n) \in I, \, x_{n+1} = f(x_1, \ldots, x_n)\}$$

in R^{n+1} has zero content.

(ii) Given a set E in R^n and a non-negative function $f: E \to R^1$, the set

$$D = \{(x_1, \ldots, x_n, x_{n+1}) | (x_1, \ldots, x_n) \in E, \, 0 \leqslant x_{n+1} \leqslant f(x_1, \ldots, x_n)\}$$

in R^{n+1} is called the *ordinate set* of f on E. Show that, if E has content and f is continuous (as well as non-negative), then D has content and

$$c(D) = \int_E f.$$

(Show that χ_D is nearly everywhere continuous.)

11. Let $a, b > 0$ and let the function f be given by

$$f(x, y) = \left(\frac{x+a}{y+b}\right)^2 \quad ((x, y) \in R^2, \, y \neq -b).$$

Find the contents of the ordinate sets (see **10**) of f on (i) the closed interval $[-a, a] \times [0, b]$ and (ii) the set

$$\{(x, y) | y \geqslant 0, \, x^2/a^2 + y^2/b^2 \leqslant 1\}.$$

12. Show that, if S is the square with vertices $(1, 0)$, $(0, 1)$, $(-1, 0)$, $(0, -1)$, then

$$\int_S \frac{x-y+1}{x+y+2} \, dx \, dy = \log 3.$$

13. Evaluate

$$\int_0^1 dy \int_y^1 e^{-y/x} \, dx.$$

(Note that the integrand is not defined on the y-axis.)

14. Evaluate

$$\int_E \frac{\sqrt{x} \, dx \, dy \, dz}{(x+y+z+1)^3},$$

where $E = \{(x, y, z) | x \geqslant 0, \, y \geqslant 0, \, z \geqslant 0, \, x+y+z \leqslant 1\}$.

15. The real function f on $[a, b]$ is integrable and, for $n = 1, 2, ...,$ the functions f_n on $[a, b]$ are defined by

$$f_n(x) = \frac{1}{(n-1)!} \int_a^x (x-t)^{n-1} f(t)\, dt \quad (a \leqslant x \leqslant b).$$

Prove that
$$\int_a^x f_n = f_{n+1}(x),$$

so that f_n is the nth repeated integral of f, i.e.

$$f_n(x) = \int_a^x dt_n ... dt_2 \int_a^{t_2} f(t_1)\, dt_1.$$

8.5. Differentiation and integration

The simplicity of the space R^1 sometimes fails to reveal the ideas underlying familiar aspects of real function theory. We saw this in chapters 3 and 7; another instance occurs in this section where we wish to extend the two relations

$$\frac{d}{dx} \int_a^x f = f(x) \quad \text{and} \quad \int_a^b f' = f(b) - f(a)$$

(which hold under suitable conditions) to multiple integration. To do so we have to introduce derivatives of *set functions*.

A set function is simply a real-valued function whose domain is a class of sets. For instance, if E is a set in R^n, $\mathscr{C}_E$ is the class of subsets of E with content and $f: E \to R^1$ is a function integrable over E, a set function $\sigma_f: \mathscr{C}_E \to R^1$ is given (see exercise 8(d) 4), by

$$\sigma_f(X) = \int_X f \quad (X \in \mathscr{C}_E).$$

Definition. *Let E be a subset of R^n and let σ be a set function on $\mathscr{C}_E$. Then σ is said to be* differentiable *at the interior point x of E and to have the* derivative $D\sigma(x)$ *(a real number) if, given $\epsilon > 0$, there is a $\delta > 0$ such that*

$$\left| \frac{\sigma(I_x)}{c(I_x)} - D\sigma(x) \right| < \epsilon \tag{8.51}$$

for every cubical interval I_x which contains x and has diameter $\rho(I_x)$ less than δ.

Suppose that U is a subset of E at all points of which σ is differentiable. Then σ is said to be uniformly differentiable *on U if, given $\epsilon > 0$, there is a $\delta > 0$ such that (8.51) holds whenever $x \in U$ and I_x is a cubical interval containing x with diameter $\rho(I_x) < \delta$.*

It is easy to see that the derivative of a set function is unique. For set functions in R^1 there is a simple connection with ordinary derivatives. (See exercise 8(e), 1.) The notion of uniform differentiability has not been used before.

Theorem 8.51. *Suppose that E is a subset of R^n and that the function $f: E \to R^1$ is integrable over E. If x is an interior point of E and f is continuous at x, then σ_f is differentiable at x and $D\sigma_f(x) = f(x)$.*

The proof is very similar to that of theorem 7.62 of C1. (See also exercise 6(a), 10.)

Theorem 8.52. *Let G be an open subset of R^n and suppose that the function $f: G \to R^1$ is continuous. If H is a compact subset of G, then σ_f is uniformly differentiable on H (and $D\sigma_f(x) = f(x)$ for $x \in H$).*

Proof. The compact set H and the closed set G' are disjoint and so $\rho(H, G') = 2\alpha > 0$ (exercise 3(f), 14). Let

$$K = \overline{\bigcup_{x \in H} B(x; \alpha)}.$$

Then K is a compact subset of G and so f is uniformly continuous on K. Thus, given $\epsilon > 0$, there is a $\delta > 0$ such that

$$|f(u) - f(v)| < \epsilon \quad \text{whenever} \quad u, v \in K \quad \text{and} \quad \|u - v\| < \delta.$$

Let $x \in H$. If I_x is any cubical interval containing x such that $\rho(I_x) < \min(\alpha, \delta)$, then $I_x \subset K$ and so, for $y \in I_x$,

$$f(x) - \epsilon < f(y) < f(x) + \epsilon.$$

It follows that

$$(f(x) - \epsilon) c(I_x) \leqslant \int_{I_x} f \leqslant (f(x) + \epsilon) c(I_x),$$

i.e.

$$\left| \frac{\sigma_f(I_x)}{c(I_x)} - f(x) \right| \leqslant \epsilon. \ |$$

The last theorem has a partial converse which is particularly important. To state it we need one more definition.

Definition. *Let E be a subset of R^n. The set function $\sigma: \mathscr{C}_E \to R^1$ is said to be* additive *if, whenever $X, Y \in \mathscr{C}_E$ and $X \cap Y$ has zero content,*

$$\sigma(X \cup Y) = \sigma(X) + \sigma(Y).$$

When the function $f: E \to R^1$ is integrable over E, then, by exercises 8(d), 3 and 5, the set function σ_f is additive.

Note that two closed intervals I, J are such that $c(I \cap J) = 0$ if and only if any common points of I and J are frontier points of both. Such intervals we have called non-overlapping.

Theorem 8.53. *Let G be an open subset of R^n and let σ be a non-negative, additive set function on $\mathcal{C}_G$. Suppose that H is a compact subset of G and σ is uniformly differentiable on H with derivative f. Then the function $f: H \to R^1$ is continuous and, if E is a set with content such that $\bar{E} \subset H^\circ$,*

$$\sigma(E) = \int_E f. \tag{8.52}$$

Proof.

(i) We first show that the relation (8.52) holds when E is a closed cubical interval J in H°. For this we do not need the hypothesis that σ is non-negative.

Given $\epsilon > 0$, there is a $\delta > 0$ such that

$$\left| \frac{\sigma(I_x)}{c(I_x)} - f(x) \right| < \epsilon$$

whenever $x \in H$ and I_x is a cubical interval containing x with $\rho(I_x) < \delta$. Let $u, v \in H$ be such that $\|u - v\| < \delta/(2\sqrt{n})$. At least one of the 2^n cubical intervals with diameter $\frac{1}{2}\delta$ and one vertex at u contains v also. If I is such an interval,

$$|f(u) - f(v)| \leqslant \left| f(u) - \frac{\sigma(I)}{c(I)} \right| + \left| \frac{\sigma(I)}{c(I)} - f(v) \right| < 2\epsilon.$$

Therefore f is (uniformly) continuous on H and the set function $\sigma_f: \mathcal{C}_{H^\circ} \to R^1$ exists.

By theorem 8.52, σ_f is uniformly differentiable on J and

$$D\sigma_f(x) = f(x).$$

Hence the set function $\qquad \tau = \sigma - \sigma_f$

is uniformly differentiable on J and $D\tau(x) = 0$. Thus, given $\epsilon' > 0$, there is a $\delta' > 0$ such that

$$\left| \frac{\tau(I)}{c(I)} \right| < \epsilon'$$

whenever I is a closed cubical interval contained in J and with diameter $\rho(I) < \delta'$. Divide J into non-overlapping cubical intervals $I_1, \ldots, I_r$ each with diameter less than δ'. Since σ_f is additive, so is τ and we have

$$|\tau(J)| = \left| \sum_{j=1}^r \tau(I_j) \right| < \epsilon' \sum_{j=1}^r c(I_j) = \epsilon' c(J).$$

But this inequality holds for all $\epsilon' > 0$ and so $\tau(J) = 0$, i.e.

$$\sigma(J) = \sigma_f(J).$$

(ii) Now let E be any set with content and with closure $\bar{E}$ in H°. Since $\rho(\bar{E}, (H^\circ)') > 0$, given ϵ, there are sets U, V, both finite unions of non-overlapping cubical intervals, such that

$$U \subset E \subset V \subset H^\circ \tag{8.53}$$

and

$$c(V) - \epsilon < c(E) < c(U) + \epsilon,$$

or, by exercises 8(d), 4 and 5,

$$c(E - U) < \epsilon, \quad c(V - E) < \epsilon.$$

If $\mu = \sup_{x \in H} f(x)$, we then have

$$\int_U f = \int_E f - \int_{E-U} f > \int_E f - \mu\epsilon \tag{8.54}$$

and

$$\int_V f < \int_E f + \mu\epsilon. \tag{8.55}$$

As σ is non-negative and additive, we have, by (8.53),

$$\sigma(U) \leqslant \sigma(E) \leqslant \sigma(V)$$

and, in view of (i), $\qquad \int_U f \leqslant \sigma(E) \leqslant \int_V f.$

By (8.54) and (8.55) we now obtain

$$\int_E f - \mu\epsilon < \sigma(E) < \int_E f + \mu\epsilon$$

and these inequalities hold for all positive ϵ. |

Exercises 8(e)

1. Let (a, b) be an open interval in R^1 and suppose that the functions $f: (a, b) \to R^1$ and $\sigma : \mathscr{C}_{(a, b)} \to R^1$ are connected by the relation

$$f(x_2) - f(x_1) = \sigma([x_1, x_2]) \quad (a < x_1 \leqslant x_2 < b).$$

Prove that σ is differentiable at the point x of (a, b) if and only if f is differentiable at x and that, when the two functions are differentiable,

$$D\sigma(x) = f'(x).$$

2. Show that, if E is a subset of R^n, σ is an additive set function on $\mathscr{C}_E$ and X is a subset of E which has zero content, then $\sigma(X) = 0$.

8.6. Transformation of integrals

Much of the theory of multiple integration is a straight-forward extension of the corresponding one-dimensional theory. The most conspicuous exception is the change of variable formula

$$\int_{\phi(E)} f = \int_E (f \circ \phi) |J\phi|$$

(in which $J\phi$ is the Jacobian of ϕ). Whereas the proofs of corresponding results in R^1 (theorem 6.92 and exercise 6(i), 1) are relatively simple, our task now is long and, in places, difficult. The reason for this difference is that a continuous function transforms intervals into intervals in R^1, but not in higher dimensions. We begin by considering the effect of a linear transformation on the content of a parallelepiped.

A parallelepiped P in R^n is determined by the n pairs of hyperplanes on which its faces lie. If these hyperplanes are $B_1, C_1, ...,$ B_n, C_n, where B_i, C_i are given by

$$a_{i1}x_1 + ... + a_{in}x_n = \begin{cases} b_i \\ c_i \end{cases} \quad (b_i < c_i; \quad i = 1, ..., n),$$

then $\xi = (\xi_1, ..., \xi_n) \in P$ if and only if

$$b_i \leqslant a_{i1}\xi_1 + ... + a_{in}\xi_n \leqslant c_i \quad (i = 1, ..., n). \tag{8.61}$$

Since the faces of P have zero content (by theorem 8.46), P has content. Also, using the inequalities (8.61), we see that, if X is an $n - m$ dimensional subspace of the form

$$x_{k_1} = \text{constant}, ..., x_{k_m} = \text{constant},$$

then points of X which are interior (exterior) points of P in R^n are also interior (exterior) points of $P \cap X$ in X. The frontier of $P \cap X$ (clearly a bounded subset of X) therefore lies on the sets

$$B_i \cap X, \quad C_i \cap X \quad (i = 1, ..., n)$$

which are hyperplanes in X. Hence $P \cap X$ has $(n - m$ dimensional) content in X. It follows from theorem 8.31 that $c(P)$ is given by

$$c(P) = \int_I \chi_P = \int_{p_n}^{q_n} dx_n ... \int_{p_1}^{q_1} \chi_P(x_1, ..., x_n) dx_1, \tag{8.62}$$

where $I = [p_1, q_1] \times ... \times [p_n, q_n]$ is any interval containing P. Moreover the order of the integrations in (8.62) is immaterial.

The inequalities (8.61) also show that, if λ is a bijective linear function on R^n to R^n, then $\lambda(P)$ is again a parallelepiped.

Lemma 1. *If P is a parallelepiped in R^n and λ is a linear function on R^n to R^n, then*

$$c(\lambda(P)) = |J|\, c(P), \tag{8.63}$$

where J is the Jacobian (i.e. the determinant) of λ.

Proof. If $J = 0$, then the rows of the matrix of λ are linearly dependent. Thus $\lambda(R^n)$ is contained in a hyperplane A of R^n (given by an equation $a_1 x_1 + \ldots + a_n x_n = 0$). Hence $\lambda(P)$ is a bounded subset of A and so has zero content. This case will not, actually, be used in the sequel.

Now suppose that $J \neq 0$. We first write (8.63) in the form

$$\int_K \chi_{\lambda(P)} = |J| \int_I \chi_P, \tag{8.64}$$

where I, K are any intervals which contain P and $\lambda(P)$ respectively. Put $Q = \lambda(P)$. Then, since $J \neq 0$ and therefore λ is bijective, (8.64) may be written as

$$\int_K \chi_Q = |J| \int_I \chi_Q \circ \lambda. \tag{8.65}$$

Next we note that λ is the composition of elementary linear functions μ, α, τ_i ($i = 1, \ldots, n-1$), where

$$\mu(x_1, \ldots, x_n) = (kx_1, x_2, \ldots, x_n) \quad (k \neq 0),$$

$$\alpha(x_1, \ldots, x_n) = (x_1 + x_2, x_2, \ldots, x_n),$$

$$\tau_i(x_1, \ldots, x_i, x_{i+1}, \ldots, x_n) = (x_1, \ldots, x_{i+1}, x_i, \ldots, x_n).$$

(It is well known that a non-singular matrix is the product of elementary matrices—see, for instance, L. Mirsky, *Linear Algebra*, chapter VI.) Since the Jacobian of the composition of two (linear) functions is the product of their Jacobians, it is sufficient to prove (8.65) for the cases $\lambda = \mu, \alpha, \tau_i$. The Jacobians of these functions are k, 1 and -1 respectively.

(i) $\lambda = \mu$. Take

$$I = [-a, a] \times \ldots \times [-a, a], \quad K = [-|k|a, |k|a] \times \ldots \times [-a, a],$$

where a is so large that $I \supset P$ and $K \supset Q$. Then, by theorem 6.92,

$$|k| \int_I \chi_Q \circ \mu = |k| \int_{-a}^a dx_n \ldots \int_{-a}^a \chi_Q(kx_1, \ldots, x_n)\, dx_1$$

$$= |k| \int_{-a}^a dx_n \ldots \int_{-ka}^{ka} \chi_Q(x_1, \ldots, x_n) \frac{1}{k}\, dx_1 = \int_K \chi_Q.$$

(ii) $\lambda = \alpha$. Let

$$I = [-a, a] \times [-b, b] \times \ldots \times [-b, b],$$

$$K = [-a+b, a-b] \times [-b, b] \times \ldots \times [-b, b],$$

where $I \supset P$ and $K \supset Q$. Then

$$\int_I \chi_Q \circ \alpha = \int_{-b}^b dx_n \ldots \int_{-a}^a \chi_Q(x_1 + x_2, x_2, \ldots, x_n)\, dx_1$$

$$= \int_{-b}^b dx_n \ldots \int_{-a+x_2}^{a+x_2} \chi_Q(x_1, x_2, \ldots, x_n)\, dx_1$$

$$= \int_{-b}^b dx_n \ldots \int_{-a+b}^{a-b} \chi_Q(x_1, x_2, \ldots, x_n)\, dx_1 = \int_K \chi_Q,$$

since $[-a+x_2, a+x_2] \supset [-a+b, a-b]$ when $x_2 \in [-b, b]$.

(iii) $\lambda = \tau_i$. If $I = K = [-a, a] \times \ldots \times [-a, a]$, where $I \supset P, Q$, then

$$\int_I \chi_Q \circ \tau_i$$

$$= \int_{-a}^a dx_n \ldots \int_{-a}^a dx_{i+1} \int_{-a}^a dx_i \ldots \int_{-a}^a \chi_Q(x_1, \ldots, x_{i+1}, x_i, \ldots, x_n)\, dx_1$$

$$= \int_{-a}^a dx_n \ldots \int_{-a}^a dx_i \int_{-a}^a dx_{i+1} \ldots \int_{-a}^a \chi_Q(x_1, \ldots, x_{i+1}, x_i, \ldots, x_n)\, dx_1$$

$$= \int_I \chi_Q,$$

since the order in the repeated integral may be changed. |

In the next lemma we no longer confine ourselves to linear transformations. We denote by $I(x; \eta)$ the closed cubical interval with centre x and sides of length 2η.

Lemma 2. *Let G be an open subset of R^n and suppose that the function $\phi : G \to R^n$ is continuously differentiable, injective and such that $J\phi(x) \neq 0$ for $x \in G$.*

(i) *If the set E has content and $\bar{E} \subset G$, then the set $\phi(E)$ also has content.*

(ii) *If H is a compact subset of G, then, as $\eta \to 0$,*

(a) $$\rho\{\phi(I(x; \eta))\} \to 0$$

and (b) $$\frac{c\{\phi(I(x; \eta))\}}{c\{I(x; \eta)\}} \to |J\phi(x)|$$

uniformly for $x \in H$.

Proof.

(i) Since the set E is bounded (it has content), $\bar{E}$ is compact and so, by theorem 3.64, $\phi(\bar{E})$ is compact. Hence the subset $\phi(E)$ of $\phi(\bar{E})$ is bounded and, by theorem 8.43 and the lemma of §8.4, to prove that $\phi(E)$ has content it is sufficient to prove that fr $\phi(E) = \phi(\text{fr } E)$.

We first note that the condition $J\phi(x) \neq 0$ ensures that the subset X of G is open if and only if $\phi(X)$ is open. (See theorems 3.23 and 7.52.) In particular, $\phi(G)$ is open.

Now let $x \in \text{fr } E$. Put $y = \phi(x)$ and let $B_y (\subset \phi(G))$ be an open ball with centre y. Then $\phi^{-1}(B_y)$ is an open set containing x and, since x is a frontier point of E, $\phi^{-1}(B_y)$ contains points of E and of $G - E$. The injectiveness of ϕ then ensures that B_y contains points of $\phi(E)$ and of $\phi(G) - \phi(E) = \phi(G - E)$. Hence $\phi(x) = y \in \text{fr } \phi(E)$ and so $\phi(\text{fr } E) \subset \text{fr } \phi(E)$. The proof that fr $\phi(E) \subset \phi(\text{fr } E)$ is similar.

(ii) (*a*) When $I(x; \eta) \subset G$,

$$\rho\{\phi(I(x; \eta))\} \leqslant 2 \sup_{u \in I(x;\eta)} \|\phi(x) - \phi(u)\| \tag{8.66}$$

and, for $u \in I(x; \eta)$, $\|x - u\| \leqslant (\sqrt{n})\eta$. Hence (see exercise 7(*c*), 4(i)) the right-hand side of (8.66) tends to 0 as η tends to 0, uniformly for $x \in H$.

(*b*) By the inverse function theorem 7.51, ϕ^{-1} is continuously differentiable on the set $\phi(G)$ and

$$D\phi^{-1}(\phi(x)) = [D\phi(x)]^{-1}.$$

Also, by theorem 3.64, $\phi(H)$ is compact and so $\|D\phi^{-1}(y)\|$ is bounded for $y \in \phi(H)$; let

$$\sup_{y \in \phi(H)} \|D\phi^{-1}(y)\| = k.$$

If therefore $D\phi(x) = \lambda_x$, $D\phi^{-1}(\phi(x)) = \lambda_x^{-1}$ and

$$\|\lambda_x^{-1}(u)\| \leqslant k \|u\|$$

whenever $x \in H$ and $u \in R^n$. The idea of the proof is that, for small η, the image of $I(x; \eta)$ under the transformation $\psi_x = \lambda_x^{-1} \circ \phi$ is, approximately, $I(\psi_x(x); \eta)$.

Take ϵ such that $0 < \epsilon < 1/(k\sqrt{n})$. By exercise 7(*c*), 4(ii), there is a $\delta > 0$ such that

$$\|\phi(x+h) - \phi(x) - \lambda_x(h)\| < \epsilon \|h\|$$

for $x \in H$ and $\|h\| < \delta$. Then, for such x and h,

$$\|\lambda_x^{-1}\{\phi(x+h) - \phi(x) - \lambda_x(h)\}\| < k\epsilon \|h\|,$$

i.e. $$\|\psi_x(x+h) - \psi_x(x) - h\| < k\epsilon \|h\|. \tag{8.67}$$

Let $x \in H$ and let η be such that $0 < \sqrt{n\eta} < \delta$. It was shown in the proof of (i) that

$$\text{fr } \psi_x(I(x; \eta)) = \psi_x(\text{fr } I(x; \eta)).$$

Now, if $x + h \in \text{fr } I(x; \eta)$, then $\eta \leqslant \|h\| \leqslant \sqrt{n\eta}$. Hence, by (8.67), points on the frontier of $\psi_x(I(x; \eta))$ lie within a distance $k\epsilon\sqrt{n\eta}$ of the frontier of $I(\psi_x(x); \eta)$. It follows (see exercise 8(f), 1) that

$$M = I(\psi_x(x); \eta - k\epsilon\sqrt{n\eta}) \subset \psi_x(I(x; \eta)) \subset I(\psi_x(x); \eta + k\epsilon\sqrt{n\eta}) = N.$$

Since $\lambda_x \circ \psi_x = \phi$,

$$\lambda_x(M) \subset \phi(I(x; \eta)) \subset \lambda_x(N).$$

By (i), $\phi(I(x; \eta))$ has content. Then, since $J\lambda_x = J\phi(x)$, lemma 1 gives

$$|J\phi(x)|(1 - k\epsilon\sqrt{n})^n (2\eta)^n \leqslant c\{\phi(I(x; \eta))\} \leqslant |J\phi(x)|(1 + k\epsilon\sqrt{n})^n (2\eta)^n,$$

i.e. $\quad (1 - k\epsilon\sqrt{n})^n |J\phi(x)| \leqslant \dfrac{c\{\phi(I(x; \eta))\}}{c\{I(x; \eta)\}} \leqslant (1 + k\epsilon\sqrt{n})^n |J\phi(x)|. \;\blacksquare$

Theorem 8.61. *Let G be an open set in R^n and let E be a set with content such that $\bar{E} \subset G$. Suppose also that the function $\phi: G \to R^n$ is continuously differentiable and injective and that $J\phi(x) \neq 0$ nearly everywhere in G. If the function $f: \phi(\bar{E}) \to R^1$ is continuous, then the integrals*

$$\int_{\phi(E)} f \quad and \quad \int_E (f \circ \phi)|J\phi|$$

exist and are equal.

In the classical notation, according to which the Jacobian $J\phi$ of the transformation $x = \phi(t)$ (or $x_i = \phi_i(t_1, ..., t_n)$ $(i = 1, ..., n)$) is denoted by

$$\frac{\partial(x_1, ..., x_n)}{\partial(t_1, ..., t_n)},$$

the relation to be proved is written

$$\int_{\phi(E)} f(x_1, ..., x_n)\, dx_1 ... dx_n = \int_E f\{\phi(t_1, ..., t_n)\} \left| \frac{\partial(x_1, ..., x_n)}{\partial(t_1, ..., t_n)} \right| dt_1 ... dt_n.$$

Proof.

(i) First suppose that the set Q on which $J\phi(x) = 0$ is empty.

In view of Tietze's extension theorem (5.52) we may suppose that f is defined and continuous not only on the compact set $\phi(\bar{E})$, but on the whole of G. Moreover f may be expressed in the form $f = f^+ - f^-$, where $f^+ = \frac{1}{2}(|f| + f), f^- = \frac{1}{2}(|f| - f)$ and these functions are non-negative and continuous. We shall therefore assume from now on that f is also non-negative.

If the set X has content and $\bar{X} \subset G$, then by lemma 2(i), $\phi(X)$ has content. For all such sets X we define (as we may by theorem 8.44) the non-negative set function σ by

$$\sigma(X) = \int_{\phi(X)} f.$$

σ is also additive, for if $\bar{X}_1, \bar{X}_2 \subset G$ and $c(X_1 \cap X_2) = 0$, then, by the lemma of §8.4, $c(\phi(X_1) \cap \phi(X_2)) = c(\phi(X_1 \cap X_2)) = 0$.

Let $\rho(\bar{E}, G') = 3\delta$ and put

$$H = \overline{\bigcup_{x \in E} B(x; \delta)}, \quad K = \overline{\bigcup_{x \in H} B(x; \delta)}.$$

Take $x \in H$ and let I_x be a closed, cubical interval containing x and with $\rho(I_x) < \delta$ so that $I_x \subset K^\circ$. Since $\phi(I_x)$ is connected and compact (theorems 3.34 and 3.64), there is a point $y \in \phi(I_x)$ such that

$$\sigma(I_x) = f(y)\,c(\phi(I_x)).$$

By lemma 2(ii) (a) and the uniform continuity of f on the compact set $\phi(K)$, $f(y) \to f(\phi(x))$ as $\rho(I_x) \to 0$, uniformly for $x \in H$. Now $J\phi$ is also uniformly continuous on $\phi(K)$ and therefore lemma 2(ii) (b) shows that

$$\frac{c(\phi(I_x))}{c(I_x)} \to |J\phi(x)|$$

as $\rho(I_x) \to 0$, uniformly for $x \in H$. It follows that

$$\frac{\sigma(I_x)}{c(I_x)} = f(y)\frac{c(\phi(I_x))}{c(I_x)} \to f(\phi(x))\,|J\phi(x)|,$$

uniformly for $x \in H$. Since E has content and $\bar{E} \subset H^\circ$, theorem 8.53 gives

$$\sigma(E) = \int_E (f \circ \phi)\,|J\phi|.$$

(ii) When Q is not empty (but has zero content), then, for each integer $k \geqslant 1$, there is a set S_k, the union of a finite number of closed intervals, such that $S_k \supset Q$ and $c(S_k) < 1/k$. Replace each interval of S_k by a similar, concentric, but open interval of twice the content. If T_k is the union of these intervals, then $c(T_k) < 2/k$.

Now $E - T_k$ has content (8(d), 5), $\overline{E - T_k}$ is a subset of the open set $G - S_k$ and

$$\phi(E - T_k) = \phi(E - (E \cap T_k)) = \phi(E) - \phi(E \cap T_k).$$

It follows from (i) that the integrals

$$\int_{\phi(E)-\phi(E\cap T_k)} f \quad \text{and} \quad \int_{E-(E\cap T_k)} (f\circ\phi)\,|J\phi|$$

exist and are equal. Also f is bounded on $\phi(E)(\subset \phi(\bar{E}))$ and $J\phi$ is bounded on $E(\subset \bar{E})$. Moreover $c(E\cap T_k)\to 0$ as $k\to\infty$ and, since $\|D\phi(x)\|$ is bounded for $x\in H$, an argument similar to that used in the lemma of §8.4 shows that $\bar{c}(\phi(E\cap T_k))\to 0$. The theorem now follows from exercise 8(d), 6. |

The change of variable technique for evaluating integrals in R^1 is familiar and the use of theorem 8.61 in higher dimensions is analogous. We start with an integral $\int_D f$ and wish to find a function ϕ such that, when $\phi(E) = D$, $\int_E (f\circ\phi)\,|J\phi|$ is more amenable to computation than the original integral. Unfortunately as it stands the theorem is not quite strong enough to cope with certain situations that occur quite frequently. Consider the polar transformation ϕ in R^2 given by

$$x = \phi_1(r, \theta) = r\cos\theta, \quad y = \phi_2(r, \theta) = r\sin\theta \qquad (8.68)$$

whose Jacobian $J\phi(r, \theta)$ is r. When D is a (bounded) closed set not meeting the half line $\{(x, y)\,|\,x\leqslant 0, y = 0\}$ (or some other half line starting at the origin), there is no difficulty. But, even when D is as simple a set as the closed unit disc (given by $x^2+y^2\leqslant 1$), theorem 8.61 is inapplicable; for a corresponding set $\bar{E}$, such as

$$[0, 1]\times[0, 2\pi],$$

cannot be enclosed in an open set G in which ϕ is injective. In this case a slightly stronger (but more cumbersome) version of theorem 8.61 is required.

Theorem 8.62. *Let G^* be an open set in R^n and let E, G be sets with content such that G is open and $E \subset \bar{G} \subset G^*$. Suppose also that the function $\phi:G^*\to R^n$ is continuously differentiable on G^*, injective on G and with Jacobian $J\phi(x) \neq 0$ nearly everywhere in G. If the function $f:\phi(\bar{E})\to R^1$ is continuous, then the integrals*

$$\int_{\phi(E)} f \quad \text{and} \quad \int_E (f\circ\phi)\,|J\phi|$$

exist and are equal.

Proof. We use the same notation as in the proof of theorem 8.61. The case $\dot{\imath} \subset G$ and $Q = \varnothing$ corresponds to part (i) of the earlier proof. In part (ii) we now let S_k cover the set $Q \cup \text{fr } G$ and have

$$\overline{E - T_k} = \overline{E \cap T'_k} \subset \overline{E} \cap T'_k \subset \overline{G} \cap T'_k = G \cap T'_k \subset G \cap S'_k = G - S_k.$$

The argument then proceeds as before, except that we use the boundedness of $D\phi$ on $\overline{G}$ instead of on H. |

Example.
$$\int_0^\infty e^{-x^2} dx = \tfrac{1}{2}\sqrt{\pi}.$$

For $u > 0$, let
$$E(u) = \int_0^u e^{-x^2} dx.$$

Then
$$E^2(u) = \int_0^u e^{-x^2} dx \int_0^u e^{-y^2} dy = \int_S e^{-x^2-y^2} dx\, dy,$$

where S is the square $[0, u] \times [0, u]$ in R^2. Let C_1, C_2 be the sectors

$$\{(x, y) | x \geqslant 0, y \geqslant 0, x^2 + y^2 \leqslant u^2\},$$

$$\{(x, y) | x \geqslant 0, y \geqslant 0, x^2 + y^2 \leqslant 2u^2\},$$

respectively, so that $C_1 \subset S \subset C_2$. Since $e^{-x^2-y^2} > 0$,

$$\int_{C_1} e^{-x^2-y^2} dx\, dy \leqslant \int_S e^{-x^2-y^2} dx\, dy \leqslant \int_{C_2} e^{-x^2-y^2} dx\, dy.$$

We now evaluate the integrals over C_1 and C_2 by using the polar transformation ϕ in R^2 given by (8.68) and appealing to theorem 8.62. The function ϕ is continuously differentiable in $G^* = R^2$ and injective in $G = (0, 2u) \times (0, 2\pi)$; also $J\phi(r, \theta) = r \neq 0$ in G. For a set E such that $\phi(E) = C_1$ we can take the interval $[0, u] \times [0, \tfrac{1}{2}\pi]$, since theorem 8.62 does not require ϕ to be injective on E. We now have

$$\int_{C_1} e^{-x^2-y^2} dx\, dy = \int_E e^{-r^2} r\, dr\, d\theta = \int_0^{\frac{1}{2}\pi} d\theta \int_0^u r\, e^{-r^2} dr = \tfrac{1}{4}\pi(1 - e^{-u^2}).$$

Similarly
$$\int_{C_2} e^{-x^2-y^2} dx\, dy = \tfrac{1}{4}\pi(1 - e^{-2u^2}).$$

Hence
$$\tfrac{1}{4}\pi(1 - e^{-u^2}) \leqslant E^2(u) \leqslant \tfrac{1}{4}\pi(1 - e^{-2u^2})$$

and so $E^2(u) \to \tfrac{1}{4}\pi$ as $u \to \infty$.

The theorems of this section, when interpreted for integrals over one-dimensional intervals, are rather weaker than the corresponding results of chapter 6; at first sight they also look different, for they lead to the formula

$$\int_{\phi(I)} f = \int_I (f \circ \phi)|\phi'|. \tag{8.69}$$

However, let $I = [\alpha, \beta]$ and put $\phi(\alpha) = a$, $\phi(\beta) = b$. The condition that ϕ is continuously differentiable and injective entails that ϕ' has constant sign. Hence, when $\phi' \geq 0$, $\phi(I) = [a, b]$ and (8.69) is just the relation (6.95) of p. 188; and, when $\phi' \leq 0$, $\phi(I) = [b, a]$ and (8.69) is

$$\int_b^a f = \int_\alpha^\beta (f \circ \phi)(-\phi'),$$

which again reduces to (6.95). The reason why $|J\phi|$ appears in theorems 8.61 and 8.62, but ϕ' (not $|\phi'|$) in theorem 6.92 is that $\int_a^b f$ is 'oriented' $\left(\text{i.e. } \int_a^b f = -\int_b^a f \right)$, whereas a multiple integral has no orientation.

An *affine* transformation is a function $\phi: R^n \to R^n$ which is of the form

$$\phi(x) = c + \lambda(x),$$

where $c \in R^n$ and the function $\lambda: R^n \to R^n$ is linear. Theorem 8.61 (with $f \equiv 1$) shows that content is invariant under an affine transformation with determinant 1 or -1 (a rigid body motion). This means, for instance, that the content of *any* rectangular parallelepiped is equal to the product of the lengths of its sides.

Finally we note that, taking again $f \equiv 1$ in theorems 8.61 and 8.62, we obtain two propositions on content which include various auxiliary results that had been needed (the lemma of §8.4 and lemmas 1 and 2(i) of this section).

Exercises 8 (f)

1. Let $0 < a < b$. Show that, if the bounded set E in R^n is such that

$$u \in E \quad \text{and} \quad \text{fr } E \subset I(u; b) - I(u; a),$$

then $$I(u; a) \subset E \subset I(u; b).$$

2. Prove that, in theorems 8.61 and 8.62, the function f need only be assumed to be bounded on $\phi(\bar{E})$ and continuous nearly everywhere in $\phi(\bar{E})$.

3. Show that, in R^2, the content of any triangle is equal to half the length of any one side multiplied by the corresponding altitude.

4. Show that, if D is the half-disc $\{(x, y) \,|\, y \geq 0, \, x^2 + y^2 \leq 1\}$,

$$\int_D \frac{(x+y)^2}{\sqrt{(1+x^2+y^2)}} \, dx \, dy = \frac{2 - \sqrt{2}}{3} \pi.$$

5. The function $f: R^2 \to R^1$ is given by $f(x, y) = x^2/p^2 + y^2/q^2$. Find the content of the ordinate set (see exercise 8(d), 10) of f on the elliptical set

$$\{(x, y) \,|\, x^2/a^2 + y^2/b^2 \leqslant 1\}.$$

6. Use the transformation $(x, y) = \phi(u, v)$ given by $x = u,\; y = v - u$ to evaluate

$$\int_D \text{arc tan } (x + y)\, dx\, dy,$$

where D is the triangle $\{(x, y) \,|\, x \geqslant 0,\, y \geqslant 0,\, x + y \leqslant 1\}$.

7. Show that the equations $u = x^3/y,\; v = y^3/x$ define a continuously different-iable, bijective transformation $(x, y) = \phi(u, v)$ on $(0, \infty) \times (0, \infty)$ to $(0, \infty) \times (0, \infty)$. Hence find the content of the set

$$\{(x, y) \,|\, x > 0,\, y > 0,\, a^2 y \leqslant x^3 \leqslant b^2 y,\, p^2 x \leqslant y^3 \leqslant q^2 x\}$$

in R^2, where $0 < a < b,\, 0 < p < q$.

8. Use the transformation in R^2 given by $x = r \cos^3 \theta,\, y = r \sin^3 \theta$ to prove that the content of the set
$$\{(x, y, z) \,|\, x^{\frac{2}{3}} + y^{\frac{2}{3}} + z^{\frac{2}{3}} \leqslant 1\}$$
is $4\pi/35$.

9. The curve in R^2 represented by

$$x = a(1 + 2\cos t + \cos 2t),\quad y = a(2\sin t + \sin 2t)\quad (0 \leqslant t \leqslant 2\pi)$$

is called a *cardioid* ($p = q = 1$, and change of origin, in exercise 8(a), 7). Show that the trace of this cardioid is given by the equation

$$r = 2a(1 + \cos \theta)\quad (0 \leqslant \theta \leqslant 2\pi)$$

in polar co-ordinates.

Find the content and centroid of the set

$$S = \{(r\cos\theta,\, r\sin\theta) \,|\, 0 \leqslant r \leqslant 2a(1 + \cos\theta),\;\; 0 \leqslant \theta \leqslant 2\pi\}$$

in R^2. (The centroid (ξ, η) of a set E with content $c(E) > 0$ is defined by the equations
$$\xi c(E) = \int_E x\, dx\, dy,\quad \eta c(E) = \int_E y\, dx\, dy.)$$

10. The set of points (x, y, z) in R^3 such that $8az \geqslant x^2 + y^2$ ($a > 0$) is denoted by S; and P is the plane with equation $x + y + 2z = 8a$. Find the content of the (bounded) set Q cut off from S by P.

11. A cylindrical hole of diameter a is drilled through a solid sphere of radius a. Find the content of the material removed if (i) the hole is symmetrically placed, (ii) the centre of the sphere lies on the edge of the hole.
(In (ii) the set whose content is to be found may be taken to be

$$\{(x, y, z) \,|\, x^2 + y^2 + z^2 \leqslant a^2;\; x^2 + y^2 \leqslant ay\}.$$

Note that the polar equation of the circle $x^2 + y^2 = ay$ in R^2 is $r = a \sin \theta$ $(0 \leqslant \theta \leqslant \pi)$.)

12. (i) Prove that, for $n \geqslant 2$, the n-dimensional polar transformation $(x_1, \ldots, x_n) = \phi(r, \theta_1, \ldots, \theta_{n-1})$ given by

$$x_1 = r \cos \theta_1 \sin \theta_2 \ldots \sin \theta_{n-2} \sin \theta_{n-1},$$
$$x_2 = r \sin \theta_1 \sin \theta_2 \ldots \sin \theta_{n-2} \sin \theta_{n-1},$$
$$x_3 = r \cos \theta_2 \sin \theta_3 \ldots \sin \theta_{n-1},$$
$$\cdots\cdots\cdots\cdots\cdots\cdots\cdots\cdots\cdots\cdots$$
$$x_{n-1} = r \cos \theta_{n-2} \sin \theta_{n-1},$$
$$x_n = r \cos \theta_{n-1}$$

has Jacobian

$$J\phi(r, \theta_1, \ldots, \theta_{n-1}) = (-1)^n r^{n-1} \sin \theta_2 \sin^2 \theta_3 \ldots \sin^{n-2} \theta_{n-1}$$

and is injective on every open interval

$$(0, a) \times (0, 2\pi) \times (0, \pi) \times \ldots \times (0, \pi).$$

(ii) Show that, in R^n, the content $c_n(a)$ of an open or closed ball of radius a is

$$\frac{\pi^m}{m!} a^{2m} \quad \text{when} \quad n = 2m \quad (m = 1, 2, \ldots),$$

$$\frac{2^{2m+1} m! \pi^m}{(2m+1)!} a^{2m+1} \quad \text{when} \quad n = 2m+1 \quad (m = 0, 1, \ldots).$$

Note that, for every a, $c_n(a) \to 0$ as $n \to \infty$.
(For $k = 0, 1, 2, \ldots$,

$$\int_0^\pi \sin^{2k} \theta \, d\theta = \frac{(2k)! \, \pi}{2^{2k}(k!)^2}, \quad \int_0^\pi \sin^{2k+1} \theta \, d\theta = \frac{2^{2k+1}(k!)^2}{(2k+1)!};$$

see, for instance, C1, 133.)

13. Let S be the portion of the closed unit ball in R^3 which lies in the positive octant (where $x, y, z \geqslant 0$). Evaluate

$$\int_S \frac{yz}{1+x} \, dx \, dy \, dz.$$

14. *Green's theorem for a disc* (cf. theorem 8.31, corollary 3). Let G be an open subset of R^2 which contains the closed unit disc B; and denote the positively described unit circle by ∂B. Prove that, if the function $f = (f_1, f_2) : G \to R^2$ is continuously differentiable, then

$$\int_{\partial B} f = \iint_B (D_1 f_2 - D_2 f_1).$$

(Show that

$$D_1 f_2(r \cos t, r \sin t) r = \frac{\partial}{\partial r} [f_2(r \cos t, r \sin t) r \cos t] - \frac{\partial}{\partial t} [f_2(r \cos t, r \sin t) \sin t]$$

and derive a similar formula for $D_2 f_1(r \cos t, r \sin t) r$.)

8.7. Functions defined by integrals

Let I, J be closed intervals in R^m and R^n respectively. If the function $f: I \times J \to R^1$ is such that, for each $y \in J$, $f(., y)$ is integrable over I, we consider the function ϕ on J given by

$$\phi(y) = \int_I f(x, y)\, dx.$$

The question we pose is what conditions on f will ensure the properties of integrability, continuity and differentiability for ϕ. Actually we have already dealt with the first of these properties. Theorem 8.48 shows that, if f is integrable over $I \times J$, then ϕ is integrable over J and

$$\int_J \phi = \int_{I \times J} f.$$

(In this context ϕ need only be defined nearly everywhere on J.)

Theorem 8.71. *If the function $f: I \times J \to R^1$ is continuous, then, for every $y \in J$, $f(., y)$ is integrable over I and the function $\phi: J \to R^1$ defined by*

$$\phi(y) = \int_I f(x, y)\, dx$$

is continuous on J.

Proof. The integrability of $f(., y)$ for each y is obvious. Next, since $I \times J$ is compact, f is uniformly continuous. Thus, given $\epsilon > 0$, there is a $\delta > 0$ such that

$$|f(x, y) - f(x', y')| < \epsilon$$

whenever

$$(x, y), (x', y') \in I \times J \quad \text{and} \quad \|x - x'\|^2 + \|y - y'\|^2 < \delta^2$$

(where $\|x - x'\|$ and $\|y - y'\|$ denote norms in R^m and R^n respectively). Therefore, if $y, y' \in J$ and $\|y - y'\| < \delta$,

$$|\phi(y) - \phi(y')| \leqslant \int_I |f(x, y) - f(x, y')|\, dx \leqslant \epsilon c(I).\; \blacksquare$$

In the next theorem, on the differentiation of ϕ, we take J to be an interval $[c, d]$ in R^1. This is the most important case; for the general case see exercise 8(g), 3.

Theorem 8.72. (Differentiation under the integral sign.) *Given the function $f: I \times [c, d] \to R^1$ ($I \subset R^m$), suppose that*

$$\phi(y) = \int_I f(x, y)\, dx \qquad (8.71)$$

exists for some y in $[c, d]$ and that $D_{m+1}f$ exists and is continuous on $I \times [c, d]$. Then $\phi(y)$ exists for all y in $[c, d]$, the function ϕ is differentiable and

$$\phi'(y) = \int_I D_{m+1}f(x, y)\,dx \quad (c \leqslant y \leqslant d). \tag{8.72}$$

In the classical notation (8.72) takes the expressive form

$$\frac{d}{dy}\int_I f(x, y)\,dx = \int_I \frac{\partial}{\partial y}f(x, y)\,dx.$$

Proof. Suppose that

$$\int_I f(x, \eta)\,dx$$

exists, and take any y in $[c, d]$. Then, by theorem 6.84,

$$f(x, y) - f(x, \eta) = \int_\eta^y D_{m+1}f(x, t)\,dt$$

for each $x \in I$; and, by theorem 8.71, the right-hand side defines a continuous function on I. Since $f(., \eta)$ is integrable over I, so therefore is $f(., y)$; in other words, (8.71) exists for all y in $[c, d]$.

By theorem 8.71, the function ψ on $[c, d]$ given by

$$\psi(y) = \int_I D_{m+1}f(x, y)\,dx$$

is continuous. Then, by theorem 8.48 (or theorem 8.31, corollary 1) and, again by theorem 6.84, we have, for $c \leqslant y \leqslant d$,

$$\int_c^y \psi(t)\,dt = \int_c^y dt \int_I D_{m+1}f(x, t)\,dx = \int_I dx \int_c^y D_{m+1}f(x, t)\,dt$$

$$= \int_I [f(x, y) - f(x, c)]\,dx = \phi(y) - \phi(c).$$

Finally, since ψ is continuous and therefore

$$\frac{d}{dy}\int_c^y \psi(t)\,dt = \psi(y),$$

it follows that $\phi'(y)$ exists and is equal to $\psi(y)$. |

The results we have so far obtained in this section have analogues for functions defined by improper integrals (in R^1). We shall state all theorems for functions ϕ given in the form

$$\phi(y) = \int_a^\infty f(x, y)\,dx;$$

other improper integrals are treated in the same way. The notion of uniform convergence once again occupies a key position.

Theorem 8.73. *If*

(i) *the function* $f:[a, \infty) \times J \to R^1$ $(J \subset R^n)$ *is continuous, and*

(ii) $\int_a^\infty f(x, y) \, dx$ *converges uniformly for* $y \in J$,

then the function $\phi: J \to R^1$ *given by*

$$\phi(y) = \int_a^\infty f(x, y) \, dx$$

is continuous on J.

Proof. By (i) and theorem 8.71, each function ϕ_n $(n = 1, 2, ...)$ defined by

$$\phi_n(y) = \int_a^{a+n} f(x, y) \, dx \quad (y \in J)$$

is continuous on J; and by (ii) the sequence of functions ϕ_n converges uniformly to ϕ. Hence ϕ is continuous on J. |

Theorem 8.74. *Suppose that the function* $f:[a, \infty) \times [c, d] \to R^1$ *satisfies the following conditions:*

(i) *for every* $X > a$, f *is integrable over* $[a, X] \times [c, d]$;

(ii) *for every* $X > a$, $\int_c^d f(x, y) \, dy$ *exists nearly everywhere in* $[a, X]$;

(iii) $\int_a^\infty f(x, y) \, dx$ *exists and converges uniformly for* $c \leqslant y \leqslant d$.

Then

$$\int_c^d dy \int_a^\infty f(x, y) \, dx \quad and \quad \int_a^\infty dx \int_c^d f(x, y) \, dy$$

exist and are equal.

Proof. By theorem 8.48, for every $X > a$,

$$\int_c^d dy \int_a^X f(x, y) \, dx \quad and \quad \int_a^X dx \int_c^d f(x, y) \, dy$$

exist and are equal.

Next, since the functions ϕ_n given by $\phi_n(y) = \int_a^{a+n} f(x, y) \, dx$ are integrable over $[c, d]$ and $\phi_n(y) \to \phi(y) = \int_a^\infty f(x, y) \, dx$ uniformly for $c \leqslant y \leqslant d$, it follows from theorem 5.22 that ϕ is integrable over $[c, d]$. To prove the theorem we show that

$$\int_a^X dx \int_c^d f(x, y) \, dy \to \int_c^d dy \int_a^\infty f(x, y) \, dx$$

as $X \to \infty$. Now, given $\epsilon > 0$, there is an X_0 such that

$$\sup_{c \leqslant y \leqslant d} \left| \int_X^\infty f(x, y) \, dx \right| < \epsilon$$

for all $X > X_0$. Then, for $X > X_0$,

$$\left| \int_c^d dy \int_a^\infty f(x, y) \, dx - \int_a^X dx \int_c^d f(x, y) \, dy \right|$$

$$= \left| \int_c^d dy \int_a^\infty f(x, y) \, dx - \int_c^d dy \int_a^X f(x, y) \, dx \right|$$

$$\leqslant \int_c^d dy \left| \int_X^\infty f(x, y) \, dx \right| \leqslant (d-c)\epsilon. \mid$$

Note that conditions (i) and (ii) are both satisfied when f is continuous on $[a, \infty) \times [c, d]$.

Theorem 8.75. (Differentiation under the integral sign.) *Given the function $f : [a, \infty) \times [c, d] \to R^1$, suppose that*

(i) $$\phi(y) = \int_a^\infty f(x, y) \, dx$$

exists for some y in $[c, d]$;

(ii) $D_2 f$ *exists and is continuous on* $[a, \infty) \times [c, d]$;

(iii) $\int_a^\infty D_2 f(x, y) \, dx$ *converges uniformly for* $c \leqslant y \leqslant d$.

Then $\int_a^\infty f(x, y) \, dx$ exists and converges uniformly for $c \leqslant y \leqslant d$, the function ϕ is differentiable and

$$\phi'(y) = \int_a^\infty D_2 f(x, y) \, dx \quad (c \leqslant y \leqslant d).$$

As in theorem 8.72, the last identity may be written

$$\frac{d}{dy} \int_a^\infty f(x, y) \, dx = \int_a^\infty \frac{\partial}{\partial y} f(x, y) \, dx.$$

Proof. Suppose that $\int_a^\infty f(x, \eta) \, dx$ exists. By (ii) and the first part of theorem 8.72, $\int_a^X f(x, y) \, dx$ exists for all $X > a$ and all y in $[c, d]$.

In view of (i) and (iii), given $\epsilon > 0$, there is an X_0 such that, whenever $X_2 > X_1 > X_0$,

$$\left| \int_{X_1}^{X_2} D_2 f(x, t) \, dx \right| < \epsilon \quad \text{for} \quad c \leqslant t \leqslant d, \quad \text{and} \quad \left| \int_{X_1}^{X_2} f(x, \eta) \, dx \right| < \epsilon.$$

Now take any y in $[c, d]$. Then, by theorem 6.84 and theorem 8.48 (or theorem 8.31, corollary 1), we have when $X_2 > X_1 > X_0$,

$$\left| \int_{X_1}^{X_2} \{f(x, y) - f(x, \eta)\} dx \right| = \left| \int_{X_1}^{X_2} dx \int_{\eta}^{y} D_2 f(x, t) dt \right|$$

$$= \left| \int_{\eta}^{y} dt \int_{X_1}^{X_2} D_2 f(x, t) dx \right|$$

$$\leqslant |y - \eta| \epsilon \leqslant (d - c) \epsilon,$$

and so

$$\left| \int_{X_1}^{X_2} f(x, y) dx \right| \leqslant \left| \int_{X_1}^{X_2} f(x, \eta) dx \right| + (d - c) \epsilon < (1 + d - c) \epsilon.$$

Therefore, by the general principle of uniform convergence (6.34), $\int_{a}^{\infty} f(x, y) dx$ exists and converges uniformly for $c \leqslant y \leqslant d$.

By theorem 8.73, the function ψ on $[c, d]$ given by

$$\psi(y) = \int_{a}^{\infty} D_2 f(x, y) dx$$

is continuous. Theorems 8.74 and 6.84 now show that, for $c \leqslant y \leqslant d$,

$$\int_{c}^{y} \psi(t) dt = \int_{c}^{y} dt \int_{a}^{\infty} D_2 f(x, t) dx = \int_{a}^{\infty} dx \int_{c}^{y} D_2 f(x, t) dt$$

$$= \int_{a}^{\infty} [f(x, y) - f(x, c)] dx = \phi(y) - \phi(c).$$

But, since ψ is continuous,

$$\frac{d}{dy} \int_{c}^{y} \psi(t) dt = \psi(y).$$

Therefore $\phi'(y)$ exists and is equal to $\psi(y)$. |

The last three theorems invite comparison with theorems 5.21–5.23 which describe the limit functions of uniformly convergent sequences.

Example. For $y \geqslant 0$,

$$\int_{0}^{\infty} e^{-xy} \frac{\sin x}{x} dx = \frac{\pi}{2} - \text{arc tan } y. \tag{8.73}$$

(i) $y > 0$. Comparing the integrand with e^{-xy} we see immediately that the integral exists.

We have

$$\frac{\partial}{\partial y} \left(e^{-xy} \frac{\sin x}{x} \right) = -e^{-xy} \sin x. \tag{8.74}$$

Also the function defined by (8.74) is continuous on R^2 and, by the M-test (6.35),

$$\int_0^\infty (-e^{-xy} \sin x)\, dx$$

converges uniformly on any interval $[c, d]$ with $c > 0$. Hence, by theorem 8.75, if $\phi(y)$ is the integral in (8.73),

$$\phi'(y) = -\int_0^\infty e^{-xy} \sin x\, dx = -\frac{1}{1+y^2}$$

for $c \leqslant y \leqslant d$. Thus there is a constant A such that

$$\phi(y) = A - \arctan y \qquad (8.75)$$

when $y \in [c, d]$. But $[c, d]$ is an arbitrary interval such that $c > 0$ and so (8.75) must, in fact, hold for all $y > 0$.

To determine A we note that

$$|\phi(y)| \leqslant \int_0^\infty e^{-xy}\, dx = \frac{1}{y}.$$

Therefore, letting $y \to \infty$ in (8.75), we obtain $0 = A - \frac{1}{2}\pi$, i.e. (8.73) holds for $y > 0$.

(ii) The fact that (8.73) holds for $y \geqslant 0$ follows at once from the example at the end of §6.9 where it was shown that

$$\int_0^\infty \frac{\sin x}{x}\, dx = \frac{\pi}{2}. \qquad (8.76)$$

But we can also use the methods of the present chapter to evaluate this integral.

We first prove that the integral in (8.73) converges uniformly for $y \geqslant 0$. For $x > 0$, e^{-xy}/x is positive and decreasing in x. Hence, by the second mean value theorem (6.93),

$$\left| \int_{X_1}^{X_2} e^{-xy} \frac{\sin x}{x}\, dx \right| \leqslant \frac{e^{-X_1 y}}{X_1} \left| \int_{X_1}^{\xi} \sin x\, dx \right| \leqslant \frac{2}{X_1}$$

when $X_2 > X_1 > 0$ and $y \geqslant 0$. Since $2/X_1 \to 0$ as $X_1 \to \infty$, the general principle of uniform convergence shows that the integral defining $\phi(y)$ converges uniformly for $y \geqslant 0$.

We now know from theorem 8.73 that ϕ is continuous on $[0, \infty)$. Therefore, by (i),

$$\phi(0) = \lim_{y \to 0+} \phi(y) = \lim_{y \to 0+} (\tfrac{1}{2}\pi - \arctan y) = \tfrac{1}{2}\pi,$$

i.e. (8.76) holds.

Exercises 8(g)

1. Suppose that the function $f: [a, \infty) \times J \to R^1$ ($J \subset R^n$) is continuous and non-negative, that

$$\phi(y) = \int_a^\infty f(x, y) dx$$

exists for all $y \in J$ and that ϕ is continuous on J. Prove that $\int_a^\infty f(x, y) dx$ converges uniformly for $y \in J$. (Cf. Dini's theorem 5.35.)

2. Let I be the closed interval $[a, b] \times [c, d]$ in R^2 and let the function $f: I \to R^1$ be continuous. For $(x, y) \in I^\circ$, put $J(x, y) = [a, x] \times [c, y]$. Prove that the function $F: I^\circ \to R^1$ defined by

$$F(x, y) = \int_{J(x,y)} f$$

is such that, for all $(x, y) \in I^\circ$,

$$D_1 D_2 F(x, y) = D_2 D_1 F(x, y) = f(x, y).$$

(Cf. exercise 8(d), 8.)

3. Let I, J be closed intervals in R^m and R^n respectively. Suppose that the function $f: I \times J \to R^1$ is such that

$$\phi(y) = \int_I f(x, y) dx$$

exists for some $y \in J$, and the partial derivatives $D_{m+1}f, \ldots, D_{m+n}f$ exist and are continuous on $I \times J$. Prove that $\phi(y)$ exists for all $y \in J$, that

$$D_k\phi(y) = \int_I D_{m+k}f(x, y) dx \quad (1 \leqslant k \leqslant n, y \in J),$$

and that ϕ is continuously differentiable on J°.

4. Show, by an example, that in theorem 8.72 the condition that

$$\int_I f(x, y) dx$$

exists for some y in $[c, d]$ may not be omitted.

5. The function ϕ on R^1 is defined by

$$\phi(x) = \int_0^{\frac{1}{2}\pi} \cos (x \sin t) dt.$$

Show that $y = \phi(x)$ is a solution of Bessel's differential equation of order 0:

$$x\frac{d^2y}{dx^2} + \frac{dy}{dx} + xy = 0.$$

6. The function ϕ on $R^1 - \{-1, 1\}$ is defined by

$$\phi(a) = \int_0^\pi \log (1 - 2a \cos x + x^2) dx.$$

Prove that $\phi'(a) = 0$ and hence evaluate $\phi(a)$ for $|a| < 1$ and for $|a| > 1$.

7. The function $f : [a, b] \times [c, d] \to R^1$ is continuous and such that $D_2 f$ exists and is continuous. Also the function $g : [c, d] \to [a, b]$ is differentiable. Show that the function $\phi : [c, d] \to R^1$ defined by

$$\phi(y) = \int_a^{g(y)} f(x, y)\, dx$$

is differentiable and that

$$\phi'(y) = \int_a^{g(y)} D_2 f(x, y)\, dx + g'(y) f\{g(y), y\}.$$

The function F on $(-\tfrac{1}{2}\pi, \tfrac{1}{2}\pi)$ is given by

$$F(y) = \int_0^y \log(1 + \tan x \tan y)\, dx.$$

Prove that $F'(y) = \log \sec y + y \tan y$ and deduce that $F(y) = y \log \sec y$ for $-\tfrac{1}{2}\pi < y < \tfrac{1}{2}\pi$.

8. Let B be an open ball in R^n and let the function $f = (f_1, \ldots, f_n) : B \to R^n$ be continuously differentiable. Prove that a necessary and sufficient condition for $\int_\gamma f$ to be independent of path in B is that, in B, $D_i f_j = D_j f_i$ $(i, j = 1, \ldots, n)$. (See exercises $8(b)$, 7 and 8.)

9. Prove that, if $0 < a < b$, then

$$\int_a^b dy \int_0^\infty e^{-xy} dx \quad \text{and} \quad \int_0^\infty dx \int_a^b e^{-xy} dy$$

exist and are equal. Hence evaluate

$$\int_0^\infty \frac{e^{-ax} - e^{-bx}}{x}\, dx.$$

(This integral may also be evaluated by the Frullani method of exercise $6(i)$, 2.)

10. Prove that

$$\int_0^1 dy \int_0^\infty \frac{1}{1 + x^2 + y^2}\, dx \quad \text{and} \quad \int_0^\infty dx \int_0^1 \frac{1}{1 + x^2 + y^2}\, dy$$

exist and are equal. Deduce that

$$\int_0^{\frac{1}{2}\pi} \frac{\arctan(\sin \theta)}{\sin \theta}\, d\theta = \tfrac{1}{2}\pi \log(1 + \sqrt{2}).$$

11. Let ϕ on R^1 be defined by

$$\phi(y) = \int_0^\infty \left(\frac{\sin xy}{x}\right)^2 dx.$$

Show that $\phi'(y) = \tfrac{1}{2}\pi$ for $y > 0$, and hence evaluate $\phi(y)$ for all y.

Use a similar method (and the formula $4 \sin^3 \theta = 3 \sin \theta - \sin 3\theta$) to evaluate

$$\int_0^\infty \left(\frac{\sin xy}{x}\right)^3 dx.$$

12. Show that

$$\phi(y) = \int_0^\infty e^{-x^2} \cosh xy \, dx$$

exists for all y and everywhere satisfies the differential equation $\phi'(y) - \frac{1}{2}y\phi(y) = 0$. Hence evaluate $\phi(y)$ for all y.

13. The function ϕ on R^1 is defined by

$$\phi(y) = \int_0^\infty \frac{\sin xy}{x(1+x^2)} \, dx.$$

Prove that, when $y > 0$, $\phi''(y) - \phi(y) = -\frac{1}{2}\pi$. Hence find $\phi(y)$ for all y.

14. The function ϕ on R^2 is defined by

$$\phi(a, b) = \int_0^\infty \frac{\arctan ax \arctan bx}{x^2} \, dx.$$

Prove that, for $a > 0$, $b > 0$,

$$D_1 D_2 \phi(a, b) = \int_0^\infty \frac{dx}{(1+a^2x^2)(1+b^2x^2)} = \frac{\frac{1}{2}\pi}{a+b}$$

and

$$\phi(a, b) = \frac{1}{2}\pi\{(a+b)\log(a+b) - a\log a - b\log b\}.$$

15. Suppose that the function $f : [a, \infty) \times [b, \infty) \to R^1$ is non-negative and, for every $Y > b$,

$$\int_a^\infty dx \int_b^Y f(x,y)\,dy, \qquad \int_b^Y dy \int_a^\infty f(x, y)\,dx$$

exist and are equal. Show that, if one of the expressions

$$\int_a^\infty dx \int_b^\infty f(x, y)\,dy, \qquad \int_b^\infty dy \int_a^\infty f(x, y)\,dx$$

exists, then so does the other and the two are equal. (Cf. theorems 4.73, 6.33.)

Evaluate

$$\int_0^\infty dx \int_0^\infty \frac{1}{x^2 + e^{y^2}} \, dy.$$

NOTES ON CHAPTER 8

§8.1. In chapters 10–14 we generally deal with paths, i.e. equivalence classes composed of piecewise continuously differentiable transits. The definition of equivalence which we use is best suited to such transits. In a context where differentiability properties are unimportant, we might call the transits $\phi : [a, b] \to R^n$, $\psi : [c, d] \to R^n$ equivalent if, simply, there exists a strictly increasing, continuous function α such that $\alpha(a) = c$, $\alpha(b) = d$ and $\phi = \psi \circ \alpha$.

In the definition on p. 248 the supposition that $\phi : [a, b] \to R^n$ ($n > 1$) is continuous could be omitted. The expression

$$\sup_{\mathscr{D}} \sum_{i=1}^m \|\phi(t_i) - \phi(t_{i-1})\|, \tag{$*$}$$

if finite, is then more appropriately called the *variation* of ϕ and denoted by $V_a^b(\phi)$. However this extension of the concept of bounded variation does not lead to an essentially new class of functions, for the proof of theorem 8.11 (in which continuity is not used) shows that (∗) is finite if and only if the components of ϕ are of bounded variation.

A space filling curve was first constructed by G. Peano (1890). The example in exercise 8(a), 11 is due to I. J. Schoenberg (*Bull. American Math. Soc.* **44** (1938), 519) who simplified an earlier construction of Lebesgue's.

§8.4. Surfaces, like transits, are naturally grouped into equivalence classes. The manner in which this is done again depends to some extent on the circumstances. For instance the two surfaces $\psi : G \to R^n$, $\psi^* : G^* \to R^n$ ($G, G^* \subset R^m$, $1 \leqslant m < n$) could be taken to be equivalent if there exists a bijective, continuously differentiable function $\alpha : G \to G^*$ such that $J\alpha > 0$ throughout G and $\psi = \psi^* \circ \alpha$. The restriction on the functions to be continuously differentiable is an indication of the considerable difficulties with which the theory of surfaces is bedevilled.

The phrase 'nearly everywhere' (meaning 'except in a set of zero content') corresponds to the phrase 'almost everywhere', used in the Lebesgue theory of measure and integration, which means 'except in a set of zero measure'. (In the notes at the end of chapter 6 we defined one-dimensional zero measure; the generalization to R^n involves no more than the replacement of length by volume.) Since a set of zero content necessarily has zero measure, 'nearly everywhere' implies 'almost everywhere'; but the opposite implication is false.

§8.6. Two results (theorem 8.31, corollary 3 and exercise 8(f), 14) have been labelled special cases of Green's theorem. With the material at our disposal only a general description of the parent theorem is possible. Suppose that the trace of a path γ in R^2 is the frontier of a bounded set E and that $f = (f_1, f_2)$ is a function on $\overline{E}$ to R^2. Green's theorem states conditions under which

$$\int_\gamma f = \iint_E (D_1 f_2 - D_2 f_1).$$

§8.7. The integral

$$\int_0^\infty \frac{\sin x}{x}\, dx$$

has now been evaluated in two different ways (in this section and in §6.9). A further method, depending on complex function theory, will be found in §14.1, example 2. G. H. Hardy made an entertaining comparison of a number of methods of calculating this integral, assigning marks for simplicity, elegance, etc. (*Mathematical Gazette* **5** (1909), 98–103 and **8** (1916), 301–3.)

FOURIER SERIES

9.1. Trigonometric series

This chapter contains salient facts about trigonometric series of the form

$$\tfrac{1}{2}a_0 + \sum_{n=1}^{\infty} (a_n \cos nx + b_n \sin nx),$$

which we shall sometimes write more shortly as

$$\sum_{n=0}^{\infty} A_n(x).$$

The topic has intrinsic importance in the expression of the periodic phenomena which occur in all branches of natural science. Moreover it will give the reader valuable practice in applying principles and theorems from earlier chapters, notably 4 and 5.

The history of the subject is of high interest and is outlined in a note at the end of the chapter. Repeatedly the settlement of a controversy or the resolution of a difficulty thrown up by trigonometric series has clarified the foundations of analysis or has engendered new ideas and methods.

The central problem of the chapter is the validity of the formula, say for x in $[-\pi, \pi]$,

$$f(x) = \tfrac{1}{2}a_0 + \sum_{n=1}^{\infty} (a_n \cos nx + b_n \sin nx). \tag{9.11}$$

The coefficients a_n, b_n are real and the function f is from R^1 to R^1. The sum of the series on the right of (9.11)—if it converges—has period 2π, so it is convenient to suppose that f is defined for all values of x and that $f(x+2\pi) = f(x)$. (If an f is assigned for which $f(\pi) \neq f(-\pi)$, then f can be redefined at one or both of these values.)

We shall always suppose that f is Riemann integrable (§6.7) over $[-\pi, \pi]$.

Let us make the assumption that the series on the right of (9.11) converges uniformly in $[-\pi, \pi]$. The convergence remains uniform when we multiply by either of the bounded functions $\cos nx$ or $\sin nx$. We can then integrate term by term over $[-\pi, \pi]$ by theorem 5.22.

Use the facts (the orthogonality of the cosine and sine functions)

$$\int_{-\pi}^{\pi} \sin mx \cos nx \, dx = 0 \quad (\text{all } m, n),$$

$$\int_{-\pi}^{\pi} \cos mx \cos nx \, dx = \int_{-\pi}^{\pi} \sin mx \sin nx \, dx = 0 \quad (m \neq n),$$

$$\int_{-\pi}^{\pi} \cos^2 nx \, dx = \int_{-\pi}^{\pi} \sin^2 nx \, dx = \pi \quad (n \neq 0)$$

and, if $n = 0$, $\qquad \int_{-\pi}^{\pi} \cos^2 nx \, dx = 2\pi.$

The result of integration term by term is

$$a_n = \frac{1}{\pi} \int_{-\pi}^{\pi} f(x) \cos nx \, dx, \quad b_n = \frac{1}{\pi} \int_{-\pi}^{\pi} f(x) \sin nx \, dx. \quad (9.12)$$

(Note the reason for the factor $\frac{1}{2}$ in the term $\frac{1}{2}a_0$ in 9.11.)

We repeat that the passage from (9.11) to (9.12) is purely formal in default of some justification such as the uniform convergence of the series in (9.11). We can, however, starting with an integrable f, define the numbers a_n, b_n from it.

Definition. *The numbers a_n, b_n given by* (9.12) *are called the* Fourier coefficients *of f. The series* (9.11) *with these values of a_n, b_n is called the* Fourier series *of f.*

From the periodicity of the integrands in (9.12), the interval $[\theta, \theta + 2\pi]$ for any θ would serve to determine a_n, b_n.

Theorem 9.11. *If we are given a trigonometric series*

$$\tfrac{1}{2}\alpha_0 + \sum_{1}^{\infty} (\alpha_n \cos nx + \beta_n \sin nx)$$

and it converges uniformly in $[\theta, \theta + 2\pi]$, then it is the Fourier series of its sum.

Proof. Let $s(x)$ be the sum. We can integrate the uniformly convergent series term by term after multiplying by $\cos nx$ or $\sin nx$. This proves that

$$\alpha_n = \frac{1}{\pi} \int_{\theta}^{\theta+2\pi} s(x) \cos nx \, dx, \quad \beta_n = \frac{1}{\pi} \int_{\theta}^{\theta+2\pi} s(x) \sin nx \, dx. \ |$$

From the definition, the Fourier series of a given function is unique. The converse does not hold without restriction on the function. If $g(x) = f(x)$ except for a finite number of values of x (more generally, except in a set of content zero), the corresponding integrals (9.12) are equal and f, g have the same Fourier series. The next theorem shows that two different *continuous* functions cannot have the same Fourier series. The reader should note this fact but he may appreciate the proof better if he defers it until he has gained more familiarity with trigonometric series (e.g. from §9.2).

Theorem 9.12. *If the R-integrable functions g_1 and g_2 are continuous at c and $g_1(c) \neq g_2(c)$, then the Fourier series of g_1 and g_2 are different.*

Proof. Suppose the conclusion false. Then all the Fourier coefficients of $f = g_1 - g_2$ vanish, and $f(c) \neq 0$, say $f(c) > 0$.

If t_m is any trigonometric polynomial

$$\tfrac{1}{2}\alpha_0 + \sum_1^m (\alpha_n \cos nx + \beta_n \sin nx)$$

and η is any number, then

$$\int_\eta^{\eta+2\pi} f t_m = 0. \tag{9.13}$$

Since f is continuous at c, we can find h (with $0 < h < \pi$), k such that $f(x) \geqslant k > 0$ for $c-h \leqslant x \leqslant c+h$.

We now take t_m to be defined by

$$t_m(x) = \{1 + \cos (x-c) - \cos h\}^m.$$

Then $t_m(x)$ may be shown to be a trigonometric polynomial. It has the properties

   (i) $t_m(x) \geqslant 1$ for $c-h \leqslant x \leqslant c+h$,

   (ii) as $m \to \infty$, $t_m(x) \to \infty$ uniformly for

$$c - \tfrac{1}{2}h \leqslant x \leqslant c + \tfrac{1}{2}h,$$

   (iii) $|t_m(x)| \leqslant 1$ for $c+h \leqslant x \leqslant c-h+2\pi$.

Then, by (i) and (ii),

$$\int_{c-h}^{c+h} f t_m \geqslant \int_{c-\frac{1}{2}h}^{c+\frac{1}{2}h} f t_m \geqslant hk \inf t_m(x) \to \infty \quad \text{as} \quad m \to \infty.$$

Also, by (iii),

$$\int_{c+h}^{c-h+2\pi} f t_m$$

is bounded. So, if m is large enough,

$$\int_{c-h}^{c-h+2\pi} f t_m$$

cannot be 0, which contradicts (9.13). ▌

Corollary. *If f is continuous and the Fourier series of f converges uniformly, its sum is f.*

Proof. If the sum is g, then from theorem 9.11 the series is the Fourier series of g. Now apply theorem 9.12. ▌

Note on complex Fourier series. Some results are more elegant if expressed not in cosines and sines but in the equivalent complex exponentials. We have

$$\frac{1}{2\pi}\int_{-\pi}^{\pi} e^{inx}\, dx = 0 \quad (n = \pm 1, \pm 2, \ldots),$$
$$= 1 \quad (n = 0).$$

The formal Fourier series of f, in general complex and equal to $g + ih$, say, is then

$$\sum_{-\infty}^{\infty} c_n e^{inx} = c_0 + \sum_{1}^{\infty} (c_n e^{inx} + c_{-n} e^{-inx}),$$

where

$$c_n = \frac{1}{2\pi}\int_{-\pi}^{\pi} f(x)\, e^{-inx}\, dx.$$

(The integral of a complex function $\int_{a}^{b} \{p(x) + iq(x)\}\, dx$ is defined to be $\int_{a}^{b} p(x)\, dx + i\int_{a}^{b} q(x)\, dx$.)

It is easy to verify that a necessary and sufficient condition for f to be real is that, if c_n is $\alpha_n + i\beta_n$, then c_{-n} is its conjugate $\alpha_n - i\beta_n$.

It will be convenient to confine ourselves in this chapter to real functions and to cosines and sines.

9.2. Some special series

Before studying general theorems about Fourier series, it is desirable to make oneself familiar with particular examples.

Example 1. *Prove that the Fourier series of the function which is equal to x^2 in $[-\pi, \pi]$ is*

$$\frac{\pi^2}{3} + 4\left\{ -\frac{\cos x}{1^2} + \frac{\cos 2x}{2^2} - \ldots + (-1)^n \frac{\cos nx}{n^2} + \ldots \right\}.$$

State the sum of the series for all values of x.
Deduce that

$$\frac{1}{1^2} + \frac{1}{2^2} + \ldots + \frac{1}{n^2} + \ldots = \frac{\pi^2}{6}.$$

Solution. By definition of Fourier series, the constant term and the coefficients of cos nx ($n \geqslant 1$), sin nx are

$$\frac{1}{2\pi}\int_{-\pi}^{\pi} x^2\,dx, \quad \frac{1}{\pi}\int_{-\pi}^{\pi} x^2 \cos nx\,dx, \quad \frac{1}{\pi}\int_{-\pi}^{\pi} x^2 \sin nx\,dx.$$

Since x^2 is an even function, the integral with sin nx is 0 for all n, and the cosine integral is most easily evaluated as twice the integral over $[0, \pi]$. Integration by parts gives the series required.

The series converges uniformly for all values of x by Weierstrass's M-test (theorem 5.32). By theorem 9.12 (corollary), the sum for $-\pi \leqslant x \leqslant \pi$ is x^2. The sum outside this range is $(x-2\pi)^2$ in $[\pi, 3\pi]$, $(x-4\pi)^2$ in $[3\pi, 5\pi]$ and so on (the parabolic arcs being seen more vividly in a diagram).

Putting $x = \pi$, we obtain the sum $\sum\limits_{1}^{\infty} (1/n^2)$. |

Example 2. *Prove that the Fourier series of the function which is equal to x^2 in $(0, 2\pi)$ is*

$$\frac{4\pi^2}{3} + \sum_{n=1}^{\infty}\left(\frac{4}{n^2}\cos nx - \frac{4\pi}{n}\sin nx\right).$$

Calculate the sum of this series for $x = 0$.

Solution. To obtain the series, we have only to work out the constant term and the coefficients of cos nx, sin nx.

$$\frac{1}{2\pi}\int_{0}^{2\pi} x^2\,dx, \quad \frac{1}{\pi}\int_{0}^{2\pi} x^2 \cos nx\,dx, \quad \frac{1}{\pi}\int_{0}^{2\pi} x^2 \sin nx\,dx.$$

Outside the interval $(0, 2\pi)$ the function which has this Fourier series has, by periodicity, the value $(x-2k\pi)^2$ when

$$2k\pi < x < 2(k+1)\pi$$

for every positive or negative integer k. The function may be given an arbitrary value for $x = 2k\pi$; whatever value is assigned, the function is discontinuous there.

The only convergence theorems so far proved (9.11 and 9.12) say nothing about the possible sum of the Fourier series of a discontinuous function. Such theorems will follow in §9.4.

This example shows a phenomenon which will be found later to be general. At a discontinuity $x = c$ of f at which the right- and left-hand limits $f(c+)$ and $f(c-)$ exist, the Fourier series of f will, if it converges, have the sum $\frac{1}{2}\{f(c+)+f(c-)\}$.

Taking here $c = 0$, we have as the sum of the series

$$\frac{4\pi^2}{3} + \sum_{n=1}^{\infty} \frac{4}{n^2},$$

which is $2\pi^2$ from example 1. As $x \to 0+$, the function x^2 in $(0, 2\pi)$ tends to 0. As $x \to 0-$, the function $(x+2\pi)^2$ in $(-2\pi, 0)$ tends to $4\pi^2$.

Fourier cosine series and sine series. We saw in example 1 that, if f is even, i.e. $f(-x) = f(x)$, its Fourier series contains cosines only. Similarly if f is odd, i.e. $f(-x) = -f(x)$, the integrals for a_n are all 0 and we have sines only.

If, then, f is defined in the interval $[0, \pi]$, there is a *Fourier cosine series* associated with the even function in $[-\pi, \pi]$ coinciding with f in $[0, \pi]$. Similarly, if f is defined in $(0, \pi)$ and we define

$$f(0) = f(\pi) = 0,$$

there is a *Fourier sine series* associated with the corresponding odd function in $[-\pi, \pi]$.

Example 3. *Let $f(x)$ be defined in $[0, \pi]$ to be x in $[0, \frac{1}{2}\pi)$ and $\pi - x$ in $[\frac{1}{2}\pi, \pi]$. Prove that the Fourier cosine series of f is*

$$\frac{\pi}{4} - \frac{2}{\pi} \left(\frac{\cos 2x}{1^2} + \frac{\cos 6x}{3^2} + \frac{\cos 10x}{5^2} + \ldots \right).$$

Prove that the Fourier sine series of f is

$$\frac{4}{\pi} \left(\frac{\sin x}{1^2} - \frac{\sin 3x}{3^2} + \frac{\sin 5x}{5^2} - \ldots \right).$$

The only feature not present in Example 1 is the necessity to divide the range of integration into $[0, \frac{1}{2}\pi]$, $[\frac{1}{2}\pi, \pi]$ in calculating the Fourier coefficients.

As in Example 1, the series are uniformly convergent to the appropriate $f(x)$.

Example 4. *Prove that the series*

$$\sum_{n=1}^{\infty} \frac{\sin nx}{n}$$

converges to $\frac{1}{2}(\pi - x)$ for $0 < x < 2\pi$.

Notes. (1) We have previously shown in the example on p. 118 that this series converges for all values of x, and that the convergence is uniform for

$$\delta \leqslant x \leqslant 2\pi - \delta$$

for any fixed positive δ.

(2) Observe that we have not here mentioned the phrase *Fourier series*. We shall solve the example from first principles.

Solution. Write

$$s_n(x) = \sum_{r=1}^{n} \frac{\sin rx}{r}.$$

Then

$$s_n'(x) = \sum_{r=1}^{n} \cos rx = -\tfrac{1}{2} + \frac{\sin (n+\tfrac{1}{2})x}{2 \sin \tfrac{1}{2}x}.$$

So

$$s_n(x) = -\tfrac{1}{2}x + \int_0^x \frac{\sin (n+\tfrac{1}{2})t}{2 \sin \tfrac{1}{2}t} dt.$$

Putting $x = \pi$ in this, we have

$$0 = -\tfrac{1}{2}\pi + \int_0^\pi \frac{\sin (n+\tfrac{1}{2})t}{2 \sin \tfrac{1}{2}t} dt.$$

By subtraction we have

$$\tfrac{1}{2}(\pi - x) - s_n(x) = \int_x^\pi \frac{\sin (n+\tfrac{1}{2})t}{2 \sin \tfrac{1}{2}t} dt.$$

The right-hand side, integrated by parts, is

$$\left[-\frac{\cos (n+\tfrac{1}{2})t}{2(n+\tfrac{1}{2}) \sin \tfrac{1}{2}t} \right]_x^\pi - \int_x^\pi \frac{\cos (n+\tfrac{1}{2})t \cos \tfrac{1}{2}t}{4(n+\tfrac{1}{2}) \sin^2 \tfrac{1}{2}t} dt.$$

If $\delta \leqslant x \leqslant 2\pi - \delta$, the modulus of this does not exceed (say)

$$\frac{1}{(2n+1) \sin \tfrac{1}{2}\delta} + \frac{\pi}{2(2n+1) \sin^2 \tfrac{1}{2}\delta}.$$

We have thus proved that the series

$$\sum_{n=1}^{\infty} \frac{\sin nx}{n}$$

converges to $\tfrac{1}{2}(\pi - x)$ for $0 < x < 2\pi$. Moreover the bound obtained for $|s_n(x) - \tfrac{1}{2}(\pi - x)|$ shows that the convergence is uniform for $\delta \leqslant x \leqslant 2\pi - \delta$. (This result had previously been deduced from theorem 5.33.) The sum of the series (obtained by periodicity outside $(0, 2\pi)$) is discontinuous at $0, \pm 2\pi \ldots$.

It is easy to verify that the series is the Fourier series of its sum, as defined by (9.12).

We add one further property of the series which will be used later (in theorem 9.51), namely the existence of a constant B such that

$$|s_n(x)| < B \quad \text{for all } x, n.$$

In proving this we may assume $0 < x < \pi$.

Let $m = \min(n, [\pi/x])$, where $[a]$ means the integral part of a. Then

$$s_n(x) = \left(\sum_1^m + \sum_{m+1}^n \right) \frac{\sin rx}{r} = t_1 + t_2, \text{ say,}$$

where the sum t_2 is empty, and so zero, if $m = n$. Then, using the inequality $\sin u \leqslant u$ valid for $u \geqslant 0$, we have

$$|t_1| \leqslant x \sum_1^m 1 = mx \leqslant \pi.$$

Also, from example (p. 91) and the inequality $\sin u \geqslant 2u/\pi$ $(0 \leqslant u \leqslant \tfrac{1}{2}\pi)$

$$|t_2| \leqslant \frac{1}{(m+1)\sin \tfrac{1}{2}x} \leqslant \frac{\pi}{(m+1)x} \leqslant 1. \ |$$

Exercises 9(a)

1. Calculate the Fourier series of $|x|$ for $-\pi < x < \pi$. What function does the series represent for other ranges of x?

Sum the series

$$\sum_{n=0}^{\infty} \frac{1}{(2n+1)^2}.$$

2. Prove that, if $a \neq 0$, the Fourier series of e^{ax} in $(-\pi, \pi)$ is

$$\frac{2\sinh a\pi}{\pi} \left\{ \frac{1}{2a} + \sum_{n=1}^{\infty} \frac{(-1)^n (a\cos nx - n\sin nx)}{n^2 + a^2} \right\}.$$

3. Prove that, if $-\pi \leqslant x \leqslant \pi$ and a is not an integer,

$$\cos ax = \frac{2a\sin a\pi}{\pi} \left\{ \frac{1}{2a^2} - \frac{\cos x}{a^2 - 1} + \frac{\cos 2x}{a^2 - 4} - \dots \right\}.$$

Hence express $\cot a\pi$ as an infinite series of partial fractions. (This result will be examined more generally, for complex a, in §14.3.)

4. Prove that, for a range of x to be specified,

$$|\sin x| = \frac{2}{\pi} - \frac{4}{\pi} \sum_{n=1}^{\infty} \frac{\cos 2nx}{4n^2 - 1}.$$

5. Expand $\cos bx$ in a sine series for $0 < x < \pi$, taking b to be (i) not an integer, (ii) an integer.

6. Prove that the function f defined by

$$f(\theta) = \tfrac{1}{2}\theta(\pi - \alpha) \quad \text{for} \quad -\alpha \leqslant \theta \leqslant \alpha(<\pi),$$
$$f(\theta) = \tfrac{1}{2}\alpha(\pi - \theta) \quad \text{for} \quad \alpha \leqslant \theta \leqslant 2\pi - \alpha,$$

is represented in $[-\alpha, 2\pi - \alpha]$ by the series

$$\sum_{n=1}^{\infty} \frac{\sin n\theta \sin n\alpha}{n^2}.$$

What locus in R^2 is represented by

$$\sum_{n=1}^{\infty} \frac{\sin nx \sin ny}{n^2} = 0?$$

7. Prove that the following three expressions all represent $x(l-x)$ for $0 \leqslant x \leqslant l$:

$$\frac{8l^2}{\pi^3} \sum_{n=1}^{\infty} \frac{\sin (2n-1)(\pi x/l)}{(2n-1)^3}, \quad \frac{2l^2}{\pi^2} \sum_{n=1}^{\infty} \frac{\sin^2 (n\pi x/l)}{n^2},$$

$$\frac{l^2}{6} - \frac{l^2}{\pi^2} \sum_{n=1}^{\infty} \frac{\cos (2n\pi x/l)}{n^2}.$$

8. Prove that the function $l^2 - x^2$ has, in the range $[-l, l]$, the Fourier series

$$4l^2 \left\{ \frac{1}{6} - \frac{1}{\pi^2} \sum_{1}^{\infty} \frac{(-1)^n}{n^2} \cos \frac{n\pi x}{l} \right\}.$$

9. The function f is defined in the range $[0, \pi]$ as

$$f(x) = \tfrac{3}{2}x \qquad \text{when} \quad 0 \leqslant x \leqslant \tfrac{1}{3}\pi,$$

$$f(x) = \tfrac{1}{2}\pi \qquad \text{when} \quad \tfrac{1}{3}\pi < x < \tfrac{2}{3}\pi,$$

$$f(x) = \tfrac{3}{2}(\pi - x) \qquad \text{when} \quad \tfrac{2}{3}\pi \leqslant x \leqslant \pi.$$

Show that

$$f(x) = \frac{6}{\pi} \sum_{1}^{\infty} \frac{\sin \tfrac{1}{3}(2n-1)\pi \sin (2n-1)x}{(2n-1)^2}.$$

10. We proved in §9.2, example 4, that

$$\left| \sum_{1}^{n} \frac{\sin rx}{r} \right| \leqslant \pi + 1.$$

Sharpen the bound to $\int_0^{\pi} \frac{\sin t}{t} dt + 1$. The exact supremum is $\int_0^{\pi} \frac{\sin t}{t} dt$; this is harder to prove (Gronwall, *Math. Annalen* **72** (1912), 228–61). Prove, for the analogous integral, that, for all T, x,

$$\left| \int_0^T \frac{\sin xt}{t} dt \right| \leqslant \int_0^{\pi} \frac{\sin t}{t} dt.$$

9.3. Theorems of Riemann. Dirichlet's integral

The first theorem of Riemann (extended later by Lebesgue) includes the result that the Fourier coefficients a_n, b_n of an R-integrable function tend to 0 as n tends to infinity.

Theorem 9.31. (Riemann–Lebesgue.) *If f is R-integrable over $[a, b]$, then*

$$\int_a^b f(x) \cos \lambda x \, dx \quad \text{and} \quad \int_a^b f(x) \sin \lambda x \, dx$$

tend to 0 as λ tends to infinity.

Proof. It is sufficient to give the proof for the cosine integral.

Given ϵ, take a dissection

$$a = x_0 < x_1 < x_2 \ldots < x_{n-1} < x_n = b,$$

such that
$$\sum_{r=1}^{n} (M_r - m_r)(x_r - x_{r-1}) < \epsilon,$$

where M_r, m_r are the supremum and infimum of f in $[x_{r-1}, x_r]$.

Define g by putting $g(x) = m_r$ for $x_{r-1} < x \leqslant x_r$. Then

$$\int_a^b f(x) \cos \lambda x \, dx = \int_a^b \{f(x) - g(x)\} \cos \lambda x \, dx + \int_a^b g(x) \cos \lambda x \, dx.$$

The first integral on the right is numerically at most

$$\int_a^b |f(x) - g(x)| \, dx \leqslant \sum_{r=1}^{n} (M_r - m_r)(x_r - x_{r-1}) < \epsilon.$$

The second integral is

$$\sum_{r=1}^{n} m_r (\sin \lambda x_r - \sin \lambda x_{r-1})/\lambda$$

which is numerically less than or equal to

$$2(\Sigma |m_r|)/\lambda$$

and this is less than ϵ if $\lambda > \lambda_0(\epsilon)$. |

Taking $\lambda = n$ and $[a, b] = [-\pi, \pi]$ we have the result for Fourier coefficients. Without restricting f it is not possible to make any assertion about the rapidity with which a_n, b_n tend to 0. As a specimen of an improved result under an additional assumption about f, we quote the following theorem.

Theorem 9.32. *If f is of bounded variation in $[-\pi, \pi]$, then, as $n \to \infty$,*

$$a_n = O\left(\frac{1}{n}\right), \quad b_n = O\left(\frac{1}{n}\right).$$

Proof. From theorem 6.43, we may suppose that f increases in $[-\pi, \pi]$. By the corollary to the second mean value theorem 6.93,

$$\int_{-\pi}^{\pi} f(x) \cos nx \, dx = f(-\pi) \int_{-\pi}^{\xi} \cos nx \, dx + f(\pi) \int_{\xi}^{\pi} \cos nx \, dx$$

$$= O\left(\frac{1}{n}\right). \, |$$

Theorem 9.33. (Dirichlet's integral.) *The n-th partial sum*

$$s_n = s_n(x) = \tfrac{1}{2}a_0 + \sum_{r=1}^{n} (a_r \cos rx + b_r \sin rx)$$

of the Fourier series of f at x is

$$\frac{1}{\pi} \int_{-\pi}^{\pi} D_n(t-x) f(t) \, dt,$$

where the Dirichlet kernel $D_n(u)$ is defined by

$$D_n(u) = \frac{\sin (n+\tfrac{1}{2})u}{2 \sin \tfrac{1}{2}u}.$$

Proof. Note that $D_n(u)$ may be left undefined at $u = 0$ or may be defined as its limit $n+\tfrac{1}{2}$.

For $r = 0, 1, 2, \ldots,$

$$a_r \cos rx + b_r \sin rx$$

$$= \left(\frac{1}{\pi} \int_{-\pi}^{\pi} f(t) \cos rt \, dt\right) \cos rx + \left(\frac{1}{\pi} \int_{-\pi}^{\pi} f(t) \sin rt \, dt\right) \sin rx$$

$$= \frac{1}{\pi} \int_{-\pi}^{\pi} \cos r(t-x) f(t) \, dt.$$

Substituting in s_n we have

$$s_n = \frac{1}{\pi} \int_{-\pi}^{\pi} \left\{\tfrac{1}{2} + \sum_{1}^{n} \cos r(t-x)\right\} f(t) \, dt$$

which leads to the result. |

There are easy variants of this formula for s_n; it is to be borne in mind that f and D_n have period 2π and so any range of integration of length 2π will serve. We have then

$$s_n(x) = \frac{1}{\pi} \int_{-\pi}^{\pi} D_n(u) f(x+u) \, du.$$

Putting in this formula $-u$ for u in $(-\pi, 0)$, we have

$$s_n(x) = \frac{1}{\pi} \int_{0}^{\pi} D_n(u) \{f(x+u) + f(x-u)\} \, du.$$

We note also that

$$\int_{0}^{\pi} D_n(u) \, du = \int_{0}^{\pi} \left(\tfrac{1}{2} + \sum_{1}^{n} \cos ru\right) du$$

$$= \tfrac{1}{2}\pi.$$

We can now state as a separate result one of the most useful conditions for convergence of a Fourier series.

Theorem 9.34. *A necessary and sufficient condition that $s_n(x)$ converges to s as $n \to \infty$ is that*

$$\int_0^\pi D_n(u)\,\phi(u)\,du$$

tends to 0 as n tends to ∞, where

$$\phi(u) = \phi(x, u) = f(x+u) + f(x-u) - 2s.$$

Proof. From the immediately preceding work the integral is $\pi(s_n(x) - s)$. |

Theorem 9.35. (Riemann's localization theorem.) *The convergence or divergence of the Fourier series of f at x depends only on the values of f in an arbitrarily small neighbourhood $(x - \delta, x + \delta)$ of x.*

Proof. From the remark following theorem 9.33 we have

$$s_n(x) = \frac{1}{\pi}\left(\int_0^\delta + \int_\delta^\pi\right) D_n(u)\{f(x+u) + f(x-u)\}\,du.$$

In $[\delta, \pi]$, $\{f(x+u) + f(x-u)\}$ cosec $\frac{1}{2}u$ being the product of two integrable functions is integrable. By the Riemann–Lebesgue theorem 9.31, the term $\int_\delta^\pi$ in the formula for $s_n(x)$ tends to 0 as n tends to ∞. So $s_n(x)$ and the integral from 0 to δ in the formula for it both converge to the same limit as n tends to infinity or both diverge. The integral involves only the values of f in $(x - \delta, x + \delta)$. |

We close this section with a modification of theorem 9.34 which is often easier to apply.

Theorem 9.36. *A necessary and sufficient condition that $s_n(x)$ converges to s is that, for some δ (where $0 < \delta \leqslant \pi$),*

$$\int_0^\delta \sin\,(n+\tfrac{1}{2})\,u\,\frac{\phi(u)}{u}\,du$$

tends to 0 as n tends to ∞.

Proof.

$$D_n(u) - \frac{\sin\,(n+\tfrac{1}{2})u}{u} = \sin\,(n+\tfrac{1}{2})u\left\{\frac{1}{2\sin\tfrac{1}{2}u} - \frac{1}{u}\right\}$$

and the expression in curly brackets, being $O(u)$ as $u \to 0$, is integrable over $[0, \delta]$. Hence, by the Riemann–Lebesgue theorem 9.31,

$$\int_0^\delta \sin (n+\tfrac{1}{2})u \, \frac{\phi(u)}{u} \, du - \int_0^\delta D_n(u) \, \phi(u) \, du$$

tends to 0 as n tends to infinity.

The proof of theorem 9.35 shows that the condition stated in the present theorem is equivalent with that of theorem 9.34. |

9.4. Convergence of Fourier series

Theorems 9.34 and 9.36 can be transformed into a variety of shapes; two are given in theorems 9.41 and 9.42. The first form is very easily deduced from theorem 9.36; the second also is in common use.

Since the value of f can be changed arbitrarily at a point without alteration of the Fourier coefficients, any condition of convergence of $s_n(x)$ to $f(x)$ must take account of values of f not only at x but in a neighbourhood of x. (Continuity of f at x would be such a condition: we shall see in theorem 9.51 that it is not, in fact, strong enough to entail convergence.)

Theorem 9.41. (Dini.) *A sufficient condition for the Fourier series of f to converge to s at x is that the integral*

$$\int_0^\pi \frac{|\phi(u)|}{u} \, du$$

exists $\left(\text{in the sense of } \lim_{\eta \to 0} \int_\eta^\pi \right)$.

Proof. Given ϵ, choose η such that

$$\int_0^\eta \frac{|\phi(u)|}{u} \, du < \epsilon.$$

Then, for all n,

$$\left| \int_0^\eta \sin (n+\tfrac{1}{2})u \, \frac{\phi(u)}{u} \, du \right| \leqslant \int_0^\eta \frac{|\phi(u)|}{u} \, du < \epsilon.$$

By the Riemann–Lebesgue theorem 9.31, there is $n_0(\epsilon)$ such that

$$\left| \int_\eta^\pi \sin (n+\tfrac{1}{2})u \, \frac{\phi(u)}{u} \, du \right| < \epsilon \quad \text{if} \quad n > n_0.$$

These two inequalities combine to give the sufficient condition of theorem 9.36. |

Corollary. *If f is differentiable at the point x, then the Fourier series of f converges at x to f(x). The conclusion remains true under the more general hypothesis that, for some $\alpha > 0$,*

$$|f(x+h)-f(x)| = O(|h|^{\alpha})$$

as h tends to 0.

Proof. With $\phi(u) = f(x+u)+f(x-u)-2f(x)$, we have as $u \to 0$,

$$|\phi(u)/u| = O(|u|^{\alpha-1})$$

and the integral in the theorem exists. |

Theorem 9.42. (Jordan.) *If f is of bounded variation in a neighbourhood of x, then the Fourier series of f converges at x to*

$$\tfrac{1}{2}\{f(x+)+f(x-)\}.$$

Proof. In theorem 9.36 write

$$s = \tfrac{1}{2}\{f(x+)+f(x-)\}.$$

(These limits exist by theorem 6.43.)

Then $\phi(u)$ has bounded variation in an interval $[0, \delta]$ say, and $\phi(u)$ tends to 0 as u tends to 0.

By theorem 6.43 $\phi(u)$ may be expressed as the difference of two increasing functions, each of which tends to 0 as $u \to 0$.

The theorem will follow if we prove that, if $\phi(u)$ is increasing, the integral in theorem 9.36 tends to 0 as $n \to \infty$.

Given ϵ, choose η so that $\phi(\eta) < \epsilon$. By the second mean value theorem 6.93 there exists a number ξ such that $0 < \xi < \eta$ and

$$\int_0^{\eta} \frac{\sin(n+\tfrac{1}{2})u}{u} \phi(u)\,du = \phi(\eta) \int_{\xi}^{\eta} \frac{\sin(n+\tfrac{1}{2})u}{u}\,du$$

$$= \phi(\eta) \int_a^b \frac{\sin u}{u}\,du,$$

where $a = (n+\tfrac{1}{2})\xi$, $b = (n+\tfrac{1}{2})\eta$. This last integral is bounded for all n, ξ, η, and so the last line is numerically less than $A\epsilon$.

By theorem 9.31, there exists n_0 such that

$$\left| \int_{\eta}^{\delta} \frac{\sin(n+\tfrac{1}{2})u}{u} \phi(u)\,du \right| < \epsilon \quad \text{if} \quad n > n_0.$$

So the integral in theorem 9.36 is numerically less than $(A+1)\epsilon$ if $n > n_0$. |

The Gibbs phenomenon. Suppose that the Fourier series of a particular function f, which has discontinuities, is known to converge to f. We shall probe the manner of approach of the continuous curve $y = s_n(x)$ to the limit curve $y = f(x)$, as n tends to infinity, in the vicinity of a discontinuity of f. The characteristic behaviour is seen most clearly if we make the simplest possible choice of f, namely

$$f(x) = -\tfrac{1}{2}\pi \quad \text{if} \quad -\pi < x < 0,$$

$$f(x) = \tfrac{1}{2}\pi \quad \text{if} \quad 0 < x < \pi,$$

f has period 2π and $f(n\pi) = 0$ for all n.

The Fourier series of f, which is found to be

$$2(\sin x + \tfrac{1}{3} \sin 3x + \tfrac{1}{5} \sin 5x + \ldots),$$

converges to $f(x)$ for all values of x, by theorem 9.42.

It will appear in theorem 9.43 that, if n is large, as x increases from the discontinuity 0, the curve $y = s_n(x)$ rises at a steep gradient to a maximum value b_n at $x = \pi/2n$. We shall prove that b_n is approximately, not $\tfrac{1}{2}\pi$ as might be expected, but the number

$$\int_0^\pi \frac{\sin t}{t}\, dt,$$

which is about 18 per cent greater than $\tfrac{1}{2}\pi$. This phenomenon of 'overshooting the mark' is named after Gibbs (see a historical note at the end of the chapter). As x increases beyond the value $\pi/2n$, the sum $s_n(x)$ will have a sequence of minima and maxima alternately below and above the line $y = \tfrac{1}{2}\pi$. Since $s_n(x) = s_n(\pi - x)$, the curve $y = s_n(x)$ for $\tfrac{1}{2}\pi \leqslant x \leqslant \pi$ is the mirror image in the line $x = \tfrac{1}{2}\pi$ of the curve for $0 \leqslant x \leqslant \tfrac{1}{2}\pi$.

Theorem 9.43. *With the notation of the last two paragraphs,*

$$\lim_{n \to \infty} b_n = \int_0^\pi \frac{\sin t}{t}\, dt.$$

Proof.

$$s_n'(x) = 2(\cos x + \cos 3x + \ldots + \cos (2n-1)x)$$

$$= \frac{\sin 2nx}{\sin x}.$$

Therefore

$$s_n(x) = \int_0^x \frac{\sin 2nt}{\sin t}\, dt.$$

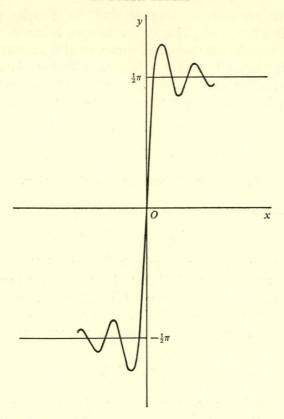

$s_n(x)$ has maxima or minima when $\sin 2nx = 0$, that is, when $2nx = r\pi$. Insertion of these values in $s_n''(x)$ shows that odd values of r give maxima and even values minima.

At the first positive maximum $x = \pi/2n$,

$$s_n(x) = 2\left\{\sin\frac{\pi}{2n} + \frac{1}{3}\sin\frac{3\pi}{2n} + \ldots + \frac{1}{2n-1}\sin\frac{(2n-1)\pi}{2n}\right\}$$

$$= \frac{\pi}{n}\left\{\frac{\sin(\pi/2n)}{\pi/2n} + \frac{\sin(3\pi/2n)}{3\pi/2n} + \ldots + \frac{\sin(2n-1)\pi/2n}{(2n-1)\pi/2n}\right\}.$$

This is seen to be an approximative sum to the required integral if we dissect $[0, \pi]$ into n equal parts and take the value of $(\sin t)/t$ at the mid-point of each part. |

The reader will expect the Gibbs phenomenon at a discontinuity to be shown by the Fourier series of a wide class of functions. In order to define such a class we need further knowledge about the Fourier series of continuous functions. This will be found in the next section, in particular in theorem 9.54; see, then, exercise 9(c), 2.

Exercises 9(b)

1. Prove that the function defined by

$$f(x) = 1/\log(\pi|x|^{-1}) \quad (x \neq 0),$$
$$f(0) = 0$$

satisfies at $x = 0$ the condition of theorem 9.42 (Jordan) but not that of theorem 9.41 (Dini).

2. Construct an example in which Dini's condition is satisfied but not Jordan's.

9.5. Divergence of Fourier series

Theorem 9.51. *The Fourier series of a continuous function may diverge at a point.*

Proof. Define

$$t(x, m, n) = 2 \sin mx \sum_{r=1}^{n} \frac{\sin rx}{r} \quad (m > n > 0).$$

We proved in §9.2, example 4, that the sum on the right-hand side is bounded for all x, n.

Hence $t(x, m, n)$ is bounded for all x, m, n.

Now $t(x, m, n)$ is the cosine polynomial

$$\frac{\cos(m-n)x}{n} + \frac{\cos(m-n+1)x}{n-1} + \dots + \frac{\cos(m-1)x}{1}$$

$$- \frac{\cos(m+1)x}{1} - \frac{\cos(m+2)x}{2} - \dots - \frac{\cos(m+n)x}{n}.$$

Although $t(x, m, n)$ is bounded, nevertheless, at $x = 0$, the sum of the first n terms (or the last n) is numerically greater than $\log n$.

Take two sequences of positive integers m_r and n_r with $n_r < m_r$, and a convergent series Σa_r of positive terms. Then, from the boundedness of $t(x, m, n)$, the series

$$\sum_{r=1}^{\infty} a_r t(x, m_r, n_r) \tag{9.51}$$

converges uniformly for all x to an even continuous function, say $f(x)$.

If, for all r,

$$m_r + n_r < m_{r+1} - n_{r+1}, \tag{9.52}$$

then the multiples of x occurring in $t(x, m_r, n_r)$ and $t(x, m_{r+1}, n_{r+1})$ are all different.

The series (9.51) is of the form $\Sigma A_\nu \cos \nu x$. Since the sequence of its partial sums, when the terms are bracketed as in (9.51), converges uniformly, we can multiply by $\cos \nu x$ and integrate term by term over $[-\pi, \pi]$.

Every integral is 0 except that of the νth term where

$$m_r - n_r \leqslant \nu \leqslant m_r + n_r$$

and we have

$$\pi A_\nu = \int_{-\pi}^{\pi} f(x) \cos \nu x \, dx.$$

The series $\Sigma A_\nu \cos \nu x$ is the Fourier series of f and its partial sums satisfy the inequality

$$s_{m_r}(0) = a_r \left(1 + \frac{1}{2} + \dots + \frac{1}{n_r} \right) > a_r \log n_r.$$

If $m_r = 2n_r$ and $n_{r+1} > 3n_r$ then (9.52) is satisfied. We have

$$s_{m_r}(0) > \tfrac{1}{2} a_r \log m_r$$

and the theorem will be proved if $a_r \log m_r$ tends to infinity with r. To ensure this we may take $a_r = 1/r^2$ and an even

$$m_r > \max (3m_{r-1}, \exp r^3). \ \blacksquare$$

The above construction, due to Fejér, defines a continuous function whose Fourier series diverges for a specified x. Continuous functions can be constructed whose Fourier series diverge for infinitely many x.

Theorem 9.52. *At a point of continuity x of f, we have $s_n(x) = o(\log n)$. This statement is the best possible.*

Proof. With the notation of theorem 9.36, we shall prove

$$\int_0^\pi \sin (n + \tfrac{1}{2}) u \frac{\phi(u)}{u} du = o(\log n).$$

Take $s = f(x)$; then $\phi(u) \to 0$ as $u \to 0$. Given ϵ, there is η (less than 1) such that $|\phi(u)| < \epsilon$ for $0 \leqslant u \leqslant \eta$. Choose n greater than $1/\eta$.

Divide the range of integration $[0, \pi]$ into the parts

$$[0, 1/n], \quad [1/n, \eta], \quad [\eta, \pi]$$

and call the corresponding integrals I_1, I_2, I_3.

In I_1, $\sin(n+\tfrac{1}{2})u < (n+\tfrac{1}{2})u$ and so

$$|I_1| \leqslant (n+\tfrac{1}{2}) \int_0^{1/n} |\phi(u)| \, du < \left(1 + \frac{1}{2n}\right)\epsilon.$$

Also
$$|I_2| \leqslant \epsilon \int_{1/n}^{\xi} \frac{du}{u} < \epsilon \log n,$$

$$I_3 = o(1) \quad \text{by theorem 9.31.}$$

Adding, we have $s_n(x) = o(\log n)$.

To say that this limitation on $s_n(x)$ is the best possible means that, however slowly $\psi(n)$ decreases to 0, there is a function f for which, at some point of continuity x, $s_n(x)$ is not $o\{\psi(n) \log n\}$.

This is already contained in the construction of theorem 9.51. We have only to impose the additional condition on m_r that $\psi(m_r) < 1/r^2$.

Then $s_{m_r}(0) > \tfrac{1}{2}a_r \log m_r > \tfrac{1}{2}\psi(m_r) \log m_r$. |

Fourier series, even if divergent, are not useless. Surprisingly, *any* Fourier series can be integrated term by term.

Theorem 9.53. *If a_n, b_n are the Fourier coefficients of f, then*

$$\tfrac{1}{2}a_0 x + \sum_{n=1}^{\infty} \frac{a_n \sin nx + b_n(1 - \cos nx)}{n} = \int_0^x f(t) \, dt.$$

Proof. Write
$$F(x) = \int_0^x f(t) \, dt - \tfrac{1}{2}a_0 x.$$

Then F is continuous and of bounded variation. Also, since

$$F(x+2\pi) - F(x) = \int_x^{x+2\pi} f(t) \, dt - a_0 \pi = 0,$$

F has period 2π.

By theorem 9.42, the Fourier series of F converges to $F(x)$ for all x. The Fourier coefficients A_n, B_n of F are

$$A_n = \frac{1}{\pi} \int_0^{2\pi} F(x) \cos nx \, dx$$

$$= \frac{1}{\pi} \left[F(x) \frac{\sin nx}{n} \right]_0^{2\pi} - \frac{1}{n\pi} \int_0^{2\pi} \{f(x) - \tfrac{1}{2}a_0\} \sin nx \, dx$$

$$= -\frac{1}{n\pi} \int_0^{2\pi} f(x) \sin nx \, dx = -\frac{b_n}{n}$$

and

$$B_n = \frac{1}{\pi}\int_0^{2\pi} F(x)\sin nx\,dx$$

$$= \frac{1}{\pi}\left[-F(x)\frac{\cos nx}{n}\right]_0^{2\pi} + \frac{1}{n\pi}\int_0^{2\pi}\{f(x)-\tfrac{1}{2}a_0\}\cos nx\,dx$$

$$= \frac{1}{n\pi}\int_0^{2\pi} f(x)\cos nx\,dx = \frac{a_n}{n}.$$

So
$$F(x) = \tfrac{1}{2}A_0 + \sum_{n=1}^{\infty} \frac{a_n\sin nx - b_n\cos nx}{n}.$$

Putting $x = 0$, we have
$$\tfrac{1}{2}A_0 = \sum_{n=1}^{\infty}\frac{b_n}{n}. \;|$$

Corollary 1. *The series* $\Sigma\,(b_n/n)$ *converges.*

Corollary 2. *An everywhere convergent trigonometric series need not be a Fourier series.*

Proof. By the example following theorem 4.54 (p. 91)

$$\sum_{n=2}^{\infty}\frac{\sin nx}{\log n}$$

converges for all x. If this were a Fourier series, then, by corollary 1,

$$\sum_{n=2}^{\infty}\frac{1}{n\log n}$$

would converge, which is false. $\;|$

Theorem 9.54. *In theorem 9.53 the series on the left-hand side converges uniformly for all x.*

Proof. Write $B_n(x) = b_n\cos nx - a_n\sin nx$.

We shall show that $B_n(x)$ is the coefficient of $\sin nt$ in the Fourier series in t of $f(x+t)$.

We have

$$\int_{-\pi}^{\pi} f(x+t)\sin nt\,dt = \int_{-\pi+x}^{\pi+x} f(t)\sin n(t-x)\,dt$$

$$= \int_{-\pi}^{\pi} f(t)\sin n(t-x)\,dt$$

$$= \cos nx\int_{-\pi}^{\pi} f(t)\sin nt\,dt - \sin nx\int_{-\pi}^{\pi} f(t)\cos nt\,dt.$$

Dividing by π, we have the result. (Similarly

$$A_n(x) = a_n \sin nx + b_n \cos nx$$

is the coefficient of cos nt in the Fourier series in t of $f(x+t)$.)

Now we have to prove that $\Sigma\, B_n(x)/n$ converges uniformly. From the first part of the proof,

$$\sum_r^s \frac{B_n(x)}{n} = \frac{1}{\pi} \int_0^{2\pi} f(x+t) \sum_r^s \frac{\sin nt}{n}\, dt$$

$$= \frac{1}{\pi}(I_1 + I_2 + I_3),$$

where the I's are the integrals taken over $[0, \delta]$, $[\delta, 2\pi - \delta]$, $[2\pi - \delta, 2\pi]$.

We know that, for all x, t,

$$|f(x+t)| \leqslant A,$$

and, from example 4 of §9.2, for all r, s, t,

$$\left| \sum_r^s \frac{\sin nt}{n} \right| \leqslant B.$$

Hence each of $|I_1|$, $|I_3|$ is at most $AB\delta$, and is less than a given ϵ by choice of δ. Now $\Sigma\,(\sin nt/n)$ converges uniformly for

$$\delta \leqslant t \leqslant 2\pi - \delta.$$

Therefore we can find r_0 such that

$$\left| \sum_r^s \frac{\sin nt}{n} \right| < \epsilon$$

for $s > r > r_0$ and all t in $[\delta, 2\pi - \delta]$. We have then $|I_2| \leqslant 2\pi A\epsilon$ and

$$\left| \sum_r^s \frac{B_n(x)}{n} \right| \leqslant 2\epsilon\left(\frac{1}{\pi} + A\right). \;\; \blacksquare$$

Exercises 9(c)

1. Deduce from theorem 9.53 an alternative proof that two different continuous functions cannot have the same Fourier series (cf. theorem 9.12).

2. Prove that the Gibbs phenomenon holds at $x = 0$ for the function $g = f_1 + f_2$, where

$$f_1(x) = \int_0^x f(t)\, dt, f \text{ being any integrable function,}$$

$$f_2(x) = -k \text{ in } (-\pi, 0), \quad k \text{ in } (0, \pi) \quad (k > 0).$$

9.6. Cesàro and Abel summability of series

Consider the infinite series

$$1-1+1-1+1- \ldots.$$

If s_n is the sum of the first n terms,

$$s_n = 1 \quad (n \text{ odd}), \qquad s_n = 0 \quad (n \text{ even}).$$

The average (arithmetic mean) of the sums s_n is

$$\sigma_n = \frac{s_1 + \ldots + s_n}{n}$$

and $\qquad \sigma_n = \frac{1}{2}\left(1 + \frac{1}{n}\right) \quad (n \text{ odd}), \qquad \sigma_n = \frac{1}{2} \quad (n \text{ even}).$

Thus σ_n tends to $\frac{1}{2}$ as n tends to ∞.

In the passage from s_n to σ_n, oscillations have been ironed out by cancellation of positive and negative terms of the original series. With increasing n the trigonometric functions $\sin nx$ and $\cos nx$ go through positive and negative values, and it is plausible that averaging the sums of a divergent trigonometric series may result in a convergent sequence.

This section will deal with series of constant terms. In §9.7 we shall discuss trigonometric series.

Definition. Let $s_n = u_1 + u_2 + \ldots + u_n$ and

$$\sigma_n = \frac{s_1 + s_2 + \ldots + s_n}{n}.$$

Then σ_n is the (C, 1) mean of the s_n and, if σ_n tends to l as n tends to infinity, we say that $\sum_{n=1}^{\infty} u_n$ is summable (C, 1) to l.

C refers to Cesàro (1859–1906). More generally a (C, k) mean can be defined; if k is a positive integer it is essentially a k-times repeated average. We shall confine ourselves to the case $k = 1$.

Theorem 9.61. If Σu_n converges, it is summable (C, 1) to the same sum.

We omit the proof which is a straightforward exercise on limits (see exercise 4(b), 9(i)).

Theorem 9.62. If Σu_n is summable (C, 1), then $s_n = o(n)$ and $u_n = o(n)$.

Proof. The first conclusion follows from

$$s_n = n\sigma_n - (n-1)\sigma_{n-1},$$

and then the second from

$$u_n = s_n - s_{n-1}. \quad |$$

Theorem 9.62, a limitation on the magnitude of u_n, states that if the terms of a series are too large, the Cesàro averaging is too weak to sum it.

It is much less clear that if the terms of a series are *too small*, the (C, 1) method will not serve any useful purpose—it will apply only to convergent series. The next theorem states a limitation of magnitude of u_n under which summability (C, 1) of u_n implies convergence. It will not be needed in the rest of the book, but it is an excellent exercise in analysis.

Theorem 9.63. (Hardy, 1910.) *If $u_n = O(1/n)$ and Σu_n is summable* (C, 1), *then it converges.*

Proof.

(*a*) We separate out two preliminary results (*b*), (*c*), in which the hypothesis $u_n = O(1/n)$ is not used.

(*b*) Let $t_n = u_1 + 2u_2 + \ldots + nu_n$. A necessary and sufficient condition that the series Σu_n, summable (C, 1), should converge is that $t_n = o(n)$.

This follows from $t_n = (n+1)s_n - n\sigma_n$.

(*c*) A necessary and sufficient condition that Σu_n should be summable (C, 1) is the convergence of

$$\Sigma \frac{t_n}{n(n+1)}.$$

This follows from

$$\frac{t_n}{n(n+1)} = \frac{n}{n+1}\sigma_n - \frac{n-1}{n}\sigma_{n-1}.$$

(*d*) Using (*b*), (*c*), we prove the theorem. Suppose that Σu_n does not converge. From (*b*), there is a positive A such that either

$$t_N > AN \quad \text{or} \quad t_N < -AN$$

(assume the former) for infinitely many N.

Now, for any n,

$$t_{n+1} = t_n + (n+1)u_{n+1} > t_n - B,$$

since $u_n = O(1/n)$. So, if $t_N > AN$, we have $t_{N+r} > AN - Br$. Then

$$t_{N+r} > \tfrac{1}{2}AN \quad \text{if} \quad r < \tfrac{1}{2}AN/B.$$

Therefore

$$\sum_{N}^{N+r} \frac{t_n}{n(n+1)} > \tfrac{1}{2}AN \sum_{N}^{N+r} \frac{1}{n(n+1)} = \tfrac{1}{2}A \frac{r+1}{N+r+1}.$$

If we take r to be the integer next less than $\tfrac{1}{2}AN/B$, we have

$$\sum_{N}^{N+r} \frac{t_n}{n(n+1)} > C,$$

where C depends on A and B but not on N. Hence the series $\Sigma t_n/n(n+1)$ diverges and by (c), Σu_n is not summable (C, 1). $\mathbf{|}$

There are many methods of 'summing' series besides that of taking arithmetic means. If the reader will turn back to theorem 5.47 (Abel's limit theorem) he will realize that the damping factors $r^n (r < 1)$ suggest another method.

Definition. *If, as $r \to 1-$,*

$$\sum_{n=0}^{\infty} u_n r^n \to l,$$

then we say that $\sum_{n=0}^{\infty} u_n$ is summable (A) *to l.*

Theorem 5.47 shows that every convergent series is summable (A).

Illustrations. The series

$$1 - 1 + 1 - 1 + \dots \quad \text{and} \quad 1 - 2 + 3 - 4 + 5 - \dots$$

are summable (A) to $\tfrac{1}{2}$ and $\tfrac{1}{4}$ respectively.

We shall prove that (A) is a more powerful method of summation than (C, 1); that is to say, any series which is summable (C, 1) is summable (A). Moreover there are series which are summable (A) but not summable (C, 1); such a series is

$$1 - 2 + 3 - 4 + 5 - \dots,$$

which has just been mentioned.

Theorem 9.64. *A series which is summable (C, 1) is summable (A) to the same sum.*

Notation. We shall take the series to be $\sum_{0}^{\infty} u_n$, starting with a term u_0, as is appropriate to summability (A). This entails a modification of the formulae for (C, 1) sums at the beginning of §9.6, where it was preferable to start with u_1. We write henceforward

$$s_n = \sum_{0}^{n} u_r \quad \text{and} \quad \sigma_n = \frac{s_0 + s_1 + \dots + s_{n-1}}{n} \quad (n \geqslant 1).$$

Proof. By theorem 9.62, $u_n = o(n)$ and so $\Sigma u_n x^n$ converges absolutely for $0 \leqslant x < 1$.

By two multiplications of absolutely convergent series we have (C1, theorem 5.7, corollary)

$$(1-x)^{-2} \sum_0^\infty u_n x^n = (1-x)^{-1} \left\{ \sum_0^\infty x^n \sum_0^\infty u_n x^n \right\}$$

$$= \sum_0^\infty x^n \sum_0^\infty s_n x^n = \sum_1^\infty n\sigma_n x^{n-1}.$$

Therefore (supposing throughout that $0 \leqslant x < 1$),

$$\sum_0^\infty u_n x^n = (1-x)^2 \sum_1^\infty n\sigma_n x^{n-1}.$$

The binomial expansion of $(1-x)^{-2}$ (or the substitution

$$u_n = 0 \quad (n \geqslant 1)$$

in the last line) gives

$$1 = (1-x)^2 \sum_1^\infty n x^{n-1}.$$

Multiplying this equation by l and subtracting from the one before, we have

$$\sum_0^\infty u_n x^n - l = (1-x)^2 \sum_1^\infty n(\sigma_n - l) x^{n-1}.$$

Suppose that l is the (C, 1) sum of $\sum_0^\infty u_n$. Then, given $\epsilon > 0$, there is an N such that $|\sigma_n - l| < \epsilon$ if $n > N$. Now

$$\sum_0^\infty u_n x^n - l = (1-x)^2 \left(\sum_1^N + \sum_{N+1}^\infty \right) n(\sigma_n - l) x^{n-1}.$$

In this, $\left| \sum_{N+1}^\infty n(\sigma_n - l) x^{n-1} \right| \leqslant \epsilon \sum_{N+1}^\infty n x^{n-1} < \epsilon(1-x)^{-2}.$

Keeping N fixed, and letting $x \to 1-$, we have

$$\limsup_{x \to 1-} \left| \sum_0^\infty u_n x^n - l \right| \leqslant \epsilon.$$

Since ϵ is arbitrary, u_n is summable (A) to l. |

9.7. Summability of Fourier series

Theorems 9.51 and 9.52 are confessions of failure to determine a function as the limit of partial sums of its Fourier series. To a physicist this would seem lamentable; the results of his observations in spectral analysis are the Fourier coefficients of the functions that he seeks. We are now in a position to show that, though the sequence of sums $s_n(x)$ may fail, the sequence of means $\sigma_n(x)$ succeeds in reconstructing a continuous f. This result of Fejér was a landmark in the theory.

Definition.

$$\sigma_n(x) = \frac{s_0(x) + \ldots + s_{n-1}(x)}{n}.$$

Note that the first suffix is 0, not 1.

Theorem 9.71. *For the Fourier series of f,*

$$\sigma_n(x) = \frac{1}{\pi} \int_0^\pi K_n(u)\{f(x+u) + f(x-u)\}\,du,$$

where the Fejér kernel $K_n(u) = \sin^2 \tfrac{1}{2}nu / 2n \sin^2 \tfrac{1}{2}u$.

Proof. Using the Dirichlet integral (theorem 9.33) for s_n, we have

$$\sigma_n(x) = \frac{1}{n\pi} \int_0^\pi \left\{ \sum_{r=0}^{n-1} \frac{\sin (r + \tfrac{1}{2})u}{2 \sin \tfrac{1}{2}u} \right\} \{f(x+u) + f(x-u)\}\,du. \quad (9.71)$$

Sum the trigonometric series by the formula

$$2 \sin \tfrac{1}{2}u \sin (r + \tfrac{1}{2})u = \cos ru - \cos (r+1)u. \;|$$

Corollary 1. $\displaystyle \int_0^\pi K_n(u)\,du = \tfrac{1}{2}\pi.$

The important property of the Fejér kernel K is that it is always positive (or zero), in contrast with the Dirichlet kernel which takes both signs. This fact yields inequalities such as the following:

If
$$m \leqslant \psi(u) \leqslant M,$$

then
$$\tfrac{1}{2}\pi m \leqslant \int_0^\pi K_n(u)\,\psi(u)\,du \leqslant \tfrac{1}{2}\pi M.$$

The following corollary is analogous to theorem 9.34.

Corollary 2. *A necessary and sufficient condition that the Fourier series of f is summable (C, 1) to s at x is that*

$$\lim_{n \to \infty} \frac{1}{n} \int_0^\pi \frac{\sin^2 \tfrac{1}{2}nu}{\sin^2 \tfrac{1}{2}u}\, \phi(u)\,du = 0,$$

where $\phi(u) = f(x+u) + f(x-u) - 2s$.

Theorem 9.72. (Fejér, 1904.) *The Fourier series of f is summable (C, 1) to $\frac{1}{2}\{f(x+)+f(x-)\}$ whenever this expression has a meaning, and in particular to $f(x)$ at every point x of continuity.*

Proof. Putting $s = \frac{1}{2}\{f(x+)+f(x-)\}$, we have $\phi(u) \to 0$ as $u \to 0$. Given ϵ, there is an $\eta\, (\leqslant \pi)$ such that $|\phi(u)| \leqslant \epsilon$ for $|u| \leqslant \eta$.

Dividing the range of integration $[0, \pi]$ at η, we have

$$\left|\frac{1}{n}\int_0^\pi \frac{\sin^2 \frac{1}{2}nu}{\sin^2 \frac{1}{2}u}\, \phi(u)\, du\right| \leqslant \frac{\epsilon}{n}\int_0^\eta \frac{\sin^2 \frac{1}{2}nu}{\sin^2 \frac{1}{2}u}\, du + \frac{1}{n}\int_\eta^\pi \frac{|\phi(u)|}{\sin^2 \frac{1}{2}u}\, du. \quad (9.72)$$

The first term on the right is less than

$$\frac{\epsilon}{n}\int_0^\pi \frac{\sin^2 \frac{1}{2}nu}{\sin^2 \frac{1}{2}u}\, du,$$

which is $\pi\epsilon$. The second term tends to 0 as n tends to ∞. |

Corollary 1. *If f is continuous in an interval (a, b) and $a < c < d < b$, then $\sigma_n(x) \to f(x)$ uniformly in the sub-interval $[c, d]$.*

Proof. Given ϵ, there is η independent of x such that, for all x in $[c, d]$,
$$|f(x+u)-f(x)| < \tfrac{1}{2}\epsilon \quad \text{if} \quad |u| < \eta.$$
Hence $|\phi(u)| < \epsilon$ if $|u| < \eta$, which is the estimate required for the first term on the right in (9.72). Also $|\phi(u)| < A$, a constant independent of x, u, and the second term on the right of (9.72) is less than B/n. |

Corollary 2. (Weierstrass's approximation theorem for trigonometric polynomials.) *If f is continuous and has period 2π, then given ϵ, there is a trigonometric polynomial $t(x)$ for which $|f(x)-t(x)| < \epsilon$ for all x.*

This is the analogue of theorem 5.52 for algebraic polynomials. See also exercise 5(f), 2.

Summability (A) of Fourier series.

Lemma. *If $0 \leqslant r < 1$, then*

(i) $\dfrac{1}{2} + \sum\limits_{n=1}^{\infty} r^n \cos n\theta = \dfrac{1-r^2}{2(1-2r\cos\theta+r^2)}$,

(ii) *for fixed r, the series converges uniformly in θ.*

Proof. (i) Writing $z = re^{i\theta}$, we have

$$\tfrac{1}{2} + \sum_1^\infty z^n = \frac{1}{1-z} - \tfrac{1}{2} = \frac{1-re^{-i\theta}}{1-2r\cos\theta+r^2} - \tfrac{1}{2}.$$

We now take real parts.

(ii) $|r^n \cos n\theta| \leqslant r^n$ and Σr^n converges.

Uniformity follows from theorem 5.32 (M-test). |

Theorem 9.73. *If $f(x)$ has Fourier series $\sum_0^\infty A_n(x)$, then, for $0 \leqslant r < 1$,*

$$\frac{1}{2\pi}\int_{-\pi}^{\pi} \frac{1-r^2}{1-2r\cos(t-x)+r^2} f(t)\, dt = \sum_0^\infty A_n(x)r^n.$$

Proof. From the lemma,

$$\frac{1-r^2}{2(1-2r\cos(t-x)+r^2)} = \tfrac{1}{2} + \sum_1^\infty r^n \cos n(t-x).$$

We can now multiply by the bounded function $f(t)$ and integrate term by term over $[-\pi, \pi]$. |

The left-hand side in theorem 9.73 is called the *Poisson integral* of f.

Theorem 9.74.

$$\lim_{r\to 1-} \frac{1}{2\pi}\int_{-\pi}^{\pi} \frac{1-r^2}{1-2r\cos(t-x)+r^2} f(t)\, dt = \tfrac{1}{2}\{f(x+)+f(x-)\},$$

whenever the right-hand side has a meaning. In particular the right-hand side is $f(x)$ at every point x of continuity.

Proof. By theorem 9.72, $\sum_0^\infty A_n(x)$ is summable (C, 1) to

$$\tfrac{1}{2}\{f(x+)+f(x-)\}.$$

By theorem 9.64, it is summable (A). Theorem 9.74 then follows from 9.73. |

Exercises 9(*d*)

1. The power series $\Sigma a_n z^n$ has radius of convergence R. Can it be summable (C, 1) for a value of z with $|z| > R$?

For what values of z is the geometric series Σz^n summable (C, 1)?

Define uniform summability (C, 1). Specify a set of z in which Σz^n is uniformly summable.

2. Prove that the summability (A) of Σa_n implies its convergence if

either (i)　$a_n \geqslant 0$　for all n,

or　　(ii)　$a_n = o(1/n)$　(Tauber).

(The much deeper result with O in place of o in (ii) was proved by Littlewood.)

3. Adapt the proof of theorem 9.64 to prove the following theorem. Suppose that (i) Σd_n is a divergent series of positive terms, (ii) $\Sigma d_n x^n$ converges for $|x| < 1$, (iii) as $n \to \infty$, $\lim (c_n/d_n) = k \ (\neq 0)$. Then, as $x \to 1-$,

$$\frac{\Sigma c_n' x^n}{\Sigma d_n x^n} \to k.$$

4. Satisfy yourself that

$$\lim_{X \to \infty} \int_0^X \left(1 - \frac{x}{X}\right) f(x)\,dx$$

is a suitable definition of (C, 1) $\displaystyle\int_0^\infty f(x)\,dx$.

Prove that $\displaystyle\int_0^\infty x^k \sin x\,dx$ exists in the (C, 1) sense if $0 \leqslant k < 1$.

Can you suggest a definition of summability (A) of $\displaystyle\int_0^\infty f(x)\,dx$?

9.8. Mean square approximation. Parseval's theorem

We saw in §9.5 that, for some functions f, the Fourier sums s_n give poor approximations to f in the sense of the metric

$$\sup_{-\pi \leqslant x \leqslant \pi} |f(x) - s_n(x)|.$$

Indeed at particular points x the difference $|f(x) - s_n(x)|$ may tend to infinity with n.

We shall show in this section that the sums s_n give good approximations to f as measured by

$$\int_{-\pi}^{\pi} \{f(x) - s_n(x)\}^2 dx.$$

Among all trigonometric polynomials t_n of degree at most n, the Fourier sum s_n minimizes the integral

$$\int (f - t_n)^2.$$

This property is shared by expansions in sets of functions other than trigonometric functions. It is easier to write out a proof under general conditions.

Orthogonal and orthonormal sets of functions.

Definition. *The functions* ϕ_n *for* $n = 1, 2, \ldots,$ *Riemann integrable over* $[a, b]$, *form an* orthogonal set *if*

$$\int_a^b \phi_m(x)\phi_n(x)\,dx = 0 \quad (m \neq n).$$

If, in addition,

$$\int_a^b \{\phi_n(x)\}^2\, dx = 1 \quad (n = 1, 2, \ldots),$$

the set (or system) ϕ_n *is orthonormal.*

Definition. *The number*

$$c_n = \int_a^b f(x)\,\phi_n(x)\, dx$$

is the n-th Fourier coefficient *of f for the orthonormal set* ϕ_n, *and*

$$\sum_1^\infty c_n \phi_n$$

is the Fourier series *of f. We write* $s_n = \sum_1^n c_r \phi_r.$

The trigonometric functions

$$\frac{1}{\sqrt{(2\pi)}}, \quad \frac{1}{\sqrt{\pi}} \cos nx, \quad \frac{1}{\sqrt{\pi}} \sin nx$$

form an orthonormal set in $[-\pi, \pi]$.

Theorems 9.81 and 9.82 will be stated for a general orthonormal set ϕ_n in $[a, b]$.

Theorem 9.81. *If* $t_n = \sum_1^n d_r \phi_r$, *then, for fixed n and arbitrary* d_r, *the integral*

$$\int_a^b \{f(x) - t_n(x)\}^2\, dx$$

attains its least value when $d_r = c_r = \int_a^b f(x)\,\phi_r(x)\, dx$ *for* $r = 1, 2, \ldots n.$

Proof. All integrals are over $[a, b]$ and all sums from 1 to n.

$$\int (f - t_n)^2 - \int (f - s_n)^2 = -2\int f t_n + \int t_n^2 + 2\int f s_n - \int s_n^2$$

$$= -2\Sigma c_r d_r + \Sigma d_r^2 + 2\Sigma c_r^2 - \Sigma c_r^2$$

$$= \Sigma (c_r - d_r)^2.$$

The sum is never negative and is 0 only when $t_n \equiv s_n$. |

Theorem 9.82. (Bessel's inequality.)

$$\sum_1^\infty c_n^2 \leqslant \int f^2.$$

Proof. We showed in the proof of theorem 9.81 that, for all n,

$$\int (f - s_n)^2 = \int f^2 - \sum_1^n c_r^2.$$

The left-hand side is not negative. $\mathbf{|}$

Corollary 1. *In the trigonometric case*

$$\tfrac{1}{2}a_0^2 + \sum_1^\infty (a_n^2 + b_n^2) \leqslant \frac{1}{\pi} \int_{-\pi}^\pi f^2.$$

Corollary 2. *As $n \to \infty$, $c_n \to 0$.*

The trigonometric case of Corollary 2 ($a_n \to 0$, $b_n \to 0$) was proved differently in theorem 9.31.

For some (not all) orthonormal sets the inequality in theorem 9.82 can be strengthened to equality. This is true for the set of trigonometric functions and we proceed to prove it. The most direct argument depends on Fejér's theorem 9.72.

Theorem 9.83. (Parseval's identity.) *If the a_n, b_n are Fourier coefficients of f, then*

$$\tfrac{1}{2}a_0^2 + \sum_{n=1}^\infty (a_n^2 + b_n^2) = \frac{1}{\pi} \int_{-\pi}^\pi f^2.$$

Proof. We have as in theorem 9.82 (for trigonometric functions)

$$0 \leqslant \int_{-\pi}^\pi f^2 - \pi \left\{ \tfrac{1}{2}a_0^2 + \sum_1^k (a_r^2 + b_r^2) \right\} = \int_{-\pi}^\pi (f - s_k)^2.$$

From theorem 9.81, if t_n is a polynomial of degree n, and $k \geqslant n$,

$$\int_{-\pi}^\pi (f - s_k)^2 \leqslant \int_{-\pi}^\pi (f - t_n)^2.$$

Theorem 9.83 will follow if, given ϵ, we can find a t_n such that

$$\int_{-\pi}^\pi (f - t_n)^2 < K\epsilon,$$

where K is a constant which may depend on f. This we set out to do.

(i) If f is continuous, by theorem 9.72, corollary 1, there is a σ_n for which

$$|f(x) - \sigma_n(x)| < \sqrt{(\epsilon/2\pi)} \quad \text{for all } x.$$

Hence we have

$$\int_{-\pi}^\pi (f - \sigma_n)^2 < \epsilon.$$

(ii) Assume only that f is integrable. With the usual notation, there is a dissection of $[-\pi, \pi]$ for which

$$\sum_{i=1}^{N} (M_i - m_i)(x_i - x_{i-1}) < \epsilon.$$

Define a continuous function ϕ such that

$$\phi(x_i) = f(x_i) \quad \text{for} \quad i = 0, 1, \ldots, N,$$

and ϕ is linear in each $[x_{i-1}, x_i]$. Then, all the integrals being over $[-\pi, \pi]$,

$$\int |f - \phi| \leqslant \sum (M_i - m_i)(x_i - x_{i-1}) < \epsilon,$$

so that, if

$$M = \sup |f(x)| \geqslant \sup |\phi(x)|,$$

$$\int (f - \phi)^2 \leqslant \int 2M |f - \phi| < 2M\epsilon.$$

Since ϕ is continuous, there is by (i) a trigonometric polynomial t for which

$$\int (\phi - t)^2 < \epsilon.$$

From the inequality $(a + b)^2 \leqslant 2(a^2 + b^2)$ we have

$$\int (f - t)^2 \leqslant 2 \int (f - \phi)^2 + 2 \int (\phi - t)^2$$

$$< (4M + 2)\epsilon. \quad |$$

Exercises 9(e)

1. If (a_n, b_n) are the Fourier coefficients of f and (c_n, d_n) those of g, prove that

$$\frac{1}{\pi} \int_{-\pi}^{\pi} fg = \tfrac{1}{2} a_0 c_0 + \sum_{1}^{\infty} (a_n c_n + b_n d_n).$$

2. By Parseval's theorem or otherwise prove that

$$\sum_{1}^{\infty} \frac{1}{n^4} = \frac{\pi^4}{90}, \quad \sum_{1}^{\infty} \frac{1}{(2n-1)^6} = \frac{\pi^6}{960}.$$

3. Prove that, if $k \neq 0$,

$$\sum_{1}^{\infty} \frac{1}{n^2(n^2 + k^2)} = \frac{3 + k^2\pi^2 - 3k\pi \coth k\pi}{6k^4}.$$

9.9. Fourier integrals

We now attempt the analysis into harmonic components of a function f defined for $-\infty < x < \infty$, which possesses no periods. This analysis will involve an integral in place of a series.

The requisite integral was discovered by Fourier himself from the following heuristic argument. Suppose that f has period $2\pi\lambda$ (where λ will later tend to infinity) and that its Fourier series is

$$\tfrac{1}{2}a_0 + \sum_{n=1}^{\infty} \left(a_n \cos\frac{nx}{\lambda} + b_n \sin\frac{nx}{\lambda} \right),$$

where
$$a_0 = \frac{1}{\pi\lambda} \int_{-\pi\lambda}^{\pi\lambda} f(t)\,dt$$

and $\quad a_n = \dfrac{1}{\pi\lambda} \displaystyle\int_{-\pi\lambda}^{\pi\lambda} f(t) \cos\dfrac{nt}{\lambda}\,dt, \quad b_n = \dfrac{1}{\pi\lambda} \displaystyle\int_{-\pi\lambda}^{\pi\lambda} f(t) \sin\dfrac{nt}{\lambda}\,dt.$

Write $\dfrac{n}{\lambda} = u_n$ and $\dfrac{1}{\lambda} = \delta u$, so that

$$a_n = \frac{\delta u}{\pi} \int_{-\pi\lambda}^{\pi\lambda} f(t) \cos u_n t\,dt \quad \text{(and } b_n \text{ likewise).}$$

Let now $\lambda \to \infty$ and we conjecture that

$$f(x) = \int_0^{\infty} \{a(u) \cos ux + b(u) \sin ux\}\,du,$$

where
$$a(u) = \frac{1}{\pi} \int_{-\infty}^{\infty} f(t) \cos ut\,dt, \quad b(u) = \frac{1}{\pi} \int_{-\infty}^{\infty} f(t) \sin ut\,dt, \quad (9.91)$$

or, finally,
$$f(x) = \frac{1}{\pi} \int_0^{\infty} du \int_{-\infty}^{\infty} f(t) \cos u(t-x)\,dt. \quad (9.92)$$

To inject rigour into Fourier's analysis would be troublesome. Instead, we start afresh and investigate the formula (9.92) on lines parallel to those which guided us in §§9.3, 9.4.

Throughout §9.9 we make the assumption that $\displaystyle\int_{-\infty}^{\infty} |f(t)|\,dt$ exists.

Then the integrals in (9.91) exist for all values of u and the inner integral in (9.92) exists for all u, x. Moreover (by theorem 6.35) the convergence of the infinite integrals in (9.91) is uniform with respect to u for all u.

The reader will prove without difficulty (exercise 9(f), 1) that $a(u)$, $b(u)$ are continuous functions of u.

We first extend theorem 9.31 (Riemann–Lebesgue).

Theorem 9.91. *If* $\displaystyle\int_{-\infty}^{\infty} |f(x)|\,dx$ *exists, then*

$$\int_{-\infty}^{\infty} f(x) \cos \lambda x\,dx \to 0 \quad \text{and} \quad \int_{-\infty}^{\infty} f(x) \sin \lambda x\,dx \to 0$$

as $\lambda \to \infty$.

Proof (for the cosine integral). Given ϵ, there is X such that

$$\int_{-\infty}^{-X} |f| + \int_{X}^{\infty} |f| < \epsilon.$$

By theorem 9.31 there is $\lambda_0 = \lambda_0(X, \epsilon) = \lambda_0(\epsilon)$ such that

$$\left| \int_{-X}^{X} f(x) \cos \lambda x \, dx \right| < \epsilon \quad \text{for} \quad \lambda > \lambda_0.$$

Hence $\left| \int_{-\infty}^{\infty} f(x) \cos \lambda x \, dx \right| < 2\epsilon \quad \text{for} \quad \lambda > \lambda_0.$ |

The next theorem is the counterpart of theorem 9.33 (Dirichlet's integral).

Theorem 9.92.

$$\int_{0}^{U} du \int_{-\infty}^{\infty} f(t) \cos u(t-x) \, dt = \int_{-\infty}^{\infty} f(t) \frac{\sin U(t-x)}{t-x} \, dt.$$

Proof. The inner integral on the left converges uniformly in u. Hence, from theorem 8.74 (the hypothesis (i) of which is satisfied from exercise 8(c), 2), the left-hand side is equal to

$$\int_{-\infty}^{\infty} f(t) \, dt \int_{0}^{U} \cos u(t-x) \, du,$$

which is the right-hand side. |

We now define $\phi(t) = f(x+t) + f(x-t) - 2s$

and follow the steps leading from theorem 9.33 to 9.34 and 9.35, obtaining two theorems.

Theorem 9.93. *A necessary and sufficient condition that*

$$\frac{1}{\pi} \int_{0}^{\infty} du \int_{-\infty}^{\infty} f(t) \cos u(t-x) \, dt = s$$

is that $\displaystyle\lim_{U \to \infty} \int_{0}^{\infty} \phi(t) \frac{\sin Ut}{t} \, dt = 0.$

Proof.

$$\int_{-\infty}^{\infty} f(t) \frac{\sin U(t-x)}{t-x} \, dt = \int_{0}^{\infty} \{f(x+t) + f(x-t)\} \frac{\sin Ut}{t} \, dt$$

$$= \int_{0}^{\infty} \phi(t) \frac{\sin Ut}{t} \, dt + 2s \int_{0}^{\infty} \frac{\sin Ut}{t} \, dt$$

and this last integral is $\frac{1}{2}\pi$ from §8.7 (example). |

Theorem 9.94. (Localization theorem.) *The convergence or divergence as $U \to \infty$ of the integral*

$$\int_0^U du \int_{-\infty}^\infty f(t) \cos u(t-x)\, dt$$

depends only on the values of f in an arbitrarily small neighbourhood $(x-\delta, x+\delta)$ of x.

Proof. Following the proof of theorem 9.35, we decompose

$$\int_0^\infty \{f(x+t)+f(x-t)\} \frac{\sin Ut}{t}\, dt$$

into $\int_0^\delta + \int_\delta^\infty$. The function $|f(x+t)+f(x-t)|/t$ is integrable in (δ, ∞), because $|f|$ is integrable in $(-\infty, \infty)$. By theorem 9.91,

$$\int_\delta^\infty \{f(x+t)+f(x-t)\} \frac{\sin Ut}{t}\, dt$$

tends to 0 as U tends to ∞.

Therefore, as $U \to \infty$,

$$\int_0^U du \int_{-\infty}^\infty f(t) \cos u(t-x)\, dt$$

and

$$\int_0^U du \int_{-\delta}^\delta f(t) \cos u(t-x)\, dt$$

both converge to the same limit or both diverge. The latter integral depends only on the values of f in $(x-\delta, x+\delta)$. |

We are now in a position to write down the criterion for integrals analogous to that of theorem 9.36 for series.

Theorem 9.95. *A necessary and sufficient condition that*

$$\frac{1}{\pi} \int_0^\infty du \int_{-\infty}^\infty f(t) \cos u(t-x)\, dt = s$$

is that, for some $\delta > 0$,

$$\lim_{U \to \infty} \int_0^\delta \phi(t) \frac{\sin Ut}{t}\, dt = 0.$$

Proof.

$$\int_\delta^\infty \phi(t) \frac{\sin Ut}{t}\, dt = \int_\delta^\infty \{f(x+t)+f(x-t)\} \frac{\sin Ut}{t}\, dt - 2s \int_\delta^\infty \frac{\sin Ut}{t}\, dt.$$

The first integral tends to 0 as $U \to \infty$ by the proof of theorem 9.94.

The integral in the second term is $\int_{U\delta}^{\infty} \frac{\sin v}{v}\, dv$, which tends to 0. |

Theorem 9.95 shows that the Fourier integral of f converges at a point x if f (restricted to an interval containing x) satisfies any known set of sufficient conditions for the convergence at x of its Fourier series, for instance those in theorem 9.41 or 9.42.

Fourier transforms. Suppose that f is defined on $[0, \infty)$. Let f_1 be the even function on $(-\infty, \infty)$ coinciding with f on $[0, \infty)$. From equations (9.91),

$$a(u) = \frac{1}{\pi} \int_{-\infty}^{\infty} f_1(t) \cos ut\, dt = \frac{2}{\pi} \int_{0}^{\infty} f(t) \cos ut\, dt,$$

$$b(u) = \frac{1}{\pi} \int_{-\infty}^{\infty} f_1(t) \sin ut\, dt = 0.$$

If now f, as well as being absolutely integrable over $[0, \infty)$, satisfies at x sufficient conditions for the convergence of Fourier series (such as Dini's or Jordan's) we have

$$f(x) = \int_{0}^{\infty} a(u) \cos ux\, du.$$

The formulae for $a(u)$ and $f(x)$ become quite symmetrical if we adjust the numerical multiplier and write

$$a(u) = \left(\frac{2}{\pi}\right)^{\frac{1}{2}} \int_{0}^{\infty} f(t) \cos ut\, dt,$$

$$f(x) = \left(\frac{2}{\pi}\right)^{\frac{1}{2}} \int_{0}^{\infty} a(u) \cos ux\, du.$$

Two functions such as $f(x)$, $a(u)$ linked by this reciprocal relation are called *cosine transforms* of each other.

Similarly an odd function gives rise to a pair of Fourier *sine transforms*

$$f(x) = \left(\frac{2}{\pi}\right)^{\frac{1}{2}} \int_{0}^{\infty} b(u) \sin ux\, du,$$

where

$$b(u) = \left(\frac{2}{\pi}\right)^{\frac{1}{2}} \int_{0}^{\infty} f(t) \sin ut\, dt.$$

It is in the shape of transforms that Fourier integrals are most readily used.

It will be observed that, though the formulae are symmetrical, the properties of the functions $f(x)$ and $a(u)$ (or $b(u)$) differ. To attain further symmetry it is necessary to use the Lebesgue integral.

Transforms according to other definitions (notably Laplace transforms) are also important in analysis and in mathematical physics.

Exercises 9(f)

1. Prove that, in (9.91), the functions $a(u)$ and $b(u)$ are continuous.

2. Find the function $b(u)$ satisfying the *integral equation*

$$f(x) = \int_0^\infty b(u) \sin ux\, du \qquad (x > 0),$$

given that, for $x > 0$, $f(x)$ is
　(i)　1　for　$0 < x < \pi$,
　　　　0　for　$x \geqslant \pi$,
　(ii)　e^{-x},
　(iii)　xe^{-kx},　where $k > 0$.

3. Prove that, if $b > 0$,

$$\int_0^\infty \frac{\cos ax}{x^2 + b^2}\, dx = \frac{\pi}{2b} e^{-ba}, \quad \text{if } a > 0.$$

What is the value if $a < 0$?

Further exercises on Fourier and Laplace transforms are in **14**(a), 16–19, when results obtained by complex integration are available.

The reader may verify that the complex-exponential version of Fourier transforms (in the sense of the note at the end of §9.1) is, with the same hypotheses on f as for cosine or sine transforms,

$$c(u) = \left(\frac{1}{2\pi}\right)^{\frac{1}{2}} \int_{-\infty}^\infty f(t) e^{-iut}\, dt,$$

$$f(x) = \left(\frac{1}{2\pi}\right)^{\frac{1}{2}} \int_{-\infty}^\infty c(u) e^{iux}\, du.$$

NOTES ON CHAPTER 9

§9.1. (*History.*) The physical problem which impelled the study of trigonometric series was that of the transverse vibration of a string, with ends fixed at the points $(0, 0)$, $(l, 0)$. Mathematicians of the 18th century were in possession of the differential equation

$$\frac{\partial^2 y}{\partial t^2} = c^2 \frac{\partial^2 y}{\partial x^2} \quad (c \text{ constant})$$

satisfied by the configuration $y = y(x, t)$ of the string. Here $y(0, t) = y(l, t) = 0$ for all t.

In 1747 d'Alembert, by the change of variables

$$\xi = x + ct, \quad \eta = x - ct,$$

transformed the equation to $\qquad \dfrac{\partial^2 y}{\partial \xi \partial \eta} = 0,$

which is satisfied by $\qquad y = \phi(\xi) + \psi(\eta)$
$$= \phi(x+ct) + \psi(x-ct),$$

where the arbitrary functions ϕ, ψ are to be determined by the initial shape and motion of the string. Suppose, following d'Alembert, that, when $t = 0$, the string is at rest in the configuration $y = f(x)$ for $0 \leqslant x \leqslant l$. Then

$$\phi(x) + \psi(x) = f(x).$$

Since the velocity at every point is 0 when $t = 0$,

$$c\phi'(x) - c\psi'(x) = 0,$$

and this can be integrated to give

$$\phi(x) = \psi(x).$$

In this way d'Alembert was led to the solution for $0 \leqslant x \leqslant l$ and all positive t,

$$y = \tfrac{1}{2}f(x+at) + \tfrac{1}{2}f(x-at).$$

At about the same time, Euler made a similar analysis of the problem. The two men differed, however, in their concept of a *function*. Was an analytical expression required (as d'Alembert held) or was a specified graph an adequate definition (Euler)? Essentially, d'Alembert's notion of a function was too narrow (needing in his hands derivatives of every order) and Euler's was too vague.

In 1753, Daniel Bernoulli used the method of separating variables to find particular solutions $y = u(x)v(t)$ of the wave equation. The equation is satisfied by $u(x) = \sin px$ or $\cos px$, and $v(t) = \sin pct$ or $\cos pct$, where p is a constant. The condition $y(0, t) = 0$ (all t) excludes $\cos px$, and the condition $y(l, t) = 0$ restricts p to the values $n\pi/l$.

Since the equation is linear, particular solutions may be added to give

$$y = \sum_{n=1}^{\infty} \sin \frac{n\pi x}{l} \left(a_n \cos \frac{n\pi ct}{l} + b_n \sin \frac{n\pi ct}{l} \right).$$

Assume, with d'Alembert and Euler, that when $t = 0$, the string is at rest in the form $y = f(x)$. We have $b_n = 0$ (all n) and

$$y(x, t) = \sum_{n=1}^{\infty} a_n \sin \frac{n\pi x}{l} \cos \frac{n\pi ct}{l}.$$

Bernoulli asserted this value of $y(x, t)$ to be the general solution and so it had to include the d'Alembert–Euler solution. This implied (for $t = 0$) that an arbitrary function $f(x)$ in $[0, \pi]$ could be expressed as

$$\sum_{n=1}^{\infty} a_n \sin nx,$$

but this was unacceptable to both d'Alembert and Euler.

The 18th century mathematicians summed a number of infinite trigonometric series, and their results, though lacking rigorous proof, were often right. As instances, Euler stated that

$$\sin x - \tfrac{1}{2} \sin 2x + \tfrac{1}{3} \sin 3x - \dots$$

had sum $\frac{1}{2}x$, but apparently believed this to hold for all x instead of for $-\pi < x < \pi$ only. Bernoulli gave the sum $\frac{1}{2}(\pi - x)$ for

$$\sin x + \tfrac{1}{2}\sin 2x + \tfrac{1}{3}\sin 3x + \ldots,$$

with the correct range of validity $0 < x < 2\pi$.

The next advance came in 1807 when Fourier proved in many special cases that a function f, which might have discontinuities, could be expanded in a convergent trigonometric series. Proofs of the possibility of expanding a general function were attempted by Fourier himself, Poisson, and Cauchy. Fourier's contributions were recognized by the attachment of his name to the trigonometric series associated with a given function. The first real proof that a wide class of functions could be expressed as trigonometric series is due to Dirichlet (1829).

Since the days of the early 19th century analysts, the mutual interaction of trigonometric series with the main branches of analysis has been fruitful. As illustrations we cite (with names and rough dates) the topics of integration and differentiation (Riemann 1850), sets of points (Cantor 1880), functions of bounded variation (Jordan 1890), summability of divergent series (Fejér 1904), integration and differentiation (a second wave, Lebesgue 1900). If theorems about trigonometric series are to be free from restrictions not inherent in the problems, it is necessary to interpret integration in the sense of Lebesgue. The Lebesgue integral is not included in this book, and in this chapter we do the best we can with integrals in the Riemann sense.

§9.4. Gibbs pointed out the phenomenon in 1899. It had however been observed by Wilbraham in 1848. There is a good chapter, with illuminating diagrams, in Carslaw, *Fourier Series and Integrals*.

§9.6. The student should read the historical account given in Hardy, *Divergent Series*. This contains a section on Fourier and his theorem (p. 29).

COMPLEX FUNCTION THEORY

10.1. Complex numbers and functions

The rest of the book will be devoted to a closer study of *complex functions*, namely functions whose domain of definition and range are both in Z.

The function value $f(z)$ will usually be called w, and we shall keep to the notation

$$z = x + iy = r(\cos\theta + i\sin\theta),$$

$$w = u + iv = \rho(\cos\phi + i\sin\phi).$$

Numbers such as a and b in $w = az + b$ will in general be complex unless common usage proclaims their reality (as for u, v, ρ, ϕ above).

The reader will be familiar with the definitions and properties of the *conjugate*, $\bar{z} = x - iy$; the *modulus*, $|z| = r = \sqrt{(x^2 + y^2)}$; and the *phase*, the angle θ for which $\cos\theta = x/r$ and $\sin\theta = y/r$, where $r \neq 0$. The phase has infinitely many values differing by multiples of 2π. We shall write $\mathrm{Ph}\,z$ (with a capital P) for any of these values. It is often convenient to have a *principal value* of $\mathrm{Ph}\,z$, and this is defined to be the value such that $-\pi < \mathrm{Ph}\,z \leqslant \pi$. The principal value will be written $\mathrm{ph}\,z$ with a small p.

We write an equation such as

$$\mathrm{Ph}(z_1 z_2) = \mathrm{Ph}\,z_1 + \mathrm{Ph}\,z_2$$

as being true *modulo* 2π, that is to say, the two sides may differ by an integral multiple of 2π. The reader may verify that

$$\mathrm{ph}(z_1 z_2) = \mathrm{ph}\,z_1 + \mathrm{ph}\,z_2 + 2k\pi,$$

where k may be 0 or 1 or -1, depending on z_1 and z_2.

Writing $w = f(z)$ as

$$u + iv = f(x + iy),$$

we see that this is equivalent to a pair of real functions of two real variables

$$u = u(x, y), \quad v = v(x, y),$$

defined for a given set of pairs (x, y). This notion of a function of z is too wide to be the most useful. For instance, functions so defined cannot in general be differentiated if we define, as seems natural,

$$f'(z) = \lim(\delta w/\delta z).$$

Illustrations. (1) $w = \frac{1}{2}(z+\bar{z})$, that is $w = x$, is a function of z in the sense that z determines w uniquely. But, in notation which explains itself, if δz has phase θ,

$$\left|\frac{\delta w}{\delta z}\right| = \left|\frac{\delta x}{\delta z}\right| = |\cos\theta|,$$

and so the modulus of $\delta w/\delta z$ takes all values between 0 and 1 according to the phase of δz, and $\delta w/\delta z$ does not tend to a limit as $\delta z \to 0$.

(2) If, however, $w = z^2$, then, as $\delta z \to 0$, $\delta w/\delta z$ tends to the limit $2z$. The reasoning is just like that which proves that, if $y = x^2$, then $\delta y/\delta x$ tends to $2x$.

Usually the domain of definition of a function of z is a *region* (defined in §3.3 to be a connected open set). Continuity of f for a given z then means that $\lim f(z+h) = f(z)$ as h tends to 0 through complex values.

Definition of derivative. *If, for a given z,*

$$\frac{f(z+h)-f(z)}{h}$$

tends to a limit as h tends to 0, then f is said to be differentiable *for that z, and the limit, written $f'(z)$, is the* derivative.

Theorem 10.1. *Necessary and sufficient conditions for f to be differentiable at a given point z are that u and v are differentiable (in the sense of 7.11) and their partial derivatives satisfy the Cauchy–Riemann equations*

$$u_x = v_y, \quad u_y = -v_x$$

at the corresponding (x, y).

Proof. We prove necessity. Let $f'(z) = p+iq$. Write $h = k+il$. By definition of $f'(z)$,

$$f(z+h)-f(z) = (p+iq)(k+il)+o\{\sqrt{(k^2+l^2)}\}.$$

Taking real and imaginary parts, we have

$$u(x+k, y+l)-u(x, y) = pk-ql+o\{\sqrt{(k^2+l^2)}\},$$

$$v(x+k, y+l)-v(x, y) = qk+pl+o\{\sqrt{(k^2+l^2)}\}.$$

Hence $u(x, y)$ and $v(x, y)$ are differentiable in the sense of (7.11) or C1 (156) and

$$u_x = p = v_y, \quad u_y = -q = -v_x.$$

To prove sufficiency, we reverse the steps of the argument. |

Notes. (1) Comparing the definition of $f'(z)$ with that of the *linear derivative* λ in (7.12) we see that the linear function $\lambda(h)$ is $f'(z)h$.

The connection between the mappings from Z to Z and from R^2 to R^2 is elaborated in a note at the end of the chapter.

(2) In complex function theory the notation u_x for the partial derivative $\partial u/\partial x$ is more expressive than the general notation $D_1 f$ which is best in chapter 7.

(3) Taking δz to be δx and $i\delta y$ in turn, we have

$$f'(z) = \frac{\partial f}{\partial x} = u_x + iv_x$$

and

$$f'(z) = -i\frac{\partial f}{\partial y} = v_y - iu_y.$$

(4) If the first partial derivatives of u and v are differentiable (which by theorem 7.84, is a sufficient condition that $u_{xy} = u_{yx}$), then the Cauchy–Riemann equations show that

$$u_{xx} = v_{xy} = -u_{yy},$$

that is to say, u (and similarly v) satisfies Laplace's equation in two variables

$$\frac{\partial^2 u}{\partial x^2} + \frac{\partial^2 u}{\partial y^2} = 0.$$

Functions which satisfy this equation are called *harmonic functions*. They play a large part in both pure mathematics and physics.

Manipulative properties of complex derivatives are in general straightforward adaptations of those of real derivatives, for instance

$$\frac{d}{dz} f\{g(z)\} = f'\{g(z)\}g'(z).$$

The following examples illustrate methods of solving some problems suggested by this section.

Example 1. *If, for all z in a region D, $f'(z) = 0$, then f is constant in D.*

Proof. By note (3) above, at every point of D,

$$u_x = u_y = v_x = v_y = 0.$$

If the segment (x, y_1), (x, y_2) is in D, the mean value theorem shows that

$$u(x, y_2) - u(x, y_1) = (y_2 - y_1)u_y(x, \eta) = 0,$$

where $y_1 < \eta < y_2$.

If a and b are any two points of D, they can, by theorem 3.32, be joined by a polygon whose sides are horizontal or vertical. The above argument applied to u and v along these sides gives $f(b) = f(a)$, as required. |

This splitting into real and imaginary parts is necessary because (§7.3) Rolle's theorem and the mean value theorem do not extend directly to a function from R^2 to R^2 (or from Z to Z). It is possible, however, to prove an inequality form of the mean value theorem on the model of theorem 7.31 (see exercise $10(a)$, 10).

Example 2. *Find $f(z)$, differentiable for all z except $z = 0$, having real part*

$$\frac{x+y}{x^2+y^2}$$

(which satisfies Laplace's equation except for $x = y = 0$).
Solution. The neatest way is to exploit conjugates,

$$2u(x, y) = (u+iv)+(u-iv) = f(z)+\overline{f(z)}.$$

Writing $2x = z+\bar{z}$ and $2iy = z-\bar{z}$, we have

$$2u = \frac{z+\bar{z}-i(z-\bar{z})}{z\bar{z}} = \frac{1+i}{z}+\frac{1-i}{\bar{z}}.$$

Hence $f(z) = \dfrac{1+i}{z}$ is a solution; and iA, where A is an arbitrary real constant, may be added to it.

10.2. Regular functions

In §10.1 we indicated that the existence of the derivative f' is necessary if complex functions f are to have interesting properties. For instance,

$$w = \frac{az+b}{cz+d} \quad \text{or} \quad w = \exp z$$

are acceptable, the former being differentiable except for $z = -(d/c)$ and the latter for all z. We shall investigate both of these functions in §§10.3–10.6.

Suppose now that $w^2 = z$. Corresponding to a given z there are two values of w (if $z \neq 0$). We were faced with this ambiguity in discussing the relation $y^2 = x$ between real numbers. We could then agree that $y^2 = x$ comprises two separate functions, namely (for $x \geqslant 0$)

$$y = +\sqrt{x} \quad \text{and} \quad y = -\sqrt{x},$$

and each of these could be investigated in its own right. From $w^2 = z$ in complex variables the two values of w cannot be so readily disentangled, and the reader should go into the following details in order to appreciate this fact.

Write

$$z = r\,(\cos\theta + i\sin\theta) \quad \text{and} \quad w = \rho(\cos\phi + i\sin\phi),$$

where $r \geqslant 0$ and $\rho \geqslant 0$. Equating moduli and phases in $w^2 = z$, we have

$$\rho^2 = r \quad \text{and} \quad 2\phi = \theta + 2n\pi,$$

where n may be any integer. The values

$$\phi = \tfrac{1}{2}\theta \quad \text{and} \quad \phi = \tfrac{1}{2}\theta + \pi$$

give two distinct values of w. Taking as an illustration the special case $r = \rho = 1$, suppose that $0 \leqslant \theta < 2\pi$. Write

$$w_1 = \cos\tfrac{1}{2}\theta + i\sin\tfrac{1}{2}\theta, \quad w_2 = \cos(\tfrac{1}{2}\theta + \pi) + i\sin(\tfrac{1}{2}\theta + \pi).$$

If z moves once round the circle $|z| = 1$, one of the two corresponding values of w moves from w_1 round to w_2 and the other from w_2 to w_1.

In tracing the values of w we have for simplicity moved z round a circle with centre $z = 0$. The phenomenon of interchange of values occurs if z follows any path which winds round the origin $z = 0$.

A point, like $z = 0$ here, about which values of w interchange is called a *branch point*. The different values of w (there being two in this illustration) generate *branches*.

In order to define a *function* of z satisfying the relation $w^2 = z$, we must restrict the domain of definition to one which contains no path winding round the origin. Such a domain is the z-plane with the negative real axis ($y = 0$, $x \leqslant 0$) deleted. We shall call this region the z-plane *cut along the negative real axis*. Any ray from the origin would serve equally well for the cut. If then w is defined to be (say) -1 when $z = 1$, the relation $w^2 = z$ determines a continuous function of z in the cut plane. We shall meet another instance of multiplicity of values in §10.6 and we shall take the problem up more systematically in §13.3.

When writing in the sequel $w = f(z)$ we shall always understand that, for any z in the region (or set of points) contemplated, w is defined uniquely.

Definition. *f is regular in a region D if $f'(z)$ exists at every point z of D.*

Definition. f is regular at the point *a if there is an open disc containing a in which f is regular.*

The reader will remark that the order of these definitions is opposite to that customary for real functions. In defining, say, continuity of a function $f: R^1 \to R^1$, we first define continuity at a point $x = a$. Afterwards we define continuity in an interval to mean continuity at each point of it. The reason for the emphasis on a region is that (as we shall prove in the next chapter) differentiability of $f(z)$ in a region entails the existence of higher derivatives as well. No such further implication holds for a real function $f(x)$.

On the other hand functions regular in a region or at a point have most properties that our knowledge of differentiable real functions would lead us to expect. Regularity is preserved under algebraic operations; for example the quotient f/g of two regular functions is regular except at points where g vanishes.

Illustration. If n is a positive integer, z^n is regular for all z, z^{-n} is regular for all z except $z = 0$.

An extensive class of regular functions is formed by those defined as power series or known to be expressible as power series. We prove this.

Theorem 10.2. If the power series $\sum_{0}^{\infty} a_n z^n$ *converges for* $|z| < R$, *then its sum* $f(z)$ *is regular for* $|z| < R$. *Moreover its derivative* $f'(z)$ *is* $\sum_{1}^{\infty} n a_n z^{n-1}$ *for* $|z| < R$.

Proof. By theorem 4.62 the series $\sum n a_n z^{n-1}$ converges for $|z| < R$. So also does the series, soon to be used, $\sum n(n-1) a_n z^{n-2}$.

Fix z, and take $\zeta = z + h$ with

$$|z| + |h| \leqslant R_1 < R.$$

We wish to prove that, as $h \to 0$,

$$\sum_{0}^{\infty} a_n \frac{(z+h)^n - z^n}{h} - \sum_{1}^{\infty} n a_n z^{n-1} \to 0. \tag{10.21}$$

The coefficient of a_n is 0 for $n = 0$ or 1, and for $n \geqslant 2$ is

$$(\zeta^{n-1} + \zeta^{n-2}z + \ldots + z^{n-1}) - n z^{n-1}$$

$$= \sum_{p=1}^{n-1} z^{p-1}(\zeta^{n-p} - z^{n-p})$$

$$= (\zeta - z) \sum_{p=1}^{n-1} z^{p-1}(\zeta^{n-p-1} + \zeta^{n-p-2}z + \ldots + z^{n-p-1}).$$

Since $|z| \leqslant R_1$ and $|\zeta| \leqslant R_1$, the modulus of the last line is at most

$$|h| \sum_{p=1}^{n-1} (n-p) R_1^{n-2} = \tfrac{1}{2} n(n-1) R_1^{n-2} |h|.$$

Since $\sum_2^{\infty} n(n-1) |a_n| R_1^{n-2}$ converges, the left-hand side of (10.21) is $O(|h|)$. |

Notes. (1) What we have proved is 'complex differentiability term by term' for power series. The proof of theorem 5.46 held only for differentiation with respect to a real variable.

(2) Since all the series obtained by any number of term by term differentiations of a power series have the same radius of convergence, theorem 10.2 shows that a function represented by a power series has within its circle of convergence derivatives of every order, and they are all regular functions.

It will appear in §11.9 that the property of possessing derivatives of every order belongs to all regular functions, whether defined by power series or in some other way.

Theorem 10.2 shows, for example, that $\exp z$, $\sin z$ and $\cos z$ are regular for all values of z.

Terminology. *Holomorphic* (and also *monogenic*) is a synonym for *regular*. Some writers use the word *analytic* in the sense in which we have used regular; we prefer to reserve it to carry a wider meaning which will be explained in §13.4.

Exercises 10(a)

1. Establish necessary and sufficient conditions that
    (i) the points z_1, z_2, z_3 are collinear;
    (ii) the triangles formed by z_1, z_2, z_3 and z_4, z_5, z_6 are similar.

2. In theorem 10.2, prove that
$$f^{(n)}(0) = n! a_n.$$

3. Prove that, if
$$0 < a_n < a_{n-1} < \ldots < a_0,$$
no root of the equation
$$a_n z^n + a_{n-1} z^{n-1} + \ldots + a_0 = 0$$
lies in the closed unit disc $|z| \leqslant 1$.

4. State the connection between $\mathrm{ph}\, z$ and $\arctan (y/x)$, the latter having its principal value θ where $\tan \theta = y/x$ and $-\tfrac{1}{2}\pi < \theta < \tfrac{1}{2}\pi$.

5. For what values of z are the following functions differentiable?

$$\bar{z}, \quad 1/(z^2+1), \quad |z|^2, \quad y^2.$$

6. If, for all z in a region, f is differentiable and $|f|$ is constant, prove that f is constant.

7. Suppose that we define $z_1 < z_2$ to mean either
    (i) $|z_1| < |z_2|$
or (ii) $|z_1| = |z_2|$ and $\operatorname{ph} z_1 < \operatorname{ph} z_2$.
Would this establish an order among the complex numbers?

8. Find conditions that the equation

$$az + b\bar{z} + c = 0$$

shall give (i) one value of z, (ii) no value, (iii) infinitely many values.

9. Prove Lagrange's identity

$$\left| \sum_{r=1}^{n} a_r b_r \right|^2 = \sum_{r=1}^{n} |a_r|^2 \sum_{r=1}^{n} |b_r|^2 - \sum_{1 \leqslant r < s \leqslant n} |a_r \bar{b}_s - a_s \bar{b}_r|^2.$$

Deduce the Cauchy–Schwarz inequality (2.13) for complex numbers.

10. (*Mean value inequality*.) If $|f'| \leqslant M$ on the segment $[z_1, z_2]$, then

$$|f(z_2) - f(z_1)| \leqslant M|z_2 - z_1|.$$

10.3. Conformal mapping

Some parts of complex function theory gain in clarity if geometrical language and diagrams are used. There will be nothing unsure in the foundations of what we do; everything can be expressed in analytical terms. But a geometrical picture can be more vivid and revealing than collections of equations.

The geometrical meaning of $f'(z)$. Suppose that f is regular in a region D. Then $w = f(z)$ maps D onto a set of points in the w-plane.

Let a be a point of D at which $f'(a) \neq 0$. Let $f(a) = b$. Then, as $z \to a$,

$$\frac{w - b}{z - a} \to f'(a).$$

If $f'(a) = k(\cos \alpha + i \sin \alpha)$, where $k > 0$, then

$$\frac{|w - b|}{|z - a|} \to k$$

and $\operatorname{Ph}(w - b) - \operatorname{Ph}(z - a) \to \alpha \pmod{2\pi}$.

Suppose that z approaches a along a curve C represented by a differentiable function $z = z(t)$ for $t_0 \leqslant t \leqslant t_1$, where $z(t_0) = a$ and $z'(t_0) \neq 0$. In the w-plane there is a corresponding curve Γ,

$$w = f\{z(t)\} \quad \text{and} \quad w'(t_0) = f'\{z(t_0)\} z'(t_0).$$

Hence $w'(t_0) \neq 0$ and Γ has a tangent at b. If the tangents to C and Γ make angles β, γ with the respective real axes, then

$$\gamma - \beta = \alpha \ (\text{mod } 2\pi).$$

It follows that, if two curves C_1 and C_2 meeting at $z = a$ are mapped into Γ_1 and Γ_2 meeting at $w = b$, then the angle between the tangents at b to Γ_1 and Γ_2 is equal to the angle between the tangents at a to C_1 and C_2, the angles being measured in the same sense.

A mapping with this conservation of angles is called *conformal*.

10.4. The bilinear mapping. The extended plane

A good initiation into the methods of complex function theory is a close study of some simple functions (or mappings or transformations).

As a first example, let us try to form a mental picture of the transformation from z to w defined by

$$w = (2+i)z + (3+2i).$$

Here w is a function of z of the same form as $y = 2x + 3$ in real numbers. This relation (*linear* in the elementary sense) is seen as a straight line in the (x, y) plane. A corresponding picture relating z and w would need four dimensions, two for each complex variable. The best that we can do is to envisage a z-plane and a w-plane and to hold together in our attention configurations in the two planes which are connected by the given relation. To illustrate this procedure we discuss relations of the simplest possible types.

(1) $w = z + b$. Any figure in the z-plane is transformed into its w-counterpart by translation through the vector b.

(2) $w = az$. Let a have modulus k and phase α. Then

$$w = kz_1 \quad \text{where} \quad z_1 = z(\cos \alpha + i \sin \alpha).$$

The former of these equations represents a change of scale by the factor k, and the latter a rotation through the angle α about the origin.

(3) $w = az + b$. The geometrical operations involved in this combination of (1) and (2) are translation, change of scale, rotation.

Note that (if $a \neq 0$) the mapping is everywhere conformal.

(4) $w = 1/z$. This defines w for all z except $z = 0$. If

$$z = r(\cos \theta + i \sin \theta),$$

then $w = \rho(\cos \phi + i \sin \phi)$ where $\rho = 1/r$ and $\phi = -\theta$.

We recall that two points A, B are *inverse* in a circle with centre O and radius c if O, A, B are in line and $OA.OB = c^2$.

Then $w = 1/z$ corresponds geometrically to (i) inversion in the circle having centre at the origin and radius 1, and then (ii) reflection in the real axis. These two operations can be performed in either order.

The point at infinity. In example (4) there are advantages in adjoining a 'point at infinity' to the complex plane. The symbol ∞ is defined to have the properties

$$a + \infty = \infty + a = \infty$$

for any (finite) complex number a, and

$$b.\infty = \infty.b = \infty$$

for any $b \neq 0$, including $b = \infty$.

We agree further that

$$\text{if} \quad a \neq 0 \quad \text{then} \quad a/0 = \infty$$

and $\qquad\qquad\qquad$ if $\quad b \neq \infty \quad$ then $\quad b/\infty = 0$.

It is not profitable to try to define $\infty + \infty$ because algebraic manipulations involving this operation would lead to inconsistencies.

Observe that in complex function theory we are content with *one* point at infinity (or one infinite value). In different branches of mathematics other conventions about infinite elements may be appropriate. The reader will see from §4.2 that if an infinite element were adjoined to the set of real numbers R^1 it would have to carry a $+$ or $-$ sign. The view is there taken that rules for algebraic manipulation would be of more nuisance than value. Such rules would have included, for instance, $\infty + \infty = \infty$, but $\infty - \infty$ could not be interpreted without violation of algebraic laws.

In plane projective geometry every straight line has a point at infinity and these points form the *line at infinity*.

A pictorial illustration of the complex number ∞ is contained in the notes at the end of the chapter.

Denoting the complex plane by Z, we shall speak of Z with the point ∞ added to it as the *extended* (or *closed*) *plane* and call it $\bar{Z}$.

Returning to $w = 1/z$ which precipitated this discussion of ∞, we see that in the extended z and w planes the mapping is bijective, the point 0 in either plane corresponding to ∞ in the other.

The bilinear mapping. Let a, b, c, d be (complex) constants such that $ad - bc \neq 0$. Then

$$w = \frac{az+b}{cz+d},$$

or $\qquad\qquad\qquad czw - az + dw - b = 0,$

is the most general algebraic relation which is linear both in z and in w. (If $ad = bc$, either w or z is constant.) This mapping is called bilinear (or homographic) and is sometimes named after Möbius (1790–1868) who first made an extensive study of it.

Suppose that $c \neq 0$, for, if $c = 0$, we are back at (3). Then

$$w = \frac{a}{c} + \frac{bc - ad}{c(cz + d)},$$

i.e.

$$w = \frac{a}{c} + \frac{bc - ad}{c} z_1, \quad z_1 = \frac{1}{z_2}, \quad z_2 = cz + d,$$

and so the transformation is composed of simpler ones of the forms (3), (4).

Solving for z in terms of w, we have

$$z = \frac{-dw + b}{cw - a}.$$

If we admit a point at infinity in each of the z and w planes, the transformation is one-one without exception since $w = \infty$ corresponds to $z = -(d/c)$ and $z = \infty$ to $w = c/a$.

10.5. Properties of bilinear mappings

This section is included, although its results are peripheral to the main lines of theory and applications, to illustrate methods of dealing with complex variables. When a proof of a result presents no striking feature it is sometimes left to the reader.

For shortness we write $w = Tz$ for

$$w = \frac{az + b}{cz + d},$$

so that T is determined by the matrix

$$\begin{pmatrix} a & b \\ c & d \end{pmatrix}$$

with determinant different from 0. The mapping $w = Tz$ is the same if each element of the matrix is multiplied by a constant k (not 0). The range of both z and w is the extended plane $\bar{Z}$.

(1) *Bilinear transformations form a group.* Two transformations T_1 and T_2 are compounded by multiplying their matrices. The associative law

$$(T_1 T_2) T_3 = T_1 (T_2 T_3)$$

can be verified directly.

The unit is

$$\begin{pmatrix} 1 & 0 \\ 0 & 1 \end{pmatrix}$$

and the inverse T^{-1} of T is

$$\begin{pmatrix} d & -b \\ -c & a \end{pmatrix}.$$

(2) *A bilinear mapping is determined by three conditions.* In particular, there is a unique $w = Tz$ which maps z_1, z_2, z_3 onto w_1, w_2, w_3 respectively. For proof, we could substitute the pairs of values in $w = Tz$ and solve for $a : b : c : d$. It is much easier to introduce an intermediate variable ζ which takes the values $0, 1, \infty$ when z is z_1, z_2, z_3 respectively. We can then write at sight

$$\zeta = k \frac{z - z_1}{z - z_3}$$

to give the right values of ζ for z_1 and z_3. The condition $\zeta = 1$ for $z = z_2$ fixes k, and we have

$$\zeta = \frac{(z - z_1)(z_2 - z_3)}{(z - z_3)(z_2 - z_1)}.$$

From the similar connection between ζ and w we find

$$\frac{(w - w_1)(w_2 - w_3)}{(w - w_3)(w_2 - w_1)} = \frac{(z - z_1)(z_2 - z_3)}{(z - z_3)(z_2 - z_1)}.$$

Corollary. If, in the language of projective geometry, we define the cross-ratio $(z_1\, z_2\, z_3\, z_4)$ of the four points to be

$$\frac{(z_1 - z_2)(z_3 - z_4)}{(z_1 - z_4)(z_3 - z_2)},$$

then *cross-ratios are invariant under bilinear transformations.*

(3) *The cross-ratio $(z_1 z_2 z_3 z_4)$ is real if and only if the four points lie on a circle or straight line.*

Proof. $\qquad \mathrm{Ph}(z_1 z_2 z_3 z_4) = \mathrm{Ph}\,\dfrac{z_1 - z_2}{z_1 - z_4} - \mathrm{Ph}\,\dfrac{z_3 - z_2}{z_3 - z_4}.$

By angle properties of circles (or lines) this is 0 or π according to the relative positions of the points. |

In the rest of this section we shall use the word *circle* to mean either circle in the conventional sense or straight line. A straight line in $\overline{Z}$ is a circle through $z = \infty$.

(4) *A bilinear transformation maps circles into circles.*

Proof. If we assume (3), the result follows from the corollary to (2). Alternatively we can check that each of the transformations (3), (4) in §10.4 turns circles into circles.

(5) *The conformal property.* The mapping $w = Tz$ is conformal except at $w = \infty$ or $z = \infty$. (We shall not try to interpret conformality at infinite values.) The preservation of angles of intersection under the mapping $w = Tz$ is a useful tool.

(6) *The transformation* $w = k(z-a)/(z-b)$. Here circles C in $\overline{Z}$ through a, b, correspond to straight lines through $w = 0$. Circles $|w| = \rho$ with centre $w = 0$ correspond to circles D

$$\left|\frac{z-a}{z-b}\right| = \frac{\rho}{|k|}$$

(circles of Apollonius for a, b). Any C circle cuts any D circle orthogonally.

We prove that a bilinear transformation maps a circle and two points inverse in it into a circle and two inverse points. This is clear for the elementary operations of translation, change of scale, rotation and reflection which are included in §10.4, (1)–(4). To see that it is true for the remaining operation of inversion, suppose that Σ is a circle and a, b are two points inverse in it. All circles Γ through a, b cut Σ orthogonally. By the conformal property of preservation of angles, the inverses Γ' of the circles Γ cut the inverse Σ' orthogonally. The two points common to all the Γ' are inverse in Σ'.

(7) *Fixed points.* A value of z for which $z = Tz$ is called a fixed point of the mapping. The quadratic equation gives two fixed points (which may coincide).

If the fixed points are distinct, say p and q, then $w = Tz$ must be the same as

$$\frac{w-p}{w-q} = k\frac{z-p}{z-q},$$

where k is some constant.

Suppose now that the fixed points coincide. In the special case when they are both 0 for the mapping

$$w = \frac{az+b}{cz+d},$$

so that $w = z$ leads to $z^2 = 0$, it is clear that $d = a$, $b = 0$, $c \neq 0$,

and hence

$$\frac{1}{w} = \frac{1}{z} + l,$$

where l is an arbitrary constant.

If the fixed points are $w = z = p$, this becomes

$$\frac{1}{w-p} = \frac{1}{z-p} + l.$$

(8) *The general bilinear mapping of the disc $|z| < 1$ onto the disc $|w| < 1$ is*

$$w = e^{i\alpha} \frac{z-a}{1-\bar{a}z},$$

where $|a| < 1$ and α is an arbitrary real number.

Proof. The frontier $|z| = 1$ will be mapped into the frontier $|w| = 1$.

From (6), the points $0, \infty$ regarded as inverse in the circle $|w| = 1$ will be mapped into points inverse in $|z| = 1$. These are of the form $a, 1/\bar{a}$, where $|a| < 1$. So the mapping is

$$w = k \frac{z-a}{1-\bar{a}z}$$

for suitable k. Any point $z = e^{i\beta}$ on $|z| = 1$ is to correspond to a point on $|w| = 1$. Therefore

$$1 = |k| \frac{|e^{i\beta} - a|}{|1 - \bar{a}e^{i\beta}|} = |k| \frac{|e^{i\beta} - a|}{|e^{-i\beta} - \bar{a}|} = |k|.$$

Hence the circles $|z| = 1$ and $|w| = 1$ correspond. By theorem 3.33 a continuous function maps a connected set into a connected set and so the disc $|z| < 1$ is mapped into either $|w| < 1$ or $|w| > 1$. Since $z = a$ goes into $w = 0$, we have $|w| < 1$. |

Exercises 10(*b*)

1. Prove that the general bilinear mapping of the upper half z-plane $(y > 0)$ onto the disc $|w| < 1$ is

$$w = e^{i\alpha} \frac{z-a}{z-\bar{a}},$$

where im $a > 0$ and α is an arbitrary real number.

2. Find the transforms under the mapping

$$w = \frac{z-2}{2z+1}$$

of (i) the circles $|z| = 1$, $|z| = 2$, (ii) the region $1 < |z| < 2$, re $z > 0$.

3. Map the unit disc $|z| < 1$ onto the region $|w-1| < 1$, taking $z = 0$ into $w = \frac{1}{2}$ and $z = 1$ into $w = 0$.

4. Prove that every bilinear mapping of the upper half plane onto itself is expressible as

$$w = \frac{az+b}{cz+d}$$

with real coefficients and $ad-bc > 0$.

5. Investigate the convergence of the sequence z_n, where $z_n = Tz_{n-1}$ $(n \geqslant 1)$, z_0 being given and T bilinear.

6. Prove that $w = (z^3+3z)/(3z^2+1)$ transforms circles through $z = \pm 1$ into circles through $w = \pm 1$. What in Z corresponds to $|w| \leqslant 1$?

7. In the mapping

$$w = \frac{z^2-6z+10}{2z-6},$$

what in Z corresponds to the region $|w| > 1$?

8. Prove that, if $n > 1$, the transformation

$$T(z) = \frac{1}{2\cos(\pi/n)-z}$$

is such that $T^n(z) = z$. Taking $n = 3$, draw diagrams to show the successive transforms T, T^2 of the disc $|z| < 1$.

9. Transform the quadrantal region $x > 0$, $y > 0$, $|z| < 1$ into the disc $|w| < 1$ (via a semicircular region, quarter plane, half plane).

10. Frame a definition of $z_n \rightarrow \infty$ as $n \rightarrow \infty$.

11. Prove that, if $p(z)$ is a polynomial, $|p(z)| \rightarrow \infty$ as $|z| \rightarrow \infty$.

10.6. Exponential and logarithm

We assume a knowledge of the exponential, logarithmic and trigonometrical functions of a real variable such as can be gleaned from C1, chapter 6. Familiarity with the power series for $\exp x$, $\sin x$, $\cos x$ (either taken as definitions of the functions or obtained from other definitions) is taken for granted.

The power series

$$\exp z = 1+z+\frac{z^2}{2!}+\ldots+\frac{z^n}{n!}+\ldots$$

converges for all complex z. Properties deduced from it such as

$$\exp(z+w) = \exp z \exp w$$

are valid for complex numbers as well as for real numbers (C1, p. 105).

The function $\exp z$ is regular for all z and (theorem 10.2) its derivative $\exp z$ is never 0. Hence the mapping $w = \exp z$ is conformal for all z. To investigate this mapping write

$$\rho(\cos\phi+i\sin\phi) = w = \exp z = \exp x \exp iy$$
$$= e^x(\cos y+i\sin y).$$

Therefore $\rho = e^x$ and $\phi = y+2n\pi$, where n is an arbitrary integer.

The known period 2π of $\cos y$ and $\sin y$ shows that $\exp z$ has period $2\pi i$ (C1, 115).

Straight lines $x = $ constant correspond to circles $|w| = $ constant and lines $y = $ constant correspond to rays $\phi = $ constant. (Note the preservation of orthogonal intersection.) Any strip of height 2π in the z-plane is mapped onto the whole w-plane.

Consider now the logarithmic function. Can we define it as the inverse of the exponential?

Write
$$w = \mathrm{Log}\, z$$

if and only if
$$\exp w = z.$$

Then $r(\cos\theta + i\sin\theta) = z = \exp w = e^u(\cos v + i\sin v)$. Since no value of w makes $\exp w$ zero, $\mathrm{Log}\, z$ must remain undefined for $z = 0$.

If $z \neq 0$,
$$\mathrm{Log}\, z = w = u + iv = \log|z| + i\,\mathrm{Ph}\,z,$$

so that $\mathrm{Log}\, z$ has infinitely many values differing by multiples of $2\pi i$.

As in §10.2, we can define a region D by deleting from Z an assigned ray from the origin, say the ray $\mathrm{ph}\, z = \alpha$. If we then specify the value of $\mathrm{Log}\, z$ at one point of D, this determines a branch of $\mathrm{Log}\, z$ continuous at all points of D. $\mathrm{Ph}\, z$ will satisfy, for a fixed integer k, the inequalities

$$\alpha + 2k\pi < \mathrm{Ph}\, z < \alpha + 2(k+1)\pi.$$

The following definition is now natural.

The *principal value* of $\mathrm{Log}\, z$, written $\log z$, is the value of $\mathrm{Log}\, z$ for which $\mathrm{Ph}\, z$ takes its principal value $\mathrm{ph}\, z$. In fact,

$$\log z = \log|z| + i\,\mathrm{ph}\, z$$

$$= u + iv, \text{ say,}$$

where
$$-\pi < v \leqslant \pi.$$

If z is real and positive, the principal value $\log z$ is the real value.

Properties. (1) If $z \neq 0$, then

$$\frac{d}{dz}\mathrm{Log}\, z = \frac{1}{z}$$

for any value of $\mathrm{Log}\, z$.

Proof. Suppose that $w = \mathrm{Log}\, z$ is defined in a region D with

$$\alpha + 2k\pi < \mathrm{Ph}\, z < \alpha + 2(k+1)\pi.$$

Let a be a point of D and $b = \text{Log}\, a$. Then, at a, $\log |z|$ and $\text{Ph}\, z$ are continuous and $w - b \to 0$ as $z - a \to 0$.

$$\frac{\text{Log}\, z - \text{Log}\, a}{z - a} = \frac{w - b}{\exp w - \exp b} \to \frac{1}{\exp b} = \frac{1}{a}. \;\big|$$

(2) $\log (1 + z) = z - \tfrac{1}{2}z^2 + \tfrac{1}{3}z^3 - \dots$ if $|z| < 1$.

Proof. The principal value $\log (1 + z)$ is uniquely defined in $|z| < 1$ because, for all such z, $\text{re}(1 + z) > 0$.

The left and right-hand sides have the respective derivatives

$$1/(1 + z) \quad \text{and} \quad 1 - z + z^2 - \dots$$

which are equal for $|z| < 1$.

From §10.1, example 1, the difference between $\log (1 + z)$ and the series $z - \tfrac{1}{2}z^2 + \dots$ is constant. This difference is 0 for $z = 0$ and hence for all z in $|z| < 1$. $\big|$

(3) $$\text{Log}\, z_1 + \text{Log}\, z_2 = \text{Log}\, z_1 z_2.$$

$$\log z_1 + \log z_2 = \log z_1 z_2 + 2k\pi i,$$

where k may be 0, 1 or -1.

Complex powers.

Definition. *The power z^ζ (where $z \neq 0$) means* $\exp (\zeta \,\text{Log}\, z)$.

In general z^ζ has infinitely many values. If ζ is an integer, z^ζ is single valued. In particular, $z^{-1} = 1/z$.

Definition. *The principal value of z^ζ is* $\exp (\zeta \log z)$.

By convention e^ζ is always interpreted as the principal value, that is to say, $\exp \zeta$.

Properties. (1) $z^{\zeta_1} z^{\zeta_2} = z^{\zeta_1 + \zeta_2}$ provided that the same value of $\text{Log}\, z$ is taken in each of the powers.

(2) $\dfrac{d}{dz} z^\zeta = \zeta z^{\zeta - 1}$, provided that the same value of $\text{Log}\, z$ is taken in each of the powers. In particular, if the powers have their principal values, the equation holds for $-\pi < \text{ph}\, z < \pi$.

Exercises 10(c)

1. For what interpretations of the many-valued expressions are the following true?

$$1/z^\zeta = z^{-\zeta}, \quad z_1^\zeta \times z_2^\zeta = (z_1 z_2)^\zeta, \quad (z^{\zeta_1})^{\zeta_2} = z^{\zeta_1 \zeta_2}.$$

2. Find all the values of i^i.

3. Describe the mappings

$$w = \frac{1}{2}\left(z + \frac{1}{z}\right), \quad w = \cosh z,$$

4. Find a regular function $u + iv$, given that

$$v = e^y(x\cos x + y\sin x).$$

5. Prove that $w = \tan z$ maps a strip of the z-plane onto the w-plane with the points $\pm i$ deleted.

6. Examine for convergence and sum the series

$$\sum_1^\infty (-1)^{n-1}\frac{\cos n\theta}{n}, \quad \sum_1^\infty (-1)^{n-1}\frac{\cos n\theta \cos^n \theta}{n}.$$

7. Prove that, if all zeros of the polynomial $p(z)$ have positive real parts, so have all zeros of $p'(z)$.

Prove that, if all zeros of $p(z)$ lie in a convex polygon P, the zeros of $p'(z)$ lie in P.

8. Discuss convergence and uniformity of convergence of

$$\sum_1^{\lceil\infty\rceil} \frac{1}{n^z} \quad \text{(principal values)}.$$

9. Prove that the function of two complex variables z, w defined (with principal values) by

$$f(z, w) = \sum_{n=1}^\infty z^n n^{w-1}$$

is continuous in each of the three sets of (z, w)
   (i) $|z| < 1$, w arbitrary,
  (ii) $|z| \leqslant 1$,   re $w < 0$,
 (iii) $|z| = 1, z \neq 1$, w real and $\leqslant 0$.

10. *Dirichlet's problem.* The problem is to find a function harmonic (§10.1) in a given region taking assigned values at points of the frontier. The general problem is difficult and outside the scope of this book. The reader has the means of solving it in a disc with continuous values on the circumference.

Use theorems 9.73, 9.74, 10.2 to find $U(r, \theta) = u(r\cos\theta, r\sin\theta)$ where $u(x, y)$ is harmonic for $x^2 + y^2 < 1$ and, as $r \to 1-$, $U(r, \theta) \to f(\theta)$ where f is continuous for $-\pi \leqslant \theta \leqslant \pi$.

Find $U(r, \theta)$ when $f(\theta) = \pi\cos\frac{1}{2}\theta$.

NOTES ON CHAPTER 10

For alternative contemporary accounts of complex function theory see the books of Ahlfors and Cartan. They start at the beginning and carry the subject further than we do in chapters 10–14.

§10.1–10.2. The Cauchy–Riemann equations commemorate the two men who first showed the power of complex function theory. Cauchy's exposition was purely analytical. Riemann, three decades later, gave full play to geometrical insight.

Complex functions and functions from R^2 to R^2. There is a natural one-to-one correspondence between the points of R^2 and those of Z. This induces a bijective correspondence $\mathscr{F}$ between the functions from R^2 to R^2 and the complex functions (i.e. functions from Z to Z). Also, the usual metrics in R^2 and Z make these spaces indistinguishable as *metric spaces*. Hence, in $\mathscr{F}$, continuous functions correspond to one another.

As *vector spaces*, R^2 and Z differ; R^2 is two dimensional while Z is one dimensional. It is therefore not surprising that $\mathscr{F}$ does not give a one-one correspondence between linear functions. In fact, if the complex function $\lambda + i\mu$ is linear, then the function $(\lambda, \mu)^T$ from R^2 to R^2 is linear; but, if $(\lambda, \mu)^T$ is linear, $\lambda + i\mu$ is not necessarily linear. When

$$\lambda(x, y) = ax + by,$$

$$\mu(x, y) = cx + dy,$$

then $\lambda + i\mu : Z \to Z$ is linear if and only if

$$a = d \quad \text{and} \quad b = -c.$$

It follows that $\mathscr{F}$ does not give a one-one correspondence between differentiable functions. If the complex function $u + iv$ is differentiable, then $(u, v)^T : R^2 \to R^2$ is differentiable; but if $(u, v)^T : R^2 \to R^2$ is differentiable, then $u + iv$ is differentiable if and only if

$$u_x = v_y \quad \text{and} \quad u_y = -v_x.$$

It is this restriction that makes all the difference between the theory of differentiable functions from R^2 to R^2 and the theory of complex differentiable functions.

The operators $\partial/\partial z$, $\partial/\partial \bar{z}$. We express some arguments, e.g. §10.1, example 2, in terms of conjugates z, $\bar{z}$, but some writers use them more extensively than we do. It is possible to treat *formally* z, $\bar{z}$ as independent variables, though in fact they are connected. On this basis,

$$\frac{\partial f}{\partial z} = \frac{1}{2} \left(\frac{\partial f}{\partial x} - i \frac{\partial f}{\partial y} \right) \quad \text{and} \quad \frac{\partial f}{\partial \bar{z}} = \frac{1}{2} \left(\frac{\partial f}{\partial x} + i \frac{\partial f}{\partial y} \right).$$

The condition for regularity of f then takes the compact form

$$\frac{\partial f}{\partial \bar{z}} = 0.$$

This usage is particularly helpful in dealing with functions of more than one complex variable.

§10.4. *The Riemann sphere.* There is a geometric model in which all complex numbers, including ∞, are depicted on a sphere. Let points $z = x + iy$ be represented as usual in a plane Z. Through the origin $O(z = 0)$ draw a line perpendicular to Z carrying a third coordinate u. Take the sphere S

$$x^2 + y^2 + (u - \tfrac{1}{2})^2 = \tfrac{1}{4}.$$

Let N (the north pole) be the point $(0, 0, 1)$ of the sphere; the tangent plane there is parallel to Z.

Let Q be the point $z = x + iy$ of the plane Z. Let QN cut the sphere in P. Then P shall be the point of S representing the number z. The point N (having no finite counterpart in Z) represents ∞.

We verify that there is a metric defined on the sphere. We have $NP . NQ = 1$. Suppose that Q_1, Q_2 are points of Z representing z_1, z_2 and that P_1, P_2 are the points of S derived by projection from N.

Then the triangles NQ_1Q_2, NP_2P_1 are similar and therefore

$$\frac{P_1P_2}{Q_2Q_1} = \frac{NP_1}{NQ_2} = \frac{NP_1 . NQ_1}{NQ_1 . NQ_2}.$$

Hence

$$\frac{P_1P_2}{|z_1 - z_2|} = \frac{1}{\{(1 + |z_1|^2)(1 + |z_2|^2)\}^{\frac{1}{2}}}.$$

The chordal distance P_1P_2 satisfies the conditions for a metric; let us call it $\rho(z_1, z_2)$.

If $z_2 = \infty$, the corresponding formula is

$$NP_1 = \frac{1}{NQ_1} = \frac{1}{\{1 + |z_1|^2\}^{\frac{1}{2}}}.$$

The reader may verify that, if $\rho(z_m, z_n) \to 0$ as $m \to \infty$, $n \to \infty$, there is a unique z_0 (possibly ∞) such that $\lim \rho(z_0, z_n) = 0$. In fact, with the chordal metric, the extended plane $\bar{Z}$ is a complete metric space.

COMPLEX INTEGRALS. CAUCHY'S THEOREM

11.1. Complex integrals

The deeper-lying properties of complex functions were first brought to light by integrating the functions along curves in Z.

We have already to hand in §8.1 and §8.2 definitions and theorems for curves and integrals in R^n, and it was noted that they are applicable to Z. It will now be helpful to restate some results and to prove new ones in notation which is most appropriate to complex functions.

Given a path γ represented by

$$z = z(t) = x(t) + iy(t) \quad (a \leqslant t \leqslant b)$$

and a continuous function $f = u + iv : G \to Z$, where G is an open set containing the trace of γ, we define $\int_\gamma f$ to be

$$\int_a^b f\{z(t)\} z'(t) \, dt,$$

proving as in 8.2 that the value is independent of the particular representation of γ. The integral is the same as

$$\int_a^b \left(u \frac{dx}{dt} - v \frac{dy}{dt} \right) dt + i \int_a^b \left(v \frac{dx}{dt} + u \frac{dy}{dt} \right) dt,$$

and it could be defined as such.

The integrals with respect to t are Riemann integrals of piecewise continuous functions. In proofs which follow, we can treat the integrands as continuous and omit any mention of the finite number of discontinuities that dx/dt and dy/dt may have; the proof would be completed by adding the results for the intervals of continuity.

If $z(a) = z(b)$, so that the initial and end points coincide, we call the path a *circuit*.

If $z(t_1) \neq z(t_2)$ for different t_1, t_2 in $a \leqslant t \leqslant b$ (except for a, b if the path is a circuit), that is to say there are no double points, the path is called *simple*. (A circuit is not assumed to be simple unless this is explicitly stated.)

We shall prove in theorem 11.11 an inequality which is in constant use.

Lemma. *If f is a complex-valued function defined on $a \leqslant t \leqslant b$, then*

$$\left| \int_a^b f(t)\, dt \right| \leqslant \int_a^b |f(t)|\, dt.$$

Proof. The simplest method is to approximate to the integrals by finite sums and use the known inequality

$$\left| \sum_1^n u_r + i \sum_1^n v_r \right| \leqslant \sum_1^n \sqrt{(u_r^2 + v_r^2)}.$$

The following alternative is more stylish. If α is a constant,

$$\mathrm{re}\left\{ e^{-i\alpha} \int_a^b f(t)\, dt \right\} = \int_a^b \mathrm{re}\, \{ e^{-i\alpha} f(t) \}\, dt \leqslant \int_a^b |f(t)|\, dt.$$

If α is chosen to be the phase of $\int_a^b f(t)\, dt$, the first term in the line above is

$$\left| \int_a^b f(t)\, dt \right|.$$

(If that integral is 0, then α is not defined, but no proof is needed.) |

Theorem 11.11. *If $|f| \leqslant M$ and γ has length l, then*

$$\left| \int_\gamma f \right| \leqslant Ml.$$

Proof. By the lemma the left-hand side is at most

$$\int_a^b |f| \cdot \left| \frac{dz}{dt} \right| dt \leqslant M \int_a^b \sqrt{\left\{ \left(\frac{dx}{dt} \right)^2 + \left(\frac{dy}{dt} \right)^2 \right\}}\, dt = Ml$$

from theorem 8.13. |

From time to time we shall need theorems about complex integrals which extend results known for real integrals. The next theorem, on term by term integration, is an instance.

Theorem 11.12. *Let each function f_n be continuous in G. Let the sequence f_n converge uniformly to f on any compact subset of G. Then f is continuous in G and, for any path γ in G,*

$$\int_\gamma f_n \to \int_\gamma f.$$

Proof. The continuity of f in G follows from theorem 5.21. If $f_n = u_n + i v_n$ and $z = x(t) + i y(t)$ is a representation of γ, then

$$\int_\gamma f_n = \int_a^b \left(u_n \frac{dx}{dt} - v_n \frac{dy}{dt} \right) dt + i \int_a^b \left(v_n \frac{dx}{dt} + u_n \frac{dy}{dt} \right) dt.$$

Theorem 5.22 is then applicable to the integrals on the right-hand side (the interval $[a, b]$ being the union of pieces in which dx/dt and dy/dt are continuous). |

11.2. Dependence of the integral on the path

Suppose that f is continuous in a region D of the z-plane. If we take any two points z_0, z in D, it is important to know in what circumstances $\int_\gamma f$ has the same value for all paths in D joining z_0 to z. This is the same as asking whether $\int f$ round every circuit in D is zero, because, if γ_1 and γ_2 have the same initial and end points, $\gamma_1 - \gamma_2$ is a circuit and $\int_{\gamma_1 - \gamma_2} = \int_{\gamma_1} - \int_{\gamma_2}$.

Theorem 11.2. *The value of* $\int_\gamma f$ *is the same for all paths* γ *in D joining any two given points* z_0 *and* z *if and only if there is a function F regular in D such that* $F'(z) = f(z)$ *for all z in D.*

Proof. Sufficiency. If γ is the path $z = z(t)$ for $a \leqslant t \leqslant b$, then

$$\int_\gamma f = \int_a^b F'\{z(t)\} \frac{dz}{dt}\, dt$$

$$= \int_a^b \frac{d}{dt} F\{z(t)\}\, dt$$

$$= F\{z(b)\} - F\{z(a)\}$$

$$= F(z) - F(z_0).$$

Necessity. Fixing z_0, we note that $\int_{z_0}^z f(\zeta)\, d\zeta$ along any path in D defines uniquely a function $F(z)$.

If $|h|$ is small enough, the straight segment $(z, z+h)$ or λ say is in D and

$$F(z+h) - F(z) = \int_\lambda f.$$

Divide by h and let h tend to 0. Since f is continuous, we have $F'(z) = f(z)$. |

Theorem 11.2 enables us to calculate easy complex integrals by the technique of the indefinite integral familiar for real functions.

A convenient notation is $\int_\gamma f = [F]_\gamma.$

Example 1. If $n \geqslant 1$, then, along *any* path joining z_0 to z_1,

$$\int_{z_0}^{z_1} z^{n-1} dz = (z_1^n - z_0^n)/n.$$

Proof. z^n/n is regular for all z and has derivative z^{n-1}. $\mid$

Example 2. Let γ be the circle with centre a and radius r defined by

$$z = a + re^{i\theta} \quad (0 \leqslant \theta \leqslant 2\pi).$$

Then, if n is a positive or negative integer,

$$\int_{\gamma} (z-a)^{n-1} dz = 0 \quad (n \neq 0),$$

$$\int_{\gamma} \frac{dz}{z-a} = 2\pi i.$$

Proof. The integral is $r^n \int_0^{2\pi} e^{ni\theta} \, i \, d\theta.$ $\mid$

Example 2 is important. If $n \geqslant 1$, there is an indefinite integral $(z-a)^n/n$ regular for all z, as in example 1. If $n \leqslant 1$, then $(z-a)^n/n$ is regular, not in the whole z-plane, but in the punctured plane $0 < |z-a|$, and this suffices to give the result.

If $n = 0$, and D is any region containing γ (say the annulus $r - \delta < |z| < r + \delta$), then there can be no function regular in D having derivative $1/(z-a)$. If there were such a function, the integral round γ would be 0, whereas it is $2\pi i$.

Example 3. (*Integration by parts.*) If, in D, the functions Φ, Ψ are regular and Φ', Ψ' are continuous, then, for any γ in D,

$$\int_{\gamma} (\Phi\Psi' + \Phi'\Psi) = [\Phi\Psi]_{\gamma}.$$

Proof. Clear.

11.3. Primitives and local primitives

The noun *primitive* is often used in the same sense as *indefinite integral* in the calculus. It will be appropriate in this chapter.

We always suppose that the function f is continuous in a region D.

Definition. *A function f has a primitive in D if there is a function F, regular in D, such that $F'(z) = f(z)$ for all z in D.*

Definition. *A function f has a local primitive in D if, given any point z_0 in D, there is a disc with centre z_0 in which f has a primitive.*

It will be observed that the step from the first definition to the second is like that in §10.2 from regularity of a function in a region to regularity at a point.

We shall need the following modification of the necessity part of theorem 11.2, in which restrictions will be placed on the region and on the paths of integration. For the detailed definition of the boundary of a rectangle, the reader may refer to theorem 8.31, corollary 3.

Theorem 11.3. *Let D be an open disc. Let* $\int_{\gamma} f = 0$ *whenever* γ *is the boundary of a rectangle, with sides parallel to the real and imaginary axes, contained in D. Then f has a primitive in D.*

Proof. Let z_0 be the centre and z any other point of D. The rectangle, with sides parallel to the axes, whose opposite corners are at z_0, z, lies in D. There are two paths from z_0 to z, say γ_1, γ_2, each along two sides of the rectangle. Then $\int_{\gamma_1} f - \int_{\gamma_2} f = \int_{\gamma_1 - \gamma_2} f = 0$, from the hypothesis. Define $F(z)$ to be the common value of $\int_{\gamma_1} f$ and $\int_{\gamma_2} f$.

If h is real and small enough for the horizontal segment $(z, z+h)$ to lie in D, then

$$F(z+h) - F(z) = \int_0^h f(z+t) \, dt.$$

Divide by h and let $h \to 0$. Then, since f is continuous,

$$\frac{\partial F}{\partial x} = f(z).$$

Similarly

$$\frac{\partial F}{\partial y} = if(z).$$

If $F(z) = U(x, y) + iV(x, y)$, these equations show that U_x, U_y, V_x, V_y are continuous and satisfy the Cauchy–Riemann equations. By theorem 7.14, U and V are differentiable. By theorem 10.1, F is differentiable and $F'(z) = f(z)$. |

Exercises 11(a)

1. Calculate the integrals round the circle $z = Re^{i\theta}$ ($0 \leqslant \theta \leqslant 2\pi$) of $x, y,$ $|z|, xy$.

2. Calculate the integrals along the line segment from 1 to $2+i$ of $x, \cos z, e^z$.

3. Evaluate the integrals round the unit circle (anticlockwise) of

$$\bar{z}, \quad |z-1|, \quad 2^z \text{ (principal value)},$$

$z^{-\frac{1}{2}}$ (the branch continuous for $0 \leqslant \theta < 2\pi$ and equal to 1 for $\theta = 0$).

4. Which of the following functions possess primitives in the disc $B(1; 1)$?

(i) $\log z$, (ii) $\sin (1-z)$, (iii) $z^{\frac{1}{2}}$ (principal value).

Give illustrations of functions which have local primitives in the punctured disc $0 < |z-1| < 1$ but not primitives therein.

5. Let Z_1 denote the plane Z from which the segment $-1 \leqslant x \leqslant 1, y = 0$ is deleted. Examine whether

$$\frac{1}{z^2-1}, \quad \log\left(1-\frac{1}{z}\right)$$

have primitives in Z_1.

6. Find primitives of the following functions, specifying the regions in which you claim the property:

$$\exp(z-1), \quad \frac{1}{z^2+4}, \quad \tan z.$$

7. Let f be regular in a region D and γ a circuit in D. Prove that $\int_\gamma \bar{f}f'$ is a pure imaginary. (It may be assumed that f' is continuous.)

8. Let γ be the semicircular path $z = Re^{i\theta}$, where $0 \leqslant \theta \leqslant \pi$. Prove that, as $R \to \infty$,

$$\int_\gamma \frac{dz}{z^2-10} = O\left(\frac{1}{R}\right).$$

Make an O estimate for
$$\int_\gamma \frac{e^{iz}}{z^2} dz.$$

9. (This is a useful fact e.g. for §14.1, example 2.) Let f be continuous in a disc punctured at $z = a$. Suppose that, as $\delta \to 0$, $\delta e^{i\theta} f(a + \delta e^{i\theta}) \to k$ uniformly for $\alpha \leqslant \theta \leqslant \beta$. If γ is the arc $z = a + \delta e^{i\theta}$, $\alpha \leqslant \theta \leqslant \beta$, prove that

$$\lim_{\delta \to 0} \int_\gamma f = ik(\beta - \alpha).$$

10. State and prove the analogue of exercise 9, when $Re^{i\theta} f(Re^{i\theta}) \to k$ as $R \to \infty$, uniformly for $\alpha \leqslant \theta \leqslant \beta$ and γ is the arc $z = Re^{i\theta}$, $\alpha \leqslant \theta \leqslant \beta$.

11. Taking
$$(P) \int_{a-i\infty}^{a+i\infty} f(z)\,dz \quad (a \text{ real})$$

to mean the principal value
$$\lim_{Y \to \infty} \int_{-Y}^{Y} f(a+iy)\,i\,dy,$$

evaluate
$$(P) \int_{1-i\infty}^{1+i\infty} \frac{dz}{z}.$$

12. By comparing the integrals

$$\int_{-\pi}^{\pi} \theta |f(e^{i\theta})|^2 d\theta, \quad \int_{-\pi}^{\pi} |f(e^{i\theta})|^2 d\theta,$$

where
$$f(z) = \sum_{n=1}^{N} (-1)^n a_n z^n,$$

or otherwise, prove that

$$\left| \sum_{m,\,n=1}^{N} {}' \frac{a_m \bar{a}_n}{m-n} \right| \leqslant \pi \sum_{n=1}^{N} |a_n|^2,$$

where the a's are complex, and Σ' means that terms with $m = n$ are omitted in the summation.

13. Prove that, if Γ is the image of the path γ by the mapping $w = f(z)$, then

$$\int_{\Gamma} g(w)\,dw = \int_{\gamma} g\{f(z)\} f'(z)\,dz.$$

11.4. Cauchy's theorem for a rectangle

We now approach the grand theorem of complex function theory. It is normally stated as the vanishing of the integral of a function f, regular in an appropriate region, round a circuit.

We prove this first in a special case.

Theorem 11.41. *Let $\bar{R}$ be a closed rectangle with sides parallel to the axes and ∂R the circuit of its boundary. If f is regular in a region containing $\bar{R}$, then*

$$\int_{\partial R} f = 0.$$

The circuit of the boundary of a rectangle with sides parallel to the axes is defined in theorem 8.31, corollary 3.

Proof. Write $I(R) = \left| \int_{\partial R} f \right|$. Divide R by lines through its centre into four rectangles of a quarter its size. The integral round ∂R is the sum of the integrals round the similarly oriented circuits of the four small rectangles, because the integrals along interior sides cancel in pairs.

We can therefore select one of the smaller rectangles, R_1 say, for which
$$I(R_1) \geqslant \tfrac{1}{4} I(R).$$

Now quadrisect R_1, finding an R_2 for which
$$I(R_2) \geqslant \tfrac{1}{4} I(R_1).$$

Repeating the process of quadrisection we arrive at a rectangle R_n for which
$$I(R_n) \geqslant 4^{-n} I(R).$$

There is, by exercise 3(d) 7, one point z_0 common to all the closed $\bar{R}_n$. Since f is regular at z_0, given ϵ, there is δ such that

$$f(z) = f(z_0) + f'(z_0)(z - z_0) + \epsilon(z)(z - z_0), \tag{11.41}$$

where $|\epsilon(z)| < \epsilon$ provided that $|z - z_0| < \delta$.

If n is large enough, R_n lies inside the disc $|z - z_0| < \delta$.

The first two terms on the right of (11.41) form a linear function of z and, from §11.2, example 1, their integral round ∂R_n is 0.

On ∂R_n the modulus of the last term of (11.41) is less than $\frac{1}{2}\epsilon s_n$, where s_n is the length of ∂R_n. Hence $I(R_n) \leqslant \frac{1}{2}\epsilon s_n^2$. But $s_n = s/2^n$, where s is the length of ∂R. We deduce

$$I(R) \leqslant 4^n I(R_n) \leqslant 4^n \tfrac{1}{2}\epsilon(s/2^n)^2 = \tfrac{1}{2}\epsilon s^2.$$

Since ϵ is arbitrary, $I(R)$ and hence $\int_{\partial R} f$ is 0. |

Note. If we assume the continuity of f' instead of only its existence, Cauchy's theorem is a straightforward consequence of Green's theorem for a rectangle (8.31, corollary 3),

$$\int (u\,dx + v\,dy) = \iint \left(\frac{\partial v}{\partial x} - \frac{\partial u}{\partial y} \right) dx\,dy,$$

where the double integral is taken over a rectangle and the line integral round its boundary. If $f = u + iv$ is regular, then

$$\int f\,dz = \int (u\,dx - v\,dy) + i \int (v\,dx + u\,dy).$$

The double integrals equivalent to these two line integrals are zero by the Cauchy–Riemann equations.

In his original proof, published in 1825, Cauchy did in fact assume the continuity of f'. It was Goursat (1858–1936) who in 1900 first dispensed with this hypothesis. We shall see in §11.8 that it is vital to have a proof which does not rest on the continuity of f'.

It will be useful later to know that the requirement in theorem 11.41 of the regularity of f may be relaxed in specified sets of points.

Theorem 11.42. *Let f be regular in an open rectangle R and continuous on its closure $\bar{R}$. Then*

$$\int_{\partial R} f = 0.$$

Proof. We may suppose without loss of generality that the origin is the centre of R. Let $z = z(t)$ for $0 \leqslant t \leqslant 1$ be a representation of the circuit ∂R. Let $0 < \alpha < 1$. Then the point $z = \alpha z(t)$ traces out the boundary ∂S of a rectangle S similar to R with linear dimensions reduced by the factor α.

By theorem 11.41,

$$\int_{\partial S} f = 0.$$

Now

$$\int_{\partial R} f - \int_{\partial S} f = \int_0^1 f\{z(t)\} z'(t) \, dt - \int_0^1 f\{\alpha z(t)\} \alpha z'(t) \, dt$$

$$= \int_0^1 [f\{z(t)\} - f\{\alpha z(t)\}] z'(t) \, dt + (1 - \alpha) \int_0^1 f\{\alpha z(t)\} z'(t) \, dt$$

$$= p(\alpha) + q(\alpha), \text{ say.}$$

Since f is uniformly continuous in the compact set $\bar{R}$, given ϵ, we can find α_0 such that

$$|f\{z(t)\} - f\{\alpha z(t)\}| < \epsilon \quad \text{if} \quad \alpha > \alpha_0.$$

Theorem 11.11 then gives $|p(\alpha)| \leqslant \epsilon l$, where l is the length of ∂R. Also

$$|q(\alpha)| \leqslant (1 - \alpha) M l,$$

where $M = \sup |f(z)|$ for z in $\bar{R}$.

Hence, as $\alpha \to 1$ (from below),

$$\int_{\partial R} f - \int_{\partial S} f \to 0.$$

Therefore $\int_{\partial R} f = 0$. $\blacksquare$

We prove a corollary, which is a weak result compared with theorem 11.42 but nevertheless provides a useful extension of theorem 11.41.

Corollary. *If, in theorem* 11.41, *at a finite number of points of R, the function f is assumed continuous instead of regular, then*

$$\int_{\partial R} f = 0.$$

Proof. Suppose that there is one exceptional point, say ζ inside R. Through ζ draw parallels to the sides of R dividing R into four rectangles. By theorem 11.42 the integral round the boundary of each of the four is 0. Adding we have $\int_{\partial R} f = 0$. The corollary follows from a finite number of repetitions of the argument. |

11.5. Cauchy's theorem for circuits in a disc

Theorem 11.5. *Let f be regular in an open disc D, except possibly at a finite number of points at which it is continuous. Then, if γ is any circuit in D,*

$$\int_{\gamma} f = 0.$$

Proof. Theorem 11.42 (corollary) shows that, for any closed rectangle $\overline{R}$ in D,

$$\int_{\partial R} f = 0,$$

whether $\overline{R}$ contains exceptional points or not.

By theorem 11.3, f has a primitive in D.

By theorem 11.2, if γ is any circuit in D,

$$\int_{\gamma} f = 0. \,|$$

The reader will realise that the circular shape of the region is inessential. It is possible to extend theorem 11.5, for instance, to a convex region (which contains the linear segment joining any two of its points). But such an extension has no air of finality.

A typical example of an integral round a circuit which is *not* zero is (from §11.2, example 2)

$$\int_{\gamma} \frac{dz}{z} = 2\pi i,$$

where γ is a circle described counterclockwise round the origin. The integrand is not regular at $z = 0$ and is, in fact, unbounded as $z \to 0$.

A survey of §11.1–11.5 may well lead the reader to conjecture correctly that, if f is regular in a region D, then $\int_{\gamma} f = 0$ provided that the circuit γ can be 'continuously shrunk to a point' in D.

A necessary preamble to such a theorem is a short account of the topological notion of *homotopy* or the continuous deformation of curves.

11.6. Homotopy. The general Cauchy theorem

Suppose that γ_0 and γ_1 are two curves in a region D with the same initial point and the same end point. We choose a representation of each curve as a mapping $\gamma_0(t), \gamma_1(t)$ of the interval $0 \leqslant t \leqslant 1$ into D. (It will be plain that the particular representations chosen do not matter in what follows.) We have then

$$\gamma_0(0) = \gamma_1(0) \quad \text{and} \quad \gamma_0(1) = \gamma_1(1).$$

We define γ_0 and γ_1 to be *homotopic in D with fixed end-points* if there is a continuous mapping $z = \delta(t, u)$ of the square $0 \leqslant t \leqslant 1$, $0 \leqslant u \leqslant 1$ into D such that $\delta(t, u)$ satisfies

$$\delta(t, 0) = \gamma_0(t), \quad \delta(t, 1) = \gamma_1(t) \quad (0 \leqslant t \leqslant 1),$$

and
$$\left. \begin{aligned} \delta(0, u) &= \gamma_0(0) = \gamma_1(0) \quad \text{for all } u, \\ \delta(1, u) &= \gamma_0(1) = \gamma_1(1) \quad \text{for all } u. \end{aligned} \right\} \tag{11.61}$$

Thus, as u increases from 0 to 1, γ_0 is continuously deformed into γ_1.

For integration we require γ_0 and γ_1 to be paths, but the mappings $z = \delta(t, u)$ for $0 < u < 1$ can be general curves.

Suppose now that, instead of having fixed end points, each of γ_0 and γ_1 is a closed curve. If there is a continuous mapping $z = \delta(t, u)$ satisfying for $0 \leqslant u \leqslant 1$, instead of (11.61),

$$\delta(0, u) = \delta(1, u) \tag{11.62}$$

we call the paths γ_0 and γ_1, *homotopic as circuits*.

If, in particular, $\delta(t, 0)$ is constant for $0 \leqslant t \leqslant 1$, so that γ_0 is a single point, we say that γ_1 is *homotopic to a point* in D.

Theorem 11.61. *Let f be regular in a region D, except at a finite number of points at which it is continuous. Let γ_0, γ be paths in D which are homotopic,*
>   *either* (i) *with fixed end points*
>   *or* (ii) *as circuits.*

Then
$$\int_{\gamma_0} f = \int_{\gamma} f.$$

Proof. Let δ be a mapping of the square $R: 0 \leqslant t \leqslant 1, 0 \leqslant u \leqslant 1$, into D as prescribed in the definition of homotopy.

We can take dissect ons of the t, u intervals

$$0 = t_0 < t_1 < \ldots < t_i < t_{i+1} < \ldots < t_m = 1,$$

$$0 = u_0 < u_1 < \ldots < u_j < u_{j+1} < \ldots < u_n = 1,$$

so fine that every rectangle

$$R_{ij} \, (t_i \leqslant t \leqslant t_{i+1}, \, u_j \leqslant u \leqslant u_{j+1})$$

is mapped by $z = \delta(t, u)$ into an open disc B_{ij} contained in D.

If $0 < j < n$, we replace the curve $z = \delta(t, u_j)$ by the polygonal path with vertices at $z_{ij} = \delta(t_i, u_j)$ for $i = 0, 1, \ldots, m$. Call this polygonal path γ_j $(0 < j < n)$. The paths γ_0 and $\gamma_n = \gamma$ are given by $z = \delta(t, u_0)$ and $z = \delta(t, u_n)$.

Write γ_{ij} for the arc of γ_j joining z_{ij} to $z_{i+1,j}$ and η_{ij} for the linear segment joining z_{ij} to $z_{i,j+1}$.

Then the segments

$$\gamma_{ij}, \, \eta_{i+1,j}, \, -\gamma_{i,j+1}, \, -\eta_{ij}$$

form a circuit lying in the disc B_{ij}.

Integrating the function f round this circuit, we have from theorem 11.5

$$\int_{\gamma_{ij}} f - \int_{\gamma_{i,j+1}} f + \int_{\eta_{i+1,j}} f - \int_{\eta_{ij}} f = 0.$$

Summing this for $0 \leqslant i \leqslant m-1$, we obtain

$$\int_{\gamma_j} f - \int_{\gamma_{j+1}} f + \int_{\eta_{m,j}} f - \int_{\eta_{0,j}} f = 0.$$

Now sum for $0 \leqslant j \leqslant n-1$. Under hypothesis (i) $\eta_{0,j}$ and $\eta_{m,j}$ both reduce to points. Under (ii) the paths $\eta_{0,j}$ and $\eta_{m,j}$ are the same. In either case we have

$$\int_{\gamma_0} f - \int_{\gamma} f = 0. \mid$$

Simply connected regions. Intuitively a simply connected region is one without holes in it. This idea can be formalized in terms of homotopy.

Definition. *A region D is* simply connected if *every closed curve in D is homotopic to a point in D.*

An alternative definition is that any two curves in D with the same fixed initial and end points are homotopic. We assume the equivalence of the two definitions.

Illustrations. The disc $|z| < 1$ is simply connected: the annulus $1 < |z| < 2$ is not—the circuit $z = re^{i\theta}(0 \leqslant \theta \leqslant 2\pi)$, where $1 < r < 2$, cannot be shrunk to a point. Nor is the *punctured disc* $0 < |z| < 1$ simply connected.

If γ_0 in theorem 11.61 is a point, we have the most useful of all versions of the theorem of Cauchy. It is the one to which we shall attach his name.

Cauchy's theorem 11.62. *Let f be regular in a region D, except at a finite number of points at which it is continuous. Then, for any circuit γ homotopic to a point in D,*

$$\int_\gamma f = 0.$$

If D is simply connected, the integral is 0 for any circuit in D.

11.7. The index of a circuit for a point

Let γ be a circuit and a a point not on γ. We seek an analytical definition of the intuitive idea of the number of times that γ winds round a. The reader is reminded that a circuit may have double points: with a complicated circuit care is needed in counting the turns.

Definition. *The* index *of γ for a is defined by the equation*

$$n(\gamma, a) = \frac{1}{2\pi i} \int_\gamma \frac{dz}{z-a}.$$

Theorem 11.7. *The index is an integer.*

Proof. The following proof has the advantage of not resting on the geometrical notion of angle.

Let γ be $z = z(t)$ for $\alpha \leqslant t \leqslant \beta$. Then $z(\beta) = z(\alpha)$. Consider the function

$$h(t) = \int_\alpha^t \frac{z'(u)}{z(u)-a} \, du.$$

It is continuous for $\alpha \leqslant t \leqslant \beta$. At every point of continuity of $z'(t)$, the function $h(t)$ has the derivative

$$h'(t) = \frac{z'(t)}{z(t)-a}.$$

Hence the derivative of $e^{-h(t)}\{z(t)-a\}$ is zero except possibly for a finite number of values of t. Therefore $e^{-h(t)}\{z(t)-a\}$, being continuous, is constant. Equating its values for α and β, we have $e^{h(\beta)} = 1$ and so $h(\beta)$ is an integral multiple of $2\pi i$. |

Properties of the index $n(\gamma, a)$.

(1) *For fixed a, $n(\gamma, a)$ is unaltered if the circuit γ is continuously deformed without passing through a.*

This is a special case of theorem 11.61.

(2) *For fixed γ, $n(\gamma, a)$ is constant for values of a in a disc containing no point of γ.*

To prove this, let γ_h be the path γ translated through the vector $-h$. The position of $a+h$ relative to γ is the same as that of a relative to γ_h. Therefore

$$n(\gamma, a+h) = n(\gamma_h, a).$$

If $|h|$ is small enough, by (1) the right-hand side is equal to $n(\gamma, a)$.

(3) *Suppose that D is simply connected, a is not in D and γ is a circuit in D. Then $n(\gamma, a) = 0$.*

Theorem 11.62 shows that the integral defining $n(\gamma, a)$ is 0.

(4) *If γ is the circle $z = e^{i\theta}$ ($0 \leqslant \theta \leqslant 2\pi$), then*

$$n(\gamma, a) = 1 \quad \text{for} \quad |a| < 1,$$
$$n(\gamma, a) = 0 \quad \text{for} \quad |a| > 1.$$

If $|a| < 1$, we can by (2) take $a = 0$. We showed in §11.2, example 2 that

$$\int_\gamma \frac{dz}{z} = 2\pi i.$$

If $|a| > 1$, we appeal to (3). |

From (4) the circular path γ defined by $z = z_0 + re^{it}$ ($0 \leqslant t \leqslant 2\pi$) determines two regions. The first D_o, containing all points a for which $|a-z_0| > r$, is unbounded. The index $n(\gamma, a) = 0$ for every point of D_o. In the other region D_i, consisting of points a such that $|a-z_0| < r$, we have $n(\gamma, a) = 1$ and $n(-\gamma, a) = -1$. We then say that γ is *positively oriented* and $-\gamma$ is *negatively oriented*.

If γ is a simple circuit of a different shape, say a rectangle, a direct proof could be constructed that again γ determines two regions in one of which $n(\gamma, a) = 0$ and in the other $n(\gamma, a) = \pm 1$ (according to the orientation). In applications of complex function theory a circuit is always an easily defined curve, e.g. one made up of straight segments and circular arcs. We shall speak of its *outside* (points a for which $n(\gamma, a) = 0$) and its *inside* (points a for which $n(\gamma, a) = 1$ if γ is positively oriented).

There is a general theorem that *any* simple closed curve divides the plane into two regions, the outside of the curve and the inside. This topological result, Jordan's curve theorem, is harder to prove. (There is a proof in Newman's *Topology of Plane Sets of Points*.)

Exercises 11(b)

1. Adapt the proof of theorem 11.41 from a rectangle to a triangle, and hence to a simple polygon.

2. Prove that theorem 11.41 holds if R contains a finite number of points ζ at which f, though not regular, satisfies the condition $\lim_{z \to \zeta} (z - \zeta)f(z) = 0$. (Suppose that there is one point ζ. Divide R into nine rectangles, the central one R_0 containing ζ; make R_0 small).

Observe that, in the illustration $\int_\gamma \frac{dz}{z} = 2\pi i$ in §11.5, the integrand $1/z$ is just too large as $z \to 0$ to satisfy the requirement in the preceding paragraph.

3. Comment on the following suggested proof of theorem 11.41.

Taking ∂R as the circuit of integration, and $0 < \lambda \leqslant 1$, write

$$I(\lambda) = \int f(\lambda z)\,dz.$$

Then

$$I'(\lambda) = \int z f'(\lambda z)\,dz = \left[\frac{z f(\lambda z)}{\lambda}\right] - \int \frac{f(\lambda z)}{\lambda}\,dz.$$

Therefore

$$I'(\lambda) = -\frac{I}{\lambda} \quad \text{and so} \quad I(\lambda) = \frac{A}{\lambda}.$$

Letting $\lambda \to 0$, we have $A = 0$. Now put $\lambda = 1$.

4. Compute directly the index of the circuit formed by $x = \pm a, y = \pm a$ for the origin.

11.8. Cauchy's integral formula

The following is a far-reaching consequence of Cauchy's theorem 11.62.

Theorem 11.81. *Let f be regular in a region D. Let γ be a circuit in D which is homotopic to a point. Let a be any point of D not lying on γ. Then*

$$n(\gamma, a)f(a) = \frac{1}{2\pi i} \int_\gamma \frac{f(z)}{z - a}\,dz.$$

Proof. In D define

$$g(z) = \frac{f(z) - f(a)}{z - a} \quad (z \neq a),$$

$$g(a) = f'(a).$$

Then g is regular except at $z = a$, and g is continuous at a. Therefore (theorem 11.62)

$$0 = \int_\gamma g(z)\,dz = \int_\gamma \frac{f(z)}{z - a}\,dz - \int_\gamma \frac{f(a)}{z - a}\,dz.$$

In the last term use the definition (§11.7) of the index $n(\gamma, a)$. |

Corollary 1. In most applications γ is a simple circuit winding once round a (e.g. the circle $z - a = re^{i\theta}$ for $0 \leqslant \theta \leqslant 2\pi$). If the circuit is positively oriented, the formula is

$$f(a) = \frac{1}{2\pi i} \int_\gamma \frac{f(z)}{z - a} \, dz.$$

Corollary 2. Suppose that f is regular in the annulus $R < |z| < S$. Let B, C be circles with centre $z = 0$ and radii r, s, described counter-clockwise, where

$$R < r < |a| < s < S.$$

Then
$$f(a) = \frac{1}{2\pi i} \int_C \frac{f(z)}{z - a} \, dz - \frac{1}{2\pi i} \int_B \frac{f(z)}{z - a} \, dz.$$

First proof. Define $g(z)$ as in theorem 11.81. The circle B can be continuously deformed into C via the paths $z = \rho e^{it}$ ($0 \leqslant t \leqslant 2\pi$), where $r \leqslant \rho \leqslant s$. By theorem 11.62

$$\int_C g \, dz = \int_B g \, dz$$

and the proof is completed as in theorem 11.81.

Second proof. Draw a common diameter of the circles B, C, not passing through a.

Apply theorem 11.81 in turn to the two simple circuits, one indicated by continuous arrows and the other by dotted arrows. Add. The two integrals along each straight segment cancel. |

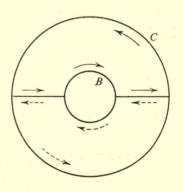

Cauchy's integral formula shows that, if the value of a regular function is known at every point of a simple circuit γ homotopic to a point, then its value is determined at every point inside γ.

In succeeding theorems, in which a will be variable, our usual notation will be to replace a by z, and to denote the variable of integration by ζ.

We observe that the only property of f required for the integral in theorem 11.81 to have a meaning is continuity on γ. This remark suggests the next theorem.

Theorem 11.82. *Let $\phi(\zeta)$ be continuous in an open set containing a path γ in the ζ-plane. If z is not on γ, define*

$$f(z) = \int_\gamma \frac{\phi(\zeta)}{\zeta - z} d\zeta.$$

Then

$$f'(z) = \int_\gamma \frac{\phi(\zeta)}{(\zeta - z)^2} d\zeta.$$

Moreover $f'(z)$ is continuous.

Observe that, in contrast with theorem 11.81, the path γ is not in general closed.

Proof. We prove from the definition that $f'(z)$ is obtained by differentiation under the integral sign.

Suppose that $|\phi(\zeta)| \leqslant M$ on γ, that γ has length l and that d is small enough for the disc with centre z and radius d to contain no point of γ. From the definition of f,

$$\frac{f(z+h) - f(z)}{h} = \int_\gamma \frac{\phi(\zeta)\,d\zeta}{(\zeta - z)(\zeta - z - h)}.$$

Therefore

$$\frac{f(z+h) - f(z)}{h} - \int_\gamma \frac{\phi(\zeta)\,d\zeta}{(\zeta - z)^2} = h \int_\gamma \frac{\phi(\zeta)\,d\zeta}{(\zeta - z)^2(\zeta - z - h)}.$$

Take $|h| < \frac{1}{2}d$ and use theorem 11.11. The modulus of the right-hand side is less than

$$\frac{|h|\,Ml}{\frac{1}{2}d^3}.$$

Let $h \to 0$ and we have the integral formula for $f'(z)$.

To prove now the continuity of f' for any z not on γ we again estimate a bound of the relevant integral. If $|h| < \frac{1}{2}d$,

$$f'(z+h) - f'(z) = \int_\gamma \left\{ \frac{1}{(\zeta - z - h)^2} - \frac{1}{(\zeta - z)^2} \right\} \phi(\zeta)\,d\zeta$$

$$= \int_\gamma \left\{ \frac{1}{\zeta - z - h} + \frac{1}{\zeta - z} \right\} \frac{h}{(\zeta - z - h)(\zeta - z)} \phi(\zeta)\,d\zeta$$

and the modulus of this is less than

$$\left(\frac{3}{d}\right) \frac{2Ml\,|h|}{d^2},$$

which tends to 0 as h tends to 0. ▐

11.9. Successive derivatives of a regular function

Combining theorems 11.81 and 11.82 we see that, if f is regular in D, then f' is continuous. Much more than this is true:

A regular function has derivatives of every order and they are all regular functions.

We must prove this fact (which will be formally stated as theorem 11.91). The reader may divine correctly that the argument of theorem 11.82 by which f' was proved continuous could be refined to prove that

$$n(\gamma, z)f''(z) = \frac{2!}{2\pi i}\int_\gamma \frac{f(\zeta)}{(\zeta-z)^3}\,d\zeta.$$

A direct extension, needing induction, to $f^{(r)}(z)$ on these lines is laborious. The following is an alternative proof, depending on integration by parts.

Theorem 11.91. *Let f be regular in a region D. Then it has a derivative of every order and every derivative is regular in D. Also, if γ is any circuit in D homotopic to a point, and z is in D but not on γ, then*

$$n(\gamma, z)f^{(r)}(z) = \frac{r!}{2\pi i}\int_\gamma \frac{f(\zeta)\,d\zeta}{(\zeta-z)^{r+1}}.$$

Proof. There are four stages (i)–(iv) in the proof.

(i) We shall prove f' continuous in D. Given z_0 in D, let γ_0 be the path

$$\zeta = z_0 + re^{it} \quad (0 \leqslant t \leqslant 2\pi),$$

where $B(z_0; 2r) \subset D$. Then, if $|z-z_0| < r$, by theorem 11.82,

$$f'(z) = \frac{1}{2\pi i}\int_{\gamma_0} \frac{f(\zeta)}{(\zeta-z)^2}\,d\zeta.$$

Also by theorem 11.82, f' is continuous inside γ_0 and so at z_0.

(ii) Next we prove the stronger result that f' is regular in D.

Since f' is continuous we can integrate the last equation by parts (§11.2, example 3) and obtain

$$f'(z) = \frac{1}{2\pi i}\int_{\gamma_0} \frac{f'(\zeta)}{\zeta-z}\,d\zeta.$$

By theorem 11.82,

$$f''(z) = \frac{1}{2\pi i}\int_{\gamma_0} \frac{f'(\zeta)}{(\zeta-z)^2}\,d\zeta$$

if z is inside γ_0, and so for $z = z_0$.

(iii) From (ii), the derivative of any function regular in D is regular in D. Therefore the derivative of f of every order exists and is regular in D.

(iv) Let now γ be any circuit in D homotopic to a point, and z any point of D not lying on γ. Then, by theorem 11.81, for any integer r,

$$n(\gamma, z)f^{(r)}(z) = \frac{1}{2\pi i}\int_\gamma \frac{f^{(r)}(\zeta)}{\zeta - z}\,d\zeta.$$

Integrating the right-hand side by parts r times, we have completed the proof. |

We emphasize again that in most applications γ is a positively oriented simple circuit enclosing z, so that $n(\gamma, z) = 1$.

Morera (1856–1909) proved in 1886 a theorem converse to Cauchy's, namely that if the integral of a continuous function round certain circuits vanishes, then the function is regular.

Theorem 11.92. (Morera.) *Suppose that f is continuous in a region D. Let z_0 be an arbitrary point of D. Suppose that there exists ρ such that*

$$\int_\gamma f = 0 \text{ round the boundary } \gamma \text{ of any rectangle, with sides parallel to}$$

the axes, contained in $B(z_0; \rho)$. Then f is regular in D.

Proof. By theorem 11.3, f has a primitive, say F, in $B(z_0; \rho)$ which is regular in that disc. By theorem 11.91, the derivative f of F is regular in $B(z_0; \rho)$, in particular at z_0, which is any point of D. |

In conclusion, the reader should satisfy himself that he can relate each of the implications in the following summary to the theorem in which it has been proved.

Write: R for the statement *f is regular in D*; L for the statement *f has a local primitive in D*; P for the statement *f has a primitive in D*;

I for the statement $\int_\gamma f = 0$ *for every circuit γ in D*.

Then, with no restriction on the region D,

$$I \Leftrightarrow P \Rightarrow L \Leftrightarrow R.$$

If D is simply connected, then $P \Leftrightarrow L$.

Proofs without Cauchy's theorem? Suppose that f is regular in D, that is to say it has a derivative f'. Using Cauchy's integral formula we proved as a first step that f' is continuous and then established the existence of derivatives of all orders. Naturally the reader will ask whether so apparently innocent a result as the continuity of f' can be proved more directly. It seems to have nothing to do with integration; can it not be proved without recourse to Cauchy's theorem? It can, and indeed the existence of all the derivatives can be proved in a few pages by a more searching analysis of the mapping by $w = f(z)$ of a region in the z-plane onto a region in the w-plane. But this was achieved only in 1960; references are given in the notes on Chapter 11. In 1958 a leading topological analyst, G. T. Whyburn, could still write that a direct proof of the continuity of f' 'remains inaccessible at present by these methods'.

Exercises 11 (c)

1. Give the values of the integrals

$$\int_\gamma \frac{z^4}{z-a}\, dz, \quad \int_\gamma \frac{e^z}{(z-a)^4}\, dz,$$

where γ is the circuit $\quad z = a + re^{i\theta} \quad (0 \leqslant \theta \leqslant 2\pi).$

2. What are the possible values of

$$\int_0^z \frac{d\zeta}{1+\zeta^2}$$

for paths joining 0 to z (not passing through $\pm i$)?

3. What are the possible values of

$$\int_\gamma \frac{dz}{z(1-z^2)},$$

where γ is a simple circuit?

4. Prove that, if γ is a circuit having index 1 for the origin,

$$\left(\frac{x^n}{n!}\right)^2 = \frac{1}{2\pi i} \int_\gamma \frac{x^n \exp zx}{n!\, z^{n+1}}\, dz.$$

Deduce that

$$\sum_{n=0}^\infty \left(\frac{x^n}{n!}\right)^2 = \frac{1}{2\pi} \int_0^{2\pi} \exp\left(2x \cos \theta\right) d\theta.$$

5. Suppose that f is regular in the disc $B(0; 1+\delta)$. If $|z| < 1$, show that

$$(1-|z|^2)f(z) = \frac{1}{2\pi i} \int_C f(\zeta) \frac{1-\zeta\bar z}{\zeta - z}\, d\zeta,$$

where C is the positively oriented unit circle. Deduce the inequality

$$(1-|z|^2)|f(z)| \leqslant \frac{1}{2\pi} \int_0^{2\pi} |f(e^{i\theta})|\, d\theta.$$

6. (*Poisson's formula.*) Let f be regular when $|z| < R_1$. Prove that, if $|a| < R < R_1$, then

$$f(a) = \frac{1}{2\pi i} \int_C \frac{R^2 - a\bar{a}}{(z-a)(R^2 - z\bar{a})} f(z)\, dz,$$

where C is the circle $z = R e^{i\phi}$ ($0 \leqslant \phi \leqslant 2\pi$). Deduce Poisson's formula that, if $r < R$,

$$f(r e^{i\theta}) = \frac{1}{2\pi} \int_0^{2\pi} \frac{R^2 - r^2}{R^2 - 2Rr \cos(\theta - \phi) + r^2} f(R e^{i\phi})\, d\phi.$$

7. Let f be regular for $|z| < R$. Prove that, if $0 < r < R$,

$$f'(0) = \frac{1}{\pi r} \int_0^{2\pi} u(\theta) \exp(-i\theta)\, d\theta,$$

where $u(\theta) = \operatorname{re} f(r \exp i\theta)$.

8. Taking, in theorem 11.82, γ to be the unit circle described positively, find f when ϕ is (i) $1/\zeta$, (ii) $\operatorname{re} \zeta$.

9. The function f is regular in the upper half plane $y > 0$, and continuous on the real axis $y = 0$. Also $f(z) = o(|z|)$ as $z \to \infty$. Prove that, if $y > 0$,

$$f(z) = \frac{1}{\pi} \lim_{X \to \infty} \int_{-X}^{X} f(\xi) \frac{y}{(\xi - x)^2 + y^2}\, d\xi.$$

10. Let l be the segment $a \leqslant x \leqslant b, y = 0$. Let D_1, D_2 be regions each having l as part of its frontier, D_1 being in the half-plane $y > 0$ and D_2 in the half-plane $y < 0$. Let f_1 be regular in D_1 and continuous in $D_1 \cup l$; f_2 regular in D_2 and continuous in $D_2 \cup l$. Let $f_1 = f_2$ at all points of l. Prove that f_1 and f_2 define a function regular in $D_1 \cup l \cup D_2$. (Use Morera's theorem 11.92.)

11. (*Schwarz's Reflection Principle.*) Suppose in exercise **10** that D_2 is the mirror image of D_1 in the line $y = 0$. Let f_1 be regular in D_1 and continuous in $D_1 \cup l$. Moreover let f_1 be real valued at all points of l. If then f_2 is defined in D_2 by

$$f_2(z) = \text{conjugate of } f_1(\bar{z}),$$

prove that f_1 and f_2 define a function regular in $D_1 \cup l \cup D_2$.

NOTES ON CHAPTER 11

§§11.1–11.3 could be written in the notation of *differential forms*. There is now an extensive theory of these forms which is of service both in analysis and mathematical physics. A good introduction is given by Harley Flanders, *Differential Forms with Applications to the Physical Sciences*.

With two real variables x, y, a differential form of the first order is,

$$\omega = P\, dx + Q\, dy,$$

where P, Q are continuous functions of (x, y) in a region D. If γ is a path $x = x(t), y = y(t)$, for $a \leqslant t \leqslant b$ then $\int_\gamma \omega$ means

$$\int_a^b [P\{x(t), y(t)\} x'(t) + Q\{x(t), y(t)\} y'(t)]\, dt.$$

A differential form which has a primitive in D is usually called *exact*. A form which has a local primitive (i.e., is locally exact) is called *closed*. We have avoided this use of the overworked word *closed*.

§§ **11.4–11.6.** *Cauchy's theorem.* Cauchy published his theorem in 1825 in his 'Mémoire sur les intégrales définies prises entre les limites imaginaires'. There is evidence that he knew the result at least ten years earlier. Ensuing discussions by writers previous to Goursat (1883) rested on the equality of a line integral $\int (u\,dx + v\,dy)$ along two paths from one point to another in virtue of the Cauchy–Riemann equations. Such an approach could not be carried through without restricting f' to be continuous.

Goursat relied on the method of quadrisection, as used in theorem 11.41. A region with a curved boundary was thereby decomposed into whole squares and part squares with curvilinear boundaries. In 1900 Goursat achieved a proof assuming only the existence of f' and not its continuity. In his approach from the interior of the region to the boundary he used the device, employed in theorem 11.42, of a family of curves $z = \alpha z(t)$ where $0 < \alpha < 1$ and $\alpha \to 1$. The success of this method depends on the region being *starlike* in the following sense.

Definition. The region R is starlike if it contains a point z_0 such that, if z_1 is any other point of R, the segment $[z_0 z_1]$ is wholly in R.

The ultimate goal of generality in Cauchy's theorem at which analysts in the early years of the 20th century could aim was that f should be regular in a region bounded by a rectifiable curve. Such a region need not be starlike and it would be difficult to frame rules for dissecting it into starlike sub-regions. So Goursat's procedure was not final.

Pringsheim (1901) initiated a method of proof which was widely adopted. The quadrisection argument establishes the theorem for a triangle and hence for a simple polygon (exercise 11 (*b*), 1). The step from a simple polygon to one which may intersect itself is not difficult intuitively. Finally a polygon can be constructed to approximate arbitrarily closely to a given rectifiable curve. The difficulties in such a programme lie in the framing of instructions for reducing a polygon with self-intersections to simple polygons and defining polygons approximating to the curved circuit of integration.

A proof, on principles derived from both Goursat and Pringsheim, was published by S. Pollard (*Proc. London Math. Soc.* **21** (1923) 456–82). Later, T. Estermann distilled the proof of the theorem for a general rectifiable boundary into a few pages of elegant analysis (*Math. Zeitschrift* **37** (1933) 556–60 and **38** (1934) 641.)

Following rapid advances in the subject of topology from (say) 1920, topological rigour demanded recognition besides analytical rigour. To illustrate that new requirement, we recall that the index of a circuit γ for a point a (§11.7) came to be *defined* as

$$\frac{1}{2\pi i} \int_\gamma \frac{dz}{z-a},$$

whereas previously it had been apprehended from a diagram as 'the number of turns that γ takes around a'.

The essence of Cauchy's theorem came to be seen as the invariance of the integral $\int_\gamma f$ for displacements of the curve γ. The topologists had, ready to hand, the theories of homology and homotopy which dealt with such displacements.

§§ **11.6–11.7.** *Topology.* M. H. A. Newman, *Elements of the Topology of Plane Sets of Points* gives a good account of ideas and methods, with special attention to the parts needed in the theory of functions.

§ **11.9.** For proofs of the existence of higher derivatives of a regular function which make no appeal to Cauchy's theorem, see E. H. Connell, *Duke Math. Journal* **28** (1961), 73–81; A. H. Read, *Journal London Math. Soc.* **36** (1961), 345–352; G. T. Whyburn, *Fundamenta Math.* **50** (1962), 305–318.

EXPANSIONS. SINGULARITIES. RESIDUES

12.1. Taylor's series. Uniqueness of regular functions

A multitude of properties of regular functions flows from Cauchy's theorem. We prove first that a function regular in a disc can be written as a Taylor's series. The expression of a regular function in a simple standard form, such as a power series, points the way to further results.

Theorem 12.11. *Let f be regular in the disc $B(a; R)$, that is $|z - a| < R$. Then, for any z in the disc,*

$$f(z) = f(a) + (z-a)f'(a) + \ldots + \frac{(z-a)^n}{n!} f^{(n)}(a) + \ldots.$$

Proof. Without loss of generality we can take $a = 0$. Given z, choose r such that $|z| < r < R$. Let C be the positively oriented circle $\zeta = re^{i\theta}$ $(0 \leqslant \theta \leqslant 2\pi)$.

Then, by theorem 11.81,

$$f(z) = \frac{1}{2\pi i} \int_C \frac{f(\zeta)}{\zeta - z} \, d\zeta.$$

In the integral substitute

$$\frac{1}{\zeta - z} = \frac{1}{\zeta} + \frac{z}{\zeta^2} + \ldots + \frac{z^n}{\zeta^{n+1}} + \frac{z^{n+1}}{\zeta^{n+1}(\zeta - z)}.$$

This gives, by use of theorem 11.91,

$$f(z) = f(0) + zf'(0) + \ldots + \frac{z^n}{n!} f^{(n)}(0) + R_n,$$

where

$$R_n = \frac{1}{2\pi i} \int_C \left(\frac{z}{\zeta}\right)^{n+1} \frac{f(\zeta)}{\zeta - z} \, d\zeta.$$

When ζ is on C, $|z/\zeta| = |z|/r < 1$, and $\sup |f(\zeta)| = M$, say. Then by theorem 11.11,

$$|R_n| < K(|z|/r)^{n+1},$$

where $K = Mr/(r - |z|)$. Hence $R_n \to 0$ as $n \to \infty$. |

Notes. (1) The proof amounts to integrating term by term the series obtained by inserting the binomial expansion of $1/(\zeta - z)$ in the integral. This series converges uniformly for ζ on C, and we could have appealed to theorem 11.12, the complex extension of theorem 5.22. Instead we have estimated the remainder R_n from first principles.

(2) Contrast theorem 12.11 with the real-variable theorem leading to Taylor's series (C1, 101) in which every derivative has to be postulated explicitly and its magnitude controlled to make sure that the remainder R_n tends to 0. With a complex function all this is implicit in the sole assumption of regularity.

(3) If we are given a power series

$$a_0 + a_1(z-a) + \ldots + a_n(z-a)^n + \ldots,$$

convergent for $|z-a| < R$, we proved in theorem 10.2 that the sum is regular for $|z-a| < R$. Theorem 10.2 (in particular note 2 and exercise 10(a), 2) shows that, if $f(z)$ is the sum, then

$$a_n = f^{(n)}(a)/n!$$

Definition of a zero. *Suppose f regular in B(a; R). If f(a) = 0, then a is called a* zero *of f. If $f^{(n)}(a)$ is the derivative of lowest order which is different from 0, then a is called a* zero of order n. *(If every derivative of f is 0 at a, then, by theorem 12.11, $f(z) \equiv 0$.)*

Observe that, if f has a zero of order n at a, then

$$f(z) = (z-a)^n g(z)$$

where g is regular at a and $g(a) \neq 0$.

Moreover there is a disc with centre a, say $B(a; \delta)$, throughout which the continuous function g is different from 0, so that a is the only zero of f in $B(a; \delta)$. Thus the zeros of finite order of a regular function f are *isolated*.

We now investigate regular functions defined not necessarily in a disc but in an arbitrarily given region.

Theorem 12.12. *If f is regular in a region D and the set of zeros of f has a limit point in D, then $f(z) \equiv 0$ in D.*

Proof. Let a be a limit point in D of the set S of zeros of f. By theorem 12.11, f has a Taylor expansion in any disc T with centre a which lies in D.

If any derivative $f^{(n)}(a)$ were different from 0, there would be a circle with centre a containing no zeros of f other than a. But this contradicts the vanishing of f in S. So $f(z) = 0$ for all z in the disc T.

We extend this result throughout D by the following construction of overlapping discs. Let b be any point of D; we wish to prove $f(b) = 0$. Join a to b by a polygon P in D (theorem 3.32). Let d be the distance between the sets P and D'. Then the disc of radius d with centre at any point of P lies in D. Take points $a_0\,(=a)$, a_1, a_2, ... $a_n\,(=b)$ in order on P with each distance $|a_r - a_{r-1}|$ less than d. Take a disc T_r of radius d with centre at each point a_r. Then a_r is in T_{r-1}. If $f \equiv 0$ in T_{r-1}, then, by the first part of the proof, since a_r is the limit of zeros of f, we have $f \equiv 0$ in T_r. Hence $f(b) = 0$. |

The following restatement of theorem 12.12 is to be noted.

Theorem 12.13. (Uniqueness theorem for regular functions.) *A function f, regular in D, is uniquely determined if we know its values at points of a set S having a limit point in D (a fortiori, if S is a linear segment or arc, however short, in D).*

This follows from theorem 12.12 because, if f and g are two regular functions having the assigned values in S, then $f - g \equiv 0$ in D.

12.2. Inequalities for coefficients. Liouville's theorem

The following simple inequalities for the coefficients in a Taylor's series are due to Cauchy.

Theorem 12.21. *If $f(z) = \sum\limits_0^\infty a_n z^n$ for $|z| < R$, and $M(r) = \sup |f(z)|$ for $|z| = r < R$, then*

$$|a_n| \leqslant \frac{M(r)}{r^n}.$$

Proof. From theorems 12.11 (note 3) and 11.91,

$$a_n = \frac{f^{(n)}(0)}{n!} = \frac{1}{2\pi i} \int_C \frac{f(\zeta)}{\zeta^{n+1}} d\zeta,$$

where C is the circle $z = re^{i\theta}$ $(0 \leqslant \theta \leqslant 2\pi)$.

By theorem 11.11,

$$|a_n| \leqslant \frac{1}{2\pi} \frac{M(r)}{r^{n+1}} 2\pi r. \; |$$

Definition. *A function regular for all z in Z is called* entire.

Some writers call an entire function an integral function.

Illustrations. Any polynomial, $\exp z$, $\sin z$, $\cos z$.

Theorem 12.22. (Liouville (1809–1882).) *An entire function, which is bounded, is constant.*

Proof. The function has a Taylor expansion, say

$$f(z) = \sum_0^\infty a_n z^n$$

valid for all z. Suppose that $|f(z)| \leqslant M$. From Cauchy's inequalities (theorem 12.21)

$$|a_n| \leqslant \frac{M}{r^n}.$$

Since r is arbitrarily large, $a_n = 0$ for $n \geqslant 1$. |

Liouville's theorem has many applications. As an illustration we deduce from it 'the fundamental theorem of algebra'.

Theorem 12.23. *Let p be a polynomial (not constant). Then there are values of z for which $p(z) = 0$.*

Proof. If the conclusion is false, $1/p$ is regular for all z. Now $|p(z)| \to \infty$ as $|z| \to \infty$ (exercise 10(*b*), 11). So we can find a (closed) disc $\overline{B}$ outside which $|1/p(z)|$ is bounded. But $|1/p(z)|$ being continuous is bounded on $\overline{B}$. So $1/p$ is a bounded entire function. By Liouville's theorem it is constant. This is a contradiction. |

Exercises 12(*a*)

In each of **1–8**, expand the given function in a Taylor's series $\Sigma a_n z^n$. Find, if you can, the general coefficient a_n or, failing that, the first three non-zero coefficients. In each example state the radius of convergence of the series.

1. The principal value of $(1+z)^k$, where k can be complex.

2. log cosh z.

3. $(\arcsin z)^2$, defined to be 0 when $z = 0$.

4. tan z.

5. $(\cos z)^{\frac{1}{2}}$, defined to be $+1$ when $z = 0$.

6. $\dfrac{1}{e^z + 1}$.

7. $\displaystyle\int_0^z e^{-\zeta^2} d\zeta$.

8. $\displaystyle\sum_1^\infty \frac{z^n}{1 - z^n}$.

9. Expand in Taylor's series $\Sigma a_n (z - \tfrac{1}{2}\pi)^n$,

$$\cot z, \quad \operatorname{cosec}^2 z.$$

10. If $\Sigma a_n z^n$ has radius of convergence 2, find the radii of convergence of

$$\Sigma a_n^k z^n, \quad \Sigma a_n z^{nk}, \quad \Sigma a_n z^{n^2},$$
$$\Sigma (a_1^k + a_2^k + \ldots + a_n^k) z^n,$$

where k is a fixed positive integer and in the fourth series $a_n > 0$.

11. For each set of values (i)–(v) corresponding to $z = 1, \frac{1}{2}, \frac{1}{3}, \frac{1}{4}, \frac{1}{5}, \frac{1}{6}, \ldots$, find whether a function exists, regular at $z = 0$, taking those values:

(i) $\quad 1, 0, \frac{1}{3}, 0, \frac{1}{5}, 0, \ldots$,

(ii) $\quad 1, \sqrt{\frac{1}{2}}, \sqrt{\frac{1}{3}}, \sqrt{\frac{1}{4}}, \sqrt{\frac{1}{5}}, \sqrt{\frac{1}{6}} \ldots$,

(iii) $\quad \frac{1}{2}, \frac{2}{3}, \frac{3}{4}, \frac{4}{5}, \frac{5}{6}, \frac{6}{7}, \ldots$,

(iv) $\quad \frac{0}{4}, \frac{1}{6}, \frac{2}{8}, \frac{3}{10}, \frac{4}{12}, \frac{5}{14}, \ldots$,

(v) $\quad 1, -\frac{1}{4}, \frac{1}{9}, -\frac{1}{16}, \frac{1}{25}, -\frac{1}{36}, \ldots$.

12. Prove that the distance from the origin to the nearest zero of $f(z) = \sum_{0}^{\infty} a_n z^n$

is at least

$$\frac{r|a_0|}{M + |a_0|},$$

where r is any number not exceeding the radius of convergence of the series and $M = M(r) = \sup |f(z)|$ for $|z| = r$.

13. In a region D, f and g are regular and $fg \equiv 0$ at all points. Prove that $f \equiv 0$ or $g \equiv 0$.

14. Prove that, if $f(z) = \sum a_n z^n$ is regular for $|z| < R$, then the average value of $|f(z)|^2$ on $|z| = r$, where $r < R$, is $\sum |a_n|^2 r^{2n}$.
  Deduce theorem 12.21.

15. Prove that, if for $n = k$ Cauchy's inequality (theorem 12.21) is an equality, that is

$$|a_k| = \frac{M(r)}{r^k},$$

then $f(z)$ is a constant multiple of z^k.

16. Let $f(z) = \sum_{0}^{\infty} a_n z^n$ be regular for $|z| < 1 + \delta$. Prove that the polynomial $p_k(z)$ of degree k which minimizes the integral

$$\frac{1}{2\pi} \int_0^{2\pi} |f(e^{i\theta}) - p_k(e^{i\theta})|^2 \, d\theta$$

is $p_k(z) = \sum_{0}^{k} a_n z^n$. Prove that the minimum value is $\sum_{k+1}^{\infty} |a_n|^2$.

17. Give an alternative proof of Liouville's theorem 12.22 by integrating $f(z)/\{(z-a)(z-b)\}$ round a circle whose radius tends to infinity.

18. The function f is entire.
  (i) Prove or refute the suggestion that, to prove f constant, boundedness may be replaced by the hypothesis

$$|f(z)| = o(|z|) \quad \text{as} \quad z \to \infty.$$

  (ii) What conclusion follows if we assume

$$|f(z)| = O(|z|^k) \quad \text{as} \quad z \to \infty?$$

19. Prove that, if $f = u + iv$ is regular in Z and either (i) $u \geqslant 0$ or (ii) $v < 7$ for all z, then f is constant. Include (i) and (ii) in a general statement.

20. Replace the hypothesis (i) or (ii) of **19** by $uv \geqslant 0$.

21. The function f is regular in Z. There is a disc $B(w_0; \delta)$ such that no value $f(z)$ is in $B(w_0; \delta)$. Prove that f is constant.

22. The functions f, g are regular and have no zero in a disc $B(z_0; r)$. There is a sequence $z_n \rightarrow z_0$ at which $f'/f = g'/g$. What follows?

12.3 Laurent's series

Suppose that f is regular in an annulus with centre a (instead of throughout a disc). Following Laurent (1813–1854) we expand $f(z)$ in a series containing both positive and negative powers of $z - a$.

A series $\sum\limits_{-\infty}^{\infty} c_n$ is said to converge if, and only if, $\sum\limits_{0}^{\infty} c_n$ and $\sum\limits_{1}^{\infty} c_{-n}$ both converge.

Theorem 12.31. *Let f be regular in the annulus $R < |z-a| < S$. Then, for any z in the annulus,*

$$f(z) = \sum_{-\infty}^{\infty} a_n(z-a)^n,$$

where $$a_n = \frac{1}{2\pi i} \int_C \frac{f(\zeta)}{(\zeta-a)^{n+1}} d\zeta.$$

and C is the circle $\zeta = a + \rho e^{i\theta}$ ($0 \leqslant \theta \leqslant 2\pi$), ρ being any number such that $R < \rho < S$.

Proof. As in theorem 12.11, we can without loss of generality take $a = 0$.

Given z such that $R < |z| < S$, let C_1, C_2 be circles, with centre $\zeta = 0$ and radii r, s, where

$$R < r < |z| < s < S.$$

By theorem 11.81, corollary 2,

$$f(z) = \frac{1}{2\pi i} \int_{C_2} \frac{f(\zeta)}{\zeta - z} d\zeta - \frac{1}{2\pi i} \int_{C_1} \frac{f(\zeta)}{\zeta - z} d\zeta,$$

where C_1 and C_2 are positively oriented.

As in theorem 12.11, the integral round C_2 yields the series $\sum\limits_{0}^{\infty} a_n z^n$, where $$a_n = \frac{1}{2\pi i} \int_{C_2} \frac{f(\zeta)}{\zeta^{n+1}} d\zeta.$$

To deal with the integral round C_1, we expand $1/(\zeta-z)$ in powers of ζ/z (since $|\zeta| < |z|$), and obtain

$$-\frac{1}{\zeta-z} = \frac{1}{z} + \frac{\zeta}{z^2} + \dots + \frac{\zeta^n}{z^{n+1}} + \frac{\zeta^{n+1}}{z^{n+1}(\zeta-z)}.$$

An argument like that of theorem 12.11 shows that the remainder term in the integral round C_1 tends to 0 as $n \to \infty$, and the consequent infinite series is $\sum\limits_{1}^{\infty} a_{-n} z^{-n}$, with the value of a_n already defined.

Since the integrand in the formula for a_n is regular for $R < |\zeta| < S$, the restriction $r < |z| < s$ on the radii of the circles C_1, C_2 can be removed (by theorem 11.61(ii)) and we can integrate round any circle C in the annulus. |

Corollary. (Cauchy's inequalities.) *If $f(z) = \sum\limits_{-\infty}^{\infty} a_n(z-a)^n$ for*

$$R < |z-a| < S$$

and $|f(z)| \leqslant M(r)$ for $|z-a| = r$, where $R < r < S$, then, for every integer n, positive or negative,

$$|a_n| \leqslant \frac{M(r)}{r^n}.$$

The proof follows that of theorem 12.21.

Theorem 12.32. (Uniqueness of Laurent expansion.)

(i) *If $f(z) = \sum\limits_{-\infty}^{\infty} a_n(z-a)^n$ and $f(z) = \sum\limits_{-\infty}^{\infty} b_n(z-a)^n$ for*

$$R < |z-a| < S,$$

then $a_n = b_n$ for all n.

(ii) *If f is regular for $R < |z-a| < S$, then $f = f_1 + f_2$, where f_1 is regular for $|z-a| < S$ and f_2 is regular for $R < |z-a|$. Moreover if $f_2 \to 0$ as $|z| \to \infty$, the expression of f as $f_1 + f_2$ is unique.*

Proof. (i) If we fix ρ with $R < \rho < S$, each of the two series for $f(\zeta)$ converges uniformly on the circle $\zeta - a = \rho e^{i\theta} (0 \leqslant \theta \leqslant 2\pi)$, by theorem 5.41 extended to cover negative powers of $\zeta - a$.

It is then legitimate to multiply by $(\zeta - a)^{-k-1}$ and integrate term by term round the circuit $\zeta - a = \rho e^{i\theta}$. Hence

$$2\pi i a_k = 2\pi i b_k \quad (k = 0, \pm 1, \pm 2, \ldots).$$

(ii) $f_1(z) = \sum\limits_{0}^{\infty} a_n(z-a)^n$ and $f_2(z) = \sum\limits_{-1}^{-\infty} a_n(z-a)^n$ satisfy the conditions imposed. To prove uniqueness, suppose that also $f = g_1 + g_2$. Then

$$h = f_1 - g_1 \quad (|z-a| < S),$$
$$= g_2 - f_2 \quad (|z-a| > R)$$

is regular for all z and (since $h \to 0$ as $|z| \to \infty$) is bounded. By Liouville's theorem 12.22, h is constant and (from $|z| \to \infty$) is 0. |

Exercises 12(b)

1. Write down the Laurent expansions of

$$\frac{1}{z-2} - \frac{1}{z-1}$$

appropriate to the respective regions $|z| < 1$, $1 < |z| < 2$, $2 < |z|$.

Expand similarly

$$\frac{z^2}{(z-1)(z-2)}, \quad \frac{1}{z^2(z-1)(z-2)}, \quad \frac{1}{(z^2+1)(z^2+4)}.$$

2. Expand in positive or negative powers of z

$$\frac{1}{1+z+z^2}.$$

3. Prove that, for all finite z except 0,

$$\cos\{u(z+z^{-1})\} = \sum_{-\infty}^{\infty} c_n z^{2n},$$

where

$$c_n = \frac{2}{\pi} \int_0^{\frac{1}{2}\pi} \cos(2u\cos\theta)\cos 2n\theta\, d\theta.$$

4. Establish the expansion

$$\exp\{\tfrac{1}{2}u(z-z^{-1})\} = \sum_{-\infty}^{\infty} J_n(u)z^n,$$

where

$$J_n(u) = \frac{1}{\pi} \int_0^{\pi} \cos(n\theta - u\sin\theta)\, d\theta.$$

Prove also that, if $n \geqslant 0$,

$$J_n(u) = \sum_{k=0}^{\infty} \frac{(-1)^k u^{n+2k}}{2^{n+2k} k!(n+k)!}$$

while, if n is a negative integer $-m$,

$$J_{-m}(u) = J_m(-u).$$

5. Discuss the possibility of expanding in power series

$$\{(z-1)(z-2)\}^{\frac{1}{2}}$$

in a suitable circle or annulus.

6. The same question for

$$\log\frac{z-a}{z-b}.$$

7. The regular function

$$w = \sum_{-\infty}^{\infty} a_n z^n$$

is a bijective mapping of the closed ring $r_1 \leqslant |z| \leqslant r_2$ onto a compact set S of the w-plane. Prove that the area (or content) of S is

$$\pi \sum_{-\infty}^{\infty} n|a_n|^2(r_2^{2n} - r_1^{2n}).$$

12.4. Singularities

Suppose that f is regular in the punctured disc $0 < |z-a| < R$. We ask whether, with the right choice of the value of f at a, the function thus extended is regular in the whole disc $|z-a| < R$.

Theorem 12.4. *A sufficient (and necessary) condition that f, regular for $0 < |z-a| < R$, can be extended to a function regular in $|z-a| < R$ is that*

$$(z-a)f(z) \to 0 \quad as \quad z \to a.$$

Proof. There is a Laurent expansion

$$f(z) = \sum_{-\infty}^{\infty} a_n(z-a)^n \quad \text{for} \quad 0 < |z-a| < R.$$

Let $$M(r) = \sup |f(z)| \quad \text{for} \quad |z-a| = r.$$

Then, from the hypothesis,

$$rM(r) \to 0 \quad \text{as} \quad r \to 0.$$

By theorem 12.21 (corollary),

$$|a_{-n}| \leqslant r^n M(r) \quad \text{if} \quad n \geqslant 1.$$

Therefore $$a_{-n} = 0 \quad \text{if} \quad n \geqslant 1.$$

If $f(a)$ is defined to be a_0, then $f(z) = \sum_{0}^{\infty} a_n(z-a)^n$ for $|z-a| < R$ and this is regular. |

Definition. *f is singular (or, has a singularity) at a if it is not regular at a.*

Definition. *If f is singular at a and regular in a punctured disc centre a, the singularity at a is* isolated.

If the condition of theorem 12.4 is fulfilled, we can assign a value to $f(a)$ so that f is regular at a. Such a point a may be called a *removable* singularity of f. We generally suppose that removable singularities have been removed.

Illustration. We shall say, without more ado, that the function

$$\frac{1 - \cos 2z}{z^2}$$

is regular for all z, on the understanding that the value 2 is assigned to it for $z = 0$.

The condition $(z-a)f(z) \to 0$ certainly holds if f is known to be continuous at a. In that case, there is, in fact, no singularity at a and it was only lack of knowledge that previously stopped us from asserting that f is regular at a. The clause in theorem 11.61 permitting a finite number of points of continuity rather than regularity was therefore only temporarily useful and now loses its significance.

If a is an isolated singularity of f, there is a Laurent expansion in a punctured disc, say

$$f(z) = \sum_{-\infty}^{\infty} a_n(z-a)^n.$$

Definition. *If the Laurent expansion of f has only a finite number of negative powers, say* $\sum_{1}^{m} a_{-n}(z-a)^{-n}$, *where* $a_{-m} \neq 0$, *then a is called a* pole of order m *of f and* $\sum_{1}^{m} a_{-n}(z-a)^{-n}$ *is its* singular (*or* principal) part.

A pole of order one is called a *simple* pole; one of order two a *double* pole.

Definition. *If there are infinitely many negative powers of $(z-a)$ in the Laurent expansion, then a is called an* essential singularity *of f.* $\sum_{1}^{\infty} a_n(z-a)^{-n}$ *is the* singular part *of f at a.*

Illustration. exp $(1/z)$ has an essential singularity at $z = 0$.

Before §13.3 we shall be concerned only with isolated singularities. We shall there encounter functions having singularities which are limit points of singularities.

Definition. *A function which is regular in a region except at poles is called* meromorphic.

12.5. Residues

We suppose that f has an isolated singularity at a and the usual Laurent expansion.

Definition. *The coefficient a_{-1} of $(z-a)^{-1}$ is called the* residue *of f at the singularity a and is written* res(f, a).

From the definition of a_{-1} (theorem 12.31)

$$2\pi i a_{-1} = \int_C f,$$

where C is a positively oriented circle containing the singularity a and no other.

This formula for a complex integral round a suitable circuit in terms of the residue of the integrand at a singularity constitutes a powerful method of calculating definite integrals. The next theorem caters for any number of singularities.

Theorem 12.5. (Residues.) *Let f be regular in a region D except for isolated singularities c (which form a countable set C). Let γ be a circuit in D which is homotopic to a point in D and is free of points of C. Then the index of γ is zero for all c except a finite number and*

$$\int_{\gamma} f = 2\pi i \sum_{c \in C} n(\gamma, c) \text{ res } (f, c).$$

Proof. Let K be a compact set in D. Then the set $K \cap C$ is finite, for if not K would contain a limit point of C which would be a non-isolated singularity.

We proceed to define the particular K which is used in the proof. The definition of homotopy requires a continuous function

$$\delta : [0, 1] \times [0, 1] \to D$$

such that $\delta(0, u) = \delta(1, u)$ for $0 \leqslant u \leqslant 1$, $\delta(t, 0)$ for $0 \leqslant t \leqslant 1$ represents γ and $\delta(t, 1)$ $(0 \leqslant t \leqslant 1)$ is a point curve γ_0 (i.e. is constant). Since $[0, 1] \times [0, 1]$ is compact, so is

$$K = \delta([0, 1] \times [0, 1]).$$

From §11.7 (1), if b is any point of $Z - K$,

$$n(\gamma, b) = n(\gamma_0, b) = 0.$$

Let C_1 be the set of singularities of f in K and C_2 those in $D - K$. We have proved that C_1 is finite and that $n(\gamma, c) = 0$ at every point of the countable set C_2.

We shall prove that $D_1 = D - C_2$ is a region.

Let $z \in D_1$. Then z cannot be a limit point of C_2 for, if it were, z would be a non-isolated singularity of f. Therefore there is a $B(z; \epsilon) \subset D - C_2$ and so D_1 is open. To prove now that D_1 is connected, let z_1, z_2 be two points of it. There is a polygon in D joining z_1, z_2. Its trace is compact and so it can go through only a finite number of points of C_2. Each such point, being isolated, can be avoided by a small detour.

In D_1, f is regular except at the points of C_1, say $c_1, c_2, \ldots, c_p$. Let $S(f, c_k)$ be the singular part of f at c_k ($k = 1, \ldots, p$) and, for z in D_1, put

$$g(z) = f(z) - \sum_{k=1}^{p} S(f, c_k).$$

Then g is regular in $D_1 - C_1$ and at each c_k it has a removable singularity. By theorem 11.62,

$$\int_{\gamma} g = 0.$$

We shall prove that, at a point $c(= c_k)$ of C_1,

$$\int_{\gamma} S(f, c) = 2\pi i n(\gamma, c) \operatorname{res}(f, c).$$

Since c is not on γ, there is a disc with centre c containing no point of γ. The power series in $1/(z-c)$

$$S(f, c) = \sum_{1}^{\infty} a_{-n}(z-c)^{-n}$$

converges uniformly on γ. Hence

$$\int_{\gamma} S(f, c) = \sum_{1}^{\infty} a_{-n} \int_{\gamma} (z-c)^{-n} dz.$$

Since $(z-c)^{-n}$ has, for $n \geqslant 2$, a primitive $(z-c)^{-n+1}/(-n+1)$, it follows that $\int_{\gamma} (z-c)^{-n} dz = 0$ for $n \geqslant 2$.

Finally, $a_{-1} \int_{\gamma} (z-c)^{-1} dz = 2\pi i\, n(\gamma, c) \operatorname{res}(f, c)$, from the definitions of index and residue.

From the definition of g and the vanishing of its integral on γ,

$$\int_{\gamma} f = 2\pi i \sum_{k=1}^{p} n(\gamma, c_k) \operatorname{res}(f, c_k).$$

Since $n(\gamma, c) = 0$ if $c \in C_2$, the right-hand side is

$$2\pi i \sum_{c \in C} n(\gamma, c) \operatorname{res}(f, c). \mid$$

12.6. Counting zeros and poles

Suppose that f has a zero of order l at a. Then, from the definition of a zero (§12.1), there is a disc $B(a; \delta)$ in which

$$f(z) = (z-a)^l g(z),$$

where $g(z)$ is regular and not zero. In this disc

$$f'(z) = l(z-a)^{l-1}g(z)+(z-a)^l g'(z).$$

Hence, for $0 < |z-a| < \delta$,

$$\frac{f'(z)}{f(z)} = \frac{l}{z-a}+\frac{g'(z)}{g(z)},$$

where g'/g is regular in $B(a;\delta)$.

Therefore f has a simple pole at $z = a$ with residue l.

Similarly, if f has a pole of order m at a, then f'/f has a simple pole there with residue $-m$.

The following theorem which enumerates the excess of the number of zeros over the number of poles of a function meromorphic in a region is particularly useful when the function is regular and we are counting its zeros.

Theorem 12.61. *If $f(\not\equiv 0)$ is meromorphic in the region D and γ is a positively oriented simple circuit in D, passing through no zero or pole of f, homotopic to a point in D, then*

$$\frac{1}{2\pi i}\int_\gamma \frac{f'(z)}{f(z)}\,dz = N-P,$$

where N is the number of zeros and P the number of poles of f enclosed by γ (a zero or pole of order l being counted l times).

Proof. The index $n(\gamma, z)$ is 1 for points z enclosed by γ and 0 for other points. The theorem follows from theorem 12.5 and the first two paragraphs of §12.6. |

Corollary. *If g is regular in D, then*

$$\frac{1}{2\pi i}\int_\gamma g(z)\frac{f'(z)}{f(z)}\,dz = \Sigma g(a_i)-\Sigma g(b_j)$$

summed over the zeros a_i and the poles b_j of f enclosed by γ.

Adapt the argument of the theorem. |

In applying theorem 12.61 we observe that

$$\frac{1}{2\pi i}\int_\gamma \frac{f'(z)}{f(z)}\,dz = \frac{1}{2\pi i}\int_\Gamma \frac{dw}{w},$$

where Γ corresponds to γ by the mapping $w = f(z)$ from the z-plane to the w-plane (exercise 11(a), 13). The w-integral is the index $n(\Gamma, 0)$ of Γ for the origin.

The reader should work out for himself the application of this principle to numerical examples. The following exercise is typical.

Exercise 12 (c), 0. *Find how many roots of the equation*

$$z^4 - 3z^3 + 7z^2 - 9z + 6 = 0$$

lie in each quadrant.
The solution is at the end of the book.

Here is another result which is useful in root-counting.

Theorem 12.62. (Rouché (1832–1910).) *Suppose that f and g are regular in D. Let γ be a simple circuit, homotopic to a point in D. If, for every z on γ, $|g(z)| < |f(z)|$, then γ encloses the same number of zeros of f as of $f+g$.*

Proof. Since $|g| < |f|$, neither f nor $f+g$ has a zero on γ. The point $w = 1 + \dfrac{g(z)}{f(z)}$, for z on γ, lies in the disc $B(1; 1)$. By §11.7 (3), if Γ is the w-image of γ, then

$$\frac{1}{2\pi i} \int_\Gamma \frac{dw}{w} = n(\Gamma, 0) = 0.$$

Now
$$\frac{1}{w} \frac{dw}{dz} = \frac{f'+g'}{f+g} - \frac{f'}{f}.$$

Integrate this round γ, and divide by $2\pi i$. The theorem then follows from theorem 12.61. |

A quick deduction from Rouché's theorem is: *A polynomial of degree n has n zeros.*

Proof. Let $f(z) = z^n$, $g(z) = a_1 z^{n-1} + \ldots + a_n$.
If γ is a circle, centre 0 and radius r, where

$$r^n > |a_1| r^{n-1} + \ldots + |a_n|,$$

the polynomial $f+g$ of degree n has the same number of zeros inside γ as z^n has. But z^n has n zeros at $z = 0$. |

A numerical illustration of Rouché's theorem follows.
Example. Show that the roots of the equation

$$z^5 - 12z^2 + 14 = 0$$

lie between the circles $|z| = 1$ and $|z| = \frac{5}{2}$. How many lie inside the circle $|z| = 2$?

Solution. On $|z| = \frac{5}{2}$, take $f = z^5$, $g = -12z^2 + 14$.
Then
$$|g| \leqslant 75 + 14 < 3125/32 = |f|.$$

So $f + g = 0$ has the same number of roots inside $|z| = \frac{5}{2}$ as $f = 0$, namely 5.

On $|z| = 1$, take $f = 14$, $g = z^5 - 12z^2$.
On $|z| = 2$, take $f = -12z^2$, $g = z^5 + 14$.

Exercises 12(c)

1. Evaluate the integrals round the unit circle, described positively, of
$$z^2 \sin \frac{1}{z}, \quad e^{-z} z^{-n}, \quad \frac{1}{z^2 (z-2)^2}.$$

2. Prove that, for all sufficiently large R, the integral of $(z-a)^{-n}(z-b)^{-n}$ round the circle $|z| = R$ is 0.

3. Prove that $z^8 + 3z^3 + 7z + 5$ has two zeros in the first quadrant. How many zeros are in $|z| < \frac{1}{2}$?

4. Prove that, if a and b are positive, the equation
$$z^4 + az + b = 0$$
has one and only one root in the first quadrant, and that its phase is greater than $\frac{1}{4}\pi$.

5. Prove that $z^5 + 3z^4 - 5z^3 + 10z^2 - 1$ has three zeros with $\operatorname{re} z > 0$.
Prove that it has two zeros in the ring $\frac{1}{4} < |z| < 1$.

6. Prove that the equation $e^{z-1} = z^n$ has $n-1$ roots inside the unit circle if $n \geqslant 2$.

7. Prove that the function $\sinh \pi z$ takes every value w_0 such that $\operatorname{re} w_0 > 0$ exactly once in the region $\operatorname{re} z > 0$, $-\frac{1}{2} < \operatorname{im} z < \frac{1}{2}$.

8. Find the number of zeros of $z(z^{2n} + 1) + 1$ having $-\pi/n < \operatorname{ph} z < \pi/n$.

9. Prove that, if $|a| < 1$, then
$$z^m \left(\frac{z-a}{1-\bar{a}z} \right)^n - a$$
has $m + n$ zeros in $|z| < 1$.

10. Prove that the function $\sin z - z$ has just two simple zeros in each of the strips
$$(2n - \tfrac{1}{2})\pi < \operatorname{re} z < (2n + \tfrac{1}{2})\pi,$$
where n is an integer other than 0; and that, apart from the triple zero at $z = 0$, these are its only zeros.

11. Prove that, however small δ is, all the zeros of the function
$$1 + \frac{1}{z} + \frac{1}{2!z^2} + \dots + \frac{1}{n!z^n}$$
lie in the disc $B(0; \delta)$ if n is large enough.

12.7. The value $z = \infty$

In §§10.4, 10.5, we paraded the advantage of adjoining a point at infinity, $z = \infty$, to the complex plane Z. The treatment of the bilinear transformation was neater in the extended plane $\overline{Z}$. In chapters 11 and 12 (so far) the value ∞ has been left on the shelf, for a good reason. Everything has been based on integration, and the inclusion of $z = \infty$ in a path of integration makes its length infinite and entails an additional limiting operation.

We assume that f is defined for all (finite) z if $|z|$ is sufficiently large.

Definitions. $f(z)$ *is said to have an* isolated singularity *for* $z = \infty$ *if* $f(1/z)$ *has an isolated singularity for* $z = 0$. *Moreover the former is a* pole *or an* essential singularity *according as the latter is.*

$f(z)$ *is said to be* regular *at* ∞ *if* $f(z)$ *is regular for* $|z| > r$ *and* $\lim_{z \to 0} f(1/z)$ *is finite.* (By theorem 12.4, the latter condition holds if $\lim_{z \to 0} zf(1/z) = 0$.)

Illustration. At $z = \infty$, the function z^2 has a double pole and $\sin(1/z)$ a simple zero; $\exp z$ has an essential singularity.

We can now state Liouville's theorem 12.22 in a more cogent form.

A function regular in $\overline{Z}$ is a constant.

Proof. From the definition of regularity at ∞, there are constants A, η such that $|f(z)| \leqslant A$ if $0 < |1/z| < \eta$. Now, regularity of f for all finite z shows that $|f(z)| \leqslant B$ if $|z| \leqslant 1/\eta$. Theorem 12.22 gives the result. |

Another useful result, extending Liouville's, follows.

Theorem 12.71. *If f is meromorphic in $\overline{Z}$, then f is a rational function p/q, where p and q are polynomials.*

Proof. The number of poles of f is finite, since an infinite set would have a limit point either at $z = a$ (finite) or $z = \infty$. f would have no Laurent expansion about this limit point and it would not be a regular point or a pole of f.

Let $a_1, a_2, \ldots, a_n$ be the finite values of z which are poles of f, and let

$$S_1(z), S_2(z), \ldots, S_n(z)$$

be the respective singular parts of f at these poles. If $z = \infty$ is a pole, the singular part there is of the form

$$S_\infty(z) = b_1 z + \ldots + b_m z^m.$$

Then $$f(z) - \sum_1^n S_r(z) - S_\infty(z)$$

has no singularities for any z in $\bar{Z}$. Therefore by the form of Liouville's theorem just stated, it is a constant A. So

$$f(z) = A + \sum_1^n S_r(z) + S_\infty(z)$$

which is a rational function. |

In further developments the best setting will be sometimes Z and sometimes $\bar{Z}$. In a discussion of singularities of a function, the behaviour at ∞ should be specified.

We shall not use the residue of a function at ∞. The definition (which needs care) is in a note at the end of the chapter.

12.8. Behaviour near singularities

Suppose that a is an isolated singularity of f.

If a is a pole of order n, the definition shows that $|f(z)|$ tends to infinity as z tends to a, and its order of magnitude is given by $f(z) \sim a_{-n}(z-a)^{-n}$. In contrast, the behaviour of $f(z)$ as z approaches an isolated essential singularity is erratic.

Theorem 12.8. (Casorati–Weierstrass.) *Let a be an isolated essential singularity of f. Given any complex number b and arbitrarily small positive numbers r, ϵ, there exists z in the punctured disc*

$$0 < |z-a| < r$$

such that $$|f(z) - b| < \epsilon.$$

(At points arbitrarily near to a the function takes values arbitrarily near to any given number.)

Proof. If there is a z in $0 < |z-a| < r$ where $f(z) = b$, the theorem is proved. If not, write

$$g = \frac{1}{f-b}.$$

Then g is regular in $0 < |z-a| < r$.

If g is bounded, then a is a removable singularity of g and $f = b+(1/g)$ has a removable singularity or a pole at a. In fact f has an essential singularity at a. Therefore g is unbounded and there is a z for which $|g(z)| > 1/\epsilon$. |

Illustration. exp $(1/z)$ assumes all values except 0 arbitrarily near to $z = 0$.

Proof. Write exp $(1/z) = b = \rho e^{i\alpha}$ say $(\rho > 0)$.
Then
$$z^{-1} = \log \rho + i(\alpha + 2n\pi).$$

To make $|z| < r$, we have only to take the integer n large enough to make
$$(\log \rho)^2 + (\alpha + 2n\pi)^2 > r^{-2}. \ |$$

When $a = \infty$ the results of this section are of daily relevance. If $f(z)$ is a polynomial
$$a_0 + a_1 z + \dots + a_n z^n,$$

where $a_n \neq 0$, then ∞ is a pole of order n.

If $f(z)$ is an infinite series in powers of z (or $z - z_0$) convergent for all z, then f is an entire function and ∞ is an isolated essential singularity.

Entire functions, a class which includes some of the most useful functions of analysis like exp z and sin z, show the Casorati–Weierstrass behaviour outside any circle $|z| = R$. The function exp z actually takes all values except 0 for arbitrarily large $|z|$.

Exercises 12(d)

In each of **1–8**, state for what values in $\bar{Z}$ the function is singular and specify the nature of the singularity

1. $\dfrac{z+1}{z-1} + \dfrac{z-1}{z+1} - \dfrac{4z^2}{z^2-1}.$

2. $\dfrac{z^4}{1+z^3}.$

3. $\dfrac{e^z}{1+z^3}.$

4. tan z.

5. tanh z.

6. exp (tan z).

7. sin $\dfrac{1}{1-z}.$

8. $\dfrac{1}{e^z - 1} - \dfrac{1}{z}$.

9. Prove that the entire function $\cos z$ assumes all values for arbitrarily large $|z|$. ($\cos z$ has no exceptional value such as 0 is for $\exp z$.)

NOTES ON CHAPTER 12

§**12.7.** *Definition of residue of f at* ∞. If f has a Laurent expansion valid for $|z| > R$, and a_{-1} is the coefficient of $1/z$ in it, then $-a_{-1}$ is defined to be the residue of f at ∞.

We have come to expect the occurrence of residues only at singularities. This definition assigning, as it does, a residue to the function $1/z$ which is regular at ∞ seems perverse. The rationale of it is that the role of the residue of f at a point z_0 is to pick out the part of the function which has a non-zero integral along a circuit round z_0. For instance, the residue at $z = 0$ of $\sum\limits_{-\infty}^{\infty} a_n z^n$ (with sum $f(z)$ for $0 < |z|$) is a_{-1} and

$$a_{-1} = \frac{1}{2\pi i} \int_C f,$$

where C is (say) a positively oriented circle with centre $z = 0$. Positive orientation is equivalent to the point 0 lying on the left as the circuit is traversed. Now a circle $|z| = R$ is a circuit round $z = \infty$, and, to keep ∞ on the left, it must be traversed clockwise. (The Riemann spherical representation §10.4 illustrates this.) Thus we arrive at the definition of the residue at ∞ as being $-a_{-1}$.

The introduction of the residue at ∞ would have the following consequence.

If a function has a finite set S of singularities in $\overline{Z}$, the sum of the residues at points of S, with ∞ included, is zero.

§**12.8.** Some writers attribute theorem 12.8 to Sochozki.

Picard's theorem. Theorem 12.8 shows that an entire function (other than a polynomial) assumes values arbitrarily close to any assigned number. We noted that the special function $\exp z$ is, by choice of z, equal to any assigned number, except 0, and does not merely take values near to it. This is a general truth for entire functions, discovered by E. Picard (1856–1941). His original theorem (1879) was that a function meromorphic in Z which omits more than two values is a constant. The proof depended on assuming, on the contrary, three values to be omitted, projecting them into 0, 1, ∞ and then drawing on knowledge which was already available about a class of functions (elliptic modular functions) which fail to take just those three values. Picard's theorem led to a vast literature in which the names of Landau, Schottky, Bloch, R. Nevanlinna and F. Nevanlinna are prominent. For a text-book account of the subject (which is impossible for us to broach) see, for instance, Hille, *Analytic Function Theory*, vol. II.

13

GENERAL THEOREMS. ANALYTIC FUNCTIONS

13.1. Regular functions represented by series or integrals

The outlook in this chapter is wider than in chapter 12, and some of the results are more general than are needed in every-day analysis. The reader who is primarily interested in applying his knowledge to special functions can read chapter 14 with occasional reference back to the present chapter.

Taylor's theorem 12.11 exhibited a regular function as the limit of a sequence of polynomials. Laurent's theorem 12.31 allowed negative as well as positive powers of the variable. The present section contains representation theorems in a much wider setting. The first, due to Weierstrass, replaces Taylor's series in powers of z by a uniformly convergent sequence of functions on which the sole restriction is regularity.

Theorem 13.11. *Suppose that* (i) *for each n, $s_n(z)$ is regular in the region D; and* (ii) *on every compact subset of D, $s_n(z)$ tends uniformly to $s(z)$ as n tends to ∞. Then* (1) *$s(z)$ is regular in D,* (2) *the derivative of any order p satisfies $s^{(p)}(z) = \lim s_n^{(p)}(z)$.*

Proof. Let a be any point of D. Choose r small enough for the closure of the disc $B(a; r)$ to lie in D. Then $s(z)$, the uniform limit of the sequence of continuous functions $s_n(z)$, is continuous in $B(a; r)$, and so at every point of D.

We shall use Morera's theorem 11.92 to prove that $s(z)$ is regular. Let γ be any circuit in $B(a; r)$. Since $s_n(z)$ tends uniformly to $s(z)$ on γ, we have, from theorem 11.12,

$$\int_\gamma s_n(z)\,dz \to \int_\gamma s(z)\,dz.$$

For each n, the left-hand side is 0 (by Cauchy's theorem 11.5). So the right-hand side is 0 and, by Morera's theorem, $s(z)$ is regular in D.

To prove (2), let $C = C(a; r)$ be the circular boundary of $B(a; r)$, positively oriented. Then, since $s_n(\zeta) \to s(\zeta)$ uniformly for ζ on C, and $(\zeta - a)^{-p-1}$ is bounded on C,

$$\lim_{n \to \infty} \int_C \frac{s_n(\zeta)}{(\zeta - a)^{p+1}} \, d\zeta = \int_C \frac{s(\zeta)}{(\zeta - a)^{p+1}} \, d\zeta.$$

Multiplying by $(p!)/2\pi i$, we have (2). |

Corollary 1. *The convergence of $s_n^{(p)}(z)$ to $s^{(p)}(z)$ is uniform on every compact subset of D.*

Proof. We refine the proof of (2). Let z be any point in the closure of the disc $B(a; \frac{1}{2}r)$. Then

$$\lim_{n \to \infty} \int_C \frac{s_n(\zeta)}{(\zeta - z)^{p+1}} = \int_C \frac{s(\zeta)}{(\zeta - z)^{p+1}} \, d\zeta,$$

since now $|\zeta - z| \geqslant \frac{1}{2}r$. This gives

$$\lim s_n^{(p)}(z) = s^{(p)}(z)$$

uniformly on the closure of the disc $B(a; \frac{1}{2}r)$, that is to say, given ϵ,

$$|s_n^{(p)}(z) - s^{(p)}(z)| < \epsilon \quad \text{if} \quad n > N(\epsilon, a) \quad \text{and} \quad z \in \overline{B(a; \frac{1}{2}r)}.$$

Let now K be a given compact subset of D. By the Heine–Borel theorem 3.7, a finite number of points a can be found such that the (open) discs $B(a; \frac{1}{2}r)$ associated with them cover K.

If N is the greatest of the corresponding numbers $N(\epsilon, a)$ defined above, we have

$$|s_n^{(p)}(z) - s^{(p)}(z)| < \epsilon \quad \text{if} \quad n > N, \quad \text{for all } z \text{ in } K. \mid$$

Corollary 2. *The hypothesis* (i) *of the theorem may be replaced by* (i)' $s_n(z)$ *is regular in* D_n, *where* $D_n \subset D_{n+1}(n = 1, 2, \ldots)$ *and* $D = \bigcup D_n$.

Proof. At each stage of the proof of the theorem we are working with a compact set, say K. If $K \subset \bigcup_1^\infty D_n$, then, by the Heine–Borel theorem 3.7, $K \subset \bigcup_1^N D_n$. Since $D_n \subset D_{n+1}$, we have $K \subset D_N$. We ignore those s_n for which $n < N$ and argue as in theorem 13.11. |

Illustration.
$$s_n(z) = \frac{z}{2z^n + 1}$$

is regular in D_n, the disc $|z| < 2^{-1/n}$. $s(z) = z$ in $|z| < 1$. This example is given by Ahlfors, *Complex Analysis* (p. 174).

Theorem 13.11, defining a regular function as the limit of a sequence (or as the sum of an infinite series) will have a parallel in which summation over the positive integers is replaced by integration with respect to a continuous variable u. We suppose u to be real, as it usually is in practice; the extension to a complex variable and a prescribed path of integration is straightforward.

We state two theorems about the regularity of a function represented by an integral, in the first over a finite interval and in the second (the analogue of theorem 13.11) over an infinite interval.

Theorem 13.12. *Let $s(u, z)$ be a continuous function of the pair (u, z) for $0 \leqslant u \leqslant 1$, z in a region D. Let $s(u, z)$ be regular in D for each u. Then*

$$f(z) = \int_0^1 s(u, z)\, du$$

is regular in D, and its derivatives are given by

$$f^{(p)}(z) = \int_0^1 \frac{\partial^p}{\partial z^p} s(u, z)\, du.$$

Proof. Since s is continuous, f is continuous in D.

As in theorem 13.11, given a disc in D we take in it a circuit with a representation $z = z(t)$. Then, since the integrand is continuous in (u, z) and the integrals along γ are real integrals with respect to t, it can be seen from theorem 8.31 that

$$\int_0^1 du \int_\gamma s(u, z)\, dz = \int_\gamma dz \int_0^1 s(u, z)\, du.$$

The left-hand side is 0, since, by Cauchy's theorem 11.5, the inner integral is 0 for each u. Therefore the right-hand side, namely $\int_\gamma f(z)\, dz$, is also 0. By Morera's theorem 11.92, f is regular in D.

To prove the formula for $f^{(p)}$, let C be a circle with centre z lying in D. From theorem 11.91

$$f^{(p)}(z) = \frac{p!}{2\pi i} \int_C \frac{f(\zeta)}{(\zeta - z)^{p+1}}\, d\zeta$$

$$= \frac{p!}{2\pi i} \int_C \frac{d\zeta}{(\zeta - z)^{p+1}} \int_0^1 s(u, \zeta)\, du.$$

The integrand of the repeated integral is continuous in (u, ζ) and we may invert the order of integration, obtaining

$$f^{(p)}(z) = \int_0^1 \frac{\partial^p}{\partial z^p} s(u, z)\, du. \qquad \rule{2pt}{10pt}$$

Theorem 13.13. *Let $s(u, z)$ be a continuous function of the pair (u, z) for $u \geqslant 0$ and for all z in D. Let $s(u, z)$ be regular in D for each u. Suppose that, on every compact subset of D, $\int_0^\infty s(u, z)\, du$ converges uniformly to $f(z)$. Then $f(z)$ is regular in D and*

$$f^{(p)}(z) = \int_0^\infty \frac{\partial^p}{\partial z^p} s(u, z)\, du.$$

Proof. The only modification needed in the proof of theorem 13.12 is to appeal to uniformity of convergence of $\int_0^\infty s(u, z)\, du$ to assert the continuity of f and to justify inverting the repeated integral in which it occurs. |

An alternative extension of theorem 13.12 caters for unboundedness of the integrand $s(u, z)$, say as $u \to 0+$, instead of integration of $s(u, z)$ over an infinite range $[0, \infty)$ in theorem 13.13.

Exercises 13(a)

1. Investigate in what regions of Z (if any), the following limits of sequences (or sums of series) represent regular functions:

(i) $\lim\limits_{n \to \infty} \dfrac{z^n - 1}{z^n + 1}$,   (ii) $\lim\limits_{n \to \infty} \dfrac{z^n - 3}{3z^n + 1}$,   (iii) $\lim\limits_{n \to \infty} \dfrac{z^n}{z^{2n} + 1}$,

(iv) $\sum\limits_0^\infty \left\{ \left(\dfrac{z}{2}\right)^n + \left(\dfrac{2}{z}\right)^n \right\}$,   (v) $\sum\limits_0^\infty \left\{ \left(\dfrac{z-1}{3}\right)^n + \left(\dfrac{2}{z}\right)^n \right\}$,   (vi) $\sum\limits_1^\infty \dfrac{z^n}{1 - z^n}$,

(vii) $\sum\limits_1^\infty \dfrac{\sin nz}{n^2}$,   (viii) $\sum\limits_1^\infty \dfrac{\sin nz}{2^n}$.

2. Investigate as in **1** these integrals:

(i) $\displaystyle\int_0^{2\pi} \dfrac{d\theta}{1 + z \sin \theta}$,   (ii) $\displaystyle\int_0^1 \exp(-zt^2)\, dt$,   (iii) $\displaystyle\int_1^\infty \exp(-zt^2)\, dt$,

(iv) $\displaystyle\int_0^1 \dfrac{\sin t}{t^z}\, dt$,   (v) $\displaystyle\int_1^\infty \dfrac{\sin t}{t^z}\, dt$,   (vi) $\displaystyle\int_0^1 \dfrac{\sin zt}{t}\, dt$,

(vii) $\displaystyle\int_1^\infty \dfrac{\sin zt}{t}\, dt$,   (viii) $\lim\limits_{T \to \infty} \displaystyle\int_{-T}^T \dfrac{e^{zt}}{a + it}\, dt \quad (a \neq 0)$.

3. Functions f, g are given which are regular in Z. From **1** (i) or otherwise construct a sequence ϕ_n such that

$$\lim_{n \to \infty} \phi_n(z) = \begin{cases} f(z) & \text{if} \quad |z| < 1, \\ g(z) & \text{if} \quad |z| > 1. \end{cases}$$

4. Show that, for every fixed N,

$$F_N(z) = \sum_{n=2N}^{\infty} \frac{1}{(n^2+z^2)(n^2z^2+1)}$$

converges uniformly on the annulus $1/N \leqslant |z| \leqslant N$. Deduce that $F_1(z)$ is regular except for certain singularities, and state the nature of the singularities.

5. In theorem 13.13 prove that the convergence of $\int_0^{\infty} \partial^p s(u, z)/\partial z^p \, du$ to $f^{(p)}(z)$ is uniform on every compact subset of D.

6. Apropos the remark following theorem 5.62, give an example of a function of z continuous in a domain D which is not the uniform limit of a sequence of polynomials in z.

13.2. Local mappings

The results of chapter 12, particularly §12.6, enable us to analyse the local properties of the correspondence effected by $w = f(z)$. We studied simple examples in chapter 10. Throughout this section we shall assume that f is not a constant.

Theorem 13.21. *Let f be regular at a, and let $f(z)-b$ have a zero of order n for $z = a$. Then, if ϵ is small enough, there is a corresponding δ such that, for every b' in the disc $|b-b'| < \delta$, the disc $|z-a| < \epsilon$ contains n zeros of $f(z)-b'$.*

Proof. The zeros of f being isolated (§12.1), we may suppose ϵ small enough for a to be the sole zero of the regular function $f-b$ in the closed disc $|z-a| \leqslant \epsilon$.

We can take $\delta > 0$ such that

$$|f(z) - b| \geqslant \delta$$

for every z on the circle $|z - a| = \epsilon$.

Let b' satisfy $|b - b'| < \delta$.

By Rouché's theorem 12.62,

$$f(z) - b \quad \text{and} \quad \{f(z) - b\} + \{b - b'\}$$

have the same number of zeros inside the circle $|z - a| = \epsilon$. The former has n zeros and therefore so has $f(z) - b'$. |

Corollary 1. *A function regular on an open set G and not constant on any of its components maps open subsets of G onto open sets.*

Proof. With the notation of the theorem, if f is regular at a, and $f(a) = b$, the z-disc $B(a; \epsilon)$ is mapped onto a w-set containing the disc $B(b; \delta)$. |

Corollary 2. If f is regular at a and f'(a) $\neq$ 0, there is a neighbourhood of a in which w = f(z) has an inverse mapping regular at b = f(a).

Proof. Since $f'(a) \neq 0$, we have $n = 1$ in theorem 13.21. By continuity of f', we can take the ϵ in the theorem small enough for f' to be different from 0 in the disc $B(a; \epsilon)$.

There is a bijective mapping of $B(b; \delta)$ onto an open subset G of $B(a; \epsilon)$ containing a. The function f restricted to G has an inverse g and, by corollary 1 and theorem 3.23, g is continuous.

Given b_1 and w_1 in the disc $B(b; \delta)$, there are unique a_1 and z_1 in G with $g(b_1) = a_1$ and $g(w_1) = z_1$. Then

$$\frac{g(w_1) - g(b_1)}{w_1 - b_1} = \frac{z_1 - a_1}{f(z_1) - f(a_1)}.$$

Let $w_1 \to b_1$. Since g is continuous, $z_1 \to a_1$ and the right-hand side tends to $1/f'(a_1)$, since $f'(a_1) \neq 0$. Hence $g'(b_1)$ exists. Since b_1 is an arbitrary point in the disc $B(b; \delta)$, g is regular at b. |

Theorem 13.21 yields a basic proof of the Maximum Principle (which is already contained in exercise 12(*a*), 14, a deduction from Cauchy's integral formula. See exercise 13(*b*), 8(ii).)

Theorem 13.22. (The maximum principle.) *If f is regular in a region D (and not constant), then $|f|$ does not have a maximum in D.*

Proof. If a is a point of D, and $b = f(a)$, then any disc Δ (in D) with centre a is mapped onto an open set G containing b. G contains points with modulus greater than $|b|$. |

Corollary. If f is regular at every point of a compact set E, then $|f|$ attains its supremum at a frontier point of E.

The maximum principle has many consequences. As illustrations we append theorems 13.23 and 13.24.

Theorem 13.23. (Schwarz's lemma.) *Suppose that f is regular and $|f(z)| \leqslant M$ for $|z| < R$, and that $f(0) = 0$. Then either*

$$|f(z)| < \frac{M|z|}{R} \quad for \quad 0 < |z| < R$$

or $$f(z) = \frac{M}{R} z e^{i\alpha} \quad (\alpha \ real).$$

Proof. We may take $M = R = 1$, and the general case will then follow by applying that result to $f(Rz)/M$.

Write $g(z) = f(z)/z$ for $z \neq 0$ and $g(0) = f'(0)$.

From theorem 12.4, g is regular for $|z| < 1$.

Given any point z_0 with $|z_0| < 1$, let $|z_0| < r < 1$. Then there is at least one value z_1 with $|z_1| = r$ such that

$$|g(z_0)| \leqslant |g(z_1)| = \frac{|f(z_1)|}{|z_1|} \leqslant \frac{1}{r}.$$

This is true, however near r is to 1, and therefore

$$|g(z_0)| \leqslant 1.$$

If there is a ζ such that $|g(\zeta)| = 1$, then, by theorem 13.22, g is constant, say $g(z) = A$, where $|A| = 1$, and $f(z) = Az$. If there is no such ζ, then $|g(z)| < 1$ for $|z| < 1$, that is to say $|f(z)| < |z|$ for $0 < |z| < 1$. |

Corollary. *With the hypothesis of the theorem either*

$$|f'(0)| < M/R \quad or \quad f(z) = \frac{M}{R} z e^{i\alpha}.$$

Proof. In the theorem, either $|g(0)| < 1$ or $g(z) = A$, where $|A| = 1$. From this the inequality for $|f'(0)|$ follows. |

The next theorem gives finality to §10.5 (8).

Theorem 13.24. *Every bijective conformal mapping of a disc onto another is bilinear.*

Proof. We lose no generality by taking the discs to be $|z| < 1$ and $|w| < 1$.

(i) We prove first that, if $z = 0$ is mapped into $w = 0$, the mapping is $w = z e^{i\alpha}$ (α real).

Denote the mapping by $w = f(z)$ and its inverse by $z = g(w)$.

Then
$$|f(z)| < 1 \quad if \quad |z| < 1.$$

By theorem 13.23 (corollary),

either $\quad |f'(0)| < 1 \quad$ or $\quad f(z) = z e^{i\alpha}$.

Similarly $\qquad |g'(0)| < 1 \quad$ or $\quad g(w) = w e^{-i\alpha}$.

Since $\qquad\qquad f'(0)g'(0) = 1$,

we must take the second alternatives.

(ii) Let the mapping $w = L(z)$ transform $|z| < 1$ onto $|w| < 1$ with $L(0) = w_0$.

From §10.5 (8) the transformation

$$\zeta = T(w) = \frac{w - w_0}{\overline{w}_0 w - 1}$$

maps $|w| < 1$ onto $|\zeta| < 1$.

The compound transformation $\zeta = T\{L(z)\}$ maps the disc $|z| < 1$ onto $|\zeta| < 1$ with $z = 0$ mapped into $\zeta = 0$. By (i) of this proof,

$$T\{L(z)\} = z e^{i\alpha}.$$

Therefore
$$L(z) = T^{-1}(z e^{i\alpha})$$

$$= \frac{w_0 - z e^{i\alpha}}{1 - \overline{w}_0 z e^{i\alpha}},$$

which is bilinear. |

Exercises 13(b)

$M(r)$ always denotes sup $|f(z)|$ for $|z| = r$.

1. Calculate $M(r)$ for the functions

    (i) exp iz,   (ii) sin z,   (iii) $1 - \dfrac{z}{3!} + \dfrac{z^2}{5!} - \ldots + \dfrac{(-1)^n z^n}{(2n+1)!} + \ldots.$

2. Prove the *minimum principle*: if f is regular and not constant in a region D, then $|f|$ can have a minimum in D only at a zero of f.

3. If f is regular in $B(0; R)$, where $R > 1$ and $M(1) < 1$, prove that there is just one point z_0 in $B(0; 1)$ for which $f(z_0) = z_0$.

4. If f is a polynomial of degree n, prove that

$$\frac{M(r_1)}{r_1^n} \geqslant \frac{M(r_2)}{r_2^n}$$

for $0 < r_1 < r_2$, with the sign of equality only if $f(z) = az^n$.

5. f is regular and bounded in $B(0; 1)$. As $r \to 1-$, $f(re^{i\theta}) \to 0$ uniformly for $\alpha \leqslant \theta \leqslant \beta$. Prove that $f \equiv 0$.

6. *The three-circles theorem of Hadamard* (1865–1963). Let f be regular for $r_1 \leqslant |z| \leqslant r_2$. By applying the maximum principle to $z^p\{f(z)\}^q$ for suitable p, q, prove that, for $r_1 < r < r_2$,

$$M(r)^{\log (r_2/r_1)} \leqslant M(r_1)^{\log (r_2/r)} M(r_2)^{\log (r/r_1)}.$$

Verify also that log $M(r)$ is a convex function of log r. (A function is convex if the arc joining any two points is below (or coincides with) the chord.)

7. Let f be regular in $B(0; R)$ and write

$$I_2(r) = \frac{1}{2\pi} \int_0^{2\pi} |f(re^{i\theta})|^2 \, d\theta \quad (0 \leqslant r < R).$$

(See also exercise 12(a), 14.) Prove that
   (i) $I_2(r)$ is a continuous, increasing function of r,
   (ii) $|f(0)|^2 \leqslant I_2(r) \leqslant \{M(r)\}^2$, and deduce theorem 13.22,
   (iii) $\log I_2(r)$ is a convex function of $\log r$.

13.3. The Weierstrass approach. Analytic continuation

We now look more closely into the foundations of complex function theory. The development in which the integral theorem $\int f \, dz = 0$ takes the central place is due to Cauchy. Other avenues of approach were followed by Riemann and by Weierstrass. As a prologue to the rest of this chapter the reader should look again at the discussion in §10.2 of the two values of w arising from $w^2 = z$ and at the uniqueness theorem 12.13 for regular functions.

Weierstrass developed complex function theory using power series as the standard representation of a regular function (in the circle of convergence). It may help the reader if the general discussion is accompanied by examples chosen to be as simple as possible.

Suppose that we are given a regular function f defined in a region D. An immediate question is—can the domain of definition of f be extended to points outside D?

Keeping theorem 12.13 in mind, let us agree on terminology and state some facts.

Definition. *A regular function f defined in a region D will be called a* function element, *written (f, D).*

Suppose that (f_0, D_0) and (f_1, D_1) are function elements, where D_0 and D_1 have common points. Then $D_0 \cap D_1$ is open; without restrictions on D_0 and D_1 it need not be connected. If D_0 and D_1 are *convex*, as we shall always in future assume, $D_0 \cap D_1$ is connected.

Suppose that $f_0(z) = f_1(z)$ for all z in $D_0 \cap D_1$.

Then $(f, D_0 \cup D_1)$, where $f = f_0$ in D_0 and $f = f_1$ in D_1, is a function element.

Definition. *With the assumptions just made, each of (f_0, D_0), (f_1, D_1) is called a* direct analytic continuation *of the other.*

It may be possible to go on with this process and form a chain of function elements

$$(f_0, D_0), (f_1, D_1), \ldots, (f_n, D_n)$$

of which each (after the first) is a direct analytic continuation of the preceding one (each D convex).

Definition. *Any two function elements of such a chain are called analytic continuations of each other.*

Example 1.

$$f_0(z) = 1 + (1-z) + (1-z)^2 + \ldots + (1-z)^n + \ldots,$$

D_0 *is the disc* $|1-z| < 1$.

This example is artificially simple because the power series defining f_0 is a geometric progression which, for z in D_0, can at once be summed to the rational function $1/z$. So the problem of extending the definition of f_0 outside D_0 can be solved by declaring that $f(z)$ shall be $1/z$ for all z except the singularity $z = 0$. But if the series for f_0 had been one without a simple formula for its sum, we could have taken

$$f_1(z) = \frac{1}{a} + \frac{a-z}{a^2} + \frac{(a-z)^2}{a^3} + \ldots + \frac{(a-z)^n}{a^{n+1}} + \ldots$$

with D_1 the disc $|a-z| < |a|$, that is to say the interior of the circle, centre a, through the origin.

If a is not real, D_1 contains points outside D_0 and $f_1 = f_0$ in $D_0 \cap D_1$.

Example 2.

$$f_0(z) = 1 - \frac{1}{2}(1-z) - \frac{1}{2.4}(1-z)^2 - \ldots - \frac{1.3.5\ldots(2n-3)}{2.4.6\ldots 2n}(1-z)^n - \ldots,$$

D_0 *is the disc* $|1-z| < 1$.

Again we have chosen a series whose sum is an elementary function, to enable us to visualize more clearly the process of continuation.

In this example let us write $g(z) = \sqrt{z}$ if, in some specified domain, $\{g(z)\}^2 = z$. Then

$$f_0(z) = \{1 - (1-z)\}^{\frac{1}{2}} = z^{\frac{1}{2}}$$

the principal value of $\sqrt{z}$ in D_0 (see exercise 12(a), 1).

Now construct a chain of function elements. We can take, for instance, the region D_n to be the interior of the circle with centre $\exp \frac{1}{4}n\pi i$ and radius 1. Each D_n overlaps the preceding one, and D_8 is the same region as D_0. As we take a sequence of values of z (say, lying on the circle $|z| = 1$) in the regions common to the pairs $D_0, D_1; D_1, D_2; \ldots; D_8, D_0$, then $\sqrt{z}$, starting at $z = 1$ with the value 1 will return to $z = 1$ with the value -1.

So the function elements (f_0, D_0) and (f_8, D_8) are different although the regions D_0 and D_8 are the same. If we pursue the process of continuation through $D_9, \ldots$ to D_{16}, making a second circumnavigation of the origin, the function element (f_{16}, D_{16}) taking the value $+1$ at $z = 1$ is the same as (f_0, D_0).

The only restriction so far laid on the region D in which a function element is defined is convexity. In writing (f, D), where D is a disc, we do not suppose, unless we say so explicitly, that D is the largest disc in which f is regular. In particular, if f is defined by a power series, D may be its disc of convergence or any disc (not necessarily concentric) inside it.

We go on to prove a simple theorem (13.3) bearing on the question whether (f, D) possesses direct analytic continuations to points outside D. We shall then give examples of function elements which have no such continuations.

Theorem 13.3. *Let D be the disc of convergence of a power series with sum f. Then there is at least one point b on the frontier of D such that (f, D) has no direct analytic continuation (f_1, D_1) for which $b \in D_1$.*

Proof. Let D be the disc $B(a; R)$. We have to prove that there is a b with $|b - a| = R$ for which there is no disc D_1, with centre b, in which a function element (f_1, D_1) continues (f, D). Suppose, on the contrary, that there is no such b. Then, by the Heine–Borel theorem 3.7, there are n elements (f_i, D_i) each continuing (f, D) such that $\bigcup_1^n D_i$ contains every point of $|z - a| = R$. The set $\bigcup_1^n D_i$ covers an annulus $R_2 < |z - a| < R_1$, where $R_2 < R < R_1$. We have to show that $f_i(z) = f_j(z)$ for $z \in D_i \cap D_j$. This follows from theorem 12.13, since, if $D_i \cap D_j \neq \varnothing$, then $D_i \cap D_j \cap D \neq \varnothing$ and

$$f_i(z) = f(z) = f_j(z)$$

for $z \in D_i \cap D_j \cap D$. The functions f_i thus define a function regular in the disc $B(a; R_1)$, and so the power series for f converges in this disc. This contradicts the assumption that R was the radius of convergence of f. |

In examples 1 and 2 above there is no direct analytic continuation into any region containing $z = 0$. In 1 this point is a pole of the function represented by the power series, in 2 a branch point (see §10.2).

Example 3. *The power series*

$$f(z) = 1 + z + z^2 + z^6 + \ldots + z^{n!} + \ldots$$

cannot be continued to any point outside its circle of convergence $|z| = 1$.

Proof. Let p/q be a rational number.

Let $z = r \exp(2\pi i \, p/q)$, where r takes values tending to 1 from below.

Then
$$f(z) = \sum_0^{q-1} z^{n!} + \sum_q^\infty r^{n!}.$$

The second sum on the right tends to ∞ as r tends to 1, and so $|f(z)|$ tends to ∞ as z tends to $\exp(2\pi i \, p/q)$ along the radius.

Let D be a region containing points of $B(0; 1)$ and of its complement. Then D contains points $\exp(2\pi i \, p/q)$. Any function (g, D) which coincides with f in $D \cap B(0; 1)$ is discontinuous at these points $\exp(2\pi i \, p/q)$. So there can be no continuation of f outside the circle. |

In example 3, the circle $|z| = 1$ is a *natural frontier* of f.

13.4. Analytic functions

In attaching the word *regular* to a differentiable function (§10.2) we said that the term *analytic function* was being reserved for a wider meaning, which we now explain.

The relation of being analytic continuations of each other is an equivalence relation between function elements. This fact serves to define an analytic function.

Definition. An analytic function *is an equivalence class of function elements.*

If, for each value of z for which it is defined, an analytic function has one value only, it may be called *one-valued* or *uniform*.

If there are values of z for which function elements have different values, the analytic function is *multiform* (or *many-valued*).

Analytic continuation along a curve. Let γ be a curve given by $z = z(t)$ for $0 \leqslant t \leqslant 1$. Let (f_0, D_0) be a function element defined in a disc D_0 containing $z(0)$. Let

$$0 = t_0 < t_1 < \ldots < t_{n+1} = 1$$

be a dissection of $[0, 1]$. Suppose that there is a chain of function elements (f_k, D_k) such that the arc $z = z(t)$ for $t_k \leqslant t \leqslant t_{k+1}$ lies in the disc $D_k (k = 0, 1, \dots n)$. We say that the function element (f_n, D_n) is obtained from (f_0, D_0) by *analytic continuation along the curve γ*.

To justify this definition we have to prove that such analytic continuation is unique in the sense of the next theorem.

Theorem 13.41. *If $(f_0, D_0), \dots, (f_n, D_n)$ and $(\phi_0, \Delta_0), \dots (\phi_m, \Delta_m)$ are chains of function elements along a given curve and $f_0(z) = \phi_0(z)$ for $z \in D_0 \cap \Delta_0$, then $f_n(z) = \phi_m(z)$ for $z \in D_n \cap \Delta_m$.*

Proof. Suppose that the dissections

$$0 = t_0 < t_1 \dots < t_{n+1} = 1$$

and

$$0 = \tau_0 < \tau_1 \dots < \tau_{m+1} = 1$$

are such that the arc $\gamma[t_i, t_{i+1}] \subset D_i (0 \leqslant i \leqslant n)$ and the arc $\gamma[\tau_j, \tau_{j+1}] \subset \Delta_j (0 \leqslant j \leqslant m)$.

The theorem is true if we prove that if the intervals $[t_i, t_{i+1}]$ and $[\tau_j, \tau_{j+1}]$ have a common point then $f_i(z) = \phi_j(z)$ for $z \in D_i \cap \Delta_j$. Suppose that there are values of i, j for which $f_i \not\equiv \phi_j$ in the non-empty region $D_i \cap \Delta_j$. Take those for which $i+j$ is least. Suppose that $t_i \geqslant \tau_j$. Then $i \geqslant 1$ and the point $z(t_i)$ lies in $D_{i-1} \cap D_i \cap \Delta_j$. Since $i+j$ is minimal, it follows that $f_{i-1} \equiv \phi_j$ in $D_{i-1} \cap \Delta_j$. But $f_{i-1} \equiv f_i$ in $D_{i-1} \cap D_i$.

Therefore $f_i \equiv \phi_j$ in the non-empty region $D_{i-1} \cap D_i \cap \Delta_j$. But f_i and ϕ_j are regular in the region $D_i \cap \Delta_j$. By theorem 12.13 $f_i \equiv \phi_j$ in $D_i \cap \Delta_j$. This contradicts the definition of i, j. |

Continuation along homotopic curves. Suppose that analytic continuation of a function element (f_0, D_0), where $z_0 \in D_0$, is possible along two different curves γ_0 and γ joining z_0 to a point z. We enquire whether both curves lead to the same function element (f, D), where $z \in D$. The nature of analytic continuation as a procedure based on overlapping regions suggests the likelihood of the two curves leading to a common function element if one curve can be continuously deformed into the other, that is to say, if γ_0 and γ are homotopic with the fixed end points z_0, z (as defined in §11.6).

We first deal with curves in a disc.

Theorem 13.42. *Let Δ be a disc containing D_0. Suppose that (f_0, D_0) can be continued along every curve in Δ. Then the analytic function thus defined is one-valued in Δ.*

Proof. Let ζ be the centre of Δ. Let z_0 be a point of D_0. Continuation along the straight segment $[z_0, \zeta]$ leads to a function element

(f, D) where $\zeta \in D$. Take a disc D_1, with centre ζ, contained in D. Then f may be supposed to be represented by its Taylor series about ζ.

By hypothesis we may pursue the continuation of (f, D_1) along any radius of Δ. By theorem 13.3, the Taylor series converges at all points of Δ. |

The extension from a disc to a more general region can now be made as in Cauchy's theorem.

Theorem 13.43. (Monodromy.) *Suppose that, in a region D, two curves γ_0, γ are homotopic with fixed end points z_0, z and that (f_0, D_0), where $z_0 \in D_0$, can be continued along all curves in D. Then continuations along γ_0 and γ lead to a common function element in a region containing z.*

Proof. We operate with the discs B_{ij} and curves γ_{ij}, η_{ij} defined in theorem 11.61. By theorem 13.42, continuation along the curve γ_{ij} is equivalent to continuation along $\eta_{ij}, \gamma_{i,j+1}, -\eta_{i+1,j}$ in order. Fix j and let i take the values of the t-dissection. Then continuation along γ_j and γ_{j+1} lead to a common function element in a region containing the final end point z. This is then true for γ_0 and γ. |

The most useful special case of theorem 13.42 is the following.

Corollary. *If D is simply connected and (f_0, D_0) can be continued along all curves in D, the analytic function thus defined is one-valued in D.*

Exercises 13(c)

1. Which of the following functions on $-1 < x < 1$ have regular extensions to regions in Z;

$$\sqrt{x^2}, \quad \sin x, \quad \tan \tfrac{1}{2}\pi x, \quad \cos |x| ?$$

2. Prove that, if

$$f(z) = \sum_{n=0}^{\infty} z^{2^n} \qquad (|z| < 1),$$

then

$$f(z) = z + f(z^2).$$

Deduce that f cannot be continued over the unit circle.

3. Assume that, if $x > 0$, then

$$\int_0^{\infty} e^{-xt^2} dt = \tfrac{1}{2}\sqrt{(\pi/x)}$$

(proved in § 8.6).

Investigate the extension to $z = x + iy$, and by letting $x \to 0+$, obtain the formula, with $y > 0$,

$$\int_0^{\infty} \cos yt^2 \, dt = \tfrac{1}{2}\sqrt{(\pi/2y)}.$$

4. Prove that, if $a_n \geqslant 0$, and $f(z) = \sum_0^\infty a_n z^n$ has radius of convergence 1, then f cannot be continued over the point 1.

5. Prove that an analytic function can take at most a countable set of values at a given point. Mention two functions for which the set of values is infinite.

6. Prove that the Riemann zeta function defined by $\zeta(z) = \sum_1^\infty n^{-z}$ is regular if $x > 1$, and $\phi(z)$ defined by $\sum_1^\infty (-1)^{n-1} n^{-z}$ is regular if $x > 0$ (n^z having its principal value).

Prove that the function $\phi(z)/(1 - 2^{1-z})$ provides the analytic continuation of $\zeta(z)$ into the punctured region $x > 0, z \neq 1$. Prove that $\zeta(z)$ has a pole at $z = 1$ with residue 1.

7. The function f is regular in a region D which contains the segment $0 \leqslant x \leqslant 1$, $y = 0$. $f(z)$ is real valued on the segment $0 < a \leqslant x \leqslant b < 1, y = 0$.

Prove that the coefficients of the Taylor expansion of f about the origin are real.

Show that this conclusion may be false if it is only known that D contains the point $z = 0$ and the segment $a \leqslant x \leqslant b, y = 0$.

NOTES ON CHAPTER 13

§ 13.2. It is interesting to compare this section with § 7.5 on mappings from R^n to R^n. Since regular functions are much more specialised than general differentiable mappings, the hypotheses in theorems can be less restrictive. For instance, the corollaries to theorem 13.21 hold without the assumption that $f'(z) \neq 0$; to prove theorem 7.52 we have to suppose that the Jacobian is not 0.

§ 13.4. *The Riemann surface.* Consider example 2 of § 13.3 which illustrates a selection of the function elements which generate a *multiform* analytic function. To visualize the mapping which it effects, it is not adequate to set up a single plane Z as the carrier of the variable z. Rather, we need *two* planes Z, one superposed on the other and having connecting bridges. The reader should compare the present section with § 10.2 in which the two branches of $\sqrt{z}$ were kept separate by the veto on analytic continuation across a fixed ray from the origin (the negative real axis).

We will now form a picture of the analytic function defined by $w^2 = z$. Corresponding to any value a of z, there are two function elements (the two square roots of z) each of which is expressible as a power series in $z - a$ convergent in a disc with centre a and radius $|a|$. Imagine these discs made of paper. The value of a function element at a given value of z may be thought of as being marked on the paper. When two function elements are direct analytic continuations of each other, we suppose their discs to be glued together. In example 2, the disc for D_8 will *not* be glued to D_0, but D_{16} is to coalesce with D_0. A model made of paper runs into difficulties because surfaces have to pass through one another. There is no trouble about this in an abstract definition of the surfaces.

So, for the 'two-valued' function $z^{\frac{1}{2}}$, we picture two planes, Z_2 being superposed on Z_1 and joined along the negative real axis in such a manner that a point in Z_1 describing (say) the circle $|z| = 1$ anticlockwise from the point 1 steps up into Z_2 on reaching the value $z = -1$ and, after going again round the circle $|z| = 1$, steps down into Z_1 on reaching $z = -1$. The purpose of this imagery is to design a structure, a *Riemann surface*, carrying the variable z on which the analytic function w is a continuous function of z.

The model would be more informative if we include $z = \infty$ and take extended planes $\overline{Z}_1$ and $\overline{Z}_2$. The points (branch-points) $z = 0$ and $z = \infty$ lie in both planes and the planes have bridges leading from each to the other along some specified line (say the negative real axis) joining 0 and ∞. Instead of extended planes, the spherical representation could be used.

APPLICATIONS TO SPECIAL FUNCTIONS

14.1. Evaluation of real integrals by residues

In this chapter we apply to special functions the methods and theorems of complex analysis (mainly to be found in chapter 12).

The theorem of residues 12.5 can be used to calculate real definite integrals of various types including many which have no indefinite integral expressible by standard functions.

We shall take four examples. The first two are of the form

$$\int_0^\infty r(x) \cos (\text{or sin}) \, mx \, dx,$$

where $r(x)$ is a rational function.

The following remark enables us to write down at sight the residue of a function at a pole of the first order.

If f/g has a simple pole at $z = a$, with $g(a) = 0$, $f(a) \neq 0$, its residue there is $f(a)/g'(a)$.

Proof.
$$\frac{f(z)}{g(z)} = \frac{a_{-1}}{z-a} + f_1(z),$$

where f_1 is regular at a. We then have for the residue

$$a_{-1} = \lim \frac{f(z)(z-a)}{g(z)-g(a)} = \frac{f(a)}{g'(a)}.$$

The remark applies only to a pole of the first order.

Example 1.

$$\int_0^\infty \frac{\cos mx}{x^2 + a^2} \, dx, \quad where \quad m > 0, a > 0.$$

Solution. Let
$$f(z) = \frac{\exp miz}{z^2 + a^2}.$$

Let $I = \int_\gamma f(z) \, dz$, where γ is the path from $-R$ to R of the real axis, and $R > a$. Let $J = \int_\Gamma f(z) \, dz$, where Γ is the semicircle $|z| = R$, $0 \leqslant \text{ph} \, z \leqslant \pi$. Then

$$I = \int_{-R}^R \frac{\cos mx + i \sin mx}{x^2 + a^2} \, dx = 2 \int_0^R \frac{\cos mx}{x^2 + a^2} \, dx.$$

The integrand f is regular for all z except simple poles at ia and $-ia$. The circuit $\gamma+\Gamma$ is seen to be homotopic to a point, and its index is 1 for ia and is 0 for $-ia$.

The remark just preceding this example shows that the residue of f at ia is $\exp(-ma)/2ia$.

The residue theorem 12.5 then gives

$$I+J = (\pi/a)\exp(-ma).$$

The modulus of the integrand in J is bounded by $1/(R^2-a^2)$, and so $|J| \leqslant \pi R/(R^2-a^2)$, and $J \to 0$ as $R \to \infty$. Collecting our results, we have proved that the integral asked for is $(\pi/2a)\exp(-ma)$.

Notes. (1) The essence of the proof that $J \to 0$ could be written

$$J = O(1/R^2)O(R) = O(1/R) \to 0.$$

(2) Where is the hypothesis $m > 0$ used? What is the value of the integral if $m \leqslant 0$?

(3) In setting up the complex integrand $f(z)$, could we not replace $\cos mx$ by $\cos mz$ instead of bringing in the exponential function?

Example 2. *Prove that*

$$\int_0^\infty \frac{\sin x}{x}\,dx = \tfrac{1}{2}\pi.$$

Preliminary survey. If we follow the method which was successful in example 1, the best upper bound that we can offer for the integrand on the large semicircle is $O(1/R)$ and for the integral the bound $O(1)$. But we need the closer estimate $o(1)$ for the integral. This calls for an argument more delicate than the boundedness of $\left|\int_\Gamma f\right|$ by the maximum of $|f|$ multiplied by the length of Γ. The extra delicacy is a reflection of the divergence of the integral

$$\int_0^\infty \frac{|\sin x|}{x}\,dx.$$

The following lemma supplies the necessary refinement.

Lemma. (Jordan's lemma.) *Let $q(z)$ be continuous for $|z| > c$, $0 \leqslant \mathrm{ph}\,z \leqslant \pi$; and let M_R be sup $|q(z)|$ on Γ, the semicircle $|z| = R$, $0 \leqslant \mathrm{ph}\,z \leqslant \pi$. Let $m > 0$. Then, as $R \to \infty$,*

$$\int_\Gamma q(z)\exp miz\,dz = O(M_R).$$

Proof. Let $I = \int_\Gamma q(z) \exp miz\, dz$. Then

$$|I| = \left| \int_0^\pi q(R \exp i\theta) \exp \{miR(\cos\theta + i \sin\theta)\} iR \exp i\theta\, d\theta \right|$$

$$\leqslant M_R \int_0^\pi \exp(-mR \sin\theta) R\, d\theta$$

$$= 2M_R \int_0^{\frac{1}{2}\pi} \exp(-mR \sin\theta) R\, d\theta.$$

But $\sin\theta/\theta$ decreases in $0 < \theta \leqslant \frac{1}{2}\pi$ and so $\sin\theta \geqslant 2\theta/\pi$. Hence

$$|I| \leqslant 2M_R \int_0^{\frac{1}{2}\pi} \exp(-2mR\theta/\pi) R\, d\theta$$

$$= 2M_R \frac{\pi}{2m} \left[-\exp(-2mR\theta/\pi) \right]_0^{\frac{1}{2}\pi} \leqslant \frac{\pi M_R}{m}. \ \blacksquare$$

Corollary. *If, as $|R| \to \infty$, $q(Re^{i\theta}) \to 0$ uniformly for $0 \leqslant \theta \leqslant \pi$, then, as $R \to \infty$,*

$$\int_\Gamma q(z) \exp miz\, dz \to 0.$$

Another useful lemma is relevant to example 2.

Lemma. *If f has a simple pole at $z = a$, with residue b, and γ is the path $z = a + \delta e^{i\theta}$ where $\alpha \leqslant \theta \leqslant \beta$, then the limit as $\delta \to 0$ of $\int_\gamma f$ is $ib(\beta - \alpha)$.*

Proof. $$f(z) = \frac{b}{z-a} + g(z),$$

where $|g(z)| \leqslant M$.

$$\int_\gamma \frac{b}{z-a} dz = \int_\alpha^\beta ib\, d\theta, \quad \int_\gamma g = O(\delta). \ \blacksquare$$

Solution of example 2. Let $f(z) = \dfrac{\exp iz}{z}$.

Let $\quad j = \int_\gamma f$, where γ is the semicircle $z = \delta e^{i\theta}$ $(0 \leqslant \theta \leqslant \pi)$,

$$J = \int_\Gamma f, \text{ where } \Gamma \text{ is the semicircle } z = Re^{i\theta}\ (0 \leqslant \theta \leqslant \pi),$$

and let $I = \left(\int_{-R}^{-\delta} + \int_\delta^R \right) \dfrac{\exp ix}{x}\, dx.$

By Cauchy's theorem 11.63,

$$I - j + J = 0,$$

the circuit being homotopic to a point and having index 0 for $z = 0$ the only singularity of f.

As $\delta \to 0$, the last lemma shows that $j \to \pi i$.

As $R \to \infty$, $J \to 0$ by Jordan's lemma.

As $\delta \to 0$ and $R \to \infty$, $I \to 2i \int_0^\infty \frac{\sin x}{x}\, dx$.

This completes the evaluation of the integral by the residue theorem. There are several alternative approaches, of which two have been given in §6.9 and §8.7; and we have referred at the end of chapter 8 to Hardy's assessment of the various methods.

Example 3.
$$\int_{-\infty}^\infty \frac{e^{ax}}{(e^x + 1)^2}\, dx, \quad where \quad 0 < a < 2.$$

This example is chosen to illustrate (i) technique for a pole of order higher than the first, (ii) exploitation of periodicity in (part of) the integrand.

Solution. Put
$$f(z) = \frac{e^{az}}{(e^z + 1)^2}.$$

Integrate anticlockwise round the rectangle whose corners are $\pm X$, $\pm X + 2\pi i$.

$f(x + 2\pi i) = e^{2a\pi i} f(x)$, so the integrals along the two horizontal sides combine to give

$$(1 - e^{2a\pi i}) \int_{-X}^X f(x)\, dx.$$

On the right vertical side $|f(z)| = O(e^{(a-2)X})$ and on the left vertical side $|f(z)| = O(e^{-aX})$. As $X \to \infty$, the integrals along both the sides tend to 0.

The singularities of f are double poles where z is an odd multiple of πi. The rectangular circuit has index 1 for πi and 0 for all others.

Put $z = \pi i + \zeta$ and expand f in its Laurent series in ζ. The residue is the coefficient of ζ^{-1} and that is all that we need to know.

We have, using a special case of theorem 5.44,

$$f(\pi i + \zeta) = e^{a\pi i}(1 + a\zeta + \ldots)(-\zeta - \tfrac{1}{2}\zeta^2 - \ldots)^{-2}$$

$$= e^{a\pi i}\left(\frac{1}{\zeta^2} + \frac{a-1}{\zeta} + \ldots\right).$$

Thus the residue is $e^{a\pi i}(a - 1)$.

Letting $X \to \infty$, we have

$$(1 - e^{2a\pi i}) \int_{-\infty}^{\infty} f(x)\,dx = 2\pi i e^{a\pi i}(a-1)$$

and hence

$$\int_{-\infty}^{\infty} f(x)\,dx = \frac{\pi(1-a)}{\sin a\pi}.$$

If $a = 1$, the last line but one says $0 = 0$. The integral is then elementary, by the substitution $e^x = u$, and its value is 1, which is in fact the limit of $\pi(1-a)/\sin a\pi$.

Example 4. $\displaystyle\int_0^\infty \frac{x^{p-1}}{x+1}\,dx, \quad where \quad 0 < p < 1.$

Observe that the conditions $p > 0$ and $p < 1$ are necessary and sufficient for the existence of the integral at 0 and ∞ respectively.

The general type, of which this is a specimen is $\displaystyle\int_0^\infty x^{p-1}q(x)\,dx$, where $q(z)$ has no singularity on the positive real axis $x \geq 0$.

The feature of this example is that z^{p-1} is multiform, $z = 0$ being the branch point. To avoid ambiguity we must specify a branch of any multiform function and keep to it.

We shall deal with this example exhaustively—by three different methods.

Method (i). In the diagram the large and small circles have radii R, δ. The angle BOC is α, where $0 < \alpha < \pi$.

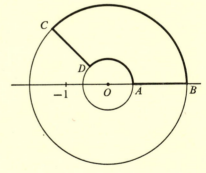

Let Γ_1 be the circuit composed of AB, arc BC (anticlockwise), CD, arc DA (clockwise).

Let Γ_2 be the circuit DC, CB (anticlockwise), BA, AD (clockwise).

Define $f_1(z) = \dfrac{z^{p-1}}{z+1}$, where z^{p-1} has its principal value. Then there is a simply connected region containing Γ_1 and excluding the origin, in which f_1 is regular. By theorem 11.62,

$$\int_{\Gamma_1} f_1 = 0.$$

We now need a function f_2 in a sector containing Γ_2 and agreeing with f_1 along CD. Define $f_2(z) = \dfrac{z^{p-1}}{z+1}$, with $z^{p-1} = r^{p-1} \exp{(p-1)\theta i}$, where $z = re^{i\theta}$ and (say) $\alpha - \eta < \theta < 2\pi + \eta$ (taking $0 < \eta < \frac{1}{2}\alpha$). The function f_2 is regular except for a simple pole at $z = -1$, where the residue is $\exp(p-1)\pi i$. By theorem 12.5,

$$\int_{\Gamma_2} f_2 = 2\pi i \exp(p-1)\pi i$$

$$= -2\pi i \exp p\pi i.$$

Observe then that, on BA,

$$f_2(x) = \frac{x^{p-1}}{x+1} \exp 2(p-1)\pi i.$$

Take

$$\int_{\Gamma_1} f_1 + \int_{\Gamma_2} f_2.$$

The integrals along CD, DC cancel. The integrals along the arcs of the outer circle are $O(R^{p-1})$ and along the small circle $O(\delta^p)$. Letting $\delta \to 0$ and $R \to \infty$, we have

$$(1 - e^{2p\pi i}) \int_0^\infty \frac{x^{p-1}}{x+1} dx = -2\pi i e^{p\pi i}$$

and hence

$$\int_0^\infty \frac{x^{p-1}}{x+1} dx = \frac{\pi}{\sin p\pi}.$$

Method (ii). By the change of variable $x = u^2$ we have

$$2 \int_0^\infty \frac{u^{2p-1}}{u^2 + 1} du.$$

The reader can verify that he can integrate the function $z^{2p-1}/(z^2 + 1)$, where z^{2p-1} is suitably defined, round the circuit composed of $-R \leqslant x \leqslant -\delta$, $\delta \leqslant x \leqslant R$ and the two semicircles with radii R, δ in the upper half plane. The residue at the pole $z = i$ is

$$(1/2i) \exp(2p-1)\tfrac{1}{2}\pi i,$$

which is $-\frac{1}{2} \exp p\pi i$.

Method (iii). The substitution $x = e^u$ gets rid of the multiformity of the integrand and yields an integral of the type of example 3.

Exercises 14(a)

1. Prove that

$$\int_{-\infty}^\infty \frac{x^4 dx}{1 + x^8} = \frac{\pi}{\sqrt{2}} \sin \frac{\pi}{8}.$$

(Choose a circuit for your complex integral which includes one pole, not four.)

2. Evaluate

$$I_n = \int_0^\infty \frac{dx}{1+x^n} \qquad (n = 2, 3, \ldots).$$

Deduce that $I_n \to 1$ as $n \to \infty$. Also prove this from the definition of I_n without calculating its value.

3. Prove that, if $a > 0$, $b > 0$, $a \neq b$,

$$\int_{-\infty}^\infty \frac{dx}{(x^2+a^2)^2(x^2+b^2)} = \frac{\pi(2a+b)}{2a^3b(a+b)^2}.$$

Is the formula valid for $a = b > 0$?

4. Prove that, if $a > 0$, $k > 0$,

$$\int_0^\infty \frac{x^3 \sin kx}{x^4+4a^4}\, dx = \tfrac{1}{2}\pi e^{-ka} \cos ka.$$

What is the value of the integral for other real values of k?

5. In example 2 of §14.1, prove that, instead of using Jordan's lemma for the semicircle $z = Re^{i\theta}$ $(0 \leqslant \theta \leqslant \pi)$ we may integrate by parts along it. Also verify that the semicircle may be replaced by other paths receding to infinity, such as the rectangular path formed with $y = 0$ by the lines $x = \pm R$, $y = R$.

6. Prove that, as $X \to \infty$,

$$\int_0^X \frac{\cos ax - \cos bx}{x^2}\, dx = \tfrac{1}{2}\pi(b-a) + O\left(\frac{1}{X^2}\right) \quad (a > 0, b > 0).$$

7. Prove that, if $m > 0$, $b > 0$,

$$\int_0^\infty \frac{(x^2-a^2)\sin mx}{x(x^2+b^2)}\, dx = \frac{\pi}{2b^2}\{(a^2+b^2)e^{-mb} - a^2\}.$$

8. By integrating $(1-e^{-z})/z$ round a suitable quadrilateral, or otherwise, prove that

$$\int_b^B \frac{e^{-az} - e^{-Az}}{z}\, dz = \int_a^A \frac{e^{-bz} - e^{-Bz}}{z}\, dz,$$

where a, A, b, B are complex, and the paths of integration are linear segments.
    Prove that, if $a > 0$, $c > 0$,

$$\int_0^\infty \frac{e^{-ax} - \cos cx}{x}\, dx = \log\frac{c}{a}, \qquad \int_0^\infty \frac{\sin cx}{x}\, dx = \frac{\pi}{2}.$$

9. By integrating a suitable function round the unit circle, prove that

$$\int_0^{2\pi} \frac{\sin^2 \theta\, d\theta}{1 - 2k \cos \theta + k^2} = \pi \qquad (-1 < k < 1).$$

Evaluate the integral for other real values of k.

10. Evaluate (n being a positive integer)

$$\int_0^{2\pi} e^{\cos \theta} \cos (n\theta - \sin \theta)\, d\theta, \qquad \int_0^{2\pi} e^{\cos \theta} \sin (n\theta - \sin \theta)\, d\theta.$$

11. Prove that, if $0 < a < 1$,

$$\int_{-\infty}^{\infty} \frac{e^{ax}}{e^x + 1}\, dx = \frac{\pi}{\sin \pi a}.$$

Deduce, or prove independently, that, if $-1 < b < 1$,

$$\int_0^{\infty} \frac{\cosh bx}{\cosh x}\, dx = \tfrac{1}{2}\pi \sec \tfrac{1}{2}\pi b.$$

12. Prove that, if $0 < a < 1$,

$$(P) \int_{-\infty}^{\infty} \frac{e^{ax}}{e^x - 1}\, dx = -\pi \cot \pi a,$$

where the principal value (P) denotes

$$\lim_{\delta \to 0} \left\{ \int_{-\infty}^{-\delta} + \int_{\delta}^{\infty} \right\}.$$

Prove that, if $-\pi < c < \pi$,

$$\int_0^{\infty} \frac{\sinh cx}{\sinh \pi x}\, dx = \tfrac{1}{2}\tan \tfrac{1}{2}c.$$

13. Prove that, if $-1 < a < 1,\ a \neq 0,\ 0 < \theta < \pi$,

$$\int_0^{\infty} \frac{x^a}{1 + 2x \cos \theta + x^2}\, dx = \frac{\pi}{\sin \theta} \frac{\sin a\theta}{\sin a\pi}. \quad \text{(Euler.)}$$

14. By integrating $(\log z)^3/(1 + z^2)$ round a suitable circuit, or otherwise, prove that

$$\int_0^{\infty} \frac{\log x}{1 + x^2}\, dx = 0 \quad \text{and} \quad \int_0^{\infty} \frac{(\log x)^2}{1 + x^2}\, dx = \frac{\pi^3}{8}.$$

15. If $f(z) = \exp \pi i z^2 / \sin \pi z$, prove that

$$f(z) - f(z - 1) = 2i \exp \{\pi i z(z - 1)\}.$$

By integrating f round the parallelogram with vertices $\pm \tfrac{1}{2} \pm R \exp \tfrac{1}{4}\pi i$, where R is large, show that

$$\int_{-\infty}^{\infty} \exp(-\pi t^2)\, dt = 1.$$

(It was long supposed that complex integration did not help in evaluating

$$\int_{-\infty}^{\infty} e^{-x^2}\, dx.$$

This integral has already been calculated on p. 285.)

16. Integrate $\exp(-z^2)$ round a sector of a circle to prove that

$$\int_0^{\infty} \cos x^2\, dx = \int_0^{\infty} \sin x^2\, dx = \frac{\sqrt{\pi}}{2\sqrt{2}}.$$

Deduce that $\dfrac{1}{\sqrt{x}} = \sqrt{\left(\dfrac{2}{\pi}\right)} \displaystyle\int_0^{\infty} \frac{1}{\sqrt{u}} \cos \text{ (or sin) } ux\, du.$

Exercises **17, 18** concern Fourier transforms (§9.9).

17. If $\phi(x) = \displaystyle\int_0^\infty \exp\left(-\tfrac{1}{2}u^2\right)\cos ux\,du$, prove that $\phi'(x) = -x\phi(x)$.

Deduce that $\exp\left(-\tfrac{1}{2}x^2\right)$ is its own cosine transform.
Prove that $x\exp\left(-\tfrac{1}{2}x^2\right)$ is its own sine transform.

18. Prove that $\operatorname{sech} x\sqrt{(\tfrac{1}{2}\pi)}$ is its own Fourier cosine transform, and that

$$\frac{1}{\exp\{x\sqrt{(2\pi)}\}-1} - \frac{1}{x\sqrt{(2\pi)}}$$

is its own sine transform.

19. *The Laplace transform.* Suppose that

$$F(z) = \int_0^\infty e^{-zt}f(t)\,dt,$$

the integral converging absolutely if re $z > c$, where $c \geqslant 0$. Suppose that f satisfies sufficient conditions for the convergence at t of its Fourier series. Using theorem 9.95, prove that, if $a > c$,

$$\tfrac{1}{2}\{f(t+)+f(t-)\} = \frac{1}{2\pi i}\lim_{U\to\infty}\int_{a-iU}^{a+iU} e^{zt}F(z)\,dz,$$

(integrating along the linear segment $(a-iU,\,a+iU)$).

14.2. Summation of series by residues

The theorem of residues 12.5 can often be used to sum series. To illustrate the method we shall prove that

$$\sum_{n=-\infty}^{\infty}\frac{1}{(z-n)^2} = \frac{\pi^2}{\sin^2 \pi z},$$

where z is any number not an integer.

The gist of the method is that the function $\pi\cot\pi\zeta$ has at each integer value of ζ a simple pole with residue 1. Therefore the desired sum is equal to the sum of the residues of the function of ζ

$$f(\zeta) = \frac{\pi\cot\pi\zeta}{(z-\zeta)^2}$$

at its poles other than the pole $\zeta = z$.

Integrate $f(\zeta)$ round the boundary ∂C_n of the square C_n with corners at $(n+\tfrac{1}{2})(\pm 1 \pm i)$. We shall prove in a moment that, as $n\to\infty$, $\displaystyle\int_{\partial C_n} f(\zeta)\,d\zeta \to 0$; assume that this has been done. Then, putting $\zeta = z+\alpha$, we have

$$f(z+\alpha) = \frac{\pi\cot\pi z}{\alpha^2} - \frac{\pi^2\operatorname{cosec}^2 \pi z}{\alpha} + \phi(\alpha),$$

where $\phi(\alpha)$ is regular at $\alpha = 0$. So the residue of $f(\zeta)$ at $\zeta = z$ is $-\pi^2 \operatorname{cosec}^2 \pi z$. The sum of the residues at $\zeta = n$ for all integers and $\zeta = z$ being zero, we have the required sum.

(What this argument proves is that the principal value

$$\lim_{N \to \infty} \sum_{n=-N}^{N} \frac{1}{(z-n)^2} = \frac{\pi^2}{\sin^2 \pi z}.$$

But, since the sums $\sum\limits_{0}^{\infty}$ and $\sum\limits_{-\infty}^{-1}$ separately converge, the principal value is otiose.)

It has still to be proved that $\displaystyle\int_{\partial C_n} f(\zeta)\, d\zeta \to 0$.

By a short calculation,

$$|\cot (x+iy)|^2 = \frac{\cosh 2y + \cos 2x}{\cosh 2y - \cos 2x}.$$

On the vertical sides of ∂C_n, if $\pi\zeta = x+iy$, then $\cos 2x = -1$ and so $|\cot \pi\zeta| \leqslant 1$. On the horizontal sides $\sup |\cot \pi\zeta|$ tends to 1 as $n \to \infty$ and so $|\cot \pi\zeta| \leqslant 2$ (say). The length of ∂C_n is $4(2n+1)$ and therefore

$$\left| \int_{\partial C_n} f(\zeta)\, d\zeta \right| \leqslant \frac{2\pi}{(n+\frac{1}{2}-|z|)^2}\, 4(2n+1).$$

This is a close estimate, but an O-argument would suffice. |

The reader would find it interesting to prove directly

$$\sum_{-\infty}^{\infty} \frac{1}{(z-n)^2} = \frac{\pi^2}{\sin^2 \pi z}$$

by showing that the difference between the left- and right-hand sides
   (i) is regular except at the values $z = n$,
  (ii) is bounded as $z \to n$,
 (iii) has period 1,
 (iv) tends to 0 as $y \to \infty$, uniformly for $-\frac{1}{2} \leqslant x \leqslant \frac{1}{2}$,
and then applying Liouville's theorem 12.22.

14.3. Partial fractions of cot z

A polynomial $p(z)$ having n zeros can be written as the product of factors

$$A \prod_{1}^{n} (z - a_r).$$

Also a rational function $q(z)/p(z)$ is expressible in partial fractions with the factors of $p(z)$ as denominators.

A vital problem of complex function theory is the resolution of a regular function which has infinitely many zeros into the product of infinitely many factors. There is the associated problem of writing a meromorphic function in partial fractions, displaying its poles (in general a countable set). Questions of convergence will arise. Trigonometric functions provide good illustrations of these problems and we shall prove the formulae

$$\sin \pi z = \pi z \prod_{1}^{\infty} \left(1 - \frac{z^2}{n^2}\right),$$

$$\pi \cot \pi z = \frac{1}{z} + \sum_{-\infty}^{\infty}{}' \left(\frac{1}{z-n} + \frac{1}{n}\right),$$

where the dash after the Σ means that the term for $n = 0$ (which is undefined) is to be omitted.

The latter formula will be proved directly. Infinite products have not yet been discussed and an investigation including the sine product as a particular case will follow in §14.4 and §14.5.

The series summed in §14.2 may be written

$$-\frac{\pi^2}{\sin^2 \pi \zeta} + \frac{1}{\zeta^2} = -\sum_{-\infty}^{\infty}{}' \frac{1}{(\zeta-n)^2}.$$

We shall integrate this equation along a path γ in the ζ-plane from 0 to z which avoids the points $\pm 1, \pm 2, \ldots$. At $\zeta = 0$ the left-hand side, ϕ say, is regular (or the singularity is removable). On the right-hand side the nth term is $O(1/n^2)$ and the series converges uniformly on the path of integration.

The function $\Phi(\zeta) = \pi \cot \pi \zeta - (1/\zeta)$, if $\Phi(0)$ is defined to be 0, is regular at $\zeta = 0$ and $\Phi'(\zeta) = \phi(\zeta)$ ($\zeta \neq \pm 1, \pm 2, \ldots$). Therefore

$$\int_{\gamma} \phi(\zeta) d\zeta = \left[\Phi(\zeta)\right]_0^z = \Phi(z)$$

and we have proved that

$$\pi \cot \pi z - \frac{1}{z} = \sum_{-\infty}^{\infty}{}' \left(\frac{1}{z-n} + \frac{1}{n}\right). \; |$$

The right-hand side has nth term $O(1/n^2)$ and converges absolutely.

If we wish we may remove negative indices by combining the terms for $+n$ and $-n$,

$$\pi \cot \pi z = \frac{1}{z} + \sum_{1}^{\infty} \frac{2z}{z^2 - n^2},$$

but the expression in first-degree partial fractions is more illuminating.

There is a corresponding formula for the cosecant

$$\frac{\pi}{\sin \pi z} = \frac{1}{z} + \sum_{-\infty}^{\infty}{}' (-1)^n \left(\frac{1}{z-n} + \frac{1}{n} \right).$$

Exercises 14(b)

Obtain the expansions in 1–3.

1. $\sec z = 2\pi \displaystyle\sum_{n=0}^{\infty} \frac{(-1)^n (n+\frac{1}{2})}{(n+\frac{1}{2})^2 \pi^2 - z^2}.$

2. $\tan z = 2z \displaystyle\sum_{n=0}^{\infty} \frac{1}{(n+\frac{1}{2})^2 \pi^2 - z^2}.$

3. $\dfrac{z}{e^z - 1} + \dfrac{z}{2} = \dfrac{z}{2} \coth \dfrac{z}{2} = 1 + 2z^2 \displaystyle\sum_{n=1}^{\infty} \frac{1}{z^2 + 4n^2\pi^2}.$

Carry out the summations in 4–7. Observe that, in many such questions, the methods of chapter 9 (particularly §9.8) provide alternatives to those of §14.2.

4. $\displaystyle\sum_{1}^{\infty} \frac{1}{64n^6 - 1} = \frac{1}{2} - \frac{\pi(1+\sqrt{3}\sinh \frac{1}{2}\pi\sqrt{3})}{12\cosh \frac{1}{2}\pi\sqrt{3}}.$

5. $\displaystyle\sum_{1}^{\infty} \frac{1}{n^2(n^2 + k^2)} = \frac{3 + k^2\pi^2 - 3k\pi \coth k\pi}{6k^4} \quad (k \neq 0).$

6. $\displaystyle\sum_{0}^{\infty} \frac{1}{a+bn^2} = \frac{1}{2a} + \frac{\pi}{2\sqrt{(ab)}} \coth \left(\pi\sqrt{\frac{a}{b}} \right) \quad (a > 0, b > 0).$

7. $\displaystyle\sum_{1}^{\infty} (-1)^n \frac{n \sin na}{x^2 - n^2} = \frac{1}{2}\pi \frac{\sin ax}{\sin \pi x} \quad (-\pi < a < \pi, x \neq n).$

Exercises **8, 9** refer to *Bernoulli numbers* which figure in a number of formulae in analysis (e.g., in theorem 14.86).

8. The Bernoulli numbers B_n are definied by

$$\frac{z}{e^z - 1} + \frac{z}{2} = \frac{z}{2} \coth \frac{z}{2} = \sum_{0}^{\infty} \frac{B_n}{n!} z^n.$$

Prove that
(i)   this series has radius of convergence 2π;
(ii) $B_0 = 1$, $B_n = 0$ if n is odd;
(iii)   the sequence of B_n satisfies

$$\sum_{r=0}^{n} 2^{2r} B_{2r} \binom{2n+1}{2r} = 2n+1;$$

(iv) $B_2 = \frac{1}{6}$, $B_4 = -\frac{1}{30}$, $B_6 = \frac{1}{42}$, $B_8 = -\frac{1}{30}$.

9. Use **3** to prove that
$$B_{2k} = (-1)^{k-1} \frac{2(2k)!\, s_{2k}}{(2\pi)^{2k}},$$

where $s_{2k} = \displaystyle\sum_{1}^{\infty} \frac{1}{n^{2k}}$, and that, for large k,

$$|B_{2k}| \sim \frac{2(2k)!}{(2\pi)^{2k}}.$$

14.4 Infinite products

The statement in §14.3 of the factors of $\sin \pi z$ was the first mention in this book of an infinite product. We must lay down the necessary definitions and prove theorems about convergence or divergence. This can be done briefly since no new principle is involved and infinite products are closely linked with infinite series.

Write

$$A_n = \prod_{r=1}^{n} (1 + a_r).$$

Definition. *If, as n tends to infinity, A_n tends to a limit A*, not 0, *we say that the product* $\prod_{r=1}^{\infty} (1 + a_r)$ converges *and has the value A.*

Definition. *Suppose that* $1 + a_r = 0$ *for at least one $r < m$ and for no $r \geqslant m$. Then the product* $\prod_{r=1}^{\infty} (1 + a_r)$ *is said to* converge to zero *if and only if* $\prod_{r=m}^{\infty} (1 + a_r)$ *converges to a non-zero limit.*

Definition. *If* $\prod_{r=1}^{\infty} (1 + a_r)$ *comes under neither of the last two definitions, it is said to* diverge.

It pays to describe as *divergent* a product for which A_n tends to 0, for instance

$$\left(1 - \frac{1}{2}\right) \left(1 - \frac{1}{3}\right) \cdots \left(1 - \frac{1}{n}\right) \cdots,$$

where $A_n = \dfrac{1}{2} \cdot \dfrac{2}{3} \cdots \dfrac{n-1}{n} = \dfrac{1}{n}$. The reason, which will become plain in the theorems that follow, is that the behaviour of the product $\Pi (1 + a_n)$ is linked with that of the series Σa_n.

Lemma. *Inequalities between products and sums. Let $a_n \geqslant 0$ and $s_n = a_1 + \ldots + a_n$. Then*

(1) $\displaystyle\prod_{1}^{n} (1 + a_r) \geqslant 1 + s_n$.

If, further, $a_r < 1$ for $1 \leqslant r \leqslant n$,

(2) $\displaystyle\prod_{1}^{n} (1 - a_r) \geqslant 1 - s_n$,

(3) $\displaystyle\prod_{1}^{n} (1 - a_r) \leqslant \dfrac{1}{1 + s_n}$.

If, further, $s_n < 1$,

$$(4) \quad \prod_1^n (1 + a_r) \leqslant \frac{1}{1 - s_n}.$$

Proof. Induction gives (1) and (2). To deduce (3) and (4) use $(1+a)(1-a) \leqslant 1$. |

Theorem. 14.41. (1) *If $a_n \geqslant 0$, then $\Pi(1+a_n)$ and Σa_n both converge or both diverge.*

(2) *If $0 \leqslant a_n < 1$, then $\Pi(1-a_n)$ and Σa_n both converge or both diverge.*

Proof. Use the appropriate inequality from the lemma. For instance, in (1) suppose Σa_n convergent.

Choose m such that

$$t = a_{m+1} + a_{m+2} + \ldots < 1.$$

By (4) of the lemma,

$$(1 + a_{m+1}) \ldots (1 + a_n) \leqslant 1/(1-t).$$

The left-hand side is a bounded increasing function of n. Hence the product converges. |

Theorem 14.42. *If $a_n \geqslant 0$, then $\Pi(1+a_n)$ is independent of the order of the factors. So is $\Pi(1-a_n)$ if $0 \leqslant a_n < 1$.*

Proof. Use monotonic sequences as for the corresponding result for series (e.g. C1, 94). |

Theorem 14.43. *A necessary condition that $\Pi(1+a_n)$ converges is that $a_n \to 0$ as $n \to \infty$.*

Proof. $1 + a_n = A_n/A_{n-1}$ and A_{n-1}, A_n both tend to A (not zero). |

Definition. (Absolute convergence.) *The product $\Pi(1+a_n)$ is said to converge absolutely if $\Pi(1+|a_n|)$ converges.*

Theorem 14.44. *If $\Pi(1+a_n)$ is absolutely convergent, then* (1) *it is convergent and* (2) *its value is independent of the order of its factors.*

Proof. Write $|a_n| = b_n$, $A_n = \prod_1^n (1 + a_r)$, $B_n = \prod_1^n (1 + b_r)$.

Then
$$\left|\frac{A_{n+p}}{A_n} - 1\right| = |(1+a_{n+1}) \ldots (1+a_{n+p}) - 1|$$
$$\leqslant (1+b_{n+1}) \ldots (1+b_{n+p}) - 1$$
$$= \frac{B_{n+p}}{B_n} - 1.$$

Also $|A_n| \leqslant B_n$. Multiplying the two inequalities we have

$$|A_{n+p} - A_n| \leqslant B_{n+p} - B_n.$$

By hypothesis the B_n form a Cauchy sequence. Hence so do the A_n, and (1) is proved.

Let $A_n \to A$ and $B_n \to B$.

Let $\Pi (1+a_n')$ be a rearrangement of $\Pi (1+a_n)$ and $\Pi (1+b_n')$ the same rearrangement of $\Pi (1+b_n)$. Let A_n' and B_n' be the respective products of the first n factors. By theorem 14.42, $B' = \lim B_n' = B$.

Given n, take q so large that A_q' contains all the factors in A_n (and so B_q' those in B_n).

Then, if $r \geqslant q$,

$$|A_r' - A_n| \leqslant B_r' - B_n \quad \text{(by the proof of (1))}$$
$$\leqslant B' - B_n = B - B_n,$$

and so

$$|A - A_r'| \leqslant |A - A_n| + |A_n - A_r'| \leqslant |A - A_n| + |B - B_n|.$$

Therefore $A_r' \to A$. |

In what follows we shall connect the logarithm of a product of factors (assumed not zero) with the series of logarithms of the separate factors. This is an easy exercise if the factors are real, but more delicate if they are complex. Each logarithm will be supposed to have its principal value.

Theorem 14.45. $\displaystyle\prod_1^\infty (1+a_n)$ *converges if and only if, for some m,* $\displaystyle\sum_{m+1}^\infty \log (1+a_n)$ *converges.*

Proof. If $\displaystyle\sum_{m+1}^\infty \log (1+a_n)$ converges, to sum l_m say, then (since the exponential function is continuous) $\displaystyle\prod_{m+1}^\infty (1+a_n)$ converges to $\exp l_m$,

and
$$\prod_1^\infty (1+a_n) = (1+a_1) \ldots (1+a_m) \exp l_m.$$

The converse needs more care because, with principal values, $\log zw$ need not be equal to $\log z + \log w$ but may differ from it by $2\pi i$. However $\log zw = \log z + \log w$ if $|\mathrm{ph}\, z| < \frac{1}{2}\pi$ and $|\mathrm{ph}\, w| < \frac{1}{2}\pi$.

Suppose that $\Pi\,(1+a_n)$ converges.

Then, given $\epsilon\,(< 1)$, there exists N such that

$$\left| \prod_{n}^{n+p} (1+a_r) - 1 \right| = |(A_{n+p} - A_{n-1})/A_{n-1}| < \epsilon \qquad (14.41)$$

for all $n > N$ and all $p > 0$.

If $|z-1| < \epsilon < 1$, then (as is clear from a diagram)

$$1 - \epsilon < |z| < 1 + \epsilon, \qquad (14.42)$$

and

$$|\mathrm{ph}\, z| < \arcsin \epsilon < \tfrac{1}{2}\pi\epsilon < \tfrac{1}{2}\pi. \qquad (14.43)$$

Now either

$$\log \prod_{n}^{n+p} (1+a_r) = \sum_{n}^{n+p} \log(1+a_r) \qquad (14.44)$$

or the two sides differ by a multiple of $2\pi i$. We prove equality for $n > N$ and $p \geqslant 0$.

Let $n > N$. Then (14.44) holds if $p = 0$. Suppose that it holds for $p = q \geqslant 0$. By (14.41) and (14.43),

$$\left| \mathrm{ph} \prod_{n}^{n+q} (1+a_r) \right| < \tfrac{1}{2}\pi, \quad |\mathrm{ph}(1+a_{n+q+1})| < \tfrac{1}{2}\pi.$$

Hence $\displaystyle \log \prod_{n}^{n+q+1} (1+a_r) = \log \prod_{n}^{n+q} (1+a_r) + \log(1+a_{n+q+1})$

$$= \sum_{n}^{n+q+1} \log(1+a_r).$$

Thus, by induction, (14.44) holds for all $p \geqslant 0$.

Finally, (14.41)–(14.44) show that, for fixed m, the function of n defined by $\displaystyle \sum_{m}^{n} \log(1+a_r)$ is a Cauchy sequence. |

Theorem 14.46. *If, for a fixed positive integer k, $\Sigma |a_n|^{k+1}$ converges, then* $\Pi\{(1+a_n)\exp(-a_n+\tfrac{1}{2}a_n^2 - \ldots + (-1)^k a_n^k/k)\}$ *converges absolutely.*

Proof. The nth factor is

$$\exp\{\log(1+a_n) - a_n + \tfrac{1}{2}a_n^2 - \ldots + (-1)^k a_n^k/k\}$$

$$= \exp\{O(|a_n|^{k+1})\}$$

$$= 1 + O(|a_n|^{k+1})$$

$$= 1 + b_n,$$

where $\Sigma |b_n|$ converges. |

Theorem 14.47. *If $\Sigma a_n, \Sigma a_n^2, \dots \Sigma a_n^k$ and $\Sigma |a_n|^{k+1}$ converge, then $\Pi (1 + a_n)$ converges.*

Proof. As in the last theorem,

$$\log (1 + z) = z - \tfrac{1}{2}z^2 + \dots + (-1)^{k-1}z^k/k + O(|z|^{k+1}).$$

In this write $z = a_1, a_2, \dots a_n, \dots$ and sum. The hypothesis shows that $\Sigma \log (1 + a_n)$ converges. Apply theorem 14.45. |

The reader should be able to define *uniformity* of convergence of products $\Pi \{1 + f_n(z)\}$ and to prove theorems which naturally arise, such as the analogue of the *M*-test (theorem 5.32).

14.5. The factor theorem of Weierstrass. The sine product

Theorem 14.51. *Given a sequence of complex numbers $a_1, a_2, \dots$, whose only limit point is ∞, there is an entire function with zeros for these values of z and no others. (a_r is repeated k times if the zero there has order k.)*

Proof. Assume for the present that no a_n is 0.
We may suppose that

$$|a_1| \leqslant |a_2| \leqslant |a_3| \dots, \quad \text{with} \quad |a_n| \to \infty.$$

If the product

$$\prod_1^\infty \left(1 - \frac{z}{a_n}\right)$$

were convergent, it would define the entire function sought. In general the product does not converge and we have to affix *convergence factors* to the factors of the product in the manner of theorem 14.46. For shortness write $E(u, 0) = 1 - u$, and

$$E(u, p) = (1 - u) \exp (u + \tfrac{1}{2}u^2 + \dots + u^p/p).$$

It is always possible to find a sequence p_n to make $\Sigma |z/a_n|^{p_n+1}$ converge for all values of z. In fact $p_n = n$ would do because, given z, if n is large enough, $|a_n| > 2|z|$ and $|z/a_n|^n < 1/2^n$.

The argument of theorem 14.46 proves that, for a fixed R, there is a constant A such that

$$|E(z/a_n, p_n) - 1| \leqslant A|z/a_n|^{p_n+1},$$

for $|z| \leqslant R$ and all n.

Then $f(z) = \prod_{n=1}^\infty E(z/a_n, p_n)$ converges for all z, and the convergence is uniform for $|z| \leqslant R$, however large R is. Theorem 13.11, shows that $f(z)$ is an entire function. It has zeros at a_n and nowhere else.

At the beginning of the proof we ruled out the possibility $a_n = 0$. If the function is to have a zero of order k for $z = 0$, then we adjoin a factor z^k. |

Notes. (1) The choice $p_n = n$ is extravagantly large. In many common applications p_n can be taken to be a suitable constant (often 1 or 2).

(2) If f and f_1 both satisfy the requirements of the theorem, then f_1/f, having removable singularities at the common zeros of f and f_1, is an entire function, say g, without zeros. The function g must be of the form $\exp h(z)$, where $h(z)$ is an entire function. For $g'(z)/g(z)$, being entire, is the derivative of an entire function $h(z)$. Then

$$\frac{d}{dz}\{g(z)\exp(-h(z))\} = \exp(-h(z))\{g'(z) - g(z)h'(z)\} = 0$$

and hence $g(z)$ is $A\exp h(z)$.

Theorem 14.52. $\sin \pi z = \pi z \prod_{n=1}^{\infty} \left(1 - \dfrac{z^2}{n^2}\right).$

Proof. In theorem 14.51, $a_n = \pm n$, and $\Sigma(1/n^{p+1})$ converges if $p = 1$. So we have

$$\sin \pi z = g(z)\pi z \prod_{-\infty}^{\infty}{}' \left\{\left(1 - \frac{z}{n}\right)\exp\frac{z}{n}\right\},$$

where the dash in $\prod_{-\infty}^{\infty}{}'$ $\left(\text{or in } \sum_{-\infty}^{\infty}{}'\right)$ denotes that the term with $n = 0$ is omitted. Here $g(z)$ is an entire function without zeros, to be found.

Writing $\qquad s_N(z) = g(z)\pi z \prod_{-N}^{N}{}' \left\{\left(1 - \frac{z}{n}\right)\exp\frac{z}{n}\right\},$

we have $\qquad s_N'(z) = \left\{\dfrac{g'(z)}{g(z)} + \dfrac{1}{z} + \sum_{-N}^{N}{}' \left(\dfrac{1}{z-n} + \dfrac{1}{n}\right)\right\} s_N(z),$

where z is not an integer.

The proof of theorem 14.51 shows that $s_N(z)$ converges uniformly to $\sin \pi z$ on any compact set. From theorem 13.11, $s_N'(z)$ converges to $\pi \cos \pi z$. Therefore

$$\pi \cot \pi z = \frac{g'(z)}{g(z)} + \frac{1}{z} + \sum_{-\infty}^{\infty}{}' \left(\frac{1}{z-n} + \frac{1}{n}\right) \qquad (z \neq 0, \pm 1, ...).$$

Collating this with the formula for $\cot \pi z$ found in §14.3, we deduce that $g'(z) = 0$, and so $g(z)$ is a constant A. As $z \to 0$, $\sin \pi z \sim \pi z$ and therefore $A = 1$.

Multiplying the factors containing $+n$ and $-n$, we have the result. |

Exercises 14(c)

Examine for convergence the products whose nth factors are given in 1–6.

1. $\dfrac{\sin (z/n)}{z/n}$.

2. $\dfrac{\alpha+n}{\beta+n}$.

3. $\cosh \dfrac{z}{n}$.

4. $1+(-1)^n \sinh \dfrac{z}{n}$.

5. $\dfrac{(1+z^n)^2}{1+z^{2n}}$.

6. $1-\left(-\dfrac{n+1}{nz}\right)^n$.

7. Prove that $\Pi\left(1+\dfrac{i}{n}\right)$ diverges and $\Pi\left|1+\dfrac{i}{n}\right|$ converges.

8. Prove that, if $|z| < 1$, the infinite product
$$(1+z)(1+z^2)(1+z^4)(1+z^8)...$$
converges to $1/(1-z)$.

9. Prove that, if $|z| < 1$, the product of the three infinite products
$$\prod_1^\infty (1+z^{2n}), \quad \prod_1^\infty (1+z^{2n-1}), \quad \prod_1^\infty (1-z^{2n-1})$$
is 1.

10. Prove that, if $|z| < 1$,
$$\prod_1^\infty (1+z^n) = \exp\left\{\dfrac{z}{1-z} - \dfrac{z^2}{2(1-z^2)} + \dfrac{z^3}{3(1-z^3)} - ...\right\}.$$

11. Express $\cos \pi z$ in factors, on the model of theorem 14.52.

12. Prove or refute the following statements for real a_n, b_n.

   (i) If $\lim\limits_{n \to \infty} \prod\limits_{r=n}^{2n} (1+a_r) = 1$, then $\prod\limits_1^\infty (1+a_n)$ converges.

   (ii) If $\lim\limits_{n \to \infty} (a_n/b_n) = 1$ and $\prod\limits_1^\infty (1+b_n)$ converges, then $\prod\limits_1^\infty (1+a_n)$ converges.

   (iii) If $\sum\limits_1^\infty a_n$ converges, then $\prod\limits_{r=1}^n (1+a_r)$ tends to a finite limit as $n \to \infty$.

13. Prove that, if p runs through the sequence of prime numbers $2, 3, 5, 7, 11, \ldots$ and $\operatorname{re} z > 1$, then

(i) $\displaystyle \prod_p \left(1 - \frac{1}{p^z}\right) = \frac{1}{\zeta(z)}$, where the Riemann zeta function $\zeta(z) = \sum_{n=1}^{\infty} \frac{1}{n^z}$,

and the zth powers have their principal values.

Prove further that

(ii) $\zeta(z)$ is never 0 for $x > 1$;

(iii) $\sum_p (1/p)$ diverges.

14. Suppose that $a_n \to \infty$. Given a sequence b_n, construct an entire function f such that $f(a_n) = b_n$.

15. *Elliptic functions.* In §14.2–14.5 we have used trigonometric functions (periodic functions) to illustrate the text. The methods and theorems can be applied to construct functions which possess *two* independent periods and to develop their properties.

Let ω_1, ω_2 be numbers such that $\operatorname{ph}(\omega_2/\omega_1) \neq 0$. Write $\Omega_{m,n} = 2m\omega_1 + 2n\omega_2$, where m, n can be any integers $(+, -$ or $0)$. Let Σ' and Π' denote a sum and a product over every pair of integers m, n except $m = n = 0$. From exercise $4(g)$, 11, $\Sigma' \Omega_{m,n}^{-3}$ converges absolutely.

(i) Prove that

$$\sigma(z) = z\Pi' \left\{ \left(1 - \frac{z}{\Omega_{m,n}}\right) \exp\left(\frac{z}{\Omega_{m,n}} + \frac{z^2}{2\Omega_{m,n}^2}\right) \right\}$$

is an entire function with a simple zero at each point $\Omega_{m,n}$.

(ii) Prove that

$$\frac{\sigma'(z)}{\sigma(z)} = \frac{1}{z} + \Sigma' \left\{ \frac{1}{z - \Omega_{m,n}} + \frac{1}{\Omega_{m,n}} + \frac{z}{\Omega_{m,n}^2} \right\}$$

is regular except for a simple pole at each point $\Omega_{m,n}$. The function σ'/σ is called ζ, but it has nothing to do with the zeta function of Riemann in **13**.

(iii) Prove that the function $\wp$, defined by

$$\wp(z) = -\zeta'(z) = \frac{1}{z^2} + \Sigma' \left\{ \frac{1}{(z - \Omega_{m,n})^2} - \frac{1}{\Omega_{m,n}^2} \right\}$$

is regular except for a double pole at each point $\Omega_{m,n}$.

In (iv) we shall outline the steps of a proof that

$$\wp(z + 2\omega_1) = \wp(z)$$

and

$$\wp(z + 2\omega_2) = \wp(z),$$

that is to say, the function $\wp$ is doubly periodic, having $2\omega_1$ and $2\omega_2$ as periods. We observe that a doubly periodic function which is regular for all finite z is bounded and therefore, by Liouville's theorem, a constant. For historical reasons, a doubly periodic function which is meromorphic in Z is called an *elliptic function.*

(iv) Prove that (if $z \neq \Omega_{m,n}$)

    (*a*) $\wp$ is an even function,

    (*b*) its derivative $\wp'$ is odd,

    (*c*) $\wp'$ is an elliptic function,

    (*d*) $\wp$ is an elliptic function.

Elliptic functions have a multitude of properties; the prevailing method of proof is by Liouville's theorem. As one illustration we ask the reader in (v) to construct the differential equation for $\wp(z)$. For anything more we must refer him to books containing systematic accounts.

(v) The differential equation for $\wp(z)$.

Establish the Laurent expansion

$$\wp(z) = z^{-2} + a_2 z^2 + a_3 z^4 + O(|z|^6),$$

with constant term 0, in a disc centre $z = 0$. Choose coefficients g_2, g_3 so that

$$\{\wp'(z)\}^2 = 4\{\wp(z)\}^3 - g_2\wp(z) - g_3 + O(|z|^2)$$

and deduce the differential equation.

14.6. The gamma function

The gamma function $\Gamma(1+z)$ extends the factorial $n!$ to any z, real or complex. We acknowledge that Jeffreys and Jeffreys are right in contending that the name 'factorial' and the symbol '!' ought to have been used for the extension and that a new notation was otiose. But we lack the courage of their convictions and continue the usage established by Euler and Legendre.

Three definitions are possible, all of which are useful:

$$\text{E (Euler)} \quad \Gamma(1+z) = \int_0^\infty t^z e^{-t}\,dt \quad (\text{re } z > -1),$$

where t^z means $\exp(z \log t)$, the principal value;

$$\text{G (Gauss)} \quad \Gamma(1+z) = \lim_{n\to\infty} \frac{n!\,n^z}{(z+1)(z+2)\ldots(z+n)},$$

where z is not a negative integer and n^z has its principal value; and, thirdly, a definition of the reciprocal as a product of the type given in theorem 14.46,

$$\text{W (Weierstrass)} \quad \frac{1}{\Gamma(1+z)} = e^{\gamma z} \prod_{n=1}^\infty \left\{\left(1+\frac{z}{n}\right) e^{-z/n}\right\},$$

where γ is Euler's constant $\left(\text{namely } \lim\left\{\sum_1^n (1/r) - \log n\right\}\right)$.

It is easy to reconcile the second and third definitions G and W. The right-hand side of W is

$$\lim_{n\to\infty} \left\{\exp z\left(1+\frac{1}{2}+\ldots+\frac{1}{n} - \log n\right)\right\} \prod_{r=1}^n \left\{\frac{z+r}{r} e^{-z/r}\right\}$$

$$= \lim_{n\to\infty} \frac{n^{-z}(z+1)\ldots(z+n)}{n!},$$

which is $1/\Gamma(1+z)$ according to the definition G.

Observe that theorem 13.13 shows that Euler's integral defines $\Gamma(1+z)$ as a regular function for re $z > -1$. By theorem 14.51, W defines $1/\Gamma(1+z)$ as a regular function for all z with simple zeros at $z = -n$. So $\Gamma(1+z)$ is meromorphic, regular except for simple poles at $z = -n$. Thus G and W will give the *analytic continuation* of the function defined by E into the whole plane Z with the negative integers removed.

We have still to reconcile the Euler and Gauss definitions. Since they both define regular functions in their domains of existence it is sufficient to prove equality for some interval on the real axis, say $0 \leqslant x \leqslant 1$. By n integrations by parts, we find that

$$\int_0^n t^x \left(1 - \frac{t}{n}\right)^n dt = \frac{n^{x+1} n!}{(x+1)(x+2) \ldots (x+n+1)}$$

and the right-hand side is the Gauss expression multiplied by the factor $n/(x+n+1)$ which has limit 1. So what we need to prove is the following theorem about the limit of an integral.

Theorem 14.61. *If* $0 \leqslant x \leqslant 1$, *then*

$$\lim_{n \to \infty} \int_0^n t^x \left(1 - \frac{t}{n}\right)^n dt = \int_0^\infty t^x e^{-t} dt.$$

Proof. We show first that, for a fixed positive t,

$$\left(1 - \frac{t}{n}\right)^n$$

increases with n.

In the straightfoward inequality (C1, 17, exercise 1(d), 8)

$$\frac{1 - a^{n+1}}{n+1} \leqslant \frac{1 - a^n}{n} \quad (0 \leqslant a \leqslant 1)$$

substitute $a^n = 1 - \dfrac{t}{n+1}$. We have then

$$1 - \frac{t}{n} \leqslant a^{n+1} \quad (0 \leqslant t \leqslant n)$$

and hence

$$\left(1 - \frac{t}{n}\right)^n \leqslant \left(1 - \frac{t}{n+1}\right)^{n+1}.$$

We know (exercise 5(c), 13 or C1, 109, exercise 6(a), 3) that

$$\lim_{n \to \infty} \left(1 - \frac{t}{n}\right)^n = e^{-t}.$$

Given ϵ, choose m such that

$$\int_m^\infty t^x e^{-t} dt < \epsilon.$$

In $[0, m]$ the increasing sequence of continuous functions

$$t^x \left(1 - \frac{t}{n}\right)^n$$

has the continuous limit $t^x e^{-t}$.

By theorem 5.35 this convergence is uniform and by theorem 5.22 we can choose $n_0 \geqslant m$ such that, if $n \geqslant n_0$,

$$0 < \int_0^m t^x e^{-t} dt - \int_0^m t^x \left(1 - \frac{t}{n}\right)^n dt < \epsilon.$$

Hence, for $n \geqslant n_0$,

$$\int_0^\infty t^x e^{-t} dt - 2\epsilon < \int_0^m t^x \left(1 - \frac{t}{n}\right)^n dt$$

$$< \int_0^n t^x \left(1 - \frac{t}{n}\right)^n dt$$

$$< \int_0^\infty t^x e^{-t} dt. \;\;|$$

In proving properties of the gamma function we can use whichever definition best serves the immediate purpose. The following results are basic:

(1) $\Gamma(1+z) = z\Gamma(z)$, $\;\;\Gamma(1) = 1$, $\;\;\Gamma(n+1) = n!$.
(Prove from either E or G.)

(2) $\Gamma(z)\Gamma(1-z) = \dfrac{\pi}{\sin \pi z}$.

$\quad \Gamma(\tfrac{1}{2}) = \sqrt{\pi}$.

Proof. Multiplying the two W products we have

$$\frac{1}{\Gamma(1+z)\Gamma(1-z)} = \prod_{n=1}^\infty \left(1 - \frac{z^2}{n^2}\right)$$

and so $\quad \dfrac{1}{z\Gamma(z)\Gamma(1-z)} = \dfrac{\sin \pi z}{\pi z}$ (from theorem 14.52).

For the value $z = \tfrac{1}{2}$ note that $\Gamma(\tfrac{1}{2})$ is not negative.

(3) *The duplication formula*

$$\Gamma(z)\Gamma(z+\tfrac{1}{2}) = \pi^{\tfrac{1}{2}} 2^{1-2z} \Gamma(2z).$$

Proof. From G, the left-hand side is

$$\lim \frac{n^{z-1}n!}{z(z+1) \dots (z+n-1)} \lim \frac{n^{z-\frac{1}{2}}n!}{(z+\frac{1}{2})(z+\frac{3}{2}) \dots (z+n-\frac{1}{2})}$$

$$= \lim \frac{n^{2z-\frac{3}{2}}(n!)^2 2^{2n}}{(2z)(2z+1)(2z+2) \dots (2z+2n-1)}.$$

From the Gauss formula for $\Gamma(2z)$ with $2n$ in place of n we find

$$\Gamma(2z) = \lim \frac{(2n)^{2z-1}(2n)!}{(2z)(2z+1)(2z+2) \dots (2z+2n-1)}.$$

Hence
$$\Gamma(z)\Gamma(z+\tfrac{1}{2}) = A\,2^{1-2z}\Gamma(2z),$$

where
$$A = \lim_{n \to \infty} \frac{(n!)^2 2^{2n}}{(2n)!\,n^{\frac{1}{2}}}.$$

The value of this limit is probably known to the reader ($A = \sqrt{\pi}$, Wallis's formula for π, e.g. C1, 135); without assuming this knowledge we have only to put $z = 1$ in the preceding line, obtaining $\Gamma(\tfrac{3}{2}) = \tfrac{1}{2}A\Gamma(2)$ and so $A = \Gamma(\tfrac{1}{2})$.

(4) The duplication formula (due to Legendre) is the most useful case of the general multiplication formula (due to Gauss)

$$\Gamma(z)\Gamma\left(z+\frac{1}{n}\right) \dots \Gamma\left(z+\frac{n-1}{n}\right) = (2\pi)^{\frac{1}{2}(n-1)}n^{\frac{1}{2}-nz}\Gamma(nz).$$

This may be proved by the same method.

(5) As an application of gamma functions we use them to evaluate the infinite product $\prod\limits_{n=1}^{\infty} u(n)$ where $u(n)$ is a rational function of n.

If the product is to converge, the degree in n of the numerator of $u(n)$ must be the same as the degree of the denominator. The method will be clear if we take the degree to be 2 and express, in terms of gamma functions, the product

$$P = \prod_{n=1}^{\infty} \frac{(n-a_1)(n-a_2)}{(n-b_1)(n-b_2)},$$

where no a or b is a positive integer.

The nth factor of P is

$$1 - \frac{a_1+a_2-b_1-b_2}{n} + O\left(\frac{1}{n^2}\right).$$

Hence a necessary and sufficient condition that the product P converges is that

$$a_1 + a_2 = b_1 + b_2.$$

The nth factor can then be written

$$\frac{\left(1 - \dfrac{a_1}{n}\right) e^{a_1/n} \left(1 - \dfrac{a_2}{n}\right) e^{a_2/n}}{\left(1 - \dfrac{b_1}{n}\right) e^{b_1/n} \left(1 - \dfrac{b_2}{n}\right) e^{b_2/n}}.$$

The W definition shows that

$$P = \frac{\Gamma(1 - b_1)\Gamma(1 - b_2)}{\Gamma(1 - a_1)\Gamma(1 - a_2)}.$$

Exercises 14(d)

Find the values of the infinite products whose nth factors are given in **1, 2.**

1. $1 - n^{-k}$ $(n \geqslant 2)$ for $k = 1, 2, 3, 4$.

2. $1 + n^{-k}$ $(n \geqslant 1)$ for $k = 1, 2, 3$.

3. Prove that

$$\prod_0^\infty \left\{ 1 + \frac{y^2}{(x+n)^2} \right\} = \frac{\{\Gamma(x)\}^2}{\Gamma(x+iy)\Gamma(x-iy)} \qquad (x > 0).$$

4. Prove that

$$|\Gamma(1+iy)|^2 = \pi y / \sinh \pi y,$$

$$|\Gamma(\tfrac{1}{2}+iy)|^2 = \pi / \cosh \pi y.$$

$$|\Gamma(iy)|^2 = \pi / y \sinh \pi y.$$

5. Prove that

$$(1-z)\left(1+\frac{z}{2}\right)\left(1-\frac{z}{3}\right)\left(1+\frac{z}{4}\right)\cdots = \frac{\sqrt{\pi}}{\Gamma(1+\tfrac{1}{2}z)\Gamma(\tfrac{1}{2}-\tfrac{1}{2}z)}.$$

What is the value of the product if the factors are rearranged so that two with a $-$ precede each factor with a $+$?

6. Find the residues of $\Gamma(1+z)$ at its poles.

14.7. Integrals expressed in gamma functions

Many common definite integrals can be evaluated by gamma functions.

Define the beta function by the equation

$$B(p, q) = \int_0^1 x^{p-1}(1-x)^{q-1} dx,$$

the integral existing if re $p > 0$, re $q > 0$.

The following properties (i) to (iv) are elementary.

  (i) $B(p, q) = B(q, p)$.

  (ii) $pB(p, q+1) = qB(p+1, q)$.

  (iii) $B(p, q) = B(p+1, q) + B(p, q+1)$.

  (iv) $pB(p, q) = (p+q)B(p+1, q)$.

Theorem 14.71. $$B(p, q) = \frac{\Gamma(p)\Gamma(q)}{\Gamma(p+q)}.$$

Proof. In order to avoid unbounded integrands we shall suppose that re $p \geqslant 1$ and re $q \geqslant 1$. The property (iv) above will then extend the theorem to any p, q with positive real part.

$$\Gamma(p)\Gamma(q) = \int_0^\infty e^{-x}x^{p-1}dx \int_0^\infty e^{-y}y^{q-1}dy$$

$$= \lim_{A\to\infty} \int_0^A dx \int_0^A e^{-(x+y)}x^{p-1}y^{q-1}dy$$

$$= \lim_{A\to\infty} \iint_{S(A)} e^{-(x+y)}x^{p-1}y^{q-1}dx\,dy,$$

where $S(A)$ is the square with opposite vertices $(0, 0)$, (A, A). The line joining the points $(A, 0)$ and $(0, A)$ divides the square of integration into two triangles, the south-west one $\Delta(A)$ having its third vertex at $(0, 0)$. In $\Delta(A)$ make the transformation $x = u$, $y = v - u$ (for which $\partial(x, y)/\partial(u, v) = 1$) and $\iint_{\Delta(A)}$ becomes an integral over $0 \leqslant u \leqslant v \leqslant A$. Expressed as a repeated integral it is

$$\int_0^A dv \int_0^v e^{-v}u^{p-1}(v-u)^{q-1}du.$$

The substitution $u = vt$ in the inner integral gives

$$\int_0^A dv \int_0^1 e^{-v}v^{p+q-1}t^{p-1}(1-t)^{q-1}dt = B(p, q)\int_0^A e^{-v}v^{p+q-1}dv.$$

If p and q are real (which is the usual case) it is simplest to use the inequalities

$$\iint_{S(\frac{1}{2}A)} \leqslant \iint_{\Delta(A)} \leqslant \iint_{S(A)},$$

since $S(\frac{1}{2}A) \subset \Delta(A) \subset S(A)$ and the integrand is positive.

As $A \to \infty$, each of the integrals over the squares tends to $\Gamma(p)\Gamma(q)$, and the integral over the triangle to $B(p, q)\Gamma(p+q)$.

If p and q are not real, we cannot assert the preceding inequalities, but we shall prove instead that the integral over the north-east triangle $(A, 0)$, (A, A), $(0, A)$ tends to 0 as $A \to \infty$.

If p_1 and q_1 are the real parts of p and q, the integrand has modulus at most $e^{-A}x^{p_1-1}y^{q_1-1}$ and the integral has modulus at most $e^{-A}A^{p_1+q_1}/p_1q_1$. This tends to 0 as $A \to \infty$ and the theorem is proved. |

To illustrate theorem 14.71 we add an alternative proof of the duplication formula (§14.6, (3)).

We have, x being real and positive,

$$\frac{\{\Gamma(x)\}^2}{\Gamma(2x)} = \int_0^1 t^{x-1}(1-t)^{x-1}dt.$$

Put $t = \frac{1}{2}(1+u)$. The right-hand side is

$$2^{1-2x}\int_{-1}^1 (1+u)^{x-1}(1-u)^{x-1}du = 2^{2-2x}\int_0^1 (1-u^2)^{x-1}\,du.$$

Now put $u^2 = v$ and we have

$$2^{1-2x}\int_0^1 (1-v)^{x-1}v^{-\frac{1}{2}}dv = 2^{1-2x}B(x, \tfrac{1}{2})$$

$$= 2^{1-2x}\frac{\Gamma(x)\,\Gamma(\tfrac{1}{2})}{\Gamma(x+\tfrac{1}{2})}.$$

This proves the formula for positive x. The left and right sides of §14.6 (3) are regular throughout the z-plane except at their poles. Theorem 12.13 shows their equality. |

Theorem 14.72 illustrates the expression of a multiple integral in gamma functions.

Theorem 14.72. (Dirichlet's integral.)

$$\iiint_\Delta x^{p-1}y^{q-1}z^{r-1}(1-x-y-z)^{s-1}dx\,dy\,dz = \frac{\Gamma(p)\,\Gamma(q)\,\Gamma(r)\,\Gamma(s)}{\Gamma(p+q+r+s)},$$

where p, q, r, s are greater than 0, *and Δ is the tetrahedron bounded by the planes* $x = 0$, $y = 0$, $z = 0$, $x+y+z = 1$.

Proof. In order to have a continuous integrand we prove the result when p, q, r, s are greater than or equal to 1. The extension to values between 0 and 1 would require a discussion of multiple integrals of unbounded functions.

The neatest method is to write

$$x+y+z = \xi, \quad y+z = \xi\eta, \quad z = \xi\eta\zeta,$$

so that $\quad\quad\quad\quad x = \xi(1-\eta), \quad y = \xi\eta(1-\zeta).$

The x, y, z tetrahedron is then seen to be mapped onto the cube in which each of ξ, η, ζ is between 0 and 1.

The Jacobian $\partial(x, y, z)/\partial(\xi, \eta, \zeta)$ is $\xi^2\eta$.

The integral then becomes

$$\int_0^1 d\xi \int_0^1 d\eta \int_0^1 \xi^{p+q+r-1}(1-\xi)^{s-1}\eta^{q+r-1}(1-\eta)^{p-1}\zeta^{r-1}(1-\zeta)^{q-1}d\zeta,$$

which is the product of three single integrals in ξ, η, ζ.

Apply theorem 14.71 to their product

$$B(p+q+r, s)\, B(q+r, p)\, B(r, q). \quad |$$

Exercises 14(e)

1. If $a > 0$, $b > 0$, $k > 0$, express in Γ functions

(i) $\quad\displaystyle\int_0^1 x^{a-1}(1-x)^{b-1}\frac{dx}{(x+k)^{a+b}},$

(ii) $\quad\displaystyle\int_{-1}^1 \frac{(1+x)^{2a-1}(1-x)^{2b-1}}{(1+x^2)^{a+b}}\,dx.$

2. Express in terms of $\Gamma(\tfrac14)$

(i) $\quad\displaystyle\int_0^{\frac12\pi} \frac{d\theta}{\sqrt{(1+\sin^2\theta)}},$

(ii) $\quad\displaystyle\int_0^1 \left(\frac{1-x^2}{1+x^2}\right)^{\frac12} dx.$

3. Prove that, if $0 < a < 1$,

$$\int_0^\infty y^{a-1}\cos y\,dy = \Gamma(a)\cos\tfrac12\pi a,$$

$$\int_0^\infty y^{a-1}\sin y\,dy = \Gamma(a)\sin\tfrac12\pi a.$$

4. Express the content of the unit ball in R^n as

$$\pi^{\frac12 n}/\Gamma(\tfrac12 n+1)$$

and show that this is the value obtained in exercise 8(f), 12.

5. Prove that, if $b > a > -1$,

$$\int_0^{\frac12\pi} \cos^a\theta\cos b\theta\,d\theta = \frac{\pi\Gamma(a+1)}{2^{a+1}\Gamma(\tfrac12 a+\tfrac12 b+1)\Gamma(\tfrac12 a-\tfrac12 b+1)}.$$

6. Prove that, if $\mathrm{re}\, z > 1$ and $\zeta(z) = \sum_1^\infty n^{-z}$,

$$\int_0^\infty \frac{t^{z-1}}{e^t-1}\, dt = \zeta(z)\Gamma(z).$$

14.8. Asymptotic formulae

The object of this section is to develop methods of approximating to $\Gamma(1+z)$ for large $|z|$ by formulae suited to numerical computation. Stirling's formula for the factorial

$$n! \sim \sqrt{(2\pi)}\, n^{n+\frac{1}{2}} e^{-n}$$

is likely to be already familiar to the reader. (A proof is outlined in C1, 148.) Our first theorem is a refinement of the formula.

Theorem 14.81. *If n is a large positive integer,*

$$n! = \sqrt{(2\pi)}\, n^{n+\frac{1}{2}} e^{-n} \left\{ 1 + \frac{1}{12n} + O\left(\frac{1}{n^2}\right) \right\}.$$

Proof. The argument is elementary and we state it without comment.

$$\int_{n-1}^n \log x\, dx - \tfrac{1}{2}\{\log(n-1) + \log n\}$$

$$= n \log n - (n-1)\log(n-1) - 1 - \tfrac{1}{2}\log(n-1) - \tfrac{1}{2}\log n$$

$$= -(n-\tfrac{1}{2})\log\left(1 - \frac{1}{n}\right) - 1$$

$$= (n-\tfrac{1}{2})\left\{ \frac{1}{n} + \frac{1}{2n^2} + \frac{1}{3n^3} + O\left(\frac{1}{n^4}\right) \right\} - 1$$

$$= \frac{1}{12n^2} + O\left(\frac{1}{n^3}\right).$$

Summing from 2 to n, we have

$$\int_1^n \log x\, dx - \sum_2^n \log r + \tfrac{1}{2}\log n = \frac{1}{12}\sum_2^n \frac{1}{r^2} + A + O\left(\frac{1}{n^2}\right)$$

where A, here and in the following lines, is a constant. It is not necessary to put suffixes $A_1, A_2, \ldots$ to denote that the constants in successive steps may differ.

The last equation is

$$(n+\tfrac{1}{2})\log n - n - \log(n!) = A - \frac{1}{12}\sum_{n+1}^\infty \frac{1}{r^2} + O\left(\frac{1}{n^2}\right).$$

Since
$$\frac{1}{(r+1)r} < \frac{1}{r^2} < \frac{1}{r(r-1)},$$

we have
$$\frac{1}{n+1} < \sum_{n+1}^{\infty} \frac{1}{r^2} < \frac{1}{n}$$

and hence
$$\sum_{n+1}^{\infty} \frac{1}{r^2} = \frac{1}{n} + O\left(\frac{1}{n^2}\right).$$

Thus we have proved that

$$\log (n!) = (n+\tfrac{1}{2}) \log n - n + A + \frac{1}{12n} + O\left(\frac{1}{n^2}\right)$$

and therefore
$$n! = An^{n+\frac{1}{2}}e^{-n} \exp\left\{\frac{1}{12n} + O\left(\frac{1}{n^2}\right)\right\}$$

$$= An^{n+\frac{1}{2}}e^{-n}\left\{1 + \frac{1}{12n} + O\left(\frac{1}{n^2}\right)\right\}.$$

To find this unknown constant A we substitute the formula we have just obtained in Wallis's formula (C1, 134)

$$\lim_{n\to\infty} \frac{(n!)^2 2^{2n}}{(2n)!n^{\frac{1}{2}}} = \sqrt{\pi},$$

and we obtain $A = \sqrt{(2\pi)}$. (Compare the evaluation of the constant in the duplication formula §14.6 (3).) |

We take up now the more difficult problem of approximating to $\Gamma(1+z)$ when the complex number z has large modulus. On account of the poles at negative integer values of z, the approximate formula must be restricted to a sector

$$-\pi + \delta \leqslant \text{ph}\, z \leqslant \pi - \delta.$$

We shall ultimately prove that the formula to be stated as theorem 14.86 holds in the smaller sector

$$-\tfrac{1}{2}\pi + \delta \leqslant \text{ph}\, z \leqslant \tfrac{1}{2}\pi - \delta.$$

The approximation will take the form of an *asymptotic expansion* (or *series*) according to Poincaré (1854–1912). The formal definition will be easier to grasp in the light of a simple illustration.

Example. *Prove that, for positive x, the function f defined by*

$$f(x) = \int_0^{\infty} \frac{e^{-xt}}{1+t^2}\, dt$$

satisfies
$$f(x) = s_n(x) + r_n(x),$$

where
$$s_n(x) = \frac{1}{x} - \frac{2!}{x^3} + \frac{4!}{x^5} - \ldots + (-1)^{n-1} \frac{(2n-2)!}{x^{2n-1}}$$

and
$$|r_n(x)| < \frac{(2n)!}{x^{2n+1}}.$$

Hence approximate to f(x) for large x.
Solution. Substitute

$$\frac{1}{1+t^2} = 1 - t^2 + \ldots + (-1)^{n-1}t^{2n-2} + (-1)^n \frac{t^{2n}}{1+t^2}$$

in the integral defining f. We have the required s_n and also the inequality

$$|r_n(x)| = \int_0^\infty \frac{e^{-xt}t^{2n}}{1+t^2}\,dt \leqslant \int_0^\infty e^{-xt}t^{2n}\,dt = \frac{(2n)!}{x^{2n+1}}.$$

The remarkable property of the decomposition

$$f(x) = s_n(x) + r_n(x)$$

is that, as $n \to \infty$, the sequence $s_n(x)$ diverges for every x. The remainder $r_n(x)$ is, however, of the order of magnitude of the $(n+1)$th term of the expansion (in this example, numerically less than it). If x is large, the terms of $s_n(x)$ initially decrease in magnitude and the value of $f(x)$ can be computed with high accuracy by stopping at the term $\frac{(2n)!}{x^{2n+1}}$ with smallest value.

We now define formally an asymptotic expansion (or series), and in theorems 14.82 and 14.83 we establish properties that we shall need.

Definition. *Let $s_n(z)$ be the sum of the first $(n+1)$ terms of the series*

$$a_0 + \frac{a_1}{z} + \ldots + \frac{a_n}{z^n} + \ldots. \tag{14.81}$$

Let $r_n(z) = f(z) - s_n(z)$. Then the series (14.81) is called an asymptotic *expansion of $f(z)$ for $\alpha \leqslant \mathrm{Ph}\,z \leqslant \beta$ if, for each fixed n,*

$$\lim_{|z| \to \infty} z^n r_n(z) = 0, \quad \text{uniformly in } \mathrm{Ph}\,z.$$

This definition applies to a power series in $1/z$ which converges for sufficiently large $|z|$, say for $|z| \geqslant R$. For then there is a constant M, depending on R only, such that for all values of $\mathrm{Ph}\,z$

$$|r_n(z)| < \frac{MR^{n+1}}{(|z|-R)|z|^n}.$$

The notation $f(z) \sim \sum_0^\infty a_n z^{-n}$ is commonly used for an asymptotic expansion. As the symbol $\sim$ has been given a different meaning in §4.1, we shall refrain from using it now.

Theorem 14.82. (Uniqueness.) (a) *For a given interval of* Phz *a function cannot have more than one asymptotic expansion.*

(b) *A series can be the asymptotic expansion of more than one function.*

Proof. (a) Suppose that, for $\alpha \leqslant$ Ph$z \leqslant \beta$, $\Sigma a_n z^{-n}$ and $\Sigma b_n z^{-n}$ are both asymptotic expansions of $f(z)$. Then, for fixed n, as $|z| \to \infty$,

$$(a_0 - b_0) z^n + (a_1 - b_1) z^{n-1} + \dots + (a_n - b_n) \to 0.$$

It follows that $a_0 = b_0, a_1 = b_1, \dots a_n = b_n$ (for every n).

(b) The reader can verify that an asymptotic expansion of $f(z)$ for $-\tfrac{1}{4}\pi \leqslant$ ph$z \leqslant \tfrac{1}{4}\pi$ is also an asymptotic expansion of $f(z) + e^{-z}$. |

Theorem 14.83. (Integration term by term.) *If, for $\alpha \leqslant$ Ph$z \leqslant \beta$,*

$$a_0 + \frac{a_1}{z} + \dots + \frac{a_n}{z^n} + \dots$$

is the asymptotic expansion of a regular function $f_0(z)$, then

$$\frac{a_2}{z} + \dots + \frac{a_n}{(n-1)z^{n-1}} + \dots$$

is the asymptotic expansion of $\int_\gamma \left\{ f_0(\zeta) - a_0 - \dfrac{a_1}{\zeta} \right\} d\zeta$, where γ is a suitable path from z to ∞ within the sector $\alpha \leqslant$ Ph$z \leqslant \beta$.

Proof. Write $\qquad f(z) = f_0(z) - a_0 - \dfrac{a_1}{z}$

and $\qquad s_n(z) = \dfrac{a_2}{z^2} + \dots + \dfrac{a_n}{z^n}, \quad f(z) = s_n(z) + r_n(z).$

From the definition of an asymptotic expansion, given ϵ, there is z_0 such that

$$|f(z) - s_n(z)| < \epsilon |z|^{-n} \quad \text{for} \quad |z| > |z_0|, \quad \alpha \leqslant \text{Ph}z \leqslant \beta.$$

Let θ be fixed in $\alpha \leqslant \theta \leqslant \beta$. Let γ_1 be the linear segment $\zeta = \rho e^{i\theta}$ from z to z' where $r = |z| > |z_0|$.

Let γ be the infinite extension of γ_1, when $r' = |z'| \to \infty$.

For ζ on γ,

$$f(\zeta) - s_n(\zeta) = O(|\zeta|^{-n}), \quad \text{where} \quad n \geq 2,$$

and

$$s_n(\zeta) = O(|\zeta|^{-2}).$$

Therefore $\displaystyle\int_\gamma \{f(\zeta) - s_n(\zeta)\} d\zeta$ and $\displaystyle\int_\gamma s_n(\zeta) d\zeta$ exist, and hence $\displaystyle\int_\gamma f(\zeta) d\zeta$ also exists. Moreover

$$\left| \int_\gamma \{f(\zeta) - s_n(\zeta)\} d\zeta \right| < \epsilon \int_r^\infty \frac{d\rho}{\rho^n} = \frac{\epsilon}{(n-1)|z|^{n-1}}.$$

This is the condition for

$$\frac{a_2}{z} + \ldots + \frac{a_n}{(n-1)z^{n-1}} + \ldots$$

to be the asymptotic expansion of $\displaystyle\int_\gamma f(\zeta) d\zeta$. |

We need a number of properties of the gamma function.

Theorem 14.84. *If* $\qquad F(z) = \dfrac{\Gamma'(1+z)}{\Gamma(1+z)},$

where z is not a negative integer, then

(1) $\quad F(z) = -\gamma + \displaystyle\sum_1^\infty \left(\frac{1}{n} - \frac{1}{z+n} \right),$

(2) $\quad F'(z) = \displaystyle\sum_1^\infty \frac{1}{(z+n)^2},$

(3) $\quad F'(z) = \displaystyle\int_0^\infty \frac{t e^{-tz}}{e^t - 1} dt \quad (\text{re } z > -1).$

Proof. (1) In the W definition of $1/\Gamma(1+z)$ the infinite product converges uniformly on any compact set not containing any of the points $-1, -2, \ldots$ (by the analogue of the M-test, since the factors are $1 + O(1/n^2)$). Writing

$$s_n(z) = e^{\gamma z} \prod_1^n \left\{ \left(1 + \frac{z}{r} \right) e^{-z/r} \right\},$$

we have, by theorem 13.11,

$$s_n'(z) \to -\frac{\Gamma'(1+z)}{\{\Gamma(1+z)\}^2} = -\frac{F(z)}{\Gamma(1+z)},$$

that is,
$$-\frac{s_n'(z)}{s_n(z)} \to F(z).$$

But
$$s_n'(z) = \left\{\gamma + \sum_{r=1}^{n} \left(\frac{1}{z+r} - \frac{1}{r}\right)\right\} s_n(z),$$

and we have proved (1).

(2) We again appeal to theorem 13.11.

(3) The reader will have observed that the series $\Sigma (z+n)^{-2}$ is of the type discussed in §14.2. We could find a formula for the sum by integrating $(z+\zeta)^{-2}\pi \cot \pi\zeta$ round a sequence of suitable circuits C_n enclosing the points $\zeta = 1, 2, \ldots n$. We can however obtain the integral in (3) much more simply as follows. By theorems 13.11 and 13.13 the series and the integral on the right-hand sides of (2) and (3) are regular if re $z > -1$. If we prove them equal for real z (say x), then theorem 12.13 shows that they are equal for re $z > -1$.

Write $f_n(t) = te^{-t(x+n)}$. Then $f_n(t) \geqslant 0$ for $t \geqslant 0$, and

$$\sum_{1}^{\infty} f_n(t) = \frac{te^{-tx}}{e^t - 1}.$$

Since this sum is continuous, by theorem 5.35 the convergence of $\sum_{1}^{\infty} f_n(t)$ is uniform in $[\delta, T]$ for $\delta > 0$. By theorem 5.22,

$$\int_{\delta}^{T} \left\{\sum_{1}^{\infty} f_n(t)\right\} dt = \sum_{1}^{\infty} \left\{\int_{\delta}^{T} f_n(t)\,dt\right\}.$$

We now see from theorem 6.33 (for $T \to \infty$) and its analogue (for $\delta \to 0$) that

$$\int_{0}^{\infty} te^{-tx} \left(\sum_{n=1}^{\infty} e^{-nt}\right) dt = \sum_{n=1}^{\infty} \frac{1}{(x+n)^2}. \quad |$$

Next we explain a widely applicable method of obtaining asymptotic expansions of functions expressed as definite integrals.

Theorem 14.85. (Watson's lemma.) *Suppose that for* $-A_1 < t < A_1$

$$f(t) = \sum_{0}^{\infty} a_n t^n,$$

and that $|f(t)| \leqslant Be^{bt}$ $(b > 0)$ *for* $t \geqslant A$, *where* $0 < A < A_1$. *Then, if* $|\text{ph}\,z| \leqslant \frac{1}{2}\pi - \delta$,

$$F(z) = \int_{0}^{\infty} e^{-zt} f(t)\,dt$$

has the asymptotic expansion $\sum_{0}^{\infty} \dfrac{a_n n!}{z^{n+1}}.$

15

We are to prove, in fact, that substitution of the power series for $f(t)$ and integration term by term gives an asymptotic expansion for $F(z)$.

Proof. Given N, there exists C such that

$$\left| f(t) - \sum_0^{N-1} a_n t^n \right| \leqslant Ct^N e^{bt} \quad \text{for all } t \geqslant 0.$$

If we now write

$$F(z) = \sum_0^{N-1} \int_0^\infty e^{-zt} a_n t^n \, dt + R_N$$

$$= \sum_0^{N-1} \frac{a_n n!}{z^{n+1}} + R_N,$$

the theorem will be proved when we have shown that $R_N z^{N+1}$ is bounded. If $z = x + iy$,

$$|R_N| = \left| \int_0^\infty e^{-zt} \left\{ f(t) - \sum_0^{N-1} a_n t^n \right\} dt \right|$$

$$\leqslant \int_0^\infty e^{-xt} Ct^N e^{bt} \, dt$$

$$= \frac{CN!}{(x-b)^{N+1}} \quad \text{if } x > b.$$

Since $|\mathrm{ph}\, z| \leqslant \frac{1}{2}\pi - \delta$, we have $x > b$ as soon as $|z| > b \operatorname{cosec} \delta$. With that restriction on z,

$$|R_N z^{N+1}| \leqslant \frac{CN! \, |z|^{N+1}}{(|z| \sin \delta - b)^{N+1}} = O(1). \;\blacksquare$$

The same method proves the following extension. *If $\alpha > -1$ and*

$$f(t) = t^\alpha \sum_0^\infty a_n t^n \text{ for } 0 < t \leqslant A_1 \text{ (with the same restrictions on } |f(t)|$$

and $|\mathrm{ph}\, z|$ *as in the theorem), then* $F(z) = \displaystyle\int_0^\infty e^{-zt} f(t) \, dt$ *has the asymptotic expansion*

$$\frac{1}{z^\alpha} \sum_0^\infty \frac{a_n \Gamma(\alpha + n + 1)}{z^{n+1}}.$$

We now establish the asymptotic formula for $\mathrm{Log}\, \Gamma(1+z)$. In it the interpretation of the Logarithm is the value arrived at by analytic continuation from the real value taken for real positive z.

The constants B_n are the Bernoulli numbers defined in exercises 14(b).

Theorem 14.86. *In the sector* $-\frac{1}{2}\pi + \delta < \mathrm{ph}\,z < \frac{1}{2}\pi - \delta$ *the function* $\mathrm{Log}\ \Gamma(1+z)$ *has the asymptotic expansion*

$$(z+\tfrac{1}{2})\log z - z + \tfrac{1}{2}\log 2\pi + \sum_{n=1}^{\infty} \frac{B_{2n}}{(2n-1)\,2n\,z^{2n-1}}.$$

Proof. Applying theorem 14.85 to the integral in (3) of theorem 14.84, we find (noting that $B_0 = 1$) that the asymptotic expansion of $F'(z) - \dfrac{1}{z}$ is

$$-\frac{1}{2z^2} + \sum_{n=1}^{\infty} \frac{B_{2n}}{z^{2n+1}}.$$

By theorem 14.83, $-F(z) + \log z + A + \dfrac{1}{2z}$ has the asymptotic expansion of

$$\sum_{n=1}^{\infty} \frac{B_{2n}}{2n z^{2n}},$$

where A is a constant of integration, to be determined.

Appealing again to theorem 14.83, we have as the asymptotic expansion of

$$\mathrm{Log}\ \Gamma(1+z) - (z+\tfrac{1}{2})\log z + z - Az - A_1$$

the series

$$\sum_{n=1}^{\infty} \frac{B_{2n}}{(2n-1)\,2n\,z^{2n-1}}.$$

The constants A and A_1 can be determined from the value found for $\log(n!)$ in theorem 14.81; we have $A = 0$ and $A_1 = \frac{1}{2}\log 2\pi$. |

Notes. (1) The result of the theorem remains valid for

$$-\pi + \delta < \mathrm{ph}\,z < \pi - \delta.$$

Our restriction to the smaller sector arises from the corresponding limitation in theorem 14.85.

(2) To determine the multiple of $2\pi i$ included in $\mathrm{Log}\ \Gamma(1+z)$ we should have to identify the branch derived by analytic continuation from the principal branch, which is real for positive z. In the calculation of $\Gamma(1+z)$ there is no difficulty, because any value of $\mathrm{Log}\ \Gamma(1+z)$ will determine it.

NOTES ON CHAPTER 14

§**14.6.** The definition of Γ commonly called Gauss's was in fact given by Euler and rediscovered by Gauss. The Weierstrass definition had been used by F. W. Newman (1848).

§**14.8.** If we substitute numerical values of the B_{2n}, the asymptotic formula for Log $\Gamma(1+z)$ begins with the terms

$$(z+\tfrac{1}{2}) \log z - z + \tfrac{1}{2} \log 2\pi + \frac{1}{12z} - \frac{1}{360z^3} + \frac{1}{1260z^5} - \frac{1}{1680z^7} + \frac{1}{1188z^9} - \dots$$

It is of great use in numerical work. It is often called Stirling's series, but a more accurate description would be de Moivre's form of Stirling's series. It is astonishing that such a formula could have been found and used correctly (for real z) as early as 1730. (See Jeffreys and Jeffreys, *Mathematical Physics*, chapters 15 and 17.)

SOLUTIONS OF EXERCISES

1(a)

2. Use theorem 1.11.

4. (ii) $(A \cup C) - (B \cup C) = (A - B) \cap C'$, and
$$(A - B) \cup C = (A - B) \cap C' \Rightarrow C = \varnothing.$$

The opposite implication is obvious.

5. (i) $(A \cup B) - B = A - B$. (ii) $(A - B) \cup B = A \cup B$.

6. $(A \cup B) - (A \cap B) = (A \cup B) \cap (A \cap B)' = (A \cup B) \cap (A' \cup B')$.

7. $(A \triangle B) \triangle C = (A \cup B \cup C) \cap (A \cup B' \cup C') \cap (A' \cup B \cup C') \cap (A' \cup B' \cup C)$ and
this expression is symmetrical in A, B, C. Hence
$$A \triangle (B \triangle C) = (B \triangle C) \triangle A = (A \triangle B) \triangle C.$$

8. (ii) $(A \cup C) \triangle (B \cup C) = (A \triangle B) \cap C'$ and $(A \triangle B) \cup C = (A \triangle B) \cap C' \Rightarrow C = \varnothing$.

9. $A \triangle B = C \triangle D \Rightarrow (A \triangle B) \triangle B = (C \triangle D) \triangle B \Rightarrow A = (B \triangle D) \triangle C$
$$\Rightarrow A \triangle C = (B \triangle D) \triangle (C \triangle C) \Rightarrow A \triangle C = B \triangle D.$$

10. $A - B$, $B - A$, $A \cap B$ are pairwise disjoint and
$$A = (A - B) \cup (A \cap B), \quad B = (B - A) \cup (A \cap B),$$
$$A \cup B = (A - B) \cup (B - A) \cup (A \cap B).$$
Hence $|A| = |A - B| + |A \cap B|, \quad |B| = |B - A| + |A \cap B|,$
$$|A \cup B| = |A - B| + |B - A| + |A \cap B|.$$

1(b)

1. Let $\{\{a\}, \{a, b\}\} = \{\{c\}, \{c, d\}\}$. If $a = b$, then $\{\{a\}, \{a, b\}\} = \{\{a\}, \{a\}\} = \{\{a\}\}$ and so $\{\{c\}, \{c, d\}\}$ has one element only. Thus $\{c\} = \{c, d\}$ and so $c = d$. Then $\{\{c\}, \{c, d\}\} = \{\{c\}\}$ and $a = c$.

If $a \neq b$, the above argument shows that $c \neq d$. Also $\{a\} = \{c\}$ or $\{a\} = \{c, d\}$. Since $\{c, d\}$ has distinct elements, $\{a\} \neq \{c, d\}$. Hence $\{a\} = \{c\}$, i.e. $a = c$. Then $\{a, b\} = \{c, d\}$ and so $b = d$.

2. (i) $x \in (A \cup B) \times C \Leftrightarrow x = (y, z)$, where $y \in A \cup B$, $z \in C \Leftrightarrow x \in A \times C$ or $x \in B \times C \Leftrightarrow x \in (A \times C) \cup (B \times C)$.

3. $A = B \Rightarrow A \times B = B \times A$. If $A \neq B$, we may suppose that A has an element x not in B. If $y \in B$, $(x, y) \in A \times B$, but $(x, y) \notin B \times A$.

Take $A = \{a\}$. Then $(A \times A) \times A = \{((a, a), a)\}$, $A \times (A \times A) = \{(a, (a, a))\}$. These are different since $a \neq (a, a)$.

4. $\{a\} \times \{a\} = \{(a, a)\}$. See **1**.

5. (i) $y \in f(\bigcup A) \Leftrightarrow \exists \, x \in \bigcup A$ such that $y = f(x) \Leftrightarrow y \in f(A)$ for some $A \in \mathscr{C} \Leftrightarrow y \in \bigcup f(A)$.

(ii) $y \in f(\bigcap A) \Leftrightarrow \exists \, x \in \bigcap A$ such that $y = f(x) \Rightarrow y \in f(A)$ for all $A \in \mathscr{C} \Leftrightarrow y \in \bigcap f(A)$.

If f is injective, $y = f(x)$ for at most one $x \in X$. Hence $y \in f(A)$ for all $A \in \mathscr{C} \Rightarrow$
$\exists\, x \in \bigcap A$ such that $y = f(x)$.

If f is not injective, there are distinct $x, x' \in X$ such that $f(x) = f(x') = y$,
say. Take $A = \{x\}$, $B = \{x'\}$. Then $f(A \cap B) = f(\varnothing) = \varnothing$, $f(A) \cap f(B) = \{y\}$.

6. Proof similar to that of **5** (ii).

7. (ii) $x \in f^{-1}(\bigcap B) \Leftrightarrow f(x) \in \bigcap B \Leftrightarrow f(x) \in B$ for all $B \in \mathscr{C} \Leftrightarrow x \in f^{-1}(B)$ for all
$B \in \mathscr{C} \Leftrightarrow x \in \bigcap f^{-1}(B)$.

9. (i) $x \in A \Rightarrow f(x) \in f(A) \Leftrightarrow x \in f^{-1}(f(A))$.

(ii) $y \in f(f^{-1}(B)) \Leftrightarrow \exists\, x \in f^{-1}(B)$ such that $y = f(x) \Leftrightarrow \exists\, x$ such that $f(x) \in B$
and $y = f(x) \Rightarrow y \in B$.

If f is injective, then $f(x) \in f(A) \Rightarrow x \in A$; hence $f^{-1}(f(A)) = A$.

If f is not injective, there are distinct $x, x' \in A$ such that $f(x) = f(x') = y$,
say. Take $A = \{x\}$. Then $f^{-1}(f(A)) = f^{-1}(\{y\}) \supset \{x, x'\}$, so that $f^{-1}(f(A)) -$
$A \neq \varnothing$.

If f is surjective, then $y \in B \Rightarrow \exists\, x$ such that $y = f(x)$; hence $f(f^{-1}(B)) = B$.

If f is not surjective, $Y - f(X) \neq \varnothing$. Take $B = Y - f(X)$. Then $f(f^{-1}(B)) =$
$f(\varnothing) = \varnothing$.

10. f_2 not surjective $\Rightarrow f_2 \circ f_1$ not surjective. f_2 not injective $\Rightarrow f_3 \circ f_2$ not injective. Hence f_2 is bijective and so are $f_2^{-1} \circ (f_2 \circ f_1) = f_1$, $(f_3 \circ f_2) \circ f_2^{-1} = f_3$.

11. The argument shows that xRy for *some* y implies xRx. But xRy need not
hold for *any* y. (E.g. $X = \{x, y\}$, $R = \{(y, y)\}$.)

1 (c)

1. By theorem 1.43, $D = C - A$ is countable. By theorem 1.42, A contains a
countably infinite subset E. Since $E \cup D$ is countable (theorem 1.45) and infinite,
$E \cup D \sim E$. Then

$$A \cup C = A \cup D = (A - E) \cup (E \cup D) \sim (A - E) \cup E = A.$$

$B \cap C$ is countable. Since B is uncountable, so is $B - C = B - (B \cap C)$; in
particular, $B - C$ is infinite. Hence, by first part,

$$B - C \sim (B - C) \cup (B \cap C) = B.$$

2. (i) $(0, 1) \sim (a, b)$: let $y = a + (b - a)x$ $(x \in (0, 1), y \in (a, b))$.

(ii) $(0, 1) \sim (a, \infty)$: let $y - a = (1/x) - 1$ $(x \in (0, 1), \ y \in (a, \infty))$. Clearly
$(-\infty, a) \sim (-a, \infty)$. $(-\tfrac{1}{2}, \tfrac{1}{2}) \sim (-\infty, \infty)$: let $y = x/(1 + x^2)$ $(x \in (-\infty, \infty),$
$y \in (-\tfrac{1}{2}, \tfrac{1}{2}))$.

(iii) $(a, b) \sim (a, b]$ etc. by **1**.

3. If $\mathscr{C}$ is countably infinite, than it is of the form $\{A_1, A_2, \ldots\}$. $A_n \sim [n - 1, n)$;
let f_n be a bijection on A_n to $[n - 1, n)$. Then $f \colon \bigcup_{n=1}^{\infty} A_n \to [0, \infty)$ given by

$$f(x) = f_n(x) \text{ if } x \in A_n$$

is a bijection. Argument similar when $\mathscr{C}$ is finite.

4. Let $x \in A_1$. Clearly $\{x\} \times A_2 \sim A_2$ and so $\{x\} \times A_2$ is countable. By theorem
1.45, $A_1 \times A_2 = \bigcup_{x \in A_1} (\{x\} \times A_2)$ is countable. Now use induction.

5. If $\mathscr{P}_0$ is the subset of $\mathscr{P}$ which consists of the sequences with infinitely many 0's, then $\mathscr{P}_0 \sim (0, 1)$. For $(p_1, p_2, \ldots) \in \mathscr{P}_0$ may be associated with the unique binary representation $0.p_1p_2\ldots$ (not using recurring 1's) of a real number in $(0, 1)$. Also $\mathscr{P} - \mathscr{P}_0$ is countable, since, for each $n \geq 1$, there is only a finite number of sequences with $p_i = 1$ for $i \geq n$. By **1**, $\mathscr{P} \sim (0, 1)$.

Define $f : \mathscr{P}^2 \to \mathscr{P}$ as follows. For $p = (p_1, p_2, \ldots)$, $q = (q_1, q_2, \ldots) \in \mathscr{P}$, $f(p, q) = (p_1, q_1, p_2, q_2, \ldots)$. Then f is bijective and so $\mathscr{P}^2 \sim \mathscr{P}$. This implies $(0, 1) \times (0, 1) \sim (0, 1)$ and, by **2**, $A_1 \times A_2 \sim R^1$. Now use induction.

6. Let S be the set of integers; A the set of all polynomials with integral coefficients, A_n the set of those of degree n. Since $A_n \sim S^{n+1}$, A_n is countable and so is $A = \bigcup\limits_{n=0}^{\infty} A_n$. Each polynomial equated to 0 has a finite number of roots.

7. $0 \leq i \leq 8$: In the decimal representation of the members of R_i replace $i+1, \ldots, 9$ by $i, \ldots, 8$ respectively. We get the set of nonary representations of the numbers in $(0, 1)$. Thus $R_i \sim (0, 1)$. $i = 9$: Some decimals of R_9 have recurring 8's; they form a countable set C. $R_9 - C \sim (0, 1)$ and so $R_9 \sim (0, 1)$.

8. Let $E_0 = E \cap \{0\}$ and, for $k = 1, 2, \ldots$, put

$$E_k = E \cap (k^{-1}, \infty), \quad E_{-k} = E \cap (-\infty, -k^{-1}).$$

Each E_i $(i \neq 0)$ is finite, for otherwise a series of distinct terms of E_i could be formed and this would diverge. Hence $E = \bigcup\limits_{-\infty}^{\infty} E_i$ is countable.

9. (i) Let $A = \{a_1, \ldots, a_n\}$. Let $\mathscr{P}$ be the set of all ordered sets $(p_1, \ldots, p_n)$ of 0's and 1's. $\mathscr{P}$ has 2^n members. With $S \in \mathscr{S}_A$ associate $(p_1, \ldots, p_n) \in \mathscr{P}$ defined by

$$p_i = \begin{cases} 1 & \text{if } a_i \in S, \\ 0 & \text{if } a_i \notin S. \end{cases}$$

Clearly $\mathscr{S}_A \sim \mathscr{P}$.

(ii) Now let $\mathscr{P}$ be the set of **5**. As in (i), $\mathscr{S}_A \sim \mathscr{P}$.

(iii) Suppose that $A \sim \mathscr{S}_A$, i.e. that there is a bijection $f : A \to \mathscr{S}_A$. For $x \in A$, put $f(x) = S_x$. Define $S^* \in \mathscr{S}_A$ by

$$x \in S^* \quad \text{if and only if} \quad x \notin S_x. \tag{†}$$

Since $S^* \in \mathscr{S}_A$, there is $a \in A$ such that $S^* = S_a$. Then (†) with $x = a$ becomes

$$a \in S^* = S_a \quad \text{if and only if} \quad a \notin S_a = S^*.$$

10. For each interval I of the collection $\mathscr{I}$ choose a rational $x_I \in I$. The x_I are distinct. Hence $\mathscr{I}$ is similar to a subset of the set of rational numbers and so is countable.

Suppose f is increasing. If c (not an end point of J) is a discontinuity, put $I_c = (f(c-), f(c+))$. When $c_1 < c_2$, $f(c_1+) \leq f(c_2-)$ and so the I_c are disjoint. Hence the set of I_c (and so of the c) is countable.

2(a)

1. (a) No; $\rho(x, -x) = 0$. (b) Yes.

2. If $\rho(\phi, \psi) = 0$, then $\int_x^{x+h} (\phi - \psi) = 0$ for all x and $0 < h \leq 1$; thus $\int_0^y (\phi - \psi) = 0$ for all y. Differentiate w.r.t. y : $\phi(y) - \psi(y) = 0$ for all y.

3. $y = x$ in (ii) gives $2\rho(x, z) \geqslant \rho(x, x) = 0$. $z = x$ in (ii) gives $\rho(x, y) \leqslant \rho(y, x)$; interchange x, y to get $\rho(y, x) \leqslant \rho(x, y)$.

4. $|\xi_{1k} - \xi_1|, \ldots, |\xi_{nk} - \xi_n| \leqslant \rho(x_k, x) \leqslant |\xi_{1k} - \xi_1| + \ldots + |\xi_{nk} - \xi_n|$.

5. Since
$$\rho(x_n, y_n) \leqslant \rho(x_n, x) + \rho(x, y) + \rho(y, y_n),$$
$$\rho(x_n, y_n) - \rho(x, y) \leqslant \rho(x_n, x) + \rho(y, y_n).$$
Similarly
$$\rho(x, y) - \rho(x_n, y_n) \leqslant \rho(x, x_n) + \rho(y_n, y).$$
Hence
$$|\rho(x, y) - \rho(x_n, y_n)| \leqslant \rho(x, x_n) + \rho(y, y_n).$$

6. To prove M3 put $\rho(x, y) = a, \rho(y, z) = b, \rho(z, x) = c$, so that $a+b-c \geqslant 0$. Then
$$\sigma(x, y) + \sigma(y, z) - \sigma(x, z) = \frac{(a+b-c) + ab(2+c)}{(1+a)(1+b)(1+c)} \geqslant 0.$$

ρ, σ are equivalent, for if $\rho(x_n, x) \to 0$, then
$$\sigma(x_n, x) = \frac{\rho(x_n, x)}{1 + \rho(x_n, x)} \to 0$$
and if $\sigma(x_n, x) \to 0$, then
$$\rho(x_n, x) = \frac{\sigma(x_n, x)}{1 - \sigma(x_n, x)} \to 0.$$

7. Example: on R^1, ρ usual metric and σ as in **6**. Then $\sigma(x, y) \leqslant \rho(x, y)$, but $\sigma(x, y)/\rho(x, y) = 1/(1 + |x - y|) \to 0$ as $x \to \infty$.

8. $\sigma \leqslant \rho \leqslant \tau \leqslant 2\sigma \leqslant 2\rho$; use **7**.

$\rho(x, a) \leqslant 1$ gives the disc D with centre a and radius 1. $\sigma(x, a) \leqslant 1$ gives the square P with centre a and sides of length 2 parallel to the axes. $\tau(x, a) \leqslant 1$ gives the square Q with centre a and sides of length $\sqrt{2}$ at 45° to the axes. $P \supset D \supset Q$.

9. The sequence (x_n) converges in (R^1, σ) if and only if x_n is constant for all sufficiently large n.

10. If $h \in C$, $h(x) \geqslant 0$ for $0 \leqslant x \leqslant 1$ and $h(\alpha) > 0$ for some $\alpha \in [0, 1]$, then $\int_0^1 h(x)dx > 0$. For $h(x) > \frac{1}{2}h(\alpha)$ in an interval $[a, b] \subset [0, 1]$ and
$$\int_0^1 h(x)dx \geqslant \int_a^b h(x)dx \geqslant \frac{1}{2}h(\alpha)(b-a).$$

This proves M1 for σ and τ. M2 is obvious and so is M3 for τ.

To prove M3 for σ adapt the argument for the usual metric in R^n: M3 is equivalent to
$$\int_0^1 \phi\psi \leqslant \left(\int_0^1 \phi^2\right)^{\frac{1}{2}} \left(\int_0^1 \psi^2\right)^{\frac{1}{2}} \tag{*}$$
and this follows from the fact that
$$\int_0^1 \{\xi\phi(x) + \psi(x)\}^2 dx \geqslant 0$$
for all real ξ.

$\tau \leqslant \sigma$ follows from (*) with $\psi \equiv 1$.

In (C, σ) and in (C, τ), $f_n \to \theta$, where $\theta(x) = 0$ for $0 \leqslant x \leqslant 1$. If (f_n) converged in (C, ρ) to f, say, then, since $\sigma(f_n, f) \leqslant \rho(f_n, f)$, f_n would have to converge to f in (C, σ) also. Thus $f = \theta$. But $\rho(f_n, \theta) = 1$ for all n, i.e. $f_n \nrightarrow \theta$ in (C, ρ).

In (C, τ), $g_n \to \theta$. If (g_n) converged in (C, σ), it would have to converge to θ. But $\sigma(g_n, \theta) = 1/\sqrt{3}$ for all n. Also convergence in (C, ρ) would imply convergence in (C, σ).

11. $\sigma_1 \leqslant \sigma_2 \leqslant \sigma_3 \leqslant n\sigma_1 \leqslant n\sigma_2$.

2(b)

2. If V is a normed vector space and ρ is defined by $\rho(x, y) = \|x-y\|$, then

$$\rho(\alpha x, \theta) = \|\alpha x - \theta\| = \|\alpha x\| = |\alpha| \|x\|.$$

Hence, if V has an element other than the zero element, then $\rho(x, y)$ takes all non-negative values.

3. The condition for $\|.\|_1$, $\|.\|_2$ to be equivalent is that $\|x_n\|_1 \to 0$ if and only if $\|x_n\|_2 \to 0$.

Suppose that there is no μ such that $\|x\|_2 \leqslant \mu\|x\|_1$ for all $x \in V$. Then, for every integer n, there is an x_n such that

$$\|x_n\|_2 > n\|x_n\|_1.$$

$x_n \neq \theta$, since otherwise $\|x_n\|_2 = 0$. Let

$$y_n = \frac{x_n}{n\|x_n\|_1}.$$

Then $\|y_n\|_1 = 1/n \to 0$, but $\|y_n\|_2 > 1$ for all n. Hence there is a μ so that $\|x\|_2 \leqslant \mu\|x\|_1$ for all x and clearly $\mu > 0$. Similarly there is a $\kappa > 0$ such that $\|x\|_1 \leqslant \kappa\|x\|_2$ for all $x \in V$.

4. If an inner product exists,

$$\|x+y\|^2 = (x+y).(x+y) = \|x\|^2 + x.y + y.x + \|y\|^2,$$

so that $\qquad 2\operatorname{re}(x.y) = \|x+y\|^2 - \|x\|^2 - \|y\|^2.$

Similarly $\qquad 2\operatorname{im}(x.y) = \|x+iy\|^2 - \|x\|^2 - \|y\|^2.$

6. $x.\theta + x.\theta = x.(\theta+\theta) = x.\theta.$

7. Take R^2 with $\|x\| = \max(|x_1|, |x_2|)$. If $x = (1, 0)$, $y = (0, 1)$,

$$\|x+y\|^2 + \|x-y\|^2 = 2, \quad 2(\|x\|^2 + \|y\|^2) = 4.$$

(Or $\|x\| = |x_1| + |x_2|$.)

8. $|x_n.y_n - x.y| \leqslant |x_n.(y_n-y)| + |(x_n-x).y|$

$$\leqslant \|x_n\| \|y_n-y\| + \|x_n-x\| \|y\|.$$

9. Proof of N 2. Theorem 2.21 clearly holds in this situation. Then

$$\|x+y\|^2 = (x+y).(x+y) = \|x\|^2 + 2\operatorname{re}(x.y) + \|y\|^2$$

$$\leqslant \|x\|^2 + 2|x.y| + \|y\|^2$$

$$\leqslant (\|x\| + \|y\|)^2.$$

2(c)

1. X contains a countable subset S. Denote the (distinct) members of S by $a, x_1, x_2, \ldots$. Let $\rho(x, x) = 0$ (all $x \in X$), $\rho(x, y) = 1$ ($x \neq y$; x, y not both in S), $\rho(x_m, x_n) = |m^{-1} - n^{-1}|$, $\rho(a, x_n) = n^{-1}$. M3 requires checking.

2. If c is a limit point, there is a sequence (x_n) such that $\rho(a, x_n) \leqslant r$ (all n) and $x_n \to c$. Since $\rho(a, x_n) \to \rho(a, c)$ (exercise 2(a), 5), $\rho(a, c) \leqslant r$.

3. If the supremum (infimum) did not belong to the set, it would be a limit point outside the set.

4. Let c be a limit point of E^*. Given ϵ, $B(c; \epsilon) - \{c\}$ contains an $x \in E^*$. Let δ be such that $B(x; \delta) \subset B(c; \epsilon) - \{c\}$. Since $B(x; \delta) - \{x\}$ contains a point of E, so does $B(c; \epsilon) - \{c\}$.

5. Each open set contains an open ball which contains a point with rational co-ordinates (cf. exercise 1(c), 10).
   No. If R^n is given the discrete metric, every set $\{x\}$ is open.

6. (i) K is the union of all open sets $G \subset E$. Every $G \subset E^\circ$ and so $K \subset E^\circ$. $K \supset E^\circ$, since E° is a G.
   (ii) H is the intersection of all closed sets $F \supset E$. Every $F \supset \bar{E}$ and so $H \supset \bar{E}$. $H \subset \bar{E}$, since $\bar{E}$ is an F.

7. For any $x \in X$, there are three exclusive and exhaustive possibilities: for every ϵ, $B(x; \epsilon) \cap E \neq \varnothing$ and $B(x; \epsilon) \cap E' \neq \varnothing$; for some ϵ, $B(x; \epsilon) \cap E = \varnothing$; for some ϵ, $B(x; \epsilon) \cap E' = \varnothing$.
   $x \in (\bar{E})' \Leftrightarrow x \notin \bar{E} \Leftrightarrow$ for some ϵ, $B(x; \epsilon) \cap E = \varnothing \Leftrightarrow$ for some ϵ, $B(x; \epsilon) \subset E' \Leftrightarrow x \in (E')^\circ$.

8. By 7 (2nd part), $\bar{E}' = (E^\circ)'$. Use (2.31).

9. (i) $x \in \overline{E_1 \cup E_2} \Leftrightarrow$ there is a sequence (x_n) in $E_1 \cup E_2$ such that $x_n \to x \Leftrightarrow$ there is a sequence (x'_m) in E_1 or in E_2 such that $x'_m \to x \Leftrightarrow x \in \bar{E}_1$ or $x \in \bar{E}_2 \Leftrightarrow x \in \bar{E}_1 \cup \bar{E}_2$. Use induction.
   For arbitrary collections, the right to left implications hold. Also, if in R^1 (usual metric), $E_n = \{1/n\}$, $\overline{\cup E_n} - \cup \bar{E}_n = \{0\}$.
   (ii) $x \in \overline{\cap E} \Rightarrow$ for every ϵ, $B(x; \epsilon)$ contains a $y \in \cap E \Rightarrow x \in \bar{E}$ for every $E \Rightarrow x \in \cap \bar{E}$.
   If, in R^1, $E_1 = \{1, \frac{1}{2}, \frac{1}{3}, \ldots\}$, $E_2 = \{-1, -\frac{1}{2}, -\frac{1}{3}, \ldots\}$, $\overline{E_1 \cap E_2} = \varnothing$, $\bar{E}_1 \cap \bar{E}_2 = \{0\}$.
   (iii) $(E_1 \cap \ldots \cap E_n)^\circ = E_1^\circ \cap \ldots \cap E_n^\circ$; $(\cap E)^\circ \subset \cap E^\circ$; $(\cup E)^\circ \supset \cup E^\circ$.

10. Use the sequence definition of limit point. For E° use 7 (2nd part) or invariance of open sets and 6 (i).

11. $C \subset D \subset E \subset \bar{D} \subset \overline{\bar{C}} = \bar{C}$.

3(a)

1. $f(x, mx^2) = m/(1 + m^2)$.

2. If (x_n), (x'_n) are sequences in $E - \{x_0\}$ with limit x_0, then $(x''_n) = (x_1, x'_1, x_2, x'_2, \ldots)$ is of same kind. Since $\lim f(x''_n)$ exists, $\lim f(x_n) = \lim f(x'_n)$. Apply theorem 3.1.

3. Given ϵ, there is δ such that $\sigma(f(x), y_0) < \epsilon$ whenever $0 < \rho(x, x_0) < \delta$ and $x \in E$. There are infinitely many such x (x_0 is a limit point of E) and so $B(y_0; \epsilon)$ contains points of $f(E)$. If f is injective, $B(y_0; \epsilon)$ contains infinitely many points of $f(E)$ and so does $B(y_0; \epsilon) - \{y_0\}$.

4. Take ϵ. There is η such that $0 < \rho(y, y_0) < \eta$ and $y \in E \Rightarrow \tau(g(y), w_0) < \epsilon$. There is δ such that $0 < \rho(x, x_0) < \delta$ and $x \in D \Rightarrow \sigma(f(x), y_0) < \eta$. If f is injective, there is at most one x for which $f(x) = y_0$. Then δ may be chosen so that $0 < \rho(x, x_0) < \delta$ and $x \in D \Rightarrow 0 < \sigma(f(x), y_0) < \eta$.

Example: $D = X = E = Y = [0, 1]$, $f(x) = 0$ $(0 \leqslant x \leqslant 1)$, $g(0) = 0$, $g(y) = 1$ $(0 < y \leqslant 1)$. $g(y) \to 1$ as $y \to 0$, but $h(x) = 0$ $(0 \leqslant x \leqslant 1)$.

5. By theorem 3.1 (necessity), if (x_n) is any sequence in $E - \{x_0\}$ with $\lim x_n = x_0$, then $\phi(x_n) \to \alpha$, $\psi(x_n) \to \beta$, $f(x_n) \to y_0$, $g(x_n) \to z_0$ as $n \to \infty$. By exercise 2(b), 8, $\phi(x_n)f(x_n) + \psi(x_n)g(x_n) \to \alpha y_0 + \beta z_0$ and (3.11) follows from theorem 3.1 (sufficiency).

3(b)

1. f continuous on X. Let $x \in \overline{f^{-1}(F)}$. There is (x_n) in $f^{-1}(F)$ such that $x_n \to x$. $f(x_n) \in F$. Since f is continuous at x, $f(x_n) \to f(x)$ and, since F is closed, $f(x) \in F$. Thus $x \in f^{-1}(F)$.

f not continuous on X. If f is discontinuous at x, there are $\epsilon > 0$ and (x_n) such that $x_n \to x$, but $\sigma(f(x_n), f(x)) \geqslant \epsilon$. Let $S = \{f(x_1), f(x_2), \ldots\}$. Then $f(x) \notin \overline{S}$ and so $x \notin f^{-1}(\overline{S})$. Thus $f^{-1}(\overline{S})$ is not closed.

2. Suppose f continuous at x_0. If N is open and $f(x_0) \in N$, there is ϵ such that $B(f(x_0); \epsilon) \subset N$. There now exists δ such that $x \in B(x_0; \delta)$ implies $f(x) \in B(f(x_0); \epsilon)$. Take $M = B(x_0; \delta)$.

Converse. Given ϵ, take $N = B(f(x_0); \epsilon)$. There is an open set M such that $x_0 \in M$ and $f(M) \subset N$. There is δ such that $B(x_0; \delta) \subset M$.

3. Define f on R^1 by $f(x) = 0$ (x rational), $f(x) = 1$ (x irrational). Take E to be the set of rational numbers.

4. Continuity of f obvious. Let y be a limit point of Y and let $y_n \to y$, where $y_n \neq y$. Then $f^{-1}(y_n) \nrightarrow f^{-1}(y)$, for otherwise $f^{-1}(y)$ would be a limit point of X.

Take $X = \{0, 1, 2, \ldots\}$, $Y = \{0, 1, \frac{1}{2}, \frac{1}{3}, \ldots\}$ and use metrics induced by R^1. Take $f(0) = 0$, $f(n) = 1/n$ $(n \geqslant 1)$.

5. ρ, σ equivalent $\Leftrightarrow (X, \rho), (X, \sigma)$ have same open sets $\Leftrightarrow I, I^{-1}$ continuous (theorem 3.23).

Let ρ be the discrete metric and σ the usual metric on R^1. Then $I : (R^1, \rho) \to (R^1, \sigma)$ is everywhere continuous, but $I^{-1} : (R^1, \sigma) \to (R^1, \rho)$ is nowhere continuous.

6. $G = [2, 3]$, $F = [0, 1)$.

7. $g((-\infty, -1]) = g((-\infty, 0)) = (0, 4/e]$.

8. If ρ is the usual metric in R^n,

$$\max_j |f_j(x_0) - f_j(x)| \leqslant \rho(f(x_0), f(x)) \leqslant \sum_j |f_j(x_0) - f_j(x)|.$$

Let $f : R^m \to R^n$ be linear and have matrix (a_{ji}). f_j is continuous at x_0, since

$$|f_j(x_0) - f_j(x)| \leqslant (|a_{j1}| + \ldots + |a_{jm}|)\|x_0 - x\|.$$

9. If ω is any one of the equivalent metrics on $X \times Y$,

$$\omega((x, y_0), (x_0, y_0)) = \rho(x, x_0), \quad \omega((x_0, y), (x_0, y_0)) = \sigma(y, y_0).$$

For the example take $f: R^2 \to R^1$ defined by

$$f(x, y) = \begin{cases} \dfrac{xy}{x^2+y^2} & \text{when } (x, y) \neq (0, 0), \\ 0 & \text{when } (x, y) = (0, 0) \end{cases}$$

and $x_0 = y_0 = 0$.

10. Let $\phi_n(x) = (1-x)^n$ for $0 \leqslant x \leqslant 1$. Then $\|f(\phi_n)\|/\|\phi_n\| = n$.

11. Let $a = (a_1, ..., a_n)$. By Cauchy's inequality, $|f(x)| = |a.x| \leqslant \|a\| \|x\|$, for all $x \in R^n$. Also $|f(a)| = \|a\|^2$. Hence $\|f\| = \|a\|$. ($a = \theta$ included.)

12. $\|g\{f(x)\}\| \leqslant \|g\| \|f(x)\| \leqslant \|g\| \|f\| \|x\| . f^{-1} \circ f$ is the identity function I on U and $\|I\| = 1$.

Let U be R^2 with $\|x\| = \max(|x_1|, |x_2|)$, let V be R^2 with norm $\|x\| = |x_1| + |x_2|$, let $f: U \to V$ be defined by $f(x) = x$. Then

$$\|f\| = \sup \frac{|x_1| + |x_2|}{\max(|x_1|, |x_2|)} = 2, \quad \|f^{-1}\| = \sup \frac{\max(|x_1|, |x_2|)}{|x_1| + |x_2|} = 1.$$

3(c)

1. If τ is metric on E induced by (X, ρ) or (Y, σ), then (E, τ) is a metric subspace of (X, ρ) [(Y, σ)] and so E is connected in (X, ρ) [(Y, σ)] if and only if (E, τ) is connected.

2. Let $G_1 \cap G_2 \cap \{x\} = \varnothing$, $G_1 \cup G_2 \supset \{x\}$. If $x \in G_1$, then $x \notin G_2$, i.e. $G_2 \cap \{x\} = \varnothing$.

Let E contain at least two points and let x be an isolated point of E. There is δ such that $B(x; \delta) \cap E = \{x\}$. Take $G_1 = B(x; \frac{1}{2}\delta)$, $G_2 = (\overline{G}_1)'$.

3. X connected $\Leftrightarrow E, E'$ not both closed (all $E \neq \varnothing, X$) $\Leftrightarrow \overline{E} \not\subset E$ or $\overline{E'} \not\subset E'$ (all $E \neq \varnothing, X$) $\Leftrightarrow \operatorname{fr} E = \overline{E} \cap \overline{E'} \neq \varnothing$ (all $E \neq \varnothing, X$).

4. Let S be set of irrational numbers and let $E(\subset S)$ contain x, y, where $x < y$. There is a rational r such that $x < r < y$. Take $G_1 = (-\infty, r) \cap S$, $G_2 = (r, \infty) \cap S$.

5. If G is open and $G \cap \overline{E} \neq \varnothing$, then $G \cap E \neq \varnothing$. For let $x \in G \cap \overline{E}$. There is δ such that $B(x; \delta) \subset G$ and $B(x; \delta) \cap E \neq \varnothing$. Hence, if $\overline{E}$ is not connected, so that there are G_1, G_2 with

$$G_1 \cap \overline{E} \neq \varnothing, \quad G_2 \cap \overline{E} \neq \varnothing, \quad G_1 \cap G_2 \cap \overline{E} = \varnothing, \quad G_1 \cup G_2 \supset \overline{E},$$

then $\quad G_1 \cap E \neq \varnothing, \quad G_2 \cap E \neq \varnothing, \quad G_1 \cap G_2 \cap E = \varnothing, \quad G_1 \cup G_2 \supset E,$

i.e. E is not connected.

No. Take $E = (0, 1) \cup (1, 2)$.

6. If C is a component of a closed set F, then $C \subset F$ and $\overline{C} \subset \overline{F} = F$. By **5**, $\overline{C}$ is connected. Since $\overline{C} \supset C$ and C is maximal, $\overline{C} = C$.

No. See **4**.

7. $f: (0, 1] \to R^2$ given by $f(x) = (x, \sin(1/x))$ is continuous. Since $(0, 1]$ is connected, by theorem 3.34, so is $E = \{(x, y) | 0 < x \leqslant 1, y = \sin(1/x)\}$. By **5** $\overline{E} = E \cup \{(x, y) | x = 0, -1 \leqslant y \leqslant 1\}$ is also connected.

3(d)

1. If (x_n) is any sequence in $E - \{x_0\}$ with limit x_0, then $\lim_{n \to \infty} f(x_n)$ exists. Apply exercise 3(a), 2.

2. $s_n \to e$ in R^1. Hence (s_n) is a Cauchy sequence in R^1 and so in Q.

$$0 < q!(e - s_q) < \frac{1}{q+1} + \frac{1}{(q+1)^2} + \ldots = \frac{1}{q} \leqslant 1. \qquad (*)$$

If $e = p/q$, $q!e$ is an integer (as well as $q!s_q$) and so is $q!(e - s_q)$. This is false by (*). Hence e is irrational and (s_n) cannot converge in Q, for if it did (s_n) would have two distinct limits in R^1.

3. Let $\phi_n(x) = (x^n + 1)^{1/n}$ $(0 \leqslant x \leqslant 2)$. (ϕ_n) converges in $B[0, 2]$ to ϕ given by

$$\phi(x) = \begin{cases} 1 & \text{for} \quad 0 \leqslant x \leqslant 1, \\ x & \text{for} \quad 1 < x \leqslant 2, \end{cases}$$

for $\rho(\phi_n, \phi) \leqslant 2(2^{1/n} - 1)$. Thus (ϕ_n) is a Cauchy sequence in $B[0, 2]$ and so in D. Since $\phi \notin D$, (ϕ_n) cannot converge in D, for if it did (ϕ_n) would have two limits in $B[0, 2]$.

4. (n) is a non-convergent Cauchy sequence since $n/(1 + n) \to 1$, but there is no real x such that $x/(1 + |x|) = 1$.

5. No. Let ρ be the usual metric on R^1 and σ the metric of **4**. Then ρ, σ are equivalent; for $f: R^1 \to R^1$ given by $f(x) = x/(1 + |x|)$ is continuous and strictly increasing, so that $f(x_n) \to f(x)$ if and only if $x_n \to x$.

6. (i) $\mathcal{E}$ is a collection of complete subsets E of X. Put $H = \bigcap_{E \in \mathcal{E}} E$. If $H = \varnothing$, then H is complete. If $H \neq \varnothing$, let (x_n) be a Cauchy sequence in H and so in E. There is an $x \in E$ such that $x_n \to x$. Limits in X are unique. Hence x is the same for all E, and $x \in H$.

(ii) We need only consider two complete sets E_1, E_2. Let (x_n) be a Cauchy sequence in $E_1 \cup E_2$. At least one of E_1, E_2, say E_1, contains an infinite subsequence (x_{ν_k}). Since E_1 is complete, there is an $x \in E_1$ such that $x_{\nu_k} \to x$. Since also (x_n) is a Cauchy sequence,

$$\rho(x, x_n) \leqslant \rho(x, x_{\nu_n}) + \rho(x_{\nu_n}, x_n) \to 0.$$

7. Let $x_n \in F_n$. If $m, n > n_0$, $\rho(x_m, x_n) \leqslant \rho(F_{n_0})$ and as $\rho(F_n) \to 0$, (x_n) is a Cauchy sequence. Since (X, ρ) is complete, there is an x such that $x_n \to x$; and since each F_n is closed, $x \in \bigcap F_n$. If $x' \in \bigcap F_n$, then $\rho(x, x') \leqslant \rho(F_n)$ for all n and so $\rho(x, x') = 0$, i.e. $x' = x$.

(i) $X = (0, 1]$, $F_n = (0, 1/n]$ (closed in X).

(ii) $X = R^1$, $F_n = (0, 1/n]$.

(iii) $X = R^1$, $F_n = [n, \infty)$. For a counter example with bounded F_n, let $X = C[0, 1]$ and let F_n be the (closed) set of continuous functions f on $[0, 1]$ such that

$$f(0) = 0, \quad 0 \leqslant f(x) \leqslant 1 \text{ in } (0, 1/n), \quad f(x) = 1 \text{ in } [1/n, 1].$$

Then $\rho(F_n) = 1$ and $\bigcap F_n = \varnothing$, since there is no continuous f such that $f(0) = 0$ and $f(x) = 1$ in $\bigcup [1/n, 1] = (0, 1]$.

8. $$\phi_n(x) = x^n \ (0 \leqslant x \leqslant 1), \quad \phi(x) = \begin{cases} 0 & (0 \leqslant x < 1), \\ 1 & (x = 1). \end{cases}$$

$\rho(\phi_n, \phi) = 1$ for all n.

9. Let (ϕ_n) be a Cauchy sequence in $L(V, W)$. For any $x \in V$,

$$\|\phi_m(x) - \phi_n(x)\| = \|(\phi_m - \phi_n)(x)\| \leqslant \|\phi_m - \phi_n\| \, \|x\|$$

and so $(\phi_n(x))$ is a Cauchy sequence in the complete space W.
Define $\phi : V \to W$ by $\phi(x) = \lim\limits_{n \to \infty} \phi_n(x)$. Then

$$\phi(\alpha x + \beta y) = \lim_{n \to \infty} \phi_n(\alpha x + \beta y) = \lim_{n \to \infty} \{\alpha \phi_n(x) + \beta \phi_n(y)\} = \alpha\phi(x) + \beta\phi(y),$$

i.e. ϕ is linear.

Since (ϕ_n) is a Cauchy sequence, $\|\phi_n\| \leqslant A$, say. (Adapt first stage of proof of theorem 3.42.) As $\|\phi(x)\| = \lim \|\phi_n(x)\| \leqslant A\|x\|$, by theorem 3.24, ϕ is continuous.

Given ϵ, there is n_0 such that $\|\phi_m - \phi_n\| < \epsilon$ for $m, n > n_0$. For any $x \in V$ and $n > n_0$,

$$\|\phi(x) - \phi_n(x)\| = \lim_{m \to \infty} \|\phi_m(x) - \phi_n(x)\| \leqslant \epsilon\|x\|.$$

Thus $\|\phi - \phi_n\| \leqslant \epsilon$ and so $\phi_n \to \phi$.

10. $[a, b]$ is complete and $f : [a, b] \to [a, b]$ is a contraction mapping since $|f(x_1) - f(x_2)| = |f'(\xi)(x_1 - x_2)| \leqslant k|x_1 - x_2|$.

11. Define Ω on $B[a, b]$ by

$$\Omega(\psi)(x) = \psi(x) - f(x, \psi(x))/M \quad (\psi \in B[a, b], \ x \in [a, b]).$$

Then $\Omega(\psi) \in B[a, b]$ and

$$|\Omega(\psi_1)(x) - \Omega(\psi_2)(x)| = |[\psi_1(x) - \psi_2(x)](1 - f_y(x, u)/M)|$$
$$\leqslant |\psi_1(x) - \psi_2(x)|(1 - m/M),$$

i.e. $\quad\quad \rho(\Omega(\psi_1), \Omega(\psi_2)) \leqslant (1 - m/M)\rho(\psi_1, \psi_2).$

Thus Ω is a contraction mapping and there is a unique ϕ such that $\Omega(\phi) = \phi$, i.e. $f(x, \phi(x)) \equiv 0$. But, if ψ is continuous, so is $\Omega(\psi)$ and the argument at the end of the proof of theorem 3.47 shows that ϕ is continuous.

12. Put $M = [x_0 - \alpha, x_0 + \alpha]$, $N = [y_0 - b, y_0 + b]$ and define Ω on $C(M, N)$ by

$$\Omega(\psi)(x) = y_0 + \int_{x_0}^x f(t, \psi(t))dt \quad (\psi \in C(M, N), \ x \in M).$$

$\Omega(\psi)$ is continuous (C1, theorem 7.61) and, for $x \in M$,

$$|\Omega(\psi)(x) - y_0| \leqslant |x - x_0|B \leqslant \alpha B < b,$$

i.e. $\Omega(\psi)(x) \in N$. Also

$$|\Omega(\psi_1)(x) - \Omega(\psi_2)(x)| \leqslant \left| \int_{x_0}^x |f(t, \psi_1(t)) - f(t, \psi_2(t))|dt \right|$$
$$\leqslant \left| \int_{x_0}^x A|\psi_1(t) - \psi_2(t)|dt \right| \leqslant \alpha A \rho(\psi_1, \psi_2)$$

and so Ω is a contraction mapping.

13. (i) $\Omega(x) - x = 1/(1 + e^x) > 0$ for all $x \in R^1$. (ii) For all distinct $x_1, x_2, \in R^1$,

$$|\Omega(x_1) - \Omega(x_2)| = \left| (x_1 - x_2) + \left(\frac{1}{1 + e^{x_1}} - \frac{1}{1 + e^{x_2}} \right) \right|$$

$$= \left| (x_1 - x_2) \left(1 - \frac{e^\xi}{(1 + e^\xi)^2} \right) \right| < |x_1 - x_2|.$$

14. Since a fixed point of Ω is a fixed point of $\Omega^{(r)}$, Ω cannot have more than one fixed point.

Let x be the unique fixed point of $\Omega^{(r)}$. Then

$$\Omega^{(r)}(\Omega(x)) = \Omega(\Omega^{(r)}(x)) = \Omega(x),$$

i.e. $\Omega(x)$ is a fixed point of $\Omega^{(r)}$. Hence $\Omega(x) = x$.

15. Take $\phi_1(x) \equiv 1$, $\phi_2(x) \equiv 0$ on $[0, 1]$. Then $\rho(\phi_1, \phi_2) = \rho(\Omega(\phi_1), \Omega(\phi_2)) = 1$.

By integration by parts,

$$\rho(\Omega(\phi), \Omega(\psi)) = \sup_{0 \leqslant x \leqslant 1} \int_0^x (x - t)\{\phi(t) - \psi(t)\}dt \leqslant \tfrac{1}{2}\rho(\phi, \psi).$$

3(e)

1. Let (X_0, σ_0) be a subspace of (Y, σ) isometric with (X, ρ). Since $\overline{X}_0$ is closed, by theorem 3.43 it is complete in (Y, σ). $(\overline{X}_0, \overline{\sigma})$ ($\overline{\sigma}$ restriction of σ to $\overline{X}_0 \times \overline{X}_0$) is a completion of (X, ρ).

2. Take $X =$ set of non-negative rational numbers, $Y = [0, \infty)$, $Y_1 = [1, \infty)$, all with metric induced by R^1.

3. For any $x \in X^+$, there is a sequence (x_n) in X convergent to x. $(\Omega(x_n))$ is a Cauchy sequence in X, since

$$\rho(\Omega(x_m), \Omega(x_n)) \leqslant k\rho(x_m, x_n),$$

and so has a limit in X^+. If also $x_n' \to x$ ($x_n' \in X$), then

$$\rho(\Omega(x_n), \Omega(x_n')) \leqslant k\rho(x_n, x_n') \to 0$$

and so $\lim \Omega(x_n') = \lim \Omega(x_n)$. Hence define Ω^+ on X^+ by

$$\Omega^+(x) = \lim \Omega(x_n).$$

The restriction of Ω^+ to X is Ω, since, when $x \in X$, a sequence from X converging to x is $(x, x, \dots)$.

Ω^+ is a contraction. If $x, y \in X^+$ and $x_n \to x$, $y_n \to y$ ($x_n, y_n \in X$),

$$\frac{\rho(\Omega(x_n), \Omega(y_n))}{\rho(x_n, y_n)} = \frac{\rho^+(\Omega(x_n), \Omega(y_n))}{\rho^+(x_n, y_n)} \to \frac{\rho^+(\Omega^+(x), \Omega^+(y))}{\rho^+(x, y)}.$$

Since l.h.s. $\leqslant k$ for all n, r.h.s. $\leqslant k$.

Ω^+ is unique. Any continuous $f: X^+ \to X^+$ which extends Ω satisfies $f(x) = \lim \Omega(x_n)$ ($x \in X^+, x_n \in X, x_n \to x$).

3(f)

1. A subsequence (x_{ν_k}) of a sequence (x_n) converges in (X, ρ) if and only if it converges in (X, σ).

2. Let E be a compact set in (X, ρ) and let (x_n) in E converge to $x \in X$. There is a subsequence which converges in E. Since every subsequence converges to x, $x \in E$. Use theorem 2.31.

3. Let $\phi_n(x) = 1$ for $x = 1/n$, 0 elsewhere in $[0, 1]$. If (ϕ_{ν_k}) is any subsequence of (ϕ_n), $\rho(\phi_{\nu_k}, \phi_{\nu_l}) = 1 (k \neq l)$ and so (ϕ_{ν_k}) does not converge.

4. If $x_n \not\to x$, there are a subsequence (x_{ν_k}) and an $\epsilon > 0$ such that $\rho(x_{\nu_k}, x) \geq \epsilon$ for all k. By compactness, (x_{ν_k}) has a convergent subsequence $(x_{\nu_{k_i}})$. Since $(x_{\nu_{k_i}})$ is a subsequence of (x_n), $x_{\nu_{k_i}} \to x$. Contradiction.

5. (i) F closed, E compact. Use theorem 3.63: $E \cap F$ is a closed subset of the compact set E.

Let $\mathscr{E}$ be a collection of compact sets E. Since each E is closed, $H = \bigcap_{E \in \mathscr{E}}$ is closed. If E^* is any member of E, $H = E^* \cap H$ is compact.

(ii) We need only consider two compact sets E_1, E_2. Let (x_n) be a sequence in $E_1 \cup E_2$. At least one of E_1, E_2, say E_1, contains an infinite subsequence (x_{ν_k}). There is a subsequence $(x_{\nu_{k_i}})$ of (x_{ν_k}) and so of (x_n) convergent in E_1 and so in $E_1 \cup E_2$.

6. (X, ρ) compact, (F_n) contracting sequence of non-empty closed sets in X. Let $x_n \in F_n$. (x_n) has a subsequence (x_{ν_k}) converging to an $x \in X$. For any n, $x_{\nu_k} \in F_n$ if $\nu_k \geq n$ and since F_n is closed, $x \in F_n$. Thus $x \in \bigcap_{n=1}^{\infty} F_n$.

7. Take $\epsilon > 0$. There is $b > a$ such that $|f(x) - l| < \frac{1}{2}\epsilon$ for $x \geq b$. When $x_1, x_2 > b$, $|f(x_1) - f(x_2)| < \epsilon$. Also f is uniformly continuous on $[a, b+1]$ and so there is $\delta (0 < \delta < 1)$ such that $|f(x_1) - f(x_2)| < \epsilon$ when $x_1, x_2 \in [a, b+1]$ and $|x_1 - x_2| < \delta$.

If $|x_1 - x_2| < \delta$, then $x_1, x_2 \in [a, b+1]$ or $x_1, x_2 \in [b, \infty)$ (possibly both). In either case, $|f(x_1) - f(x_2)| < \epsilon$.

8. f not uniformly continuous: if $\delta > 0$, $(x+\delta)^2 - x^2 \to \infty$ as $x \to \infty$.

g, h, k uniformly continuous: $|g(x_1) - g(x_2)| \leq |x_1 - x_2|$; for h use **7**; for k use argument similar to that in **7**, since $|x_1^{\frac{1}{2}} - x_2^{\frac{1}{2}}| = |x_1 - x_2|/(x_1^{\frac{1}{2}} + x_2^{\frac{1}{2}}) < \epsilon$ if $|x_1 - x_2| < 1$ and $x_1, x_2 > b$, say.

9. There is δ such that $|f(x_1) - f(x_2)| < \epsilon$ when $x_1, x_2 \in [a, b]$ and $|x_1 - x_2| < \delta$. Let $a = x_0 < x_1 < \ldots < x_{n-1} < x_n = b$, where $x_i - x_{i-1} < \delta (i = 1, \ldots, n)$. Define h on $[a, b]$ by putting

$$h(x_i) = f(x_i) \quad (i = 0, 1, \ldots, n)$$

and letting h be linear in each $[x_{i-1}, x_i]$. If $x \in [x_{i-1}, x_i]$ and $x = (1-\alpha)x_{i-1} + \alpha x_i$ $(0 \leq \alpha \leq 1)$, then

$$|h(x) - f(x)| = |(1-\alpha)f(x_{i-1}) + \alpha f(x_i) - f(x)|$$
$$\leq (1-\alpha)|f(x_{i-1}) - f(x)| + \alpha|f(x_i) - f(x)| < \epsilon.$$

10. f is uniformly continuous, for

$$|f(\phi_1)(x) - f(\phi_2)(x)| = \left| \int_0^x (\phi_1 - \phi_2) \right| \leq \int_0^x |\phi_1 - \phi_2| \leq \rho(\phi_1, \phi_2)$$

and so $\rho(f(\phi_1), f(\phi_2)) \leq \rho(\phi_1, \phi_2)$.

g is continuous. For let $\psi \in C[0, 1]$. Then $|\psi(x)| \leqslant A$, say, where $A > 0$. If $\rho(\psi, \chi) < \delta \leqslant A$,

$$|g(\psi)(x) - g(\chi)(x)| = \left| \int_0^x (\psi - \chi)(\psi + \chi) \right| \leqslant \delta(2A + \delta)$$

and so $\rho(g(\psi), g(\chi)) \leqslant 3A\delta$. But g is not uniformly continuous. For if $\psi_n(x) = n$, $\chi_n(x) = n + 1/n \ (0 \leqslant x \leqslant 1)$, $\rho(\psi_n, \chi_n) = 1/n$, while $\rho(g(\psi_n), g(\chi_n)) \geqslant 2$.

11. There are sequences (x_n), (y_n) in E such that $\rho(x_n, y_n) \to \rho(E)$. (x_n) has a subsequence (x_{ν_k}) which converges to a point $x \in E$. (y_{ν_k}) has a subsequence $(y_{\nu_{k_i}})$ which converges to a point $y \in E$. Since $x_{\nu_{k_i}} \to x, \rho(x, y) = \rho(E)$.

Take $X = B[0, 1]$ with usual metric, and $E = \{\phi_1, \phi_2, ...\}$, where

$$\phi_n(x) = \begin{cases} 1 - 1/n & \text{when } x = 1/n, \\ 0 & \text{elsewhere in } [0, 1]. \end{cases}$$

Since $\rho(\phi_m, \phi_n) = 1 - 1/n \ (m < n)$, a sequence of distinct points of E cannot be a Cauchy sequence. Hence E has no limit points and so E is closed. $\rho(E) = 1$, but there are no m, n such that $\rho(\phi_m, \phi_n) = 1$.

12.
$$\inf_{z \in E} \rho(x, z) \leqslant \inf_{z \in E} \{\rho(x, y) + \rho(y, z)\} = \rho(x, y) + \inf_{z \in E} \rho(y, z),$$

i.e.
$$\rho(x, E) - \rho(y, E) \leqslant \rho(x, y).$$

Similarly
$$\rho(y, E) - \rho(x, E) \leqslant \rho(y, x)$$

and so
$$|\rho(x, E) - \rho(y, E)| \leqslant \rho(x, y).$$

$\rho(x, E) = 0 \Leftrightarrow$ every $B(x; \epsilon)$ contains a point of $E \Leftrightarrow x \in \bar{E}$.

13. There are sequences (x_n) in A, (y_n) in B such that $\rho(x_n, y_n) \to \rho(A, B)$. (x_n) has a subsequence (x_{ν_k}) converging to a point $a \in A$ and (y_{ν_k}) has a subsequence converging to a point $b \in B$. Then $\rho(a, b) = \rho(A, B)$.

In R^n take A compact, B closed. For a suitable finite, closed interval $I, \rho(A, B) = \rho(A, B \cap I)$.

14. (i) A compact, B closed. Suppose $\rho(A, B) = 0$. Take (x_n) in A, (y_n) in B such that $\rho(x_n, y_n) \to 0$. (x_n) has a subsequence (x_{ν_k}) converging to an $x \in A$. Then $y_{\nu_k} \to x$ and, since B is closed, $x \in B$. Thus $A \cap B \neq \varnothing$.

(ii) In $B[0, 1]$ take $A = \{\phi\}$, where $\phi(x) = 0$ for $0 \leqslant x \leqslant 1$, $B = \{\phi_1, \phi_2, ...\}$, where

$$\phi_n(x) = \begin{cases} 1 + 1/n & \text{when } x = 1/n, \\ 0 & \text{elsewhere in } [0, 1]. \end{cases}$$

A is compact, B is closed (cf. **12**), $\rho(A, B) = 1$, $\rho(\phi, \phi_n) > 1$ for all n.

15. $X = R^1$, $A = \{2, 3, ...\}$, $B = \{2 + \frac{1}{2}, 3 + \frac{1}{3}, ...\}$.

3(g)

1. Let E be a set in (X, ρ). Suppose $E \subset \bigcup_{n=1}^{p} B(c_n; 1)$. If a is an arbitrary point of X and $r = \max_{1 \leqslant n \leqslant p} \rho(a, c_n)$, then $E \subset B(a; r+1)$.

Let X be an infinite set, σ the discrete metric on X. Then X is bounded in (X, σ), but not totally bounded.

2. Let E be a set in (X, ρ). E can be covered by a finite number of open balls with radius $1/k$. Hence there are infinitely many members of $(x_{k-1\ n})$ lying in one such open ball.

If $m, n \geqslant n_0$, $x_{m, m}$ and $x_{n, n}$ lie in an open ball of radius $1/n_0$ and so $\rho(x_{m, m}, x_{n, n}) < 2/n_0$.

3. Suppose $f: (X, \rho) \to (Y, \sigma)$ continuous, (X, ρ) compact. Let $\epsilon > 0$. Given $c \in X$, there is $\delta(c)$ such that $\sigma(f(x), f(c)) < \epsilon$ when $x \in B(c; \delta(c))$. $\{B(c; \frac{1}{2}\delta(c)) | c \in X\}$ is an open covering of X. Let $\{B(c_i; \delta(c_i)) | 1 \leqslant i \leqslant n\}$ be a finite subcovering and put $\delta = \min\{\delta(c_1), ..., \delta(c_n)\}$. Let $\rho(u, v) < \frac{1}{2}\delta$. If $u \in B(c_i; \frac{1}{2}\delta(c_i))$, then $u, v \in B(c_i; \delta(c_i))$ and so $\sigma(f(u), f(v)) < 2\epsilon$.

4. The set of $(x - \delta(x), x + \delta(x))$ $(\alpha \leqslant x \leqslant \beta)$ is an open covering of $[\alpha, \beta]$. There is a finite subcovering corresponding to $x_1, ..., x_n$, say. Suppose

$$\alpha \leqslant x_1 < x_2 < ... < x_{n-1} < x_n \leqslant \beta.$$

Put $\delta(x_i) = \delta_i$, $(x_i - \delta_i, x_i + \delta_i) = I_i$. We may suppose that no I_j contains an I_i $(i \neq j)$.

For $i = 1, ..., n-1$, I_i and I_{i+1} have a common point y_i such that $x_i < y_i < x_{i+1}$. (Otherwise there would exist z such that $x_i + \delta_i \leqslant z \leqslant x_{i+1} - \delta_{i+1}$ and $z \in I_j$, where $j < i$ or $j > i+1$; and then $I_i \subset I_j$ or $I_{i+1} \subset I_j$.) Hence $f(x_i) \leqslant f(x_{i+1})$. But $f(\alpha) \leqslant f(x_1)$, $f(x_n) \leqslant f(\beta)$. Hence $f(\alpha) \leqslant f(\beta)$.

4(a)

1. O. (a) Yes. (b) No: $x_n = 2v_n$.

o. (a) No: $x_n = (\frac{1}{3})^n$, $v_n = (\frac{1}{2})^n$. (b) Yes.

$\sim$. (a) No: $x_n = (\frac{1}{2})^n + (\frac{1}{3})^n$, $v_n = (\frac{1}{2})^n$. (b) No: $x_n = 1 + (1/n)$, $v_n = 1$.

2. (i) $\sum\limits_{n+1}^{\infty} \dfrac{1}{r^k} < \int\limits_{n}^{\infty} \dfrac{1}{x^k} dx = \dfrac{1}{(k-1)n^{k-1}}$. $A = \sum\limits_{1}^{\infty} x_r$.

(ii) $\sum\limits_{1}^{n} \dfrac{1}{r} < 1 + \int\limits_{1}^{n} \dfrac{1}{x} dx = 1 + \log n$.

(iii) If $-1 < k < 0$, $\sum\limits_{1}^{n} r^k < 1 + \int\limits_{1}^{n} x^k dx$. If $k \geqslant 0$, $\sum\limits_{1}^{n} r^k \leqslant n \cdot n^k$.

3. $u_n - v_n = \{(u_n/v_n) - 1\} v_n \to 0$. Counter example: $u_n = n+1$, $v_n = n$.

4. $(x_{n+1} - x_n)$ increases. Hence the bounded sequence (x_n) either decreases or eventually increases.

5. (i) For $|x| < \frac{1}{2}$, say.

$$|\log(1+x) - x| = |-\tfrac{1}{2}x^2 + \tfrac{1}{3}x^3 - ...| \leqslant \tfrac{1}{2}|x|^2(1 + |x| + |x|^2 + ...) < |x|^2.$$

(ii) $\alpha \geqslant 0$: $\binom{\alpha}{n} \to 0$ and so $\left| \sum\limits_{0}^{\infty} \binom{\alpha}{n} x^n - (1 + \alpha x) \right| \leqslant K(|x|^2 + |x|^3 + ...)$.

$\alpha < 0$: by above,

$$(1+x)^\alpha = \dfrac{1}{1 - \alpha x + O(x^2)} = 1 + \alpha x - O(x^2) + \sum\limits_{2}^{\infty} \{\alpha x - O(x^2)\}^n.$$

6. The inequality follows from the fact that, in $[r-\frac{1}{2}, r+\frac{1}{2}]$, the curve $y = 1/x$ lies between the tangent at $(r, 1/r)$ and the chord joining $(r-\frac{1}{2}, 1/(r-\frac{1}{2}))$, $(r+\frac{1}{2}, 1/(r+\frac{1}{2}))$.

$$\sum_{n+1}^{2n} \frac{1}{r} = \int_{n+\frac{1}{2}}^{2n+\frac{1}{2}} \frac{1}{x} dx + O\left(\frac{1}{n^2}\right) = \log\left(2 - \frac{1}{2n+1}\right) + O\left(\frac{1}{n^2}\right).$$

7. $0 \leqslant \sum_1^\infty \dfrac{1}{n^2 + x^2} - \dfrac{\pi}{2x} \leqslant \dfrac{1}{x^2 + 1}$ and $\dfrac{1}{x^2 + 1} \Big/ \dfrac{\pi}{2x} \to 0$ as $x \to \infty$.

4(b)

1. If the upper and lower limits are finite, x_n/v_n is bounded by theorem 4.21. Converse obvious.

2. $\limsup x_n = \infty \Leftrightarrow \sup x_n = \infty \Leftrightarrow$ given $K > 0$, there is n such that $x_n > K \Leftrightarrow$ there is (v_k) such that $x_{v_k} \to \infty$.

3. (i) Nothing to prove when $\liminf x_n = -\infty$. Otherwise (x_n), (y_n) bounded below and $\inf x_r \leqslant x_s \leqslant y_s$ for all $s \geqslant n$, so that $\inf\limits_{r \geqslant n} x_r \leqslant \inf\limits_{s \geqslant n} y_s$; let $n \to \infty$.

4. $\inf x_r + \inf y_r \leqslant x_s + y_s$ (all s), so $\inf x_r + \inf y_r \leqslant \inf(x_s + y_s)$. $x_s + y_s \leqslant x_s + \sup y_r$ (all s), so $\inf(x_s + y_s) \leqslant \inf(x_s + \sup y_r) = \inf x_s + \sup y_r$.
For $<$ everywhere take

$$(x_n) = (1, -1, 1, -1, \ldots), \quad (y_n) = (-1, 1, 0, 0, -1, 1, 0, 0, \ldots).$$

5. $(x_n) = (1, 2, 3, 1, 2, 3, \ldots)$, $(y_n) = (3, 1, 2, 3, 1, 2, \ldots)$.

6. (i) (a) Let $\lambda = \liminf x_n$. There is a subsequence (x_{μ_k}) such that $x_{\mu_k} \to \lambda$. Then $x_{\mu_k} y_{\mu_k} \to \lambda y$. If $x_{v_l} y_{v_l} \to c$, then, since $y > 0$, $x_{v_l} = (x_{v_l} y_{v_l})/y_{v_l} \to c/y$. As $c/y \geqslant \lambda$, $c \geqslant \lambda y$. Hence no subsequence of $(x_n y_n)$ has a limit $< \lambda y$ and so $\lambda y = \liminf (x_n y_n)$.
(b) If $\liminf x_n = \infty$, then $x_n \to \infty$ and $x_n y_n \to \infty$. If $\liminf x_n = -\infty$, there is (x_{v_k}) such that $x_{v_k} \to -\infty$; then $x_{v_k} y_{v_k} \to -\infty$.
(ii) $x_n = n$ (n odd), $x_n = 1$ (n even), $y_n = -1/n$ make (a), (d) false.
(iii) If $y_n \to 0$, then $x_n y_n \to 0$. If $y_n \to y < 0$, use (i) (a), (c) and (4.23).

7. Let $\liminf x_n = \lambda$. Given $\epsilon > 0$, take $\delta > 0$ so that $|f(\lambda) - f(x)| < \epsilon$ when $|\lambda - x| \leqslant \delta$. Since $x_n > \lambda - \delta$ for $n > n_0$, say, $f(x_n) \geqslant f(\lambda - \delta) > f(\lambda) - \epsilon$ for $n > n_0$. Similarly $f(x_n) < f(\lambda) + \epsilon$ for infinitely many n.

8. First inequality: Let $\liminf (u_n/v_n) = \lambda$. Nothing to prove if $\lambda = -\infty$. If λ finite and $\epsilon > 0$, $u_n \geqslant (\lambda - \epsilon) v_n$ for $n > N$, say. Then,

$$\frac{s_n}{t_n} = \frac{s_N}{t_n} + \frac{u_{N+1} + \ldots + u_n}{t_n} \geqslant \frac{s_N}{t_n} + (\lambda - \epsilon)\frac{t_n - t_N}{t_n} > \lambda - 3\epsilon$$

for $n > N'(>N)$, say, since $t_n \to \infty$. Thus $\liminf (s_n/t_n) \geqslant \lambda$. If $\lambda = \infty$, $u_n/v_n \to \infty$ and, by similar argument, $s_n/t_n \to \infty$.

9. (i) $u_n = x_n$, $v_n = 1$ in **8**. (ii) $u_n = \log x_n$, $v_n = 1$ in **8**; use **7**. (iii) For $n > 1$, replace x_n by x_n/x_{n-1} in (ii).

10. (i) $u_n = n^k$, $v_1 = 1$, $v_n = n^{k+1} - (n-1)^{k+1}$ ($n > 1$) in **8**. (ii) $x_n = n$ in **9**(iii). (iii) $x_n = n^n/n!$ in **9** (iii) gives $\lim n/(n!)^{1/n} = \lim [1 + (1/n)]^n = e$.

11. If $\lim_{\delta \to 0} g_1(\delta) = l_1$, $\lim_{\delta \to 0} g_2(\delta) = l_2$, then $\lim_{\delta \to 0} \max \{g_1(\delta), g_2(\delta)\} = \max (l_1, l_2)$.

12. Let f be upper semi-continuous in $[a, b]$. If f is not bounded above, there is a sequence (x_n) in $[a, b]$ such that $f(x_n) \to \infty$. (x_n) has a subsequence (x_{ν_k}) with limit $c \in [a, b]$. Then $\limsup_{x \to c} f(x) = \infty > f(c)$ (suitably modified if $c = a$ or b). Similar argument shows that $\sup f(x)$ is attained.

13. By the analogue of theorem 4.21 (ii)(c), $\underline{D}f(c) > 0$ implies that f increases at c in the sense of exercise $3(g)$, 4.

If $\underline{D}f(x) \geqslant 0$ and $\epsilon > 0$, $\underline{D}\{f(x) + \epsilon x\} > 0$. Thus $f(x) + \epsilon x$ increases, i.e. $f(\alpha) + \epsilon\alpha \leqslant f(\beta) + \epsilon\beta$ when $a \leqslant \alpha < \beta \leqslant b$. True for all $\epsilon > 0$.

Counter example: f on $[0, 2]$ defined by $f(x) = x$ $(0 \leqslant x \leqslant 1)$, $f(x) = x - 1$ $(1 < x \leqslant 2)$.

4(c)

1. $\Sigma(1/n^2)$ converges, $\Sigma(1/n)$ diverges.

2. Σu_n, Σv_n, where

$$u_n = \begin{cases} (\frac{1}{2})^n & (n \text{ odd}), \\ (\frac{1}{3})^n & (n \text{ even}), \end{cases} \quad v_n = \begin{cases} 2^n & (n \text{ odd}), \\ 3^n & (n \text{ even}). \end{cases}$$

3. By 4(b), 7 and 9,

$$\limsup a^{(u_1 + \cdots + u_n)/n} = a^{\limsup (u_1 + \cdots + u_n)/n} \leqslant a^{\limsup u_n} < 1.$$

4. (i) converges, (ii) diverges.

5. $\left| \dfrac{n!}{(1 - in)^n} \right|^{1/n} = \dfrac{(n!)^{1/n}}{n} \left| \dfrac{n}{1 - in} \right| \to \dfrac{1}{e}$, by 4$(b)$, 10.

6. $|x| < 1$.

7. $|a| < 1$ and all b, $|a| = 1$ and $|b| < 1$.

4(d)

1. If $\liminf (u_n/v_n) = \lambda$, $\limsup (u_n/v_n) = \Lambda$, then $\frac{1}{2}\lambda v_n < u_n < 2\Lambda v_n$ for $n > N$, say.

2.   (i) Convergence for $p > 1$ and all q, $p = 1$ and $q > 1$.

(ii) $u_n = n^{-\log\log n} < n^{-2}$ $(n > N)$; Σu_n converges.

(iii) $u_n = n^{-\log\log\log n}$; convergence.

(iv) $u_n = e^{-(\log\log n)^2} > e^{-\log n} = n^{-1}$ $(n > N)$; divergence.

(v) $n^{-1-(1/n)}/n^{-1} \to 1$; divergence.

(vi) $u_n = O(1/n^2)$.

(vii) $u_n/n^{-a} = \left(1 + \dfrac{a}{n}\right)^n \to e^a$; convergence for $a > 1$.

(viii) $\dfrac{u_n}{u_{n+1}} = 1 + \dfrac{p}{2}\dfrac{1}{n} + O\left(\dfrac{1}{n^2}\right)$; convergence for $p > 2$.

(ix) $a > 1 : (e^{(1/n)\log a} - 1)/(1/n) \to \log a > 0.$ $0 < a < 1 : (1 - e^{(1/n)\log a})/(1/n)$ $\to -\log a > 0.$ Hence, when $0 < a < 1$ or $a > 1$, series converges for $k > 1$. When $a = 1$, series converges for all k.

(x) $\{e^{(\log n)/n} - 1\}/\{(\log n)/n\} \to 1$; convergence for $k > 1$.

3. $\left(1 - \dfrac{k \log n}{n}\right)^n \Big/ \dfrac{1}{n^k} = n^k \exp\left\{n \log\left(1 - \dfrac{k \log n}{n}\right)\right\}$

$$= n^k \exp\left\{n\left[-\dfrac{k \log n}{n} + O\left(\dfrac{\log^2 n}{n^2}\right)\right]\right\} = \exp\left\{O\left(\dfrac{\log^2 n}{n}\right)\right\} \to 1.$$

4. $\dfrac{u_{n+1}}{u_n} = a\left(1 + \dfrac{1}{n}\right)^n \to ae.$ Σu_n converges for $0 < a < 1/e$, diverges for $a > 1/e$.

For $a = 1/e$, $\dfrac{u_{n+1}}{u_n} = \dfrac{1}{e}\left(1 + \dfrac{1}{n}\right)^n$. We show that

$$\dfrac{1}{e}\left(1 + \dfrac{1}{n}\right)^n > \left(1 + \dfrac{1}{n}\right)^{-1} = \dfrac{1}{n+1}\Big/\dfrac{1}{n}, \qquad (*)$$

so that Σu_n diverges by theorem 4.41. To prove $(*)$ it is sufficient to show that $\left(1 + \dfrac{1}{n}\right)^{n+1}$ decreases (to e). We have

$$\dfrac{d}{dx}\left(1 + \dfrac{1}{x}\right)^{x+1} = \left[\log\left(1 + \dfrac{1}{x}\right) - \dfrac{1}{x}\right]\left(1 + \dfrac{1}{x}\right)^{x+1} < 0.$$

5. $(p-1)(u_1 + pu_p + p^2 u_{p^2} + \dots + p^n u_{p^n})$

$$\geqslant (u_1 + \dots + u_{p-1}) + (u_p + \dots + u_{p^2-1}) + \dots + (u_{p^n} + \dots + u_{p^{n+1}-1}),$$

$(p-1)(u_p + pu_{p^2} + p^2 u_{p^3} + \dots + p^{n-1} u_{p^n})$

$$\leqslant (u_2 + \dots + u_p) + (u_{p+1} + \dots + u_{p^2}) + \dots + (u_{p^{n-1}+1} + \dots + u_{p^n}).$$

Σn^{-k} convergent $\Leftrightarrow \Sigma (2^{k-1})^{-n}$ convergent $\Leftrightarrow k > 1$. Use induction.

6. $\Sigma(u_n/s_n)$ diverges, for, when p is sufficiently large,

$$\dfrac{u_{n+1}}{s_{n+1}} + \dots + \dfrac{u_{n+p}}{s_{n+p}} > \dfrac{s_{n+p} - s_n}{s_{n+p}} > \tfrac{1}{2}.$$

$k > 1$. The inequality is equivalent to $f(x) = (k-1)x^k - kx^{k-1} + 1 \geqslant 0$ for $0 < x \leqslant 1$ $(x = s_{n-1}/s_n)$. This holds since $f(1) = 0$ and $f'(x) < 0$ for $0 < x \leqslant 1$.

7. If $\sum_1^\infty u_r = s$, $\sum_1^n u_r = s_n$, take $v_n = u_n/\sqrt{(s - s_n)}$.

8. $v_n = u_n/s_n$, where $s_n = \sum_1^n u_r$ (use **6**).

9. $u_n = \dfrac{1}{q_n} - \dfrac{1}{q_{n+1}}$ (use **6**).

10. Take $n_0 = 0$ and $n_1, n_2, \dots$ so that

$$u_{1,1} + \dots + u_{1,n_1} > 1 \quad \text{and} \quad u_{2,n}/u_{1,n} < \tfrac{1}{2} \quad \text{for} \quad n > n_1,$$

$$u_{2,n_1+1} + \dots + u_{2,n_2} > 1 \quad \text{and} \quad u_{3,n}/u_{2,n} < \tfrac{1}{2} \quad \text{for} \quad n > n_2, \text{ etc.}$$

Define $u_n = u_{k,n}$ for $n_k < n \leqslant n_{k+1}$. Clearly Σu_n diverges. Also, for every k, $u_n/u_{k,n} \to 0$ as $n \to \infty$, for, when $n > n_{k+l}$, $u_n/u_{k,n} < 1/2^l$.

To construct the convergent series, take $n_0 = 0$ and n_1, n_2, ... such that

$$u_{2,n}/u_{1,n} > 2 \quad \text{for} \quad n > n_1 \quad \text{and} \quad u_{2,n_1+1} + u_{2,n_1+2} + \ldots < 1/2,$$

$$u_{3,n}/u_{2,n} > 2 \quad \text{for} \quad n > n_2 \quad \text{and} \quad u_{3,n_2+1} + u_{3,n_2+2} + \ldots < 1/2^2, \text{ etc.}$$

Again define $u_n = u_{k,n}$ for $n_k < n \leqslant n_{k+1}$.

4(e)

1. (i) Divergence, since $a^{1/n} \to 1$.

(ii) Convergence for $x = k\pi$ ($k = 0, \pm 1, \pm 2, \ldots$), otherwise divergence. We may take $0 < |x| < \pi$. Then there are infinitely many n such that $|nx - (2r + \frac{1}{2})\pi| \leqslant \frac{1}{2}|x|$, for some integer r, and so lim sup $\sin nx \geqslant \cos \frac{1}{2}|x| > 0$. Also $\cos 1 \leqslant \cos (\sin nx) \leqslant 1$. Hence $(\sin nx)/\cos(\sin nx) \not\to 0$.

(iii) $\Sigma(2n)!/n^{2n}$ converges by theorem 4.32.

(v) Absolute convergence for $|x| \neq 1$ (theorem 4.32), convergence for $x = -1$, divergence for $x = 1$.

(vi) $\dfrac{d}{dx}(x^{1/x} - 1) = \dfrac{1}{x^2}(1 - \log x)x^{1/x} < 0$ for $x > e$. Use theorem 4.53. See also $4(d)$, $2(x)$.

(vii) Convergence for $q < 0$ and all p, $q = 0$ and $p < 0$.

(viii) Use theorem 4.54.

(ix) $\Sigma \dfrac{\sin nx}{\log n} \cos \dfrac{x}{n}$ converges for all x by theorems 4.53, 4.54. $\Sigma \dfrac{\cos nx}{\log n} \sin \dfrac{x}{n}$ converges for $x \neq 2k\pi$ ($k = 0, \pm 1, \pm 2, \ldots$) and diverges for $x = 2k\pi$, since $\dfrac{\sin (2k\pi/n)}{\log n} \sim \dfrac{2k\pi}{n \log n}$. Thus Σu_n converges for $x \neq 2k\pi$.

(x) For $\alpha \neq 2k\pi$ ($k = 1, 2, \ldots$), Σu_n converges. For $\alpha = 2k\pi$, Σu_n diverges since $(2k\pi)^{1/n} \to 1$ and $1/(2k\pi) < 1$.

2. (i) All x, by theorem 4.53. (ii) $x = k\pi$ ($k = 0, \pm 1, \pm 2, \ldots$) only; use $\sin^2 nx = \frac{1}{2}(1 - \cos 2nx)$.

3. Series converges for all x, by theorem 4.53. Also

$$\Sigma \left(1 + \frac{1}{2} + \ldots + \frac{1}{n}\right) \frac{\sin^2 nx}{n} = \Sigma \left(1 + \frac{1}{2} + \ldots + \frac{1}{n}\right) \frac{1 - \cos 2nx}{n}$$

diverges when $x \neq k\pi$. $|\sin nx| \geqslant \sin^2 nx$.

4. (i) If $\Sigma a_n n^k$ converged for $k > 1$, $a_n n^k$ would be $o(1)$, i.e. a_n would be $o(n^{-k})$ and so Σa_n would converge absolutely.

(ii) Write $a_n/n^\beta = (a_n n^{-\alpha})n^{\alpha-\beta}$ and use theorem 4.54.

Let A be the set of α for which $\Sigma(a_n/n^\alpha)$ converges. $A \neq \varnothing$, since $a_n = o(1)$ and so $\alpha \in A$ when $\alpha > 1$. Also A is bounded below, by (i). Let $\alpha_0 = \inf A$, so that $\alpha \notin A$ when $\alpha < \alpha_0$. By (ii), $\alpha \in A$ when $\alpha > \alpha_0$. (α_0 may or may not belong to A.)

5. Let $a_n - l = b_n$. Then $s_n = \sum_1^n (-1)^r b_r + l\sum_1^n (-1)^r$, where $\sum_1^n (-1)^r b_r \to \sigma$, say, and $l\sum_1^n (-1)^r = -l$ (n odd) or 0 (n even).

6. Extension of theorem 4.53: If $s_n = \sum_1^n a_r$ and $\sum_1^n |v_r - v_{r+1}|$ are bounded and $v_n \to 0$, then $\Sigma a_n v_n$ converges. Both extensions follow from

$$a_{n+1}v_{n+1} + \ldots + a_{n+p}v_{n+p}$$
$$= (s_{n+1} - s_n)(v_{n+1} - v_{n+2}) + \ldots + (s_{n+p-1} - s_n)(v_{n+p-1} - v_{n+p}) + (s_{n+p} - s_n)v_p.$$

Since $f''(x) > 0$, f' strictly increases. If $f'(x_0) \geq 0$ and $\xi > x_0$, then $f'(\xi) > 0$, $f'(x) \geq f'(\xi)$ for $x \geq \xi$, and $f(x) \to \infty$ as $x \to \infty$. Hence $f'(x) < 0$ for $x \geq 1$, i.e. f strictly decreases. Since $f(x) \to 0$, $f(x) > 0$ for $x \geq 1$. As $-f'$ is positive and decreasing and $\int_1^\infty (-f')$ exists, $\sum_1^\infty \{-f'(n)\}$ converges.

Put $v_n = f(n) \sin \{\log f(n)\}$. Then $v_n \to 0$ and $\Sigma |v_{n+1} - v_n|$ converges since $|v_{n+1} - v_n| \leq 2|f'(\xi)| \leq 2|f'(n)|$.

7. If $\liminf (x_n/v_n) = \lambda > 0$, then, for $n > N$, say, $u_n > \frac{1}{2}\lambda v_n$ and so Σu_n diverges.

  (i) $u_n = (-1)^n/\sqrt{n}$, $v_n = 1/n$.

  (ii) $u_n = 1/\sqrt{n}$ when $n = k^4$ ($k = 1, 2, 3, \ldots$), $u_n = 1/n^2$ otherwise; $v_n = 1/n$.

8. To prove that $\sum_{r=0}^n u_r \tau_{n-r} \to 0$ as $n \to \infty$, let $\Sigma |u_n| = \sigma$, $\sup|\tau_n| = \tau$. Given $\epsilon > 0$, there is N such that $|\tau_n| < \epsilon$ and $\sum_{n+1}^\infty |u_r| < \epsilon$ for $n \geq N$. For $n > 2N$,

$$|u_0\tau_n + \ldots + u_n\tau_0| \leq |u_0\tau_n + \ldots + u_{n-N}\tau_N| + |u_{n-N+1}\tau_{N-1} + \ldots + u_n\tau_0|$$
$$\leq \epsilon(|u_0| + \ldots + |u_{n-N}|) + \tau(|u_{n-N+1}| + \ldots + |u_n|)$$
$$< (\sigma + \tau)\epsilon.$$

Example. $u_n = v_n = (-1)^n/\sqrt{(n+1)}$. Then

$$-w_{2n-1} = \frac{1}{\sqrt{(2n)}} + \frac{1}{\sqrt{2}.\sqrt{(2n-1)}} + \ldots + \frac{1}{\sqrt{n}.\sqrt{n}} + \ldots + \frac{1}{\sqrt{(2n)}} > 2.$$

9. Let $\gamma_n = \left(1 + \frac{1}{2} + \ldots + \frac{1}{n}\right) - \log n$, so that $\gamma_n \to \gamma$ (Euler's constant). Then

$$s_n = \{\log(2p_n - 1) + \gamma_{2p_n-1}\} - \frac{1}{2}\{\log(p_n - 1) + \gamma_{p_n-1}\} - \frac{1}{2}\{\log q_n + \gamma_{q_n}\}.$$

10. Since the terms of the rearranged series tend to 0, we need only consider s_{2n}. Let p, q be the number of positive and negative terms in s_{2n}. Let $k > 1$. Take $\epsilon > 0$. If $n > N$, say, $p > n$ and $|ra_r - l| < \epsilon$ for $r > 2q$. Then

$$s_{2n} - (a_1 - a_2 + \ldots - a_{2n})$$
$$= (a_{2n} + a_{2n+1} + \ldots + a_{2p-1}) + (a_{2q+2} + a_{2q+4} + \ldots + a_{2n}) - (a_{2n} + a_{2n+2} + \ldots + a_{2p-2})$$
$$= (l + \epsilon_1)\left(\frac{1}{2n} + \ldots + \frac{1}{2p-1}\right) + \frac{1}{2}(l + \epsilon_2)\left(\frac{1}{q+1} + \ldots + \frac{1}{n}\right) - \frac{1}{2}(l + \epsilon_3)\left(\frac{1}{n} + \ldots + \frac{1}{p-1}\right),$$

where $|\epsilon_1|, |\epsilon_2|, |\epsilon_3| \leq \epsilon$.

11. By the absolute convergence of $\Sigma(-1)^n/n^k$ with $k > 1$ and by **9**, the series converges if $k \geq 1$. For $k < 1$, the series diverges. Obvious when $k \leq 0$. For $0 < k < 1$,

$$s_{3n} = \sum_{r=1}^n \left\{\frac{1}{(4r-3)^k} + \frac{1}{(4r-1)^k} - \frac{1}{(2r)^k}\right\} > \frac{2^{1-k}-1}{2^k} \sum_{r=1}^n \frac{1}{r^k} \to \infty.$$

$$4(f)$$

1. (i) All z.

(ii) $z = 0$ only.

(iii) $R = 1$, since $n^{1/n} < (n \log^2 n)^{1/n} < n^{3/n}$ for $n > N$, say, and $n^{1/n} \to 1$. Convergence for $|z| \leqslant 1$.

(iv) $R = 1$. When $z = \cos \theta + i \sin \theta$, series is

$$\sum_2^\infty \left(\frac{\cos n\theta}{\log^2 n} + i \frac{\sin n\theta}{\log^2 n} \right).$$

$\Sigma(\sin n\theta)/\log^2 n$ converges for all θ, $\Sigma(\cos n\theta)/\log^2 n$ except for $\theta = 2k\pi$ ($k = 0, \pm 1, \pm 2, \dots$). Power series converges for $|z| \leqslant 1$ except $z = 1$.

(v) $|z| \leqslant 1$.

(vi) $|z| < 1$.

(vii) $\alpha = 0$: all z. $\alpha \neq 0$: $|z| \leqslant 1$, $z \neq 1$.

(viii) $|z| \leqslant 1$, $z \neq 1$,

(ix) $R = 2$. Find real and imaginary parts of $(\cos 3n\theta + i \sin 3n\theta)/(1 - in)$; convergence for $|z| \leqslant 2$ except $z = 2, -1 \pm i\sqrt{3}$.

(x) $\alpha = k\pi$ ($k = 0, \pm 1, \pm 2, \dots$): all z. $\alpha \neq k\pi$: $R = 1$, since

$$0 < \limsup |\sin n\alpha| \leqslant 1$$

(see exercise $4(e)$, 1 (ii)). When $z = \cos \theta + i \sin \theta$, series is

$$\sum_1^\infty \frac{1}{2} \left\{ \frac{\sin n(\alpha + \theta) + \sin n(\alpha - \theta)}{n} + i \frac{\cos n(\alpha - \theta) - \cos n(\alpha + \theta)}{n} \right\}.$$

Real component converges for all θ, imaginary component except for $\theta = 2k\pi \pm \alpha$. Power series converges for $|z| \leqslant 1$ except $z = \cos \alpha \pm i \sin \alpha$.

2. (i) If $\alpha \neq 0, 1, 2, \dots$ and $n > \alpha$,

$$\left| \binom{\alpha}{n} \right| \Big/ \left| \binom{\alpha}{n+1} \right| = \frac{n+1}{n-\alpha} = 1 + \frac{1+\alpha}{n} + O\left(\frac{1}{n^2} \right).$$

Use theorem 4.42.

(ii) Since $[1 + (1/n)]^k = 1 + k/n + O(1/n^2)$, (*) holds if $0 < k < 1 + \alpha$. As $1/n^k \to 0$, inequality (4.41) shows that $\left| \binom{\alpha}{n} \right| \to 0$. Also $\left| \binom{\alpha}{n} \right|$ decreases and $\binom{\alpha}{n} = (-1)^n \left| \binom{\alpha}{n} \right|$. Use theorem 4.53.

(iii) $\left| \binom{\alpha}{n} \right| \geqslant 1$.

3. $R = 1$. If the series is $\Sigma a_n z^n$,

$$\frac{a_n}{a_{n+1}} = \frac{(n+1)(c+n)}{(a+n)(b+n)} = 1 + \frac{1+c-(a+b)}{n} + O\left(\frac{1}{n^2} \right).$$

The argument of **2** shows that, if $c - (a+b) > 0$, $\Sigma a_n z^n$ converges absolutely for $|z| = 1$ and, if $-1 < c - (a+b) \leqslant 0$, it converges conditionally for $|z| = 1$, $z \neq 1$. When $c - (a+b) \leqslant -1$, $\Sigma a_n z^n$ diverges for $|z| = 1$, since $a_n \nrightarrow 0$.

Obvious if < -1. If $c = a+b-1, 0 < a_n/a_{n+1} \leqslant 1+(A/n^2)$ for $n \geqslant N$ and, for $n > N$,

$$\log \frac{a_N}{a_n} \leqslant A \left\{ \frac{1}{N^2} + \frac{1}{(N+1)^2} + ... + \frac{1}{(n-1)^2} \right\} \leqslant B,$$

i.e.

$$a_n \geqslant a_N e^{-B}.$$

4. (i) If $R = S$, $T_1 \geqslant R$. When $a_n = 1$, $b_n = -1$ (all n), $R = S = 1$, $T_1 = \infty$.
(ii) If $a_n = 1+(-1)^n$, $b_n = 1-(-1)^n$, then $R = S = 1$, $T_2 = \infty$.
(iii) If $\Sigma a_n x^n = 1+2x+2x^2+...$, $\Sigma b_n x^n = 1-2x+2x^2-2x^3+...$, $R = S = 1$, $T_3 = \infty$.

5. $\Sigma a_n z^n$ converges for $|z| < 1$, since $|a_n z^n| \leqslant a_0 |z|^n$. Let $|z| < 1$.

$$\left| (1-z) \left(\sum_0^\infty a_n z^n \right) - a_0 \right| = \left| \sum_1^\infty (a_n - a_{n-1}) z^n \right| \leqslant \sum_1^\infty (a_{n-1} - a_n) |z|^n.$$

If $a_0 = a_1 = a_2 = ...$, r.h.s. $= 0 < a_0$. If $a_{r-1} > a_r$,

$$\text{r.h.s.} < \sum_1^\infty (a_{n-1} - a_n) = a_0 - \lim a_n \leqslant a_0.$$

Lastly, $\sum_0^\infty a_n z^n = 0$ implies $|-a_0| < a_0$.

4(g)

1. (i) Non-existent, $-1, 1$. (ii) Non-existent, non-existent, 0. (iii) All 0.
(iv) 0, non-existent, non-existent. (v) Non-existent, 0, 0. (vi) 0, 0, non-existent.

2. $x_{1n} = n$, $x_{mn} = 0$ otherwise.

3. Suppose x_{mn} increases, $\lim_{n \to \infty} x_{mn} = y_m$, $\lim_{m \to \infty} y_m = y$. Since y_m increases, $x_{mn} \leqslant y_m \leqslant y$. Hence, as $m \to \infty$, $x_{mn} \to z_n \leqslant y$. Also z_n increases and so $z_n \to z \leqslant y$. Similarly $y \leqslant z$.

4. Sufficiency. For $n, n' > n_0$, $|x_{nn} - x_{n'n'}| < \epsilon$. By completeness of Z, $x_{nn} \to x$, say. If $m, n, n' > n_0$, $|x_{mn} - x| \leqslant |x_{mn} - x_{n'n'}| + |x_{n'n'} - x| < 2\epsilon$.

5. Let sup $x_{mn} = x$, $x_{\mu\nu} > x - \epsilon$. For $m \geqslant \mu$, $n \geqslant \nu$, $x - \epsilon < x_{mn} \leqslant x$.

6. $s_{mn} = \sum_{i=1}^m \sum_{j=1}^n u_{ij} \leqslant \sum_{i=1}^m \sum_{j=1}^n v_{ij} \leqslant \sum_{i,j=1}^\infty v_{ij}$; use **5**.

7. Adapt the proof for $\sum_{n=1}^\infty u_n$ (C1, 90 and 95) using **6**.

8. $\sum_{m=1}^\infty \left(\sum_{n=1}^\infty \frac{1}{m^k n^l} \right)$ exists if and only if $k, l > 1$; use theorem 4.74.

9. $k > 2$: $(m+n)^{-k} \leqslant (2mn)^{-\frac{1}{2}k}$; use **8**.

$$k = 2: \qquad \sum_{n=1}^\infty (m+n)^{-2} \geqslant \int_1^\infty (m+x)^{-2} dx = (m+1)^{-1}$$

and so $\sum_{m=1}^\infty \left(\sum_{n=1}^\infty (m+n)^{-2} \right)$ diverges; use theorem 4.74.

10.
$$s_{mn} = \sum_{i=1}^{m} \sum_{j=1}^{n} u_{ij}, \quad t_{mn} = \sum_{i=1}^{m} \sum_{j=1}^{n} |u_{ij}|,$$

$$s = \sum_{m,n=1}^{\infty} u_{mn}, \quad t = \sum_{m,n=1}^{\infty} |u_{mn}|.$$

Take $\epsilon > 0$. Then $|s_{mn} - s| < \epsilon$, $|t_{mn} - t| < \epsilon$ if $m, n \geqslant n_0$, say. Let $q > 2n_0$ and let $p = [\frac{1}{2}q]$ (the greatest integer $\leqslant \frac{1}{2}q$). Then

$$\left| s_{qq} - \sum_{r=1}^{q} v_r \right| = \left| \sum_{\substack{m,n \leqslant q \\ m+n>q}} u_{mn} \right| \leqslant \sum_{\substack{m,n \leqslant q \\ m+n>q}} |u_{mn}| \leqslant t_{qq} - t_{pp} < 2\epsilon$$

and so $\left| s - \sum_{r=1}^{q} v_r \right| < 3\epsilon$.

11. If $\omega = \alpha + i\beta$, $\omega' = \alpha' + i\beta'$, $\alpha\beta' \neq \alpha'\beta$. $|m\omega + n\omega'|^2 = am^2 + 2bmn + cn^2$, where $a > 0, c > 0, ac - b^2 = (\alpha\beta' - \alpha'\beta)^2 > 0$. Then $A(m+n)^2 \geqslant |m\omega + n\omega'|^2 \geqslant Bmn$, where $A = \max(a, c) > 0$, $B = 2(\sqrt{(ac)} + b) > 0$. Use **8, 9**.

12. Let $|z| < 1$. $\sum_{n=1}^{\infty} |z^{mn}| = |z|^m/(1 - |z|^m) \leqslant |z|^m/(1 - |z|)$ and so $\sum_{m=1}^{\infty} \left(\sum_{n=1}^{\infty} |z^{mn}| \right)$
exists. Hence $\sum_{m,n=1}^{\infty} |z^{mn}|$ exists.

Let $\sum_{m,n=1}^{\infty} z^{mn} = s$. Since
$$z^{n^2}(1 + z^n)/(1 - z^n) = z^{n^2} + 2(z^{n^2+n} + z^{n^2+2n} + \ldots),$$

$$\sum_{m=1}^{p} \left(\sum_{n=1}^{\infty} z^{mn} \right) + \sum_{n=1}^{p} \left(\sum_{m=1}^{\infty} z^{mn} \right) = s_{pp} + \sum_{n=1}^{p} \frac{z^{n^2}(1 + z^n)}{1 - z^n}.$$

The second term is $\sum_{n=1}^{p} z^n/(1 - z^n)$. Also, as $p \to \infty$, the first three terms tend to s; hence so does the last.

13. Let $|z| = r < \min(R, 1)$, $r < \rho < R$. $\limsup |a_n|^{1/n} = \limsup |b_n|^{1/n} = R^{-1}$ and so $|a_n|, |b_n| < \rho^{-n}$ for $n > N$, or $|a_n|, |b_n| < K\rho^{-n}$ for all n. Hence $\Sigma |b_n| |z^{mn}|$ converges and

$$\sum_{n=1}^{\infty} |b_n| |z^{mn}| \leqslant K \frac{r^m/\rho}{1 - r^m/\rho} \leqslant \frac{K}{\rho - r} r^m,$$

so that $\sum_{m=1}^{\infty} \left(\sum_{n=1}^{\infty} |a_m| |b_n| |z^{mn}| \right)$ also converges. By theorem 4.74,

$$\sum_{m=1}^{\infty} \left(\sum_{n=1}^{\infty} a_m b_n z^{mn} \right) = \sum_{n=1}^{\infty} \left(\sum_{m=1}^{\infty} a_m b_n z^{mn} \right)$$

and this is the desired result.

The disc $|z| < \min(R, 1)$ cannot be replaced by $|z| < R$. For if $R < \infty$ and $|z| > 1$, $|z|^n > R$ for large n and for these $f(z^n)$, $g(z^n)$ are not defined; while if $R = \infty$, $\Sigma a_n g(z^n)$, $\Sigma b_n f(z^n)$ may diverge if $|z| > 1$ (e.g. $a_n = b_n = 1/n!$, $z = x > 1$).

5(a)

1. If $M(X) = \sup_{x \in X} \sigma(f_n(x), f(x))$, then $M(X \cup Y) = \max(M(X), M(Y))$.

2. $f_n g_n \to fg$ uniformly if (f_n), (g_n) are uniformly bounded (i.e. $|f_n(x)|$, $|g_n(x)| \leqslant K$ for all x, n), since

$$fg - f_n g_n = f(g - g_n) + g_n(f - f_n).$$

If (f_n), (g_n) not uniformly bounded, counter example is

$$f_n(x) = g_n(x) = x + (1/n) \quad \text{on} \quad (-\infty, \infty).$$

3. $f(x) = 0$. Convergence uniform, for f_n has maximum $\left(\dfrac{n}{n+1}\right)^n \dfrac{1}{n+1}$ at $\dfrac{n}{n+1}$;
or, given ϵ, $|f_n(x)| \leqslant \epsilon$ on $[1-\epsilon, 1]$ and $|f_n(x)| \leqslant (1-\epsilon)^n < \epsilon$ on $[0, 1-\epsilon]$ for $n > n_0$.
  $g(x) = 0$. Convergence not uniform, for $\sup |g_n(x) - g(x)| \geqslant g_n(1/n) = \sin 1 \nrightarrow 0$.
  $h(x) = x^2$. Convergence uniform, for $|h_n(x) - h(x)| = x^2/(1 + nx) \leqslant 100/n$.

4. $f = g = 0$. f_n has maximum $\frac{1}{4}$ at $(\frac{1}{2})^{1/n}$. g_n has maximum $\sqrt{(n/2e)}$ at $\sqrt{(1/2n)}$; or $g_n(1/n) = e^{-1/n} \geqslant e^{-1}$.

5. $f = 0$. In every interval $\sup f_n(x) = 1$ for infinitely many n.

6. Given q_0, there is n_0 such that all r_n with $n > n_0$ have denominator $> q_0$; then $\sup |f_n(x) - f(x)| < 1/q_0$.

7. (f_n) converges (to 0) on Z; convergence uniform on any bounded subset of Z.
  (g_n) converges on $\{0, \pm 2\pi i, \pm 4\pi i, \ldots\} \cup \{z \,|\, \mathrm{re}\, z < 0\}$; convergence uniform on

$$\{0, \pm 2\pi i, \pm 4\pi i, \ldots\} \cup \{z \,|\, \mathrm{re}\, z \leqslant -\delta\} \quad (\delta > 0).$$

$$|\sin n(x + iy)| = |\sin nx \cosh ny + i \cos nx \sinh ny|$$

$$= \sqrt{(\sin^2 nx + \sinh^2 ny)}.$$

Hence $(h_n(z))$ diverges if $\mathrm{im}\, z \neq 0$, (h_n) converges uniformly on $\{z \,|\, \mathrm{im}\, z = 0\}$.

8. No; take $s_n(x) = nx/(1 + n^2 x^2)$. Then s_n has maximum $1/2$ at $1/n$. If $M(\delta) = \sup_{0 \leqslant x \leqslant \delta} |f(x)|$ $(0 < \delta \leqslant 1)$, then $M(\delta) \to 0$ as $\delta \to 0$ and

$$\sup_{0 \leqslant x \leqslant 1} |f(x) s_n(x)| \leqslant \max\left(\tfrac{1}{2} M(\delta), \frac{n\delta}{1 + n^2\delta^2} M(1)\right) < \epsilon$$

by choice first of δ, then of $n > 1/\delta$.

9. The partial sums are $1 - x^n$, $(1 - x^n)(1 - x)$, $[1 - (-x)^n](1 - x)(1 + x)^{-1}$. Σv_n, Σw_n converge uniformly.

10. $f(x, y) = \dfrac{x}{y}$. $|f_n(x, y) - f(x, y)| = \dfrac{|y - x|}{y(1 + ny)} < \dfrac{A}{n}$.

11. $\displaystyle\int_a^b \phi(x) \sin nx \, dx = \left[-\phi(x) \frac{\cos nx}{n}\right]_a^b + \int_a^b \phi'(x) \frac{\cos nx}{n} \, dx$.

$$f(\phi) = 0 \quad \text{and} \quad |f_n(\phi) - f(\phi)| \leqslant \frac{2M}{n} + \frac{M(b-a)}{n}.$$

12. Let $f_n \to f$. Take $\epsilon > 0$. There exists n_0 such that $\sigma(f_n(x), f(x)) < \epsilon$ for all $n > n_0$ and $x \in X_0$. Take $m, n > n_0$. There exists $x \in X_0$ such that $\sigma(f_m(x), y_m) < \epsilon$, $\sigma(f_n(x), y_n) < \epsilon$. Hence

$$\sigma(y_m, y_n) \leqslant \sigma(y_m, f_m(x)) + \sigma(f_m(x), f(x)) + \sigma(f(x), f_n(x)) + \sigma(f_n(x), y_n) < 4\epsilon.$$

Since (Y, σ) is complete, $\lim y_n$ exists.

5(b)

1. Sum is $x/(1-x^2)$ for $0 \leqslant x < 1$, 0 for $x = 1$. Use theorem 5.21.

2. 'Only if': $(f_n(-1))$ diverges and sum function discontinuous at 1. 'If': When $[a, b] \subset (1, \infty)$, $\sup\limits_{a \leqslant x \leqslant b} |f_n(x)| = a^n/(1+a^{2n}) \to 0$; other cases treated similarly.

3. If $\phi(1) \neq 0$, sum function discontinuous at 1. If $\phi(1) = 0$, use method of 5(a), 8.

4. (a) $n^2 x e^{-nx^2}$, (b) $(-1)^n nx e^{-nx^2}$ on $[0, 1]$.

5. (i) $f_n(x) = 1/x$ on $[1, n]$, 0 on (n, ∞).

(ii) $g_n(x) = 1/n$ on $[0, n]$, 0 on (n, ∞); or $g_n(x) = \dfrac{n}{(x-n)^2 + n^2}$ on $[0, \infty)$.

6. $f = 0$. If $0 < p \leqslant 1$, $f_n(x)$ increases with x and $f_n(1) = n/(1+n^2)$; if $p > 1$, $f_n(x)$ has maximum $An^{1-(2/p)}$ where $n^2 x^p(p-1) = 1$. Convergence uniform if $0 < p < 2$. Yes, no.

7. $f = f' = 0$. $f_n' \to f'$ pointwise on $[0, 1]$, but (f_n') does not converge uniformly since $f_n'(1/n) = n/4$.

8. $f = f' = 0$ and $|f_n - f|$ has maximum $1/(2n)$ at $1/n$. $f_n'(0) = 1$ for all n.

9. $f(x) = 1$ for $0 \leqslant x \leqslant 1$, x for $1 \leqslant x \leqslant 2$. $\sup\limits_{0 \leqslant x \leqslant 1} |f_n(x) - f(x)| \leqslant 2(2^{1/n} - 1)$.

5(c)

1. (i) $\sup\limits_{-\infty < x < \infty} \left| \dfrac{x}{n(1+nx^2)} \right| = \dfrac{1}{2n^{\frac{3}{2}}}$; uniform convergence (u.c.) on $(-\infty, \infty)$ (M-test).

(ii) Convergence for $x \neq 0$; u.c. for $|x| \geqslant \delta$ (>0) and on no larger set since $\sup\limits_{x \neq 0} |\operatorname{cosech} nx| = \infty$ (use theorem 5.31).

(iii) Convergence on $(-\infty, -2) \cup [0, \infty)$; u.c. on $(-\infty, -2-\delta] \cup [\delta, \infty)$ and on no larger set since $\sup\limits_{x < -2} \left| \dfrac{x}{(1+x)^n} \right| = 2$ and sum discontinuous at 0.

(iv) Convergence for $x \leqslant 0$; u.c. for $x \leqslant -\delta$ and not for $x \leqslant 0$ since $\sup\limits_{x \leqslant 0} |e^{nx} \sin nx| \geqslant e^{-1} \sin 1$.

(v) Convergence for $x \neq 2k\pi$; u.c. on closed interval not including $2k\pi$ (theorem 5.33). Convergence not uniform on open interval with $2k\pi$ as end point, for when $x = 2k\pi \pm \pi/(8n)$,

$$\sum_{n+1}^{2n} \frac{\cos rx}{\sqrt{(r+1)} + \sqrt{r}} \geqslant \sum_{n+1}^{2n} \frac{1/\sqrt{2}}{\sqrt{(r+1)} + \sqrt{r}} \geqslant A\sqrt{n}.$$

(vi) Convergence and u.c. on $[0, \infty)$ (theorem 5.34).

2. For any n, $\sup_{z \in Z^*} |1/(z^2+n^2)| = \infty$.

3. For $|x| \leqslant \frac{1}{2}$ use M-test. For $\frac{1}{2} \leqslant x \leqslant 1$, $\Sigma n^{-p} \sin nx$ u.c.; then use theorem 5.34. For $-1 \leqslant x \leqslant -\frac{1}{2}$ series is $-\Sigma y^n n^{-p} \sin n(y+\pi)$, where $y = -x$.

4. For $z \in \overline{B(0;1)} - B(1;\delta)$,

$$\left| \sum_0^n z^r \right| = \left| \frac{1-z^{n+1}}{1-z} \right| \leqslant \frac{2}{|1-z|} \leqslant \frac{2}{\delta}.$$

Apply theorem 5.33 to the real and imaginary parts of $\Sigma a_n z^n$.

5.
$$\sup_{-\infty < x < \infty} \left| \frac{d}{dx} (n^2 + n^3 x^2)^{-1} \right| = An^{-\frac{3}{2}}, \quad \sup_{x \geqslant 1+\delta} \left| \frac{d}{dx} n^{-x} \right| = (\log n) n^{-1-\delta}.$$

Use theorem 5.23 B.

6. (i) If $0 \leqslant \alpha < 2$, $u_n(x)$ has maximum $An^{-\alpha}$ where $(2-\alpha)n^2 x^2 = \alpha$; if $\alpha \geqslant 2$, $u_n(x)$ increases with x and $u_n(1) = (1+n^2)^{-1}$. Thus Σu_n is u.c. on $[0,1]$ if $\alpha > 1$.
Let $0 \leqslant \alpha < 1$. Then $\sum_{n+1}^{2n} u_r(1/n) = \sum_{n+1}^{2n} 1/(2n^\alpha) \geqslant \frac{1}{2}$ and convergence non-uniform by theorem 5.31.

(ii) Consider $[1, \infty)$. If $0 \leqslant \alpha < 2$ and n large enough, $u_n(x)$ decreases in $[1, \infty)$ and $u_n(1) = (1+n^2)^{-1}$. If $\alpha \geqslant 2$, $u_n(x)$ increases in $[1, \infty)$: for $\alpha = 2$, $\sup u_n(x) = n^{-2}$; for $\alpha > 2$, $\sup u_n(x) = \infty$.

7. $\sum_1^N x^n(1-x) < 1$; use theorem 5.33. $M_n > [e(n+1)\log(n+1)]^{-1}$.

8. $u_n(x) = (-1)^n x^n(1-x)$. See $5(a)$, 9.

9. (i) For u.c. use theorem 5.33.
(ii) For any x, $u_n(x) = 1/(1+n^2)$ if n is sufficiently large. For any interval I, $\sup_{x \in I} u_n(x) = 1$ for infinitely many n.

10. Uniform convergence by M-test.
x irrational. Each term is continuous at x and so is s.
$x = p/q$.

$$s\left(\frac{p}{q} \pm h\right) = \sum_1^\infty \frac{[n(p/q) \pm nh]}{n^3}.$$

Each series is u.c. (in h) for $0 < h < 1$. Each term tends to a limit as $h \to 0+$. By $5(a)$, 12 and theorem 5.21 we can take limits term by term. $\lim [n(p/q) \pm nh]$ differ (by 1) if and only if $n = qr$ (r integer).

11. In $[0, a]$, $\sin(\pi x^n) \leqslant \pi x^n \leqslant \pi a^n$; u.c. by M-test.

$$\int_0^1 \sin(\pi x^n) \, dx \leqslant \pi \int_0^1 x^n \, dx = \frac{\pi}{n+1}.$$

So $\sum_1^\infty \int_0^1 \frac{1}{n} \sin(\pi x^n) \, dx$ converges to l, say. Since $\int_0^a \leqslant \int_0^1$, $\sum_1^\infty \int_0^a \leqslant l$. Thus, by theorem 5.22, $\int_0^a f = \int_0^a \sum_1^\infty = \sum_1^\infty \int_0^a \leqslant l$.

Since $\int_0^a f$ increases with a, $\int_0^1 f = \lim_{a\to 1-} \int_0^a f \leqslant l$.

Also $\sum_1^N \int_0^1 = \int_0^1 \sum_1^N$ increases with N and so $\int_0^1 f \geqslant \int_0^1 \sum_1^N = \sum_1^N \int_0^1 > l - \epsilon$ if $N > N_0$.

12.
$$\sum_{r=0}^{\infty} u_r(x) = \begin{cases} \sum_{r=0}^{\infty} w_r & \text{if } x = 0, \\ F(n) & \text{if } x = 1/n. \end{cases}$$

Σu_r is u.c. on $\{0, 1, \frac{1}{2}, \ldots\}$ and each u_r is continuous at 0.

13. $v_r(n) = \binom{n}{r}\left(\frac{z}{n}\right)^r$, $p(n) = n$.

14. Given ϵ, take $a = a_0 < a_1 < \ldots < a_k = b$ such that

$$|f(x) - f(x')| < \epsilon \quad \text{if} \quad a_{i-1} \leqslant x, x' \leqslant a_i \quad (i = 1, \ldots, k).$$

If $n \geqslant N$, say, $|f_n(a_i) - f(a_i)| < \epsilon$ $(i = 1, \ldots, k)$. Let $x \in [a_{j-1}, a_j]$. Then

$$|f_n(a_{j-1}) - f(x)| < 2\epsilon, \quad |f_n(a_j) - f(x)| < 2\epsilon.$$

Hence $\qquad\qquad\qquad f(x) - 2\epsilon < f_n(a_{j-1}) < f(x) + 2\epsilon$

and $\qquad\qquad\qquad\quad f(x) - 2\epsilon < f_n(a_j) < f(x) + 2\epsilon$.
Since f_n is monotonic,
$$f(x) - 2\epsilon < f_n(x) < f(x) + 2\epsilon.$$

Thus, for $n \geqslant N$, $\sup_{a\leqslant x\leqslant b} |f_n(x) - f(x)| \leqslant 2\epsilon$.

5(d)

1. (i) For $z \neq 0$, $(\sin z)/z = 1 - z^2/3! + \ldots \to 1$ as $z \to 0$, by theorem 5.42.
(ii) For $z \neq a$, $(\cos z - \cos a)/(z - a) = -2\sin\frac{1}{2}(z-a)\sin\frac{1}{2}(z+a)/(z-a) \to -\sin a$ as $z \to a$, by (i) and by continuity of sine (as sum of power series).
(iii) $(e^{z^2} - 1)/(e^z - 1)^2 \to 1$ as $z \to 0$.

2. (i) $\cos(\sin z) = 1 - \frac{1}{2}z^2 + \frac{5}{24}z^4 + \ldots$ (all z).
(ii) $e^{1/(1+z)} = e(1 - z + \frac{3}{2}z^2 - \frac{1}{6}z^3 + \frac{61}{24}z^4 + \ldots)$ $(|z| < 1)$.
(iii) $\sin(\cos z) = \sin 1 \cos(-\frac{1}{2}z^2 + \frac{1}{24}z^4 - \ldots) + \cos 1 \sin(-\frac{1}{2}z^2 + \frac{1}{24}z^4 - \ldots)$

$$= \sin 1 - \frac{1}{2}(\cos 1)z^2 + (\frac{1}{24}\cos 1 - \frac{1}{8}\sin 1)z^4 + \ldots \text{ (all } z).$$

3. For $x \neq 0$, $f(x) = \exp[(1/x)\log(1 + x)]$. Since $f(0) = e$, for $|x| < 1$, $f(x) = \exp(1 - \frac{1}{2}x + \frac{1}{3}x^2 - \frac{1}{4}x^3 + \ldots)$. $1 - \frac{1}{2}x + \frac{1}{3}x^2 - \frac{1}{4}x^3 + \ldots$ has derivatives of all orders at 0 and so therefore has f. For small $|x|$, $f(x) = e(1 - \frac{1}{2}x + \frac{11}{24}x^2 - \frac{7}{16}x^3 + \ldots)$. Thus $f'''(0) = -\frac{21}{8}e$.

4. (i) $\Sigma a_n x^n, \Sigma b_n x^n, \Sigma c_n x^n$ have radii of convergence $\geqslant 1$. Then $(\Sigma a_n x^n)(\Sigma b_n x^n) = \Sigma c_n x^n$ for $|x| < 1$. Let $x \to 1-$ and use theorem 5.47.
(ii) If $a_n = b_n = (-1)^{n+1}/\sqrt{(n+1)}$, then $|c_n| > 1$.

5. (i) $\Sigma a_n x^n$ is u.c. for $0 \leqslant x \leqslant 1$. Hence $\int_0^1 (\Sigma a_n x^n)\,dx = \Sigma \int_0^1 a_n x^n\,dx$.

(ii) $\Sigma a_n x^{n+1}/(n+1)$ has radius of convergence $\geqslant 1$ and so has $\Sigma a_n x^n$. Hence, for $0 < \xi < 1$, $\int_0^\xi (\Sigma a_n x^n)\,dx = \Sigma a_n \xi^n/(n+1) \to \Sigma a_n/(n+1)$ as $\xi \to 1-$, by theorem 5.47.

6. By **5**, sum of series is $\int_0^1 \dfrac{dx}{1+x^4} = \dfrac{1}{2\sqrt{2}} \int_0^1 \left(\dfrac{x+\sqrt{2}}{x^2+\sqrt{2}x+1} - \dfrac{x-\sqrt{2}}{x^2-\sqrt{2}x+1} \right) dx$.

7. Coefficient of x^n:

$$\sum_{r=1}^{n-1} \frac{(-1)^{n-r+1}}{n-r} \frac{(-1)^{r+1}}{r} = \frac{(-1)^n}{n} \sum_{r=1}^{n-1} \left(\frac{1}{n-r} + \frac{1}{r} \right) = \frac{(-1)^n}{n} 2s_{n-1}.$$

Valid for $x = 1$ as s_{n-1}/n decreases and tends to 0.

8. $\alpha \geqslant 0$: $\Sigma \left| \binom{\alpha}{n} \right|$ converges, by $4(f)$, 2. Hence, by M-test, $\Sigma \binom{\alpha}{n} x^n$ is u.c. for $-1 \leqslant x \leqslant 1$; the sum is $(1+x)^\alpha$ for $-1 < x < 1$ and, by continuity, for $x = \pm 1$.

$\alpha > -1$: $\Sigma \binom{\alpha}{n} x^n$ converges, by $4(f)$, 2. $\Sigma \binom{\alpha}{n} x^n$ is u.c. in $[0, 1]$ by theorem 5.34 and in $[-1+\delta, 0]$ by M-test.

9. By **8**, $1 - \dfrac{1}{2}x^2 + \dfrac{1.3}{2.4}x^4 - \dots$ converges uniformly to $(1+x^2)^{-\frac{1}{2}}$ for $0 \leqslant x^2 \leqslant 1$, i.e. $|x| \leqslant 1$. Integrate.

10. Put $\sum_0^n \binom{\alpha}{r}\binom{\beta}{n-r} = c_n$. For $|x| < 1$, $\Sigma c_n x^n = \Sigma \binom{\alpha}{n} x^n . \Sigma \binom{\beta}{n} x^n = (1+x)^{\alpha+\beta}$.

By theorem 5.43, $c_n = \binom{\alpha+\beta}{n}$.

5(e)

1. If f is a continuous nowhere differentiable function on $[a, b]$, define g_k $(k = 1, 2, \dots)$ on $[a, b]$ by

$$g_1(x) = \int_a^x f(t)\,dt, \quad g_k(x) = \int_a^x g_{k-1}(t)\,dt \quad (k = 2, 3, \dots).$$

Then $g_n^{(n)} = f$ and so g_n is the desired function.

2. Since $\Sigma u_n/n^2$ is u.c., the sum g is continuous in $[0, 1] - \{r_1, r_2, \dots\}$; at r_n the left and right limits exist, but are unequal. Each u_n is integrable and so is g. If $h(x) = \int_0^x g$, h is continuous everywhere, but is not differentiable at r_n, for the left and right derivatives differ. Finally, $f(x) = h(x) + \sum_1^\infty r_n/n^2$.

3. Take $f(x) = 1/x$ for $0 < x \leqslant 1$.

4. $F \subset B(0; r)$, say. Take $R > r$, $F_0 = R^k - B(0; R)$, $F_1 = F \cup F_0$. Let f_1 be the extension of f to F_1 with $f_1(x) = 0$ for $x \in F_0$. Apply theorem 5.52 to f_1.

5. Given ϵ, choose δ so that $\sigma(f(x), f(x')) < \epsilon$ when $x, x' \in E$ and $\rho(x, x') < \delta$. If $x, x' \in E$ and $\rho(x, \xi) < \frac{1}{2}\delta$, $\rho(x', \xi) < \frac{1}{2}\delta$, then $\sigma(f(x), f(x')) < \epsilon$. By exercise $3(d)1$, $\lim_{x \to \xi} f(x)$ exists; define $f^*(\xi)$ to be this.

Let $u, u' \in \bar{E}$ be such that $\rho(u, u') < \frac{1}{2}\delta$. There are $x, x' \in E$ such that

$$\rho(x, x') < \delta, \quad \sigma(f^*(u), f^*(x)) < \epsilon, \quad \sigma(f^*(u'), f^*(x')) < \epsilon.$$

Then $\sigma(f^*(u), f^*(u')) < 3\epsilon$.

5 (f)

1. $|f(x)| \leq M$, say, for $a \leq x \leq b$. Take $\epsilon > 0$. There exists a polynomial p such that $|f(x) - p(x)| < \epsilon$ for $a \leq x \leq b$. Then

$$\int_a^b f^2 = \int_a^b f(f-p) + \int_a^b fp = \int_a^b f(f-p) \leq (b-a) M\epsilon.$$

Hence $\int_a^b f^2 = 0$.

If $f \not\equiv 0$, there exists $\xi \in [a, b]$ such that $f(\xi) = \eta \neq 0$. By continuity, $f^2(x) > \frac{1}{2}\eta^2$ for $\alpha \leq x \leq \beta$, say. Then

$$\int_a^b f^2 \geq \int_\alpha^\beta f^2 \geq \frac{1}{2}\eta^2(\beta - \alpha) > 0.$$

2. The set A of trigonometric polynomials is an algebra in $C[\alpha, \beta]$ which contains the constant functions. If $\beta - \alpha < 2\pi$, A separates points for, when $0 < |x_1 - x_2| < 2\pi$, at least one of $\cos x_1 - \cos x_2$, $\sin x_1 - \sin x_2$ is not 0. If $\beta - \alpha \geq 2\pi$, clearly A does not separate points. Then only continuous functions with period 2π can be approximated.

3. The even polynomials on $[a, b]$ separate points if and only if $[a, b] \subset (-\infty, 0]$ or $[a, b] \subset [0, \infty)$.

4. For each x, $y_n(x) \to y(x)$, where $y(x) = \frac{1}{2}\{y^2(x) + 1 - x^2\}$. By first part, $y(x) = 1 - |x|$. Convergence uniform by theorem 5.35. y_n is a polynomial (of degree 2^n).

6 (a)

1. Consider $\mathscr{D}'$ with one point of division added to those of $\mathscr{D}$.

2. When $k \geq 0$, $S(\mathscr{D}, kf, g) = kS(\mathscr{D}, f, g)$ and $S(\mathscr{D}, -f, g) = -s(\mathscr{D}, f, g)$.

3. $S(\mathscr{D}, f, g_1 + g_2) = S(\mathscr{D}, f, g_1) + S(\mathscr{D}, f, g_2)$ and $\mathscr{D}$ may be chosen so that all three sums lie within ϵ of the corresponding upper integrals.

4. Follows from 3.

5. $S(\mathscr{D}, f_1, g) \leq S(\mathscr{D}, f_2, g)$.

6. $f(x) = 1$ for x rational, -1 for x irrational.

7. For x rational, $f(x) = 1$, $h(x) = 0$; for x irrational, $f(x) = 0$, $h(x) = 1$; $g(b) > g(a)$.

8. If m, M are the infimum and supremum of f on $[a, b]$, then $mh(x) \leqslant f(x)h(x) \leqslant Mh(x)$ for $a \leqslant x \leqslant b$ and so

$$m \int_a^b h \, dg \leqslant \int_a^b fh \, dg \leqslant M \int_a^b h \, dg.$$

9. Given ϵ, there is δ such that $|f(u)-f(v)| < \epsilon$ if $u, v \in [a, b]$ and $|u-v| < \delta$. When $\mu(\mathscr{D}) < \delta$, $S(\mathscr{D})-s(\mathscr{D}) \leqslant \epsilon\{g(b)-g(a)\}$ and

$$s(\mathscr{D}) \leqslant \int_a^b f \, dg \leqslant S(\mathscr{D}).$$

10. (i) $|F(v)-F(u)| \leqslant K\{g(v)-g(u)\}$, where K is the supremum of $|f|$ in $[a,b]$.
(ii) If $m(u, v)$, $M(u, v)$ are the infimum and supremum of f in $[u, v]$,

$$m(u, v)\frac{g(v)-g(u)}{v-u} \leqslant \frac{F(v)-F(u)}{v-u} \leqslant M(u, v)\frac{g(v)-g(u)}{v-u}.$$

11. There is a dissection $\mathscr{D}'$ of $[a+\epsilon, b]$, say $a+\epsilon = x_1 < x_2 < \ldots < x_n = b$ such that $S(\mathscr{D}')-s(\mathscr{D}') < \epsilon$. If $\mathscr{D}$ is the dissection $a = x_0 < x_1 < \ldots < x_n = b$ of $[a, b]$, then $S(\mathscr{D})-s(\mathscr{D}) < (M-m)\epsilon+\epsilon$, where m, M are the infimum and supremum of f on $[a, b]$.

12. Let c be a common right discontinuity. Any $\mathscr{D}$ has points x_{i-1}, x_i such that $x_{i-1} \leqslant c < x_i$. Then

$$S(\mathscr{D})-s(\mathscr{D}) \geqslant (M_i-m_i)\{g(x_i)-g(x_{i-1})\} \geqslant |f(c+)-f(c)|\{g(c+)-g(c)\} > 0.$$

13. For $n \geqslant N$, say, $|f_n(x)-f(x)| < \epsilon$ $(a \leqslant x \leqslant b)$. Thus

$$\overline{\int_a^b} f \, dg - \underline{\int_a^b} f \, dg \leqslant \int_a^b (f_N+\epsilon) \, dg - \int_a^b (f_N-\epsilon) \, dg = 2\epsilon\{g(b)-g(a)\}.$$

Also, for $n \geqslant N$, $\left| \int_a^b f \, dg - \int_a^b f_n \, dg \right| \leqslant \epsilon\{g(b)-g(a)\}$.

14. If $f(x) \not\equiv 0$, let $\sup\limits_{a \leqslant x \leqslant b} f(x) = M > 0$. There is an interval $[\alpha, \beta]$ in which $f(x) > M-\epsilon$. Hence

$$M^n\{g(b)-g(a)\} \geqslant \int_a^b f^n \, dg \geqslant \int_\alpha^\beta f^n \, dg \geqslant (M-\epsilon)^n\{g(\beta)-g(\alpha)\}$$

and, for n sufficiently large,

$$\{g(b)-g(a)\}^{1/n} < 1+\epsilon, \quad \{g(\beta)-g(\alpha)\}^{1/n} > 1-\epsilon.$$

6 (b)

2.

$$g(x) = \begin{cases} 0 & (x = 0), \\ \sum\limits_{r=1}^n |u_r| & \left(1-\dfrac{1}{n} < x \leqslant 1-\dfrac{1}{n+1}\right), \\ \sum\limits_{r=1}^\infty |u_r| & (x = 1); \end{cases} \qquad f\left(1-\frac{1}{n}\right) = \begin{cases} 1 & (u_n \geqslant 0), \\ -1 & (u_n < 0), \end{cases}$$

also f is linear in each interval $\left[1-\dfrac{1}{n}, 1-\dfrac{1}{n+1}\right]$ and $f(1) = 0$.

3. $g_n(x) = x^n$ on $[0, 1]$, f monotonic on $[0, 1]$ and discontinuous at 1.

4. f, g_n have no common left or right discontinuities and nor have f, g, by uniform convergence. Thus $f \in R(g; a, b)$. Then use integration by parts and exercise 6(a), 13.

6 (c)

1. Necessity obvious. To prove sufficiency adapt the proof of the completeness of R^1 given in §4.2 (p. 81); or *use* the completeness of R^1 as follows. The sequence

$$c_n = \int_a^{a+n} f\,dg \quad (n = 1, 2, \ldots)$$

converges. Also

$$\int_{[X]}^X f\,dg \to 0 \quad \text{as} \quad X \to \infty.$$

2. Use theorem 6.31; or write $f = (|f|+f)-|f|$ and note that $\int_a^\infty (|f|+f)\,dg$ exists, since $0 \leqslant |f(x)|+f(x) \leqslant 2|f(x)|$.

3. Integrate by parts.

4. To prove convergence use integration by parts. Also

$$\int_{(n-\frac{1}{2})\pi}^{(n+\frac{1}{2})\pi} \left|\frac{\cos x}{x+\sin x}\right| dx \geqslant \frac{1}{(n-\frac{3}{2})\pi} \int_{(n-\frac{1}{2})\pi}^{(n+\frac{1}{2})\pi} |\cos x|\,dx \geqslant \frac{2}{(n-2)\pi}.$$

5.
$$\frac{(n+1)^{\alpha+1}-n^{\alpha+1}}{n^\alpha} = \sum_{r=1}^\infty \binom{\alpha+1}{r} \frac{1}{n^{r-1}} \to \alpha+1$$

as $n \to \infty$, since the power series $\Sigma \binom{\alpha+1}{r} x^{r-1}$ has radius of convergence 1 and is therefore continuous at 0.

(i) Let $\alpha \geqslant 0$. Since $f(n+1) \leqslant f(x) \leqslant f(n)$ for $n \leqslant x \leqslant n+1$,

$$\frac{(n+1)^{\alpha+1}-n^{\alpha+1}}{\alpha+1} f(n+1) \leqslant \int_n^{n+1} x^\alpha f(x)\,dx \leqslant \frac{(n+1)^{\alpha+1}-n^{\alpha+1}}{\alpha+1} f(n)$$

and so, for $n \geqslant N$, say

$$\tfrac{1}{4}(n+1)^\alpha f(n+1) \leqslant \tfrac{1}{2}n^\alpha f(n+1) \leqslant \int_n^{n+1} x^\alpha f(x)\,dx \leqslant 2n^\alpha f(n).$$

For $X > N$, $\quad \tfrac{1}{4}\sum_{N+1}^{[X]} n^\alpha f(n) \leqslant \int_N^X x^\alpha f(x)\,dx \leqslant 2\sum_N^{[X]} n^\alpha f(n).$

(ii) If $\alpha < 0$, $x^\alpha f(x)$ decreases and (i) may be applied to $x^0(x^\alpha f(x))$.

6. Let $f(x) \geqslant 0$. $\int_{a+h}^b f\,dg$ is bounded and increases as h decreases and so

$\lim\limits_{h\to 0+} \int_{a+h}^b f\,dg$ exists. In general, write $f = (|f|+f)-|f|$.

If $f \in R(g; a, b)$ and g is discontinuous at a, then f is continuous at a. Thus

$$\int_a^{a+h} f\,dg \to f(a)\{g(a+)-g(a)\} \text{ as } h \to 0+.$$

7.
$$\int_{0+}^{\infty} \frac{\log x}{(x+1)^2} \, dx = \lim_{X \to \infty} \left(\lim_{\delta \to 0+} \int_{\delta}^{X} \frac{\log x}{(x+1)^2} dx \right) = 0.$$

8. $0 < \alpha+1 < \beta$ and $\beta < \alpha+1 < 0$.

9. By the M-test, there is uniform convergence on $[\alpha, \beta]$ and so, by theorem 5.22 (or exercise 6(a), 13),

$$\sum_{n=0}^{\infty} \int_{\alpha}^{\beta} x^{n+c-1} \log (1/x) \, dx = \int_{\alpha}^{\beta} \frac{x^{c-1} \log (1/x)}{1-x} \, dx.$$

Applying the analogue of theorem 6.33 we first get

$$\sum_{n=0}^{\infty} \int_{0}^{\beta} x^{n+c-1} \log (1/x) \, dx = \int_{0}^{\beta} \frac{x^{c-1} \log (1/x)}{1-x} \, dx$$

and then

$$\sum_{n=0}^{\infty} \frac{1}{(n+c)^2} = \sum_{n=0}^{\infty} \int_{0}^{1} x^{n+c-1} \log (1/x) \, dx = \int_{0}^{1} \frac{x^{c-1} \log (1/x)}{1-x} \, dx.$$

10. On $(0, Y]$, $e^{-x/y} \leqslant e^{-x/Y}$. Also $\sup\limits_{y>0} \int_{X}^{\infty} e^{-x/y} \, dx = \sup\limits_{y>0} y e^{-X/y} = \infty.$

11.
$$\int_{2}^{X} \frac{\cos x}{x+\sin y} \, dx = \left[\frac{\sin x}{x+\sin y} \right]_{2}^{X} + \int_{2}^{X} \frac{\sin x}{(x+\sin y)^2} \, dx$$

and consider each term.

12. (ii) $\sup\limits_{y \geqslant c} \int_{X}^{\infty} \frac{y \, dx}{1+x^2 y^2} = \frac{\pi}{2} - \arctan Xc.$

(iii) $\sup\limits_{y>0} \int_{X}^{\infty} \frac{y \, dx}{1+x^2 y^2} = \frac{\pi}{2}.$

6 (d)

1. If $|f'(x)| \leqslant K$ for $a < x < b$ and $a \leqslant s < t \leqslant b$, $|f(t)-f(s)| \leqslant K(t-s)$.

2. $|\, |f(t)| - |f(s)| \,| \leqslant |f(t)-f(s)|$. If $|f(x)|, |g(x)| \leqslant K$ in $[a, b]$, then

$$|f(t)g(t)-f(s)g(s)| = |f(t)\{g(t)-g(s)\}+g(s)\{f(t)-f(s)\}|$$
$$\leqslant K\{|f(t)-f(s)|+|g(t)-g(s)|\}.$$

3. $P(\mathscr{D}, f)+Q(\mathscr{D}, f) = V(\mathscr{D}, f), \quad P(\mathscr{D}, f)-Q(\mathscr{D}, f) = f(b)-f(a);$

$P(\mathscr{D}, f) = \frac{1}{2}\{V(\mathscr{D}, f)+f(b)-f(a)\}, \quad Q(\mathscr{D}, f) = \frac{1}{2}\{V(\mathscr{D}, f)-f(b)+f(a)\};$

$P_a^b(f) = \frac{1}{2}\{V_a^b(f)+f(b)-f(a)\}, \quad Q_a^b(f) = \frac{1}{2}\{V_a^b(f)-f(b)+f(a)\}.$

4. For any $\mathscr{D}$, $V(\mathscr{D}, F) \leqslant \int_{a}^{b} |f| \, dg$. Take $\mathscr{D}'$ so that $S(\mathscr{D}', f, g) - s(\mathscr{D}', f, g) < \epsilon.$
For each i, there are λ_i, μ_i such that

$$\int_{x_{i-1}}^{x_i} f \, dg = \lambda_i \{g(x_i)-g(x_{i-1})\},$$

$$\int_{x_{i-1}}^{x_i} |f| \, dg = \mu_i \{g(x_i)-g(x_{i-1})\}.$$

Then $0 \leqslant \mu_i - |\lambda_i| \leqslant M_i - m_i$ and so

$$0 \leqslant \int_a^b |f|\, dg - \sum_{i=1}^n \left| \int_{x_{i-1}}^{x_i} f\, dg \right| \leqslant \sum_{i=1}^n (M_i - m_i)\{g(x_i) - g(x_{i-1})\} < \epsilon.$$

Thus
$$V_a^b(F) \geqslant V(\mathscr{D}', F) > \int_a^b |f|\, dg - \epsilon.$$

For the last part, use **3**.

5. $f(x) = \begin{cases} 0 & (x=0), \\ x\cos(\pi/2x) & (0 < x \leqslant 1); \end{cases}$ $f_n(x) = \begin{cases} 0 & (0 \leqslant x < 1/n), \\ f(x) & (1/n \leqslant x \leqslant 1). \end{cases}$

6 (e)

1. By exercises 6(d), 3, 4, G is of bounded variation and $p_G(x) = \int_a^x g^+$,

$q_G(x) = \int_a^x g^-$. Hence, by theorem 6.21 B, $f \in R(p_G; a, b) \cap R(q_G; a, b)$ and

$$\int_a^b f\, dp_G = \int_a^b fg^+, \qquad \int_a^b f\, dq_G = \int_a^b fg^-.$$

2. Since $s(\mathscr{D}, f, p_g) \leqslant \Sigma f(\xi_i)\{p_g(x_i) - p_g(x_{i-1})\} \leqslant S(\mathscr{D}, f, p_g)$, by exercise 6(a), 9, $\Sigma f(\xi_i)\{p_g(x_i) - p_g(x_{i-1})\} \to \int_a^b f\, dp_g$ as $\mu(\mathscr{D}) \to 0$. Replace p_g by q_g and subtract.

3. For every n, $f_n \in R(p_g; a, b) \cap R(q_g; a, b)$. Now use exercise 6(a), 13.

4. Choose f, g so that $\int_a^b f\, dg \neq 0$, $g(b) - g(a) = 0$;   e.g. $f(x) = \cos x$, $g(x) = \sin x$ on $[0, \pi]$.

5. Let $|g(x)| \leqslant K$ for $x \geqslant a$ and let $0 \leqslant f(x) < \epsilon$ for $x \geqslant X$. By exercise 6(a), 12, f and p_g, f and q_g have no common left or right discontinuities. Hence, by theorem 6.54, if $X_2 > X_1 \geqslant X$,

$$\left| \int_{X_1}^{X_2} f\, dg \right| = \left| f(X_2)g(X_2) - f(X_1)g(X_1) - \int_{X_1}^{X_2} g\, df \right| < 3K\epsilon.$$

7. (ii) $\int_\pi^{n\pi} \left| \dfrac{\sin x}{x} \right| d(\cos x) = \sum_{r=1}^{n-1} (-1)^{r+1} \int_{r\pi}^{(r+1)\pi} \dfrac{\sin^2 x}{x}\, dx = \sum_{r=1}^{n-1} (-1)^{r+1} u_r$ and $\Sigma(-1)^{r+1} u_r$ converges by the alternating series test. Also, for $0 < \alpha < \pi$,

$$\left| \int_{n\pi}^{n\pi + \alpha} \left| \frac{\sin x}{x} \right| d(\cos x) \right| \leqslant \frac{1}{n}.$$

6 (f)

1. We need only consider the cases where $f_n^*(x)$ and $f_{n+1}^*(x)$ are defined by different formulae. For instance let $f_n^*(x) = f_n(x)$, so that $|f_n(x) - h_n(x)| \leqslant K$. If $f_{n+1}^*(x) = h_{n+1}(x) + K$,

$$f_{n+1}^*(x) - f_n^*(x) \geqslant (h_{n+1}(x) + K) - (h_n(x) + K) = h_{n+1}(x) - h_n(x) \geqslant 0.$$

If $f_{n+1}^*(x) = h_{n+1}(x) - K$, so that $f_{n+1}(x) - h_{n+1}(x) < -K$,

$$f_{n+1}^*(x) - f_n^*(x) > f_{n+1}(x) - f_n(x) \geqslant 0.$$

2. *Necessity.* (i) gives $\phi(b) - \phi(a) = 0$.   (ii) $\int_a^b f_{x,\delta}\,d\phi = \int_a^x + \int_x^{x+\delta} + \int_{x+\delta}^b = 0$.

$$\int_a^x = \phi(x) - \phi(a); \qquad \left|\int_x^{x+\delta}\right| \leqslant V_x^{x+\delta}(\phi); \qquad \int_{x+\delta}^b = 0.$$

Since ϕ is continuous at x, $V_x^{x+\delta}(\phi) \to 0$ as $\delta \to 0$ (theorem 6.44) and so $\phi(x) - \phi(a) = 0$.

Sufficiency. Take $\epsilon > 0$. By exercise $6(e)$, 2, there is δ such that, for any dissection $\mathscr{D}: a = x_0 < x_1 < \ldots < x_{n-1} < x_n = b$ with $\mu(\mathscr{D}) < \delta$,

$$\left| \Sigma f(\xi_i)\{\phi(x_i) - \phi(x_{i-1})\} - \int_a^b f\,d\phi \right| < \epsilon. \tag{*}$$

But, since ϕ has only countably many discontinuities, $x_1, \ldots, x_{n-1}$ can be chosen to be points of continuity of ϕ, so that the sum in (*) is 0. Thus $\left|\int_a^b f\,d\phi\right| < \epsilon$.

3. $g = g^* + \phi$, where

$$g^*(x) = \begin{cases} 0 & \text{for} \quad a \leqslant x < c, \\ 1 & \text{for} \quad c \leqslant x \leqslant b \end{cases}$$

and ϕ is a function of the kind described in **2**. Note that ϕ may be chosen so that

$$g(x) = \begin{cases} 0 & \text{for} \quad a \leqslant x \leqslant c, \\ 1 & \text{for} \quad c < x \leqslant b. \end{cases}$$

6 (g)

1. Follow the proof of theorem 6.71 with two changes.
   (i) In the notation of that proof,

$$0 \leqslant S(\mathscr{D}) - S(\mathscr{D}') \leqslant (M-m)\{g(x_i) - g(x_{i-1})\}.$$

   (ii) Since g is uniformly continuous, we may choose δ so that, if $0 < v - u < \delta$, then $0 \leqslant g(v) - g(u) < \epsilon/n_0$. It follows that

$$0 \leqslant S(\mathscr{D}_\delta) - S(\mathscr{D}^*) \leqslant (M-m)n_0(\epsilon/n_0) = \epsilon(M-m).$$

2. On $[-1, 1]$ let

$$f(x) = \begin{cases} 1 & \text{for} \quad -1 \leqslant x < 0, \\ 0 & \text{for} \quad 0 \leqslant x \leqslant 1; \end{cases} \qquad g(x) = \begin{cases} 0 & \text{for} \quad -1 \leqslant x \leqslant 0, \\ 1 & \text{for} \quad 0 < x \leqslant 1. \end{cases}$$

If 0 is a point of division of $\mathscr{D}$, $S(\mathscr{D}, f, g) = 0$; otherwise $S(\mathscr{D}, f, g) = 1$.

3. Adapt the proof of theorem 6.71. If $\mathscr{D}'$ is obtained from $\mathscr{D}$ by adding new points of division in the subinterval $[x_{i-1}, x_i]$ of $\mathscr{D}$, then

$$0 \leqslant V(\mathscr{D}') - V(\mathscr{D}) \leqslant V_{x_{i-1}}^{x_i}(f).$$

   Choose $\mathscr{D}_0$ so that $V(\mathscr{D}_0, f) > V_a^b(f) - \epsilon$. Let n_0 be the number of subintervals of $\mathscr{D}_0$. Since v_f is continuous (theorem 6.44), there is δ such that $V_u^v(f) < \epsilon/n_0$

if $0 < v-u < \delta$. Let $\mathscr{D}_\delta$ be a dissection with $\mu(\mathscr{D}_\delta) < \delta$ and let $\mathscr{D}^*$ be the dissection with all points of division of $\mathscr{D}_0$ and $\mathscr{D}_\delta$. Then

$$0 \leqslant V(\mathscr{D}^*,f)-V(\mathscr{D}_\delta,f) \leqslant n_0(\epsilon/n_0) = \epsilon$$

and so $V_a^b(f) \geqslant V(\mathscr{D}_\delta,f) \geqslant V(\mathscr{D}^*,f)-\epsilon \geqslant V(\mathscr{D}_0,f)-\epsilon \geqslant V_a^b(f)-2\epsilon$.

Similarly, if $V(\mathscr{D},f)$ is unbounded, $V(\mathscr{D}_\delta,f) \to \infty$ as $\delta \to 0$.

4. On $[-1,1]$ let $f(x) = 0$ if $x \neq 0$, $f(0) = 1$. If 0 is a point of division of $\mathscr{D}$, then $V(\mathscr{D},f) = 2$; otherwise $V(\mathscr{D},f) = 0$.

6. For any dissection of $[a, b]$,

$$\sum_{i=1}^n f(x_{i-1})(x_i-x_{i-1})+ \sum_{i=1}^n x_i\{f(x_i)-f(x_{i-1})\} = bf(b)-af(a).$$

If $f(x_i) = t_i$, this can be written

$$\sum_{i=1}^n f(x_{i-1})(x_i-x_{i-1})+ \sum_{i=1}^n f^{-1}(t_i)(t_i-t_{i-1}) = b\beta - a\alpha.$$

Since f is uniformly continuous, $\max(t_i-t_{i-1}) \to 0$ as $\max(x_i-x_{i-1}) \to 0$. Use theorem 6.72 (i).

6 (h)

1. $[a, b]$ may be divided into a finite number of subintervals $[a_k, b_k]$ such that f is continuous in $[a_k, b_k]$ except at one of the end points. Use exercise 6(a), 11, theorems 6.12, 6.16.

3. Let $\xi_1, ..., \xi_l$ be the limit points of E. Outside $\cup(\xi_k-\epsilon, \xi_k+\epsilon)$ there is only a finite number of points of E. These may be covered by intervals of total length less than ϵ. Hence E may be covered by intervals of total length less than $(l+1)\epsilon$.

4. By 3, $c(E^{(n-1)}) = 0$. The proof of 3 is easily adapted to show that, if D is bounded and $c(D^{(1)}) = 0$, then $c(D) = 0$. Hence $c(E^{(n-2)}) = ... = c(E) = 0$.

5. Let $[a, b] \supset E_1 \cup E_2$. Since χ_{E_1}, χ_{E_2} are integrable over $[a, b]$, so are

$$\chi_{E_1 \cup E_2} = \tfrac{1}{2}\{(\chi_{E_1}+\chi_{E_2})+|\chi_{E_1}-\chi_{E_2}|\}, \chi_{E_1 \cap E_2} = \tfrac{1}{2}\{(\chi_{E_1}+\chi_{E_2})-|\chi_{E_1}-\chi_{E_2}|\}.$$

Also $$\chi_{E_1 \cup E_2}+\chi_{E_1 \cap E_2} = \chi_{E_1}+\chi_{E_2}.$$

6. By 5, $0 \leqslant c(E_1 \cup E_2) \leqslant c(E_1)+c(E_2) = 0$. By induction, $c(E_1 \cup ... \cup E_n) = 0$. For the second part, let $r_1, r_2, ...$ be the rational numbers in $[a, b]$ and put $E_n = \{r_n\}$.

7. $h(x) = f(x)-g(x) = 0$ except in a set of zero content. Thus (see theorem 6.83) $\int_a^b h$ exists and is 0. Hence

$$\overline{\int_a^b} f \leqslant \overline{\int_a^b} g+ \int_a^b h = \overline{\int_a^b} g \leqslant \overline{\int_a^b} f+ \int_a^b (-h) = \overline{\int_a^b} f.$$

8. If $E = \{1, \tfrac{1}{2}, \tfrac{1}{3}, ...\}$, $c(E) = 0$; and, if f is assigned any bounded set of values on E, then f is bounded on $[0, 1]$ and continuous on $[0, 1]-E$.

9. Let

$$f_n(x) = \begin{cases} 1/q & \text{if } x = p/q, \text{ where } p, q \text{ are coprime and } 1 \le q \le n, \\ 0 & \text{otherwise.} \end{cases}$$

then $f_n \to f$ uniformly on $[a, b]$ and each f_n is integrable since $f_n(x) = 0$ except at a finite number of points.

10. Take $\epsilon > 0$. Choose $\mathscr{D}$ so that

$$\int_a^b f' - \epsilon < s(\mathscr{D}, f') \le S(\mathscr{D}, f') < \int_a^b f' + \epsilon.$$

Since
$$f(b) - f(a) = \Sigma f'(\xi_i)(x_i - x_{i-1}) \quad (x_{i-1} < \xi_i < x_i),$$
$$s(\mathscr{D}, f') \le f(b) - f(a) \le S(\mathscr{D}, f').$$

Hence
$$\left| \{f(b) - f(a)\} - \int_a^b f' \right| < \epsilon.$$

11. f is not differentiable at 0, since $\{f(h) - f(0)\}/h = \sin(\log h)$ oscillates as $h \to 0+$. For $0 < x \le 1$, $f'(x) = \sin(\log x) + \cos(\log x)$; so f' is bounded and continuous on $(0, 1]$, therefore integrable over $[0, 1]$. Also f is continuous on $[0, 1]$ and so, by theorem 6.84, $\int_0^1 f' = f(1) - f(0) = 0$.

12. $\{f(h) - f(0)\}/h = h \sin(\pi/2h^2) \to 0$ as $h \to 0+$. For $0 < x \le 1$,

$$f'(x) = 2x \sin(\pi/2x^2) - \pi(1/x)\cos(\pi/2x^2).$$

Thus f' is not bounded on $[0, 1]$. But

$$\int_\delta^1 f' = f(1) - f(\delta) = 1 - \delta^2 \sin(\pi/2\delta^2) \to 1 \quad \text{as} \quad \delta \to 0+.$$

13. Let $f' = g = g^+ - g^-$. Then g, g^+, g^- are continuous and

$$f(x) - f(a) = \int_a^x g = \int_a^x g^+ - \int_a^x g^- = \phi(x) - \psi(x).$$

6 (i)

1. $a, b \in [c, d]$. Define F on $[c, d]$, G on $[\alpha, \beta]$ by $F(x) = \int_a^x f$, $G = F \circ \phi$. Then justify the following steps:

$$\int_\alpha^\beta (f \circ \phi)\phi' = \int_\alpha^\beta G' = G(\beta) - G(\alpha) = F(b) - F(a) = \int_a^b f.$$

2. Assume $0 < a < b$. When $0 < \delta < X$, by theorem 6.92 (or **1**) and exercise 6(a), **8**,

$$\int_\delta^X \frac{f(ax) - f(bx)}{x} \, dx = \int_{a\delta}^{b\delta} \frac{f(t)}{t} \, dt - \int_{aX}^{bX} \frac{f(t)}{t} \, dt = (\lambda_\delta - \Lambda_X) \log \frac{b}{a},$$

where λ_δ, Λ_X lie between the infimum and supremum of f in $[a\delta, b\delta]$ and $[aX, bX]$ respectively.

3. In the first part f is decreasing. Define $f*$ by $f*(a) = f(a+)$, $f*(x) = f(x)$ for $a < x \leqslant b$. Then $f*$ is decreasing and, by theorems 6.83, 6.93 (i),

$$\int_a^b fg = \int_a^b f*g = f*(a) \int_a^\xi g = f(a+) \int_a^\xi g.$$

4. Adapt the proof of theorem 6.93 (i): begin by considering $\int_a^b f d(g+v_g)$, which has an increasing integrator.

For the last part take f on $[a, b]$ to be continuous and decreasing with $f(a) = 2$, $f(b) = 1$; $g(x) = 0$ for $a \leqslant x < b$, $g(b) = 1$. Then $\int_a^b f dg = 1$ and $f(a) \int_a^\xi dg = 0$ for $a \leqslant \xi < b$, 2 for $\xi = b$.

5. Let $\left| \int_a^x f \right| \leqslant K$. If g decreases, then $g(x) \geqslant 0$ and

$$\left| \int_{X_1}^{X_2} fg \right| = \left| g(X_1) \int_{X_1}^\xi f \right| \leqslant 2K|g(X_1)| \to 0 \quad \text{as} \quad X_1 \to \infty.$$

6. If g decreases and $|g(x)| \leqslant K$, then $g(x)+K \geqslant 0$ and

$$\left| \int_{X_1}^{X_2} f(g+K) \right| = \left| \{g(X_1)+K\} \int_{X_1}^\xi f \right| \leqslant 2K \left| \int_{X_1}^\xi f \right| \to 0 \quad \text{as} \quad X_1 \to \infty.$$

7. $\alpha \leqslant 0$. $(2/\pi)x^\alpha \leqslant \sin(x^\alpha) \leqslant x^\alpha$ $(x \geqslant 1)$; use comparison principle.

$\alpha > 0$. $\int_1^X \sin(x^\alpha) dx = \int_1^{X^\alpha} u^\beta \sin u\, du$, where $\beta = (1/\alpha)-1$. Since $X^\alpha \to \infty$ as $X \to \infty$, $\int_1^\infty \sin(x^\alpha) dx$ exists if and only if $\int_1^\infty u^\beta \sin u\, du$ exists. When $0 < \alpha \leqslant 1$, $\beta \geqslant 0$ and $\int_{2n\pi}^{(2n+1)\pi} u^\beta \sin u\, du \geqslant 2$. When $\alpha > 1$, $\beta < 0$ and the second mean value theorem (or **5**) gives convergence.

8. (b) Use (a) and the fact that $x/(x+1)$ is bounded and monotonic.

10. For $y > 0$,

$$\left| \int_X^{X_2} e^{-xy} \sin(x/y) dx \right| = \left| e^{-X_1 y} \int_X^\xi \sin(x/y) dx \right| \leqslant 2y e^{-X_1 y} \leqslant 2X_1^{-1} e^{-1}.$$

7 (a)

1. Let $|D_i f_j(x)| < M$ for $x \in B(\xi; \delta)$. With the notation of thoerem 7.14, if $\|h\| < \delta$,

$$|f_j(\xi+h)-f_j(\xi)| \leqslant \sum_{i=1}^m |f_j(\xi+k^i)-f_j(\xi+k^{i-1})|$$

$$\leqslant \sum_{i=1}^m |h_i|M \leqslant mM\|h\|$$

and so $\qquad \|f(\xi+h)-f(\xi)\| \leqslant nmM\|h\|.$

3. Let $\|u\| = 1$. $|f(\xi+tu)-f(\xi)-Df(\xi)(tu)|/\|tu\| \to 0$ as $t \to 0$. Since $\|tu\| = |t|$ and $Df(\xi)(tu) = tDf(\xi)(u)$, this implies $\{f(\xi+tu)-f(\xi)\}/t \to Df(\xi)(u)$.

4. By theorem 2.21, $|Df(\xi)(u)| = (\nabla f(\xi))^T . u \leqslant \|\nabla f(\xi)\|$ whenever $\|u\| = 1$, and equality holds if and only if $u = u_0$. Now $|Df(\xi)(u_0)| = \sup_{\|u\|=1} |Df(\xi)(u)|$ $= \|Df(\xi)\|$. Use 3.

5. Let $u^2 + v^2 = 1$. Since $\{f(tu, tv) - f(0, 0)\}/t = u^2 v$, $D_{(u, v)} f(0, 0)$ exists. Also $|f(tu, tv) - f(0, 0)| = |u^2 vt| \leqslant |t|$ and so f is continuous at $(0, 0)^T$.

If f is differentiable at $(0, 0)^T$, then, since $D_1 f(0, 0) = D_2 f(0, 0) = 0$, $Df(0, 0) = \Theta$. But
$$|f(tu, tv) - f(0, 0) - \Theta(tu, tv)|/|t| = |u^2 v| \nrightarrow 0 \quad \text{as} \quad t \to 0.$$

6. Since $f(w, 0) = f(0, w) = 0$ for all w, $D_1 f(0, 0)$, $D_2 f(0, 0)$ exist and are 0. But, when $w \neq 0$, $D_1 f(0, w)$, $D_2 f(w, 0)$ do not exist. f is differentiable at $(0, 0)^T$, since
$$|f(h, k) - f(0, 0) - \Theta(h, k)| = |hk| \leqslant \tfrac{1}{2}(h^2 + k^2) = \tfrac{1}{2}\|(h, k)\|^2.$$

7. If λ, μ, ν are linear functions on R^m to R^n, R^p, R^{n+p} such that $\nu(h) = (\lambda(h), \mu(h))^T$, then
$$\|f(\xi + h) - f(\xi) - \nu(h)\|^2 = \|\phi(\xi + h) - \phi(\xi) - \lambda(h)\|^2 + \|\psi(\xi + h) - \psi(\xi) - \mu(h)\|^2.$$

8.
$$D_1 f(a, b) = \phi'(a)\psi(b), \quad D_2 f(a, b) = \phi(a)\psi'(b)$$
and $f(a+h, b+k) - f(a, b) - hD_1 f(a, b) - kD_2 f(a, b)$
$$= \{\phi(a+h) - \phi(a)\}\psi(b+k) + \{\psi(b+k) - \psi(b)\}\phi(a)$$
$$- h\phi'(a)\psi(b) - k\phi(a)\psi'(b)$$
$$= \{h\phi'(a) + o(|h|)\}\psi(b+k) + \{k\psi'(b) + o(|k|)\}\phi(a)$$
$$- h\phi'(a)\psi(b) - k\phi(a)\psi'(b)$$
$$= o(\|(h, k)\|), \quad \text{since} \quad \psi(b+k) - \psi(b) = o(1), \quad \psi(b+k) = O(1).$$

9. If $g : R^1 \to R^1$ is a continuous, nowhere differentiable function (theorem 5.51), define $f : R^2 \to R^1$ by $f(x, y) = g(x)$ for all $(x, y)^T \in R^2$.

10. $D_i f_j$ $(i = 1, 2; j = 1, 2, 3)$ is clearly continuous everywhere. Hence f is everywhere differentiable. Also
$$\|Df(x, y)(h, k)\| = (h^2 + k^2 \sin^2 x)^{\frac{1}{2}} \leqslant \|(h, k)\| \quad \text{for} \quad \text{all} \;\; (h, k)^T \in R^2$$
and
$$\|Df(x, y)(h, 0)\| = |h| = \|(h, 0)\|.$$

7 (b)

1. Let $g : R^n \to R^1$ be given by $g(x) = \|x\| = \sqrt{(x_1^2 + \ldots + x_n^2)}$. $D_1 g, \ldots, D_n g$ exist and are continuous except at θ. Hence g is differentiable except at θ and $\|f\| = g \circ f$ is differentiable at ξ if $Df(\xi)$ exists and $f(\xi) \neq \theta$.

2. $\|f\|^2 = f.f$ is differentiable by theorem 7.21. If $\|f\|^2$ is constant, $\Theta = D\|f\|^2(\xi) = 2f(\xi) . Df(\xi)$.

3. (i) For fixed $t > 0$ define $\psi : X \to R^p$ by $\psi(x) = g(tx) = t^\alpha g(x)$. Then $D_i \psi(x) = t D_i g(tx) = t^\alpha D_i g(x)$, so that $D_i g(tx) = t^{\alpha-1} D_i g(x)$.

(ii) For fixed $x \in X$ define $\phi : (0, \infty) \to R^p$ by $\phi(t) = g(tx)$.

If $f : (0, \infty) \to R^n$ is defined by $f(t) = tx$, then $\phi = g \circ f$ and, by theorem 7.22 (see note), $\phi'(t) = Dg(f(t))(f'(t)) = Dg(tx)(x)$.

Necessity. $\phi(t) = t^\alpha g(x)$ and so $\alpha t^{\alpha-1} g(x) = \phi'(t) = Dg(tx)(x)$. Put $t = 1$.

Sufficiency. $t\phi'(t) = tDg(tx)(x) = Dg(tx)(tx) = \alpha g(tx) = \alpha\phi(t)$. Thus, for $t > 0$, $(d/dt)(t^{-\alpha}\phi(t)) = 0$, i.e. $t^{-\alpha}\phi(t) = C$. Since $t = 1$ gives $C = \phi(1) = g(x)$, $g(tx) = \phi(t) = t^\alpha g(x)$.

4. $(f \circ p)(r, \phi) = (r^{1-\alpha} \cos \phi, r^{1-\alpha} \sin \phi)^T$ and so $J(f \circ p)(r, \phi) = (1-\alpha)r^{1-2\alpha}$. Since also $Jp(r, \phi) = r$ and

$$J(f \circ p)(r, \phi) = Jf(r \cos \phi, r \sin \phi)Jp(r, \phi),$$

$Jf(x, y) = Jf(r \cos \phi, r \sin \phi) = (1-\alpha)r^{-2\alpha}$.

5. If $f(x) = (y_1, y_2, y_3)^T$,

$$x_1 = y_1 + y_2 + y_3, \quad x_2 = \frac{y_2 + y_3}{y_1 + y_2 + y_3}, \quad x_3 = \frac{y_3}{y_2 + y_3}$$

provided that $y_1 + y_2 + y_3 \neq 0$, $y_2 + y_3 \neq 0$. Hence

$$Y = \{(y_1, y_2, y_3)^T | y_1 + y_2 + y_3 \neq 0, \quad y_2 + y_3 \neq 0\}.$$

$f^{-1}: Y \to R^3$ is differentiable, since all partial derivatives exist and are continuous on Y. If $x = f^{-1}(y)$,

$$Jf^{-1}(y) = 1/Jf(x) = 1/(x_1^2 x_2) = 1/[(y_1 + y_2 + y_3)(y_2 + y_3)].$$

7 (c)

1. Let $h = v - u$ and define $\phi: [0, 1] \to R^1$ by $\phi(t) = f(u + th)$ $(0 \leq t \leq 1)$. Then ϕ is continuous at 0, 1 and differentiable in $(0, 1)$ with $\phi'(t) = Df(u + th)(h)$. Apply mean value theorem to ϕ.

2. If $v = u + 2\pi$, $f(v) - f(u) = 0$. But, for all $\xi \in R^1$,

$$\|Df(\xi)(2\pi)\| = \|(-2\pi \sin \xi, 2\pi \cos \xi)^T\| = 2\pi.$$

3. Let $\Delta = f(v) - f(u) - Df(u)(v - u)$. Nothing to prove if $\Delta = 0$. If $\Delta \neq 0$, let ω be unit vector such that $\|\Delta\| = \omega.\{f(v) - f(u) - Df(u)(v - u)\}$. As in proof of theorem 7.31, there exists τ in $(0, 1)$ such that $\omega.\{f(v) - f(u)\} = \omega.Df(u + \tau h)(h)$, where $h = v - u$ and so $u + \tau h \in S$. Then $\|\Delta\| = \|\omega.\{Df(u + \tau h)(h) - Df(u)(h)\}\| \leq \|Df(u + \tau h) - Df(u)\| \|h\|$.

4. Let $\rho(K, G') = 2\delta(> 0)$. If $H = \bigcup_{x \in K} B(x; \delta)$, then $\overline{H}$ is a compact subset of G. Thus f, Df are uniformly continuous on $\overline{H}$. (i) Now immediate. For (ii), if $x \in K$ and $\|h\| < \delta$, by 3,

$$\|f(x+h) - f(x) - Df(x)(h)\|/\|h\| \leq \sup_{w \in B(x; \|h\|)} \|Df(w) - Df(x)\|.$$

5. Each sequence $(D_i f_j^\nu)$ converges uniformly on every compact subset of G, since

$$|D_i f_j^\mu(x) - D_i f_j^\nu(x)| \leq \|Df^\mu(x) - Df^\nu(x)\|.$$

Suppose $(f^\nu(\xi))$ converges. If x is any point of G other than ξ, ξ and x may be joined by a polygonal path in G with segments parallel to the axes. Integrating the continuous functions $D_i f_j^\nu$ along these segments we find (theorem 5.22) that each sequence $(f_j^\nu(x))$ converges and so $(f^\nu(x))$ converges. The same argument shows that the convergence is uniform in every closed ball in G.

Let K be any compact subset of G and put $\rho(K, G') = 2\delta$ (> 0). By the Heine–Borel theorem, there are finitely many open balls $B(x^k; \delta)$ which cover K. (f^ν) converges uniformly on each $\overline{B(x^k; \delta)}$ and so on K.

Take $x \in G$. Theorem 5.23 (A or B) applied to straight line segments through x parallel to the axes shows that the $D_i f_j(x)$ exist and equal $\lim D_i f_j^\nu(x)$. Since the $D_i f_j^\nu$ are continuous, so are the $D_i f_j$. Hence f is continuously differentiable. Since

$$\|Df(x) - Df^\nu(x)\|^2 \leq \sum_{i,j} |D_i f_i(x) - D_i f_j^\nu(x)|^2,$$

$Df^\nu(x) \to Df(x)$.

7 (d)

1. (i) When $m = n = 1$, existence of $D_{(i)} f(\xi, \eta)$ equivalent to existence of $D_i f(\xi, \eta)$ ($i = 1, 2$) and this does not even imply continuity.

(ii) All partial derivatives exist in the open ball (see p. 215) and are continuous at $(\xi, \eta)^T$ (as on p. 211). Apply theorem 7.14.

2. $D_2 f(x, y) = x + y^{-1} > 0$ for all $x, y > 0$
$D_1 f(x, y) = y - x^{-1} = 0 \Leftrightarrow y = x^{-1}$; $f(x, x^{-1}) = 0 \Leftrightarrow x = e^{\frac{1}{2}}$. If solution through $(e^{\frac{1}{2}}, e^{-\frac{1}{2}})^T$ is $y = \phi(x)$,

$$\phi'(x) = -\frac{\phi(x) - x^{-1}}{x + (\phi(x))^{-1}}.$$

Since $\phi(e^{\frac{1}{2}}) = e^{-\frac{1}{2}}$ and $\phi'(e^{\frac{1}{2}}) = 0$, $\phi''(e^{\frac{1}{2}}) = -\frac{1}{2} e^{-\frac{3}{2}}$.

3. Let $\xi \neq 0$. By the mean value theorem, $f(\xi + y) - f(y) = \xi f'(y + \alpha\xi)$, where $0 < \alpha < 1$. As $y \to \pm \infty$, $\xi f'(y + \alpha\xi) \to \pm \infty$ or $\mp \infty$ according as $\xi \gtrless 0$. In either case there is an η such that $f(\xi + \eta) - f(\eta) = f(\xi)$.
If $F(x, y) = f(x + y) - f(x) - f(y)$, $D_2 F(\xi, \eta) = f'(\xi + \eta) - f'(\eta) \neq 0$, since $\xi \neq 0$. Last part: $f(x) = 1 + x^2$.

4. E.g. $(a, 0, 0)^T$, $(0, a, 0)^T$ when $a \neq 0$.

5. $J_{(2)} f(x, y, z) = 3xy^5$ and $f(0, y, z), f(x, 0, z)$ cannot be θ.

6. $\begin{pmatrix} -\frac{1}{5} & -\frac{1}{5} & 0 \\ \frac{1}{5} & -\frac{24}{5} & 0 \end{pmatrix}$.

7. E.g. when $u, v, w \leq -1$. $J_{(2)} = 1 + e^{u_0 + v_0 + w_0} > 0$.

$$\frac{2}{1 + e^{u+v+w}} \begin{pmatrix} -x & ye^v & -ze^{v+w} \\ -xe^{w+u} & -y & ze^w \\ xe^u & -ye^{u+v} & -z \end{pmatrix}.$$

8. E.g. $(x, 0, 0, v)^T$ satisfies the equations if $g(v) = v - \sin(\cos v) = 0$; and $g(0) < 0, g(v) \to \infty$ as $v \to \infty$.
$J_{(2)} = -e^u \cos v \neq 0$, for $\cos v = 0 \Rightarrow x = 0 \Rightarrow v = 0$.

7 (e)

1. Counter-example: In R^1, $f(x) = x^3$, $\xi = 0$.
If the local inverse ϕ is differentiable,

$$I = D(\phi \circ f)(\xi) = D\phi(f(\xi)) \circ Df(\xi)$$

and so
$$1 = J\phi(f(\xi)) \cdot Jf(\xi).$$

2. $Jf(x, y) = \sin(y-x)$. For all δ, δ',

$$f(\alpha+\delta, \alpha+2n\pi+\delta') = f(\alpha+\delta', \alpha+2n\pi+\delta),$$
$$f(\alpha+\delta, \alpha+(2n+1)\pi+\delta) = (0, 0)^T.$$

3.

$$\frac{1}{2xyz}\begin{pmatrix} -x^2 & xy & zx \\ xy & -y^2 & yz \\ zx & yz & -z^2 \end{pmatrix}.$$

4. $(1+x^2+y^2)^2$.

7. Counter example: if $f : R^2 \to R^2$ is given by

$$f(x, y) = (e^x\cos y, e^x\sin y)^T,$$

then $f(R^2) = R^2 - \{\theta\}$.

7(f)

1. Jacobian $x^2 \sin y$.

2. If $\Psi : R^2 \to R^1$ is given by $\Psi(u, v) = 2e^{u+v} - u + v$, $\Psi(f(x, y), g(x, y)) \equiv 0$. Also, on all open sets, $\Psi \not\equiv 0$; for, given u, $\Psi(u, v)$ strictly increases with v.

3. The origin in R^6 satisfies the equations and $J_{(2)} = 27u^2v^2w^2 + 1 > 0$. Hence there exists $[-\alpha, \alpha]^3 \times [-\beta, \beta]^3$ in which there is a unique solution. Also

$$J\phi = -\frac{6(y-z)(z-x)(x-y)}{27u^2v^2w^2+1} \not\equiv 0$$

in $(-\alpha, \alpha)^3$.

4. If $f(G)$ has no interior point, define $\Psi : R^n \to R^1$ by $\Psi(y) = 0$ for $y \in f(G)$, $\Psi(y) = 1$ for $y \notin f(G)$.

If $f(G)$ has an interior point, then, whenever $\Psi(f(x)) = 0$ for $x \in G$, $\Psi(y) = 0$ in an open set containing that point.

5. In the definition of functional dependence we may clearly replace the condition $\Psi \not\equiv 0$ in any open set by $(C) : \Psi \not\equiv 0$ in any open set containing a point of $f(G)$. Then $(C) \Leftrightarrow$ for all $y \in f(G)$, y is not an interior point of $\{z | \Psi(z) = 0\} \Leftrightarrow$ for all $y \in f(G)$, $y \in$ support of $\Psi \Leftrightarrow f(G) \subset$ support of Ψ.

6. Let Ψ be as in the definition of functional dependence of $f_1, ..., f_n$. Define Ψ^* from R^{n+1} to R^1 by

$$\Psi^*(y_1, ..., y_n, y_{n+1}) = \Psi(y_1, ..., y_n).$$

Then $\Psi^* \not\equiv 0$ in any open set in R^{n+1}, since otherwise $\Psi \equiv 0$ in an open subset of R^n. Clearly $\Psi^*(f_1(x), ..., f_n(x), g(x)) = 0$ for $x \in G$.

7. If $g(x) = \Phi(f_1(x), ..., f_n(x))$ for $x \in G$, define Ψ from R^{n+1} to R^1 by

$$\Psi(y_1, ..., y_n, y_{n+1}) = \Phi(y_1, ..., y_n) - y_{n+1}.$$

Then $\{y | \Psi(y) = 0\}$ has no interior point; for if $\Psi(y_1, ..., y_n, y_{n+1}) = 0$, then $\Psi(y_1, ..., y_n, y'_{n+1}) \neq 0$ when $y'_{n+1} \neq y_{n+1}$.

8. $f(R^1) = \{(0, 0)^T\} \cup \{(t, \pm t^2 \sin t^{-\frac{1}{2}})^T | t > 0\}$ and this set clearly has no interior point.

f_1 is not functionally dependent on f_2, for $\Phi(x^2)$ is even (all Φ), while $x^4\sin(1/x)$ is odd.

f_2 is not functionally dependent on f_1, for $\Phi(x^4\sin(1/x))$ has the same value for $x = 1/(n\pi)$ $(n = \pm 1, \pm 2, ...)$, but x^2 has not.

9. Take $\Phi = g \circ f^{-1}$.

10. $h = 2(f^2+g^2)^{\frac{1}{2}}\cos[f+(f^2+g^2)^{\frac{1}{2}}]^{\frac{1}{2}}$.

11. $h = \frac{1}{3}(3fg-f^3)$.

7(g)

1. $D_i f(\xi)$ $(i = 1, ..., m)$ exists. The real function ϕ_1 defined by $\phi_1(t) = f(t, \xi_2, ..., \xi_m)$ has a maximum or minimum at ξ_1 and so $D_1 f(\xi) = \phi_1'(\xi_1) = 0$. Also $D_i f(\xi) = 0$ $(i = 2, ..., m)$ and so $Df(\xi) = \Theta$.

2. Since $\bar{G}$ is compact, f assumes its bounds (say k, K) on $\bar{G}$. If f is constant in $\bar{G}$, then $Df(x) = \Theta$ for all $x \in G$. If f is not constant, $k \neq K$ and f assumes at least one of k, K in G. At such a point f has a maximum or minimum.

3. $(2x+y, x+2y) = (0, 0) \Rightarrow x = y = 0 \Rightarrow x^2+xy+y^2 \neq 1$.
By theorem 7.71, critical points $\pm(1, 1)^T$, $\pm(1/\sqrt{3}, 1/\sqrt{3})^T$.
Largest value 1, least value 1/3.

4. $(2x, 2y, 2z) \neq (0, 0, 0)$ if $x^2+y^2+z^2 = 2$. Lagrange multiplier λ such that $\lambda(x+y+z) = 0$. $\lambda = 0$ gives critical points $\pm(\sqrt{(2/3)}, \sqrt{(2/3)}, \sqrt{(2/3)})^T$. $x+y+z = 0$ gives $\pm(0, 1, -1)^T$, $\pm(-1, 0, 1)^T$, $\pm(1, -1, 0)^T$. Largest value 2, least value -2.

5. Theorem 7.71 applicable. Lagrange equations

$$2x-y-z+\lambda(2x-2) = 0,$$
$$2y-z-x+\lambda(2y+2) = 0,$$
$$2z-x-y+\lambda(2z+6) = 0$$

yield $\lambda(x+y+z+3) = 0$. $\lambda = 0 \Rightarrow x = y = z$; $3x^2+6x+9 > 0$. $x+y+z = -3$ leads to

$$x = \frac{2\lambda-3}{2\lambda+3}, \quad y = -1, \quad z = -\frac{6\lambda+3}{2\lambda+3}.$$

Substitution in condition gives $\lambda = -\frac{9}{2}$ or $\frac{3}{2}$. Hence critical points $(2, -1, -4)^T$, $(0, -1, -2)^T$. Former gives supremum, therefore maximum; latter gives infimum, therefore minimum.

6. Box has length x, width y, height z.
(i) Consider $V = xyz$ under constraint $2(yz+zx+xy) = a^2$, $x, y, z \geq 0$. $(y+z, z+x, x+y) \neq (0, 0, 0)$ if $x, y, z > 0$. Critical point $x = y = z = a/\sqrt{6}$.
$V = 0$ when one of $x, y, z = 0$. Also $xy \leq 2a^2$, $xz \leq 2a^2$ and so $V \leq a^4/(4x)$. Thus V is small if x (or y, or z) is large. Hence sup V exists and is attained at $(\xi, \eta, \zeta)^T$, where $\xi, \eta, \zeta > 0$ and $2(\eta\zeta+\zeta\xi+\xi\eta) = a^2$. So $(\xi, \eta, \zeta)^T$ is the critical point $\xi = \eta = \zeta = a/\sqrt{6}$.
(ii) V under constraint $2yz+2zx+xy = a^2$, $x, y, z \geq 0$. $\xi = \eta = a/\sqrt{3}$, $\zeta = a/(2\sqrt{3})$.

7. Rank of
$$\begin{pmatrix} 2x & 2y & 2z \\ 1 & 1 & 1 \end{pmatrix}$$

is 2 unless $x = y = z$, and $x = y = z$ does not satisfy conditions. Critical points $(1, 0, 0)^T$, $(-\frac{1}{3}, \frac{2}{3}, \frac{2}{3})^T$. Largest value 1, least value $\frac{5}{9}$.

8. (i) Stationary points $\pm(\sqrt{2}, \sqrt{2}, \sqrt{2})^T$, both in $E = \{(x, y, z)^T | x^2 + y^2 + z^2 \leqslant 9\}$.

(ii) $(2x, 2y, 2z)$ has rank 1 when $x^2 + y^2 + z^2 = 9$. Lagrange's equations $2 - yz + 2\lambda x = 0$, etc. Subtract in pairs:

$$(y - z)(x + 2\lambda) = 0, \quad (z - x)(y + 2\lambda) = 0, \quad (x - y)(z + 2\lambda) = 0.$$

Then $x = y = z = \pm\sqrt{3}$ or $y = z = -2\lambda$, $x = (2\lambda^2 - 1)/\lambda$, etc. By substitution in constraint, $\lambda^2 = 1$ or $1/12$. Hence critical points are $\pm(\sqrt{3}, \sqrt{3}, \sqrt{3})^T$, $\pm(-1, 2, 2)^T$, $\pm(5/\sqrt{3}, 1/\sqrt{3}, 1/\sqrt{3})^T$ and 8 more by cyclic permutation.

E is compact and so $f(x, y, z) = 2(x + y + z) - xyz$ assumes its bounds on E. If a bound occurs in interior, then point is stationary point without constraint; if in $\{(x, y, z)^T | x^2 + y^2 + z^2 = 9\}$, then point is critical point of f under $x^2 + y^2 + z^2 = 9$. Hence bounds of f on E are taken at some of the 16 points of (i), (ii). Evaluation shows spuremum 10 (at $(-1, 2, 2)^T$ etc.), infimum -10 (at $(1, -2, -2)^T$ etc.).

9. Theorem 7.71 applicable in (i) and (ii). In (i), critical point $(\frac{1}{2}, 0, \frac{1}{2})^T$, which satisfies $4x^2 + y^2 < 16$. In (ii), critical points $(2, 0, -1)^T$, $(-2, 0, 3)^T$, $(-\frac{1}{2}, \pm\sqrt{15}, \frac{3}{2})^T$.

By argument of **8**, bounds of $x^2 + y^2 + z^2$ on $\{(x, y, z)^T | x + z = 1; 4x^2 + y^2 \leqslant 16\}$ taken at some of the five critical points of (i), (ii). Supremum at $(-\frac{1}{2}, \pm\sqrt{15}, \frac{3}{2})^T$; infimum at $(\frac{1}{2}, 0, \frac{1}{2})^T$.

7 (h)

1. (i) $\alpha f(x) = Df(x)(x) = \sum_{j=1}^{m} x_j D_j f(x)$. Hence

$$\alpha^2 f(x) = \alpha \sum_{i=1}^{m} x_i D_i f(x) = \sum_{i=1}^{m} x_i \left(\sum_{j=1}^{m} x_j D_i D_j f(x) + D_i f(x) \right).$$

(ii) Take $\xi = (\xi_1, \ldots, \xi_m)^T$ with $\xi_1 \neq 0$. Multiply first row of $Hf(\xi)$ by ξ_1 and add to it ξ_2(2nd row) $+ \ldots + \xi_m$(mth row). By exercise 7(b), 3, all elements of first row are now 0. Hence $\xi_1 Hf(\xi) = 0$, i.e. $Hf(\xi) = 0$. Since Hf is continuous, $Hf(\xi) = 0$ also when $\xi_1 = 0$.

2. When $(x, y) \neq (0, 0)$,
$$D_1 f(x, y) = \frac{y(x^4 + 4x^2 y^2 - y^4)}{(x^2 + y^2)^2}.$$

Since $f(x, 0) = 0$ for all x, $D_1 f(0, 0) = 0$. $D_1 f$ is continuous at points $(x, y)^T \neq (0, 0)^T$. $|D_1 f(x, y)| \leqslant 3|y|$ and so $D_1 f$ is continuous at $(0, 0)^T$ also. $D_2 f(x, y) = -D_1 f(y, x)$ has same properties. Hence f is differentiable.

All second order partial derivatives exist in $R^2 - \{\theta\}$. Also
$$D_2 D_1 f(0, 0) = \lim_{y \to 0} \{D_1 f(0, y) - D_1 f(0, 0)\}/y = \lim_{y \to 0} (-y)/y = -1.$$

Similarly $D_1 D_2 f(0, 0) = 1$, $D_1^2 f(0, 0) = D_2^2 f(0, 0) = 0$.

3. Both sides equal $\Delta(h, k)/(hk)$; e.g. if $g(t) = f(\xi + th, \eta + k) - f(\xi, \eta + k)$, there exists α in $(0, 1)$ such that $\Delta(h, k) = g(1) - g(0) = g'(\alpha)$. Again, by mean value theorem,

$$\{D_1 f(\xi + \alpha h, \eta + k) - D_1 f(\xi + \alpha h, \eta)\}/k = D_2 D_1 f(\xi + \alpha h, \eta + \beta' k)$$

and r.h.s. $\rightarrow D_2 D_1 f(\xi, \eta)$ as $(h, k)^T \rightarrow (0, 0)^T$. Thus, given ϵ, there exists δ such that

$$|\{D_2 f(\xi+h, \eta+\beta k) - D_2 f(\xi, \eta+\beta k)\}/h - D_2 D_1 f(\xi, \eta)| < \epsilon$$

if $0 < |h|, |k| < \delta$. Take $0 < |h| < \delta$ and let $k \rightarrow 0$. Then

$$|\{D_2 f(\xi+h, \eta) - D_2 f(\xi, \eta)\}/h - D_2 D_1 f(\xi, \eta)| \leqslant \epsilon.$$

4. Let g be a real function differentiable at 0 only (e.g. $g(x) = x^2$ when x is rational, 0 when x is irrational). Define f by $f(x, y) = g(x) + g(y)$. Then $D_1 f(0, y) = D_2 f(x, 0) = g'(0)$ for all x, y; and $D_2 D_1 f(0, 0) = D_1 D_2 f(0, 0) = 0$. All three sets of conditions involve the existence of $D_1 f$, $D_2 f$ in an open set containing $(0, 0)^T$, but $D_1 f(x, y)$ exists only when $x = 0$, $D_2 f(x, y)$ only when $y = 0$.

5. The continuity of the kth order partial derivatives implies the differentiability of the partial derivatives of order $k-1$. Hence f and its partial derivatives up to order $p-1$ are differentiable. Then, by induction and the chain rule, for $k = 1, ..., p$, $g^{(k)}(t)$ ($|t| < r/\|h\|$) exists and equals $D^{(k)} f(\xi)(h)^k$. By C1, 81, there exists $\alpha \in (0, 1)$ such that

$$g(1) = g(0) + g'(0) + ... + g^{(p-1)}(0)/(p-1)! + g^{(p)}(\alpha)/p!.$$

6. By 5, for $\|h\| < \delta$, say, $f(\xi+h) - f(\xi) = \frac{1}{2} D^2 f(\xi+\alpha h)(h)^2$.

(i) If $H_k f(\xi) > 0$ ($k = 1, ..., m$), by continuity, $H_k f(\xi+h) > 0$ when $\|h\| < \delta'(\leqslant \delta)$. Hence $f(\xi+h) - f(\xi) > 0$ for $0 < \|h\| < \delta'$.

(ii) Similar.

(iii) Take any $h \in R^m$. When $0 < t\|h\| < \delta$,

$$\Delta(t) = f(\xi+th) - f(\xi) = \frac{1}{2} D^2 f(\xi+\alpha th)(th)^2.$$

As $t \rightarrow 0$, $\Delta(t)/t^2 \rightarrow \frac{1}{2} D^2 f(\xi)(h)^2$.

If f has a minimum at ξ, $\Delta(t) \geqslant 0$ for all sufficiently small t and so $\frac{1}{2} D^2 f(\xi)(h)^2 = \lim \Delta(t)/t^2 \geqslant 0$. Since this holds for all h, $D^2 f(\xi)(.)^2$ is positive semi definite. If f has a maximum at ξ, then $D^2 f(\xi)(.)^2$ is negative semi-definite. But, if (iii) holds, $D^2 f(\xi)(.)^2$ is indefinite.

7. f has stationary points $(0, 0, 0)^T$, $(-1, -1, -1)^T$, $(-1, 1, 1)^T$, $(1, -1, 1)^T$, $(1, 1, -1)^T$. At $(0, 0, 0)^T$, $H_1 f = 2$, $H_2 f = 4$, $H_3 f = 8$. Elsewhere $H_1 f = 2$, $H_2 f = 0$, $H_3 f = -24$. Minimum at $(0, 0, 0)^T$; no other maxima or minima.

Infimum of f in $x^2 + y^2 + z^2 \leqslant a^2$ taken at $(0, 0, 0)^T$ or on $x^2 + y^2 + z^2 = a^2$. f under constraint $x^2 + y^2 + z^2 = a^2$ equivalent to $a^2 + 2xyz$ under same constraint. Critical points $(\pm a/\sqrt{3}, \pm \sqrt{a}/\sqrt{3}, \pm a/\sqrt{3})^T$, $(\pm a, 0, 0)^T$, $(0, \pm a, 0)^T$, $(0, 0, \pm a)^T$. $f(\pm a, 0, 0) = a^2$, $f(\pm a/\sqrt{3}, \pm a/\sqrt{3}, \pm a/\sqrt{3}) = a^2 \pm 2a^3/(3\sqrt{3})$. When $a^2 \leqslant \frac{27}{4}$, infimum 0; when $a^2 > \frac{27}{4}$, infimum $a^2 - 2a^3/(3\sqrt{3})$.

8. Stationary points $(0, 0)^T$, $(4, 4)^T$, $(-4, 4)^T$. Maximum at $(0, 0)^T$; neither maximum nor minimum at $(4, 4)^T$, $(-4, 4)^T$. If f attained its infimum on K at an interior point, this would have to be a minimum.

8 (a)

1. (1) $\phi \sim \phi$. (2) $\phi \sim \psi \Rightarrow \phi = \psi \circ \alpha$, where α satisfies (i), (ii), (iii). Since α strictly increases, α^{-1} exists, so that $\psi = \phi \circ \alpha^{-1}$. Clearly α^{-1} satisfies (i), (ii), (iii). (3) $\phi \sim \psi$ and $\psi \sim \chi \Rightarrow \phi = \psi \circ \alpha$ and $\psi = \chi \circ \beta \Rightarrow \phi = \chi \circ (\beta \circ \alpha)$. If α, β satisfy (i), (ii), (iii), so does $\beta \circ \alpha$.

2. Define $\alpha: [0, 3] \to [a, b]$ by

$$\alpha(t) = \begin{cases} a + \frac{1}{2}(b-a)t & (0 \leqslant t \leqslant 1), \\ \frac{1}{4}(3a+b) + \frac{1}{4}(b-a)t & (1 \leqslant t \leqslant 3) \end{cases}$$

and put $\psi = \phi \circ \alpha$. Then $\psi \sim \phi$, but ψ is not differentiable at 1.

3. Let $\phi^i, \psi^i : [i-1, i] \to R^n$ be (equivalent) representations of γ_i $(i = 1, ..., m)$ and define $\phi, \psi : [0, m] \to R^n$ by

$$\phi(t) = \phi^i(t), \quad \psi(t) = \psi^i(t) \quad (i-1 \leqslant t \leqslant i).$$

There exist $\alpha^i : [i-1, i] \to [i-1, i]$ such that $\phi^i = \psi^i \circ \alpha^i$, $\alpha^i(i-1) = i-1$, $\alpha^i(i) = i$ and (ii), (iii) are satisfied. Since $\alpha^i(i) = \alpha^{i+1}(i)$, we can define $\alpha : [0, m] \to [0, m]$ by $\alpha(t) = \alpha^i(t)$ $(i-1 \leqslant t \leqslant i)$. For $i-1 \leqslant t \leqslant i$, $i-1 \leqslant \alpha(t) = \alpha^i(t) \leqslant i$ and so $\phi(t) = \phi^i(t) = \psi^i(\alpha^i(t)) = \psi(\alpha^i(t)) = \psi(\alpha(t))$. Since this holds for all i, $\phi = \psi \circ \alpha$. Also α satisfies (ii), (iii).

4. Let ϕ^1, ϕ^2 be the restrictions of ϕ to $[a, u], [u, b]$. Define $\alpha^1 : [0, 1] \to [a, u]$, $\alpha^2 : [1, 2] \to [u, b]$ by

$$\alpha^1(t) = a + (u-a)t \quad (0 \leqslant t \leqslant 1), \qquad \alpha^2(t) = u + (b-u)(t-1) \quad (1 \leqslant t \leqslant 2).$$

Define ψ^1 on $[0, 1]$, ψ^2 on $[1, 2]$, ψ on $[0, 2]$ by $\psi^i = \phi \circ \alpha^i$ $(i = 1, 2)$, $\psi(t) = \psi^i(t)$ for $i-1 \leqslant t \leqslant i$ $(i = 1, 2)$. Then $\psi^1 \sim \phi^1$, $\psi^2 \sim \phi^2$ and clearly $\psi \sim \phi$. Since ψ represents $\gamma_1 + \gamma_2$ and ϕ represents γ, $\gamma_1 + \gamma_2 = \gamma$.

5. If ϕ on $[a, b]$ is any representation of γ and $\mathscr{D}$ is the dissection of $[a, b]$ with no intermediate points, $\|q - p\| = \Lambda(\mathscr{D}, \phi) \leqslant \Lambda_a^b(\phi)$.

6. Trace $\{(x, 0) \mid -1 \leqslant x \leqslant 1\}$. $\Lambda(\gamma_1) = 2$. $\Lambda(\gamma_2) = 6$. ϕ_1^3 not of bounded variation.

7. $\Lambda(\gamma) = \int_0^{2q\pi} 2\left(\frac{p}{q}+1\right) a \left|\cos\frac{p}{2q}t\right| dt = 4p\left(\frac{p}{q}+1\right) a \int_0^{q\pi/p} \cos\frac{p}{2q}t \, dt = 8(p+q)a.$

8. $\Lambda(\gamma) = \int_0^2 \sqrt{(2+t^2)} dt = \sqrt{6} + \log(\sqrt{2} + \sqrt{3}).$

9. Let $a \leqslant u < v \leqslant b$. By theorem 8.12, $s(v) - s(u) = \Lambda_u^v(\phi)$ and, by (8.11), $0 \leqslant \Lambda_u^v(\phi) \leqslant V_u^v(\phi_1) + ... + V_u^v(\phi_n)$. R.h.s. $\to 0$ as $v - u \to 0$, by theorem 6.44.

10. Adapt the proof of exercise 6(g), 3, using **9**.

11. Since $|f| \leqslant 1$, the series defining ϕ, ψ converge uniformly.

If $\alpha_1 = 0$, then $0 \leqslant \tau \leqslant \frac{1}{3}$ and $f(\tau) = 0$; if $\alpha_1 = 1$, then $\frac{2}{3} \leqslant \tau \leqslant 1$ and $f(\tau) = 1$. In either case, $f(\tau) = \alpha_1$.

$$3^2\tau = \text{even number} + 0 \cdot (2\alpha_2)(2\beta_2)...(\text{in ternary scale}).$$

Hence $f(3^2\tau) = \alpha_2$. In general $f(3^{2n-2}\tau) = \alpha_n$ and so $\phi(\tau) = \xi$. Similarly $\psi(\tau) = \eta$.

8 (b)

1. Since $\phi \sim \psi$, $\phi = \psi \circ \alpha$, where α has usual properties. There is a dissection $a = a_0 < a_1 < \ldots < a_r = b$ such that ϕ, α are continuously differentiable in each $[a_{j-1}, a_j]$ and ψ is so in each $[c_{j-1}, c_j]$, where $c_j = \alpha(a_j)$. By theorem 6.92,

$$\int_{c_{j-1}}^{c_j} (f_i \circ \psi) \psi_i' = \int_{a_{j-1}}^{a_j} \{[(f_i \circ \psi) \psi_i'] \circ \alpha\} \alpha'$$

$$= \int_{a_{j-1}}^{a_j} [(f_i \circ \psi) \circ \alpha](\psi_i' \circ \alpha) \alpha' = \int_{a_{j-1}}^{a_j} (f_i \circ \phi) \phi_i'.$$

Second part similar, for $(|\psi_i'| \circ \alpha) \alpha' = |\phi_i'|$ since $|\alpha'| = \alpha'$.

2. In an interval $[a_{j-1}, a_j]$ of continuous differentiability of ϕ,

$$\int_{-a_j}^{-a_{j-1}} f_i(\phi(-t))(-\phi_i'(-t)) \, dt = \int_{a_j}^{a_{j-1}} f_i(\phi(t)) \phi_i'(t) \, dt.$$

3. ϕ on $[a, b]$ represents γ, $\mathscr{D}$ is a dissection of $[a, b]$. Then

$$|\Sigma f\{\phi(\tau_j)\} . \{\phi(t_j) - \phi(t_{j-1})\}| \leqslant \Sigma \|f\{\phi(\tau_j)\}\| \, \|\phi(t_j) - \phi(t_{j-1})\|$$

$$\leqslant \Sigma \|f\{\phi(\tau_j)\}\| \{s_\phi(t_j) - s_\phi(t_{j-1})\} \leqslant M \Lambda(\gamma).$$

As $\mu(\mathscr{D}) \to 0$, first term $\to \left| \int_\gamma f \right|$, third term $\to \int_\gamma \|f\| \, ds$.

4. $\gamma_1 : x = t, y = 0 \ (0 \leqslant t \leqslant 1); \quad \int_{\gamma_1} = 0.$

$\gamma_2 : x = 1, y = t \ (0 \leqslant t \leqslant 1); \quad \int_{\gamma_2} = (1/\sqrt{2}) \arctan(1/\sqrt{2}).$

$\gamma_3 : x = t, y = t \ (0 \leqslant t \leqslant 1); \quad \int_{\gamma_3} = 1 - (1/\sqrt{2}) \arctan \sqrt{2}.$

$$\int_\gamma = \int_{\gamma_1 + \gamma_2 - \gamma_3} = \pi/(2\sqrt{2}) - 1.$$

5. Trace $= \{(x, y, z) | x^2 + y^2 = 2ax\} \cap \{(x, y, z) | x^2 + y^2 + z^2 = 4ax\}$. $\int_\gamma = 8a^3\pi.$

6. $\int_\gamma = \frac{1}{16}\pi^2 abc$, $\int_\sigma = -\frac{1}{12}\pi abc.$

7. ϕ on $[a, b]$ represents γ; $a = a_0 < a_1 < \ldots < a_r = b$ are such that ϕ is continuously differentiable in each $[a_{j-1}, a_j]$. Then

$$\int_\gamma f = \int_a^b (f \circ \phi) . \phi' = \sum_{j=1}^r \int_{a_{j-1}}^{a_j} \left(\sum_{i=1}^n f_i\{\phi(t)\} \phi_i'(t) \right) dt$$

$$= \sum_{j=1}^r \int_{a_{j-1}}^{a_j} \frac{d}{dt} g\{\phi(t)\} \, dt = \sum_{j=1}^r [g\{\phi(t_j)\} - g\{\phi(t_{j-1})\}]$$

$$= g(v) - g(u).$$

8. By hypothesis, g is uniquely defined. If $x \in D$, $B(x; \delta) \subset D$ for some δ. For $|t| < \delta$, $x + te^i \in D$, where e^i has ith component 1, others 0. If $|h| < \delta$, let $\gamma_i(h)$ be the segment from x to $x + he^i$. Then

$$g(x+he^i) = \int_{\gamma_x + \gamma_i(h)} f = g(x) + \int_0^h f_i(x + te^i)\,dt,$$

i.e.

$$\frac{g(x+he^i) - g(x)}{h} = \frac{1}{h}\int_0^h f_i(x + te^i)\,dt.$$

Since f is continuous, r.h.s. $\to f_i(x)$ as $h \to 0$.

9. $l = c \sinh(a/c)$. $\mathrm{sech}^2 x < 1$ $(x > 0)$ $\Rightarrow$ $\tanh(a/c) < (a/c)$ $\Rightarrow$ $(d/dc)c \sinh(a/c) < 0$ for $c > 0$. Hence $c \sinh(a/c)$ decreases as c increases. Since $c \sinh(a/c) \to \infty$ as $c \to 0+$ and $\to a$ as $c \to \infty$, the equation in c, $c \sinh(a/c) = l$ $(> a)$ has a unique solution.

10. 3514/25.

8 (c)

1. To simplify notation, take $a_i = 0$, $b_i = 1$ $(i = 1, \dots, n)$. Let $\mathscr{D}_r$ be the dissection of $[0, 1] \times \dots \times [0, 1]$ into r^n congruent subintervals. If $M = \sup |f|$,

$$S(\mathscr{D}_r) - s(\mathscr{D}_r) = \frac{1}{r^n} \sum_{i_1, \dots, i_n = 1}^{r} \left\{ f\left(\frac{i_1}{r}, \dots, \frac{i_n}{r}\right) - f\left(\frac{i_1 - 1}{r}, \dots, \frac{i_n - 1}{r}\right) \right\}$$

$$= \frac{1}{r^n} \left\{ \sum_{\text{some } i_k = n} f\left(\frac{i_1}{r}, \dots, \frac{i_n}{r}\right) - \sum_{\text{some } i_k = 0} f\left(\frac{i_1}{r}, \dots, \frac{i_n}{r}\right) \right\} \leqslant \frac{2nr^{n-1}M}{r^n}$$

$$= \frac{2nM}{r} \to 0 \quad \text{as} \quad r \to \infty.$$

2. There is a dissection $\mathscr{D}'$ of I into subintervals $I_1, \dots, I_r$ such that

$$S(\mathscr{D}', f) - s(\mathscr{D}', f) = \sum_{l=1}^{r} (M_l - m_l) V(I_l) < \epsilon.$$

Let $\mathscr{D}$ be the dissection of K into the subintervals $K_l = I_l \times J$. Since $V(K_l) = V(I_l)V(J)$, $S(\mathscr{D}, \phi) - s(\mathscr{D}, \phi) < \epsilon V(J)$.

Similarly $\psi : R^m \times J \to R^1$, defined by $\psi(x, y) = g(y)$, is integrable over K. Hence $h = \phi\psi$ is integrable over K. Also $\int_I h(x, y)\,dx = g(y)\int_I f(x)\,dx$. Use theorem 8.31.

5. Inner integral is elliptic, not expressible by elementary functions. But integrand continuous, so interchange of order of integration legitimate. Integral $= \frac{2}{3}(\tan\alpha - \sec\alpha - \cos\alpha + 2)$.

6. Define $\phi : [0, 1] \to R^1$ by $\phi(x) = 0$ (x irrational), $\phi(x) = 1/q$ ($x = p/q, p, q$ coprime). By exercise 6(h), 9, $\int_0^1 \phi$ exists and is 0. Let $\mathscr{D}'$ be a dissection of $[0, 1]$ into subintervals $[a_{i-1}, a_i]$ such that $S(\mathscr{D}', \phi) < \epsilon$. Let $f^* : [0, 1] \times [0, 1] \to R^1$ be given by $f^*(x, y) = \phi(x)$; let $\mathscr{D}$ be the dissection of I into subintervals $[a_{i-1}, a_i] \times [0, 1]$. Then $s(\mathscr{D}, f) = 0$ and $S(\mathscr{D}, f) \leqslant S(\mathscr{D}, f^*) = S(\mathscr{D}', \phi) < \epsilon$.

By exercise 6(h), 9, $\int_0^1 f(x, y)\,dx = 0$ for every y.

By illustration (ii), p. 141, $\int_0^1 f(x, y)\,dy$ does not exist for any x.
$h = f+g$, where g is defined like f, but with x, y interchanged.

7. $\int_0^\pi f(x, y)\,dx = 0$ for every y; $\int_0^1 f(x, y)\,dy$ exists for $x = \frac{1}{2}\pi$ only.
Let $\mathscr{D}$ be a dissection of $J = [0, \frac{1}{4}\pi] \times [0, 1]$ into subintervals $J_1, \ldots, J_r$. Then

$$S(\mathscr{D}, f) - s(\mathscr{D}, f) \geqslant \sum_{l=1}^r \left(\frac{1}{\sqrt{2}} - 0\right) V(J_l) = \frac{\pi}{4\sqrt{2}}.$$

Hence f is not integrable over J, and so not over I.

8. Any $[x, x+\delta] \times [y, y+\delta] \subset I$ contains a point of P_k, where $p_k > 1/\delta$. If $a = m/p$ (p prime), then the lines $x = a$, $y = a$ each contain $p-1$ points of P; if a not of this form, the lines contain no points of P.

For every $x, f(x, y) = 0$ for all except a finite number of y. Thus $\overline{\int_0^1} f(x, y)\,dy = 0$,

$\int_0^1 dx \overline{\int_0^1} f(x, y)\,dy = 0$. Since P is dense in I, $\overline{\int_I} f = 1$, $\underline{\int_I} f = 0$.

9. If $0 < y < 1$, $\int_0^1 f(x, y)\,dx = 1$. Hence $\int_0^1 dy \int_0^1 f(x, y)\,dx = 1$ ($y = 0, 1$ need

not be considered). $\int_0^1 dx \int_0^1 f(x, y)\,dy = -1$.

8 (d)

1. Let $I_1 \supset E$; let I_1, I_2 be non-overlapping with common face. Then $f_E = 0$ in $I_2 - (I_1 \cap I_2)$, so integral of f_E over I_2 is 0. Use (8.33).

2. Call interval I; $I_\epsilon = [a_1+\epsilon, b_1-\epsilon] \times \ldots \times [a_n+\epsilon, b_n-\epsilon]$, ($\epsilon > 0$, small). Then $I_\epsilon \subset I \subset \bar{I}$, $c(I_\epsilon) \leqslant \underline{c}(I) \leqslant \overline{c}(I) \leqslant c(\bar{I})$.

3. Let $m = \inf f$, $M = \sup f$, I any closed interval containing E. Then

$$0 = mc(E) = \int_I m\chi_E \leqslant \underline{\int_I} f_E \leqslant \overline{\int_I} f_E \leqslant \int_I M\chi_E = Mc(E) = 0.$$

4. (i) $f \equiv 1$ is not integrable over a set without content.
Let $A \subset E \subset I$, where A has content, I is a closed interval. If $\mathscr{D}$ is a dissection of I into subintervals I_j,

$$S(\mathscr{D}, f_A) - s(\mathscr{D}, f_A) = (\Sigma_1 + \Sigma_2 + \Sigma_3)(M_j - m_j) c(I_j),$$

where Σ_1 ranges over the $I_j \subset A^\circ$, Σ_2 over the $I_j \subset I - \bar{A}$, Σ_3 over the I_j with points of fr A. Then $\Sigma_1 \leqslant S(\mathscr{D}, f_E) - s(\mathscr{D}, f_E)$, $\Sigma_2 = 0$, $\Sigma_3 \leqslant \{S(\mathscr{D}, \chi_A) - s(\mathscr{D}, \chi_A)\}(\sup f_E - \inf f_E)$. $\Sigma_1, \Sigma_3 \to 0$ as $\mu(\mathscr{D}) \to 0$.
(ii) $f_{E-X} = f_E - f_X$.

5. (i) Replace function χ_E of exercise 6(h), 5 by f_E.
(ii) f integrable over $E_1 \cap E_2$, by first part, and so over $E_1 - E_2 = E_1 - (E_1 \cap E_2)$, by 4 (ii).
(iii) Put $f = f^+ - f^-$, where $f^+ = \frac{1}{2}(|f| + f) \geqslant 0$, $f^- = \frac{1}{2}(|f| - f) \geqslant 0$.

6. Let $m = \inf f$, $M = \sup f$ in E, I a closed interval containing E. As in **3**,

$$m\underline{c}(X_k) \leqslant \underline{\int_I} f_{X_k} \leqslant \overline{\int_I} f_{X_k} \leqslant M\bar{c}(X_k)$$

and so integrals $\to 0$. Also, since $f_E = f_{E-X_k} + f_{X_k}$,

$$\overline{\int_I} f_E \leqslant \overline{\int_I} f_{E-X_k} + \overline{\int_I} f_{X_k}, \quad \underline{\int_I} f_E \geqslant \underline{\int_I} f_{E-X_k} + \underline{\int_I} f_{X_k}$$

and so $\qquad \overline{\int_I} f_E \leqslant \lim \inf \int_I f_{E-X_k} \leqslant \lim \sup \int_I f_{E-X_k} \leqslant \underline{\int_I} f_E.$

7. (i) If $m = \inf f$, $M = \sup f$, I a closed interval containing E,

$$\int_I m\chi_E \leqslant \int_I f_E \leqslant \int_I M\chi_E.$$

(ii) Since E is compact and connected, so is $f(E) \subset R^1$. Hence $f(E)$ is the finite closed interval $[m, M]$.

8. For $x, y \in I$, put $\|x - y\| = \delta$, $\min(x_i, y_i) = u_i$, $\max(x_i, y_i) = v_i$ $(i = 1, ..., n)$. Then $0 \leqslant v_i - u_i \leqslant \delta$. If $M = \sup |f|$ in I, by **5**,

$$\left| \int_{J_x} f - \int_{J_y} f \right| = \left| \int_{(J_x - J_y) \cup (J_y - J_x)} f \right|$$

$$\leqslant \int_{J_v - J_u} |f|$$

$$\leqslant Mc(J_v - J_u)$$

$$= O(\delta) \quad \text{as} \quad \delta \to 0.$$

9. Let D be closed disc with centre (a, b), radius r. Then χ_D is continuous except on path given by $x = a + r\cos t$, $y = b + r\sin t$ $(0 \leqslant t \leqslant 2\pi)$. Hence $c(D)$ exists and, if $I = [a-r, a+r] \times [b-r, b+r]$,

$$c(D) = \int_I \chi_D = \int_{a-r}^{a+r} dx \int_{b-r}^{b+r} \chi_D(x, y)\, dy = \int_{a-r}^{a+r} 2\sqrt{(r^2 - (x-a)^2)}\, dx = \pi r^2.$$

Since $c(\text{fr } D) = 0$, $c(D^\circ) = c(D)$.

10. (i) There is a dissection of I into subintervals I_l such that $\Sigma(M_l - m_l)c(I_l) < \epsilon$. If $J_l = I_l \times [m_l, M_l + \epsilon]$ (note m_l may equal M_l), $E \subset \cup J_l$ and

$$\Sigma c(J_l) = \Sigma(M_l - m_l + \epsilon)c(I_l) < \epsilon + \epsilon c(I).$$

(ii) Let

$$D_0 = \{(x_1, ..., x_{n+1}) | (x_1, ..., x_n) \in E^\circ, 0 < x_{n+1} < f(x_1, ..., x_n)\},$$

$$D_1 = \{(x_1, ..., x_{n+1}) | (x_1, ..., x_n) \in \bar{E}, 0 \leqslant x_{n+1} \leqslant f(x_1, ..., x_n)\}.$$

D_0 is open, D_1 is closed. Since χ_D is constant in D_0 and in D_1', χ_D is continuous on $D_0 \cup D_1'$.

Let I be a closed interval containing E, S the set in (i) corresponding to f_E. If $q > \sup f$,
$$(D_0' \cap D_1) \subset (I \times \{0\}) \cup S \cup (\text{fr } E \times [0, q])$$

and sets on right have zero content. (fr E has content 0 in R^n.) Hence χ_D is integrable over $J = I \times [0, q]$ and

$$c(D) = \int_J \chi_D = \int_I dx_1 \ldots dx_n \int_0^q \chi_D(x_1, \ldots, x_{n+1}) \, dx_{n+1}$$

$$= \int_I dx_1 \ldots dx_n \int_0^{f_E(x_1, \ldots, x_n)} 1 \, dx_{n+1} = \int_I f_E = \int_E f.$$

11. In both parts the subsets of R^2 have content. (i) $\dfrac{4a^3}{3b}$, (ii) $\dfrac{a^3}{b} \left(\dfrac{8}{3} - \dfrac{\pi}{2} \right)$.

13. Let E be triangle with vertices $(0, 0)$, $(1, 0)$, $(1, 1)$. Since integrand is bounded on E and continuous except when $x = 0$,

$$\int_0^1 dy \int_y^1 e^{-y/x} dx = \int_E e^{-y/x} dx \, dy = \int_0^1 dx \int_0^x e^{-y/x} dy = \frac{e-1}{2e}.$$

14. E has content and integrand f continuous on E. If

$$\Delta_x = \{(y, z) | y \geqslant 0, z \geqslant 0, y + z \leqslant 1 - x\},$$

$$\int_E f = \int_0^1 dx \int_{\Delta_x} f \, dy \, dz = \int_0^1 dx \int_0^{1-x} dy \int_0^{1-x-y} f \, dz = \frac{4}{5} - \frac{\pi}{4}.$$

15. ϕ on $I = [a, b] \times [a, b]$ given by $\phi(t, u) = f(t)$ is integrable over I (exercise 8(c), 2). Hence $(x - t)^{n-1} f(t)$ is integrable over I and so over triangle with vertices (a, a), (b, a), (b, b). Then

$$(n-1)! \int_a^x f_n(u) \, du = \int_a^x du \int_a^u (u - t)^{n-1} f(t) \, dt = \int_a^x dt \int_t^x (u - t)^{n-1} f(t) \, du.$$

8 (e)

1. σ differentiable at x: as $h \to 0+$,

$$\frac{f(x+h) - f(x)}{h} = \frac{\sigma([x, x+h])}{c([x, x+h])}, \quad \frac{f(x) - f(x-h)}{h} = \frac{\sigma([x-h, x])}{c([x-h, x])} \to D\sigma(x).$$

f differentiable at x: $|f(x+t) - f(x) - t f'(x)| \leqslant \epsilon t$ if $|t| < \delta$. Hence, when $0 \leqslant h, k < \delta$, $h + k > 0$,

$$\left| \frac{\sigma([x-h, x+k])}{c([x-h, x+k])} - f'(x) \right|$$

$$\leqslant \frac{|f(x+k) - f(x) - k f'(x)| + |f(x) - f(x-h) - h f'(x)|}{h + k} \leqslant \epsilon.$$

2. $c(X \cap X) = c(X) = 0$, so $\sigma(X) = \sigma(X \cup X) = \sigma(X) + \sigma(X)$.

8 (f)

1. (i) If there exists $x \in I(u; a)$ such that $x \notin E$, then the segment from u to x (which lies in $I(u; a)$) contains a frontier point of E. Impossible.

(ii) If there exists $x \in (R^n - I(u; b)) \cap E$, consider the half line λ from u through x. Since E is bounded, there exists $y \in \lambda - I(u; b)$ such that $y \notin E$. Hence the segment from x to y (which lies outside $I(u; b)$) contains a frontier point of E. Impossible.

(Clearly the cubes could be replaced by convex sets S_1, S_2 with $S_1 \subset S_2$ and $E \cap S_1 \neq \varnothing$.)

2. If X is the set of points of discontinuity, proceed as in (ii) of proof of theorem 8.61 (8.62), but let S_k cover $Q \cup X$ ($Q \cup X \cup \mathrm{fr}\, G$).

3. Let AB be side in question, C opposite vertex. A rigid body motion takes A, B, C to $(a, 0), (b, 0), (p, q)$, where $q > 0$. Let $a \leqslant p \leqslant b$. Since segment from $(p, 0)$ to (p, q) has content 0,

$$\text{content} = \int_a^p \frac{q(x-a)}{p-a}\, dx + \int_p^b \frac{q(x-b)}{p-b}\, dx = \tfrac{1}{2}q(b-a).$$

Cases $p < a, p > b$ similar.

5. $\tfrac{1}{4}\pi ab(a^2/p^2 + b^2/q^2)$.

6. Clearly ϕ is continuously differentiable and injective in R^2. $E = \phi^{-1}(D)$ $= \{(u, v) | 0 \leqslant u \leqslant v \leqslant 1\}, J\phi \equiv 1$. Integral $= \tfrac{1}{4}\pi - \tfrac{1}{2}$.

7. $J\phi^{-1}(x, y) = 8xy$, so $J\phi(u, v) = \tfrac{1}{8}(uv)^{-\frac{1}{2}}$. Content $\tfrac{1}{2}(b-a)(q-p)$.

9. $c(S) = \int_0^{2\pi} d\theta \int_0^{2a(1+\cos\theta)} r\, dr = 6\pi a^2$; centroid $(\tfrac{5}{3}a, 0)$.

10. If D is the disc with centre $(-2a, -2a)$ and radius $\sqrt{40}a$, then

$$c(Q) = \int_D \left(\frac{(8a-x-y)}{2} - \frac{x^2+y^2}{8a} \right) dx\, dy = 100\pi a^3.$$

11. (i) If, in R^2, $D = \{(x, y) | x^2 + y^2 \leqslant \tfrac{1}{4}a^2\}$,

$$\text{content} = 2\int_D \sqrt{(a^2 - x^2 - y^2)}\, dx\, dy = \tfrac{1}{6}(8 - 3\sqrt{3})\pi a^3.$$

(ii) If, in R^2, $E = \{(r, \theta) | 0 \leqslant r \leqslant a\sin\theta, 0 \leqslant \theta \leqslant \pi\}$,

$$\text{content} = 2\int_E \sqrt{(a^2 - r^2)}\, r\, dr\, d\theta = \tfrac{2}{3}a^3 \int_0^\pi (1 - |\cos^3\theta|)\, d\theta$$

$$= \tfrac{4}{3}a^3 \int_0^{\frac{1}{2}\pi} (1 - \cos^3\theta)\, d\theta = \tfrac{2}{9}(3\pi - 4)a^3.$$

12. (i) Given $x_1, \ldots, x_n$, $x_1^2 + \ldots + x_n^2 = r^2$ determines r; then x_n determines $\theta_{n-1}, \ldots, x_3$ determines θ_2, x_2 and x_1 determine θ_1.

Let J_n be determinant with $r = 1$, so that $J = r^{n-1}J_n$. Multiply first row of J_n by $\cos\theta_{n-1}/\sin\theta_{n-1}$ and subtract from last row. This becomes $0, \ldots, 0$, $-1/\sin\theta_{n-1}$, and $\sin\theta_{n-1}$ factor of all elements in first $n-1$ columns. $J_n = -\sin^{n-2}\theta_{n-1}J_{n-1}$. Since $J_2 = 1$, result follows.

(ii) Use theorem 8.62 and note that, if $I_k = \int_0^\pi \sin^k\theta\, d\theta$,

$$I_{2k-1}I_{2k} = \pi/k, \quad I_{2k}I_{2k+1} = 2\pi/(2k+1).$$

13. $(19 - 24 \log 2)/36$.

14. $\displaystyle\iint_B D_1 f_2 = \int_0^{2\pi} dt \int_0^1 D_1 f_2(r\cos t, r\sin t) r\, dr$

$\displaystyle = \int_0^{2\pi} dt \int_0^1 \left\{ \frac{\partial}{\partial r}[f_2(r\cos t, r\sin t) r\cos t] - \frac{\partial}{\partial t}[f_2(r\cos t, r\sin t)\sin t] \right\} dr$

$\displaystyle = \int_0^{2\pi} f_2(\cos t, \sin t)\cos t\, dt - \int_0^1 dr \int_0^{2\pi} \frac{\partial}{\partial t}[f_2(r\cos t, r\sin t)\sin t]\, dt$

$\displaystyle = \int_0^{2\pi} f_2(\cos t, \sin t)\cos t\, dt,$

by theorem 6.84. Similarly

$$\iint_B D_2 f_1 = \int_0^{2\pi} f_1(\cos t, \sin t)\sin t\, dt.$$

8 (g)

1. For every $X > a$, $\phi_X = \displaystyle\int_a^X f(x, .)\, dx$ is continuous on the compact set J. Adapt the proof of theorem 5.35.

2. Since $\phi = \displaystyle\int_a^x f(u, .)\, du$, $\psi = f(., y)$ are continuous,

$$D_2 F(x, y) = \frac{\partial}{\partial y} \int_b^y \left(\int_a^x f(u, v)\, du \right) dv = \int_a^x f(u, y)\, du,$$

$$D_1 D_2 F(x, y) = \frac{\partial}{\partial x} \int_a^x f(u, y)\, du = f(x, y).$$

3. Let $\phi(\eta)$ exist, where $\eta = (\eta_1, ..., \eta_n)$. Take any $y = (y_1, ..., y_n)$ in J. Then
$f(x, y) - f(x, \eta)$

$$= \int_{\eta_1}^{y_1} D_{m+1} f(x, t, y_2, ..., y_n)\, dt + \int_{\eta_2}^{y_2} D_{m+2} f(x, \eta_1, t, ..., y_n)\, dt$$

$$+ ... + \int_{\eta_n}^{y_n} D_{m+n} f(x, \eta_1, \eta_2, ..., t)\, dt$$

and r.h.s. is continuous on I (theorem 8.71). Hence $f(., y)$ is integrable over I. Formula for and continuity of $D_k\phi$ follow from theorems 8.71, 8.72.

4. Define f on $[a, b] \times [c, d]$ by $f(x, y) = 1$ (x rational), 0 (x irrational). Then $D_2 f\ (\equiv 0)$ is continuous, but $f(., y)$ is not integrable for any y.

6. When showing $\phi'(a) = 0$ distinguish between $a = 0$, $a \neq 0$. Since $\phi(0) = 0$, $\phi(a) = 0$ for $|a| < 1$. For $|a| > 1$,

$$\phi(a) = \phi(1/a) + \int_0^\pi \log a^2\, dx = 2\pi \log |a|.$$

7. If ψ on $[c, d] \times [c, d]$ is given by $\psi(y, z) = \int_a^z f(x, y)\,dx$, then

$$D_1\psi(y, z) = \int_a^z D_2 f(x, y)\,dx, \quad D_2\psi(y, z) = f(z, y);$$

and
$$\phi'(y) = (d/dy)\psi\{y, g(y)\}$$
$$= D_1\psi\{y, g(y)\} + g'(y)D_2\psi\{y, g(y)\}.$$

8. Since B is a region, independence of path of $\int_\gamma f$ is equivalent to existence of g such that $f = \operatorname{grad} g$ ($8(b)$, 7, 8).

(N) If $f = \operatorname{grad} g$, then $D_i f_j = D_i D_j g$, $D_j f_i = D_j D_i g$; $D_i D_j g = D_j D_i g$, since both are continuous.

(S) Let $c = (c_1, ..., c_n)$ be centre of B. For any $x = (x_1, ..., x_n) \in B$, let γ_x be the path from c to x consisting of the segments determined by the points

$$(c_1, ..., c_n), (c_1, ..., c_{n-1}, x_n), ..., (c_1, x_2, ... x_n), (x_1, ..., x_n).$$

Define $g : B \to R^1$ by

$$g(x) = \int_{\gamma_x} f = \int_{c_1}^{x_1} f_1(t, x_2, ..., x_n)\,dt + ... + \int_{c_n}^{x_n} f_n(c_1, ..., c_{n-1}, t)\,dt.$$

Then $D_1 g(x) = f_1(x_1, ..., x_n)$ and, by theorem 8.72, for $i \geq 2$,

$$D_i g(x) = \int_{c_1}^{x_1} D_i f_1(t, x_2, ..., x_n)\,dt + ... + \int_{c_{i-1}}^{x_{i-1}} D_i f_{i-1}(c_1, ..., c_{i-2}, t, x_i, ..., x_n)\,dt$$

$$+ f_i(c_1, ..., c_{i-1}, x_i, ..., x_n) = f_i(x_1, ..., x_n),$$

since $D_i f_1 = D_1 f_i, ..., D_i f_{i-1} = D_{i-1} f_i$.

9. $\log(b/a)$.

11. (i) For $y > 0$, $\phi'(y) = \int_0^\infty \dfrac{\sin 2xy}{x}\,dx = \frac{1}{2}\pi$, since integral converges uniformly for $y \geq \delta > 0$ (see example of §6.9). Hence $\phi(y) = \frac{1}{2}\pi y + A$ ($y > 0$). But ϕ is continuous on R^1 and so $0 = \phi(0) = \lim_{y \to 0+} (\frac{1}{2}\pi y + A) = A$. Since ϕ is even, $\phi(y) = \frac{1}{2}\pi|y|$ (all y).

(ii) Call integral $\psi(y)$. Then $\psi''(y) = \frac{3}{4}\pi$ ($y > 0$) and $\psi(y) = \frac{3}{8}\pi y^2$ ($y \geq 0$), $-\frac{3}{8}\pi y^2$ ($y < 0$).

12. $|\sinh xy| \leq \cosh xy \leq e^{x|y|} \leq e^{\frac{1}{2}x^2}$ for $x \geq 2|y|$. This gives existence of $\phi(y)$ and uniform convergence of $\int_0^\infty x e^{-x^2} \sinh xy\,dx$ (integral for $\phi'(y)$) on every finite interval.

Differential equation implies $e^{-\frac{1}{4}y^2}\phi(y) = A$. Since $\phi(0) = \frac{1}{2}\sqrt{\pi}$, $\phi(y) = \frac{1}{2}\sqrt{\pi}\,e^{\frac{1}{4}y^2}$.

13. Integrals for $\phi(y)$, $\phi'(y)$ converge uniformly for all y, integral for $\phi''(y)$ for $y \geq \delta > 0$. Differential equation shows $\phi(y) = A e^y + B e^{-y} + \frac{1}{2}\pi$, $\phi'(y) = A e^y - B e^{-y}$ for $y > 0$. These also hold for $y = 0$, since ϕ, ϕ' continuous on R^1. $\phi(0) = \phi'(0) = 0$ give $A = 0$, $B = -\frac{1}{2}\pi$. Thus $\phi(y) = \frac{1}{2}\pi(1 - e^{-y})$ for $y \geq 0$, $\phi(y) = -\frac{1}{2}\pi(1 - e^y)$ for $y < 0$.

14. For any a, if $b > 0$, $D_2\phi(a, b) = \displaystyle\int_0^\infty \frac{\arctan ax}{x(1+b^2x^2)}\, dx$, since integral converges

uniformly for $b \geqslant \beta > 0$. Similarly $D_1 D_2\phi(a, b) = \displaystyle\int_0^\infty \frac{dx}{(1+a^2x^2)(1+b^2x^2)}$ if

$b > 0$; integral $= \frac{1}{2}\pi/(a+b)$ if also $a > 0$.

For $a > 0$, $b > 0$, $D_2\phi(a, b) = \frac{1}{2}\pi \log(a+b) + A$, where A independent of a. But $D_2\phi(a, b) \to 0$ as $a \to 0$ and so $D_2\phi(a, b) = \frac{1}{2}\pi\{\log(a+b) - \log b\}$. Hence $\phi(a, b) = \frac{1}{2}\pi\{(a+b)\log(a+b) - a - b\log b + B\}$, where B independent of b. As $b \to 0$, $\phi(a, b) \to 0$ (and $b \log b \to 0$). Hence $B = a - a\log a$.

15. Use theorem 4.73 as in proof of theorem 6.33. Integral $= (\pi/2)^{\frac{3}{2}}$.

9 (a)

1. $\dfrac{\pi}{2} - \dfrac{4}{\pi}\left(\cos x + \dfrac{\cos 3x}{3^2} + \dfrac{\cos 5x}{5^2} + \ldots\right).$

5. (i) $\dfrac{2}{\pi} \displaystyle\sum_{n=1}^\infty \frac{1 + (-1)^{n+1}\cos b\pi}{n^2 - b^2}\, n \sin nx,$

(ii) $\dfrac{4}{\pi} \displaystyle\sum_{n=1}^\infty \frac{2n+1}{(2n+1)^2 - b^2}\sin(2n+1)x$ if b is even.

10. In example 4, t_1 is a lower approximative sum to the integral

$$\int_0^m \frac{\sin tx}{t}\, dt.$$

For the last part, the analogous integral is $\displaystyle\int_0^{Tx} \frac{\sin u}{u}\, du$. The integrals over

$[0, \pi]$, $[\pi, 2\pi]$, $[2\pi, 3\pi]$, ... have alternate sign and decreasing magnitude.

9 (b)

2. $x^\alpha \sin(1/x)$ $(0 < \alpha < 1)$.

9 (c)

2. By theorem 9.54, given ϵ, there is $(-\delta, \delta)$ in which osc. Fourier series of f_1 is $< \epsilon$. Osc. Fourier series of f_2 in $(-\delta, \delta)$ is $2k\lambda$, where λ is the Gibbs ratio $\dfrac{2}{\pi} \displaystyle\int_0^\pi \frac{\sin t}{t}\, dt$.

9 (d)

1. No. $|z| > R$ and $\limsup |a_n|^{1/n} = 1/R$ contradict $a_n z^n = o(n)$. Theorem 9.62. Σz^n summable $(C, 1)$ for $|z| \leqslant 1$, $z \neq 1$. Uniformly summable in $\{z \mid |z| \leqslant 1, |z-1| > \delta\}$.

2. (i) $s_n(x) = \sum_1^n a_r x^r$ increases both with n and with x.

From theorem 4.73, since $\lim\limits_{x\to 1-}\lim\limits_{n\to\infty}$ exists, so does $\lim\limits_{n\to\infty}\lim\limits_{x\to 1-}$.

(ii) Write $n|a_n| = c_n$. Given ϵ, take m to make $c_n < \epsilon$ for $n > m$. Use $1 - x^k \leqslant k(1-x)$.

$$\left| \sum_0^m a_n - \sum_0^\infty a_n x^n \right| \leqslant (1-x)\sum_0^m c_n + \frac{\epsilon}{m+1}\sum_{m+1}^\infty x^n.$$

Take $x = 1-(1/m)$ and let $m \to \infty$.

3. We can take c_n real and $k = 1$. Given ϵ, choose N such that

$$d_n(1-\epsilon) \leqslant c_n \leqslant d_n(1+\epsilon) \quad \text{for} \quad n > N.$$

Then
$$C(x) = \sum_0^\infty c_n x^n = \sum_0^N + \sum_{N+1}^\infty \leqslant \sum_0^N |c_n|x^n + (1+\epsilon)\sum_0^\infty d_n x^n.$$

Divide by $D(x) = \sum_0^\infty d_n x^n$ and let $x \to 1-$.

$$\lim\sup\{C(x)/D(x)\} \leqslant 1+\epsilon.$$

Similarly, $\lim\inf \geqslant 1-\epsilon$.

4. Integrate $\displaystyle\int_0^X \left(1-\frac{x}{X}\right) x^k \sin x\,dx$ by parts.

$$\text{Sum } (A) = \lim_{\delta\to 0+} \int_0^\infty e^{-\delta x} f(x)\,dx.$$

9 (e)

1. $(f+g)^2$ and $(f-g)^2$.

2. Use Parseval on §9.2, example 1, and exercise 9(a), 7.

3. Use **1** and exercise 9(a), 2.

9 (f)

1. $\displaystyle \pi|a(u_1)-a(u_2)| \leqslant \left(\int_N^\infty + \int_{-\infty}^{-N}\right)|f| + \left|\int_{-N}^N f(t)(\cos u_1 t - \cos u_2 t)\,dt\right|$.

Given ϵ, there is N making first two $\int$'s on right side less than ϵ.

Last term $\leqslant N|u_1-u_2|\displaystyle\int_{-N}^N |f| < \epsilon$ if $|u_1-u_2|$ small enough.

2. (iii) $\displaystyle \frac{2}{\pi}\cdot\frac{2kx}{(k^2+x^2)^2}$.

3. The transformed integral $\displaystyle\int_0^\infty e^{-bu}\cos xu\,du$ is elementary, with value $b/(x^2+b^2)$.

10 (a)

1. (i) Collinear $z_r \Leftrightarrow \exists$ real A_1, A_2, A_3 with $\Sigma A_r = \Sigma A_r z_r = 0$.
From the last, $\Sigma A_r \bar{z}_r = 0$. Hence condition

$$\begin{vmatrix} 1 & 1 & 1 \\ z_1 & z_2 & z_3 \\ \bar{z}_1 & \bar{z}_2 & \bar{z}_3 \end{vmatrix} = 0.$$

(ii)

$$\begin{vmatrix} 1 & 1 & 1 \\ z_1 & z_2 & z_3 \\ z_4 & z_5 & z_6 \end{vmatrix} = 0.$$

2. See Note (2). Constant term in $f^{(n)}(z)$ is $n!\,a_n$.

3. If $|z| \leqslant 1, z \neq 1$,

$$|(1-z)\Sigma a_r z^r| = |a_0 - (a_0 - a_1)z - \ldots - a_n z^{n+1}|$$

$$\geqslant a_0 - |(a_0 - a_1)z + \ldots + a_n z^{n+1}| > a_0 - \{(a_0 - a_1) + \ldots + a_n\} = 0,$$

since the z^r $(r = 1, \ldots, n)$ have not all the same phase.

4. If $x > 0$, $\qquad\qquad$ ph $z = \arctan(y/x)$.

If $x < 0$, $\quad$ ph $z = \arctan(y/x) \pm \pi$ for $y \geqslant 0$ or < 0.

If $x = 0$, $\qquad$ ph $z = \pm\frac{1}{2}\pi$ for $y > 0$ or < 0.

5. None; all except $\pm i$; only 0; real z.

6. $u^2 + v^2 = k \Rightarrow uu_x + vv_x = uu_y + vv_y = 0 \Rightarrow u = v = 0$ $\quad$ or $\quad$ $u_x v_y - u_y v_x = 0$.
C–R equations.

7. These rules give a 'dictionary-order', but do not satisfy the axioms of an ordered field (see an Algebra text, e.g. Birkhoff & Maclane, *Survey*).
If for example, $a = -1-i$, $b = 1-i$, $c = 2i$; then $a < b$ but $a+c > b+c$.

8. Solve with the conjugate $a\bar{z} + \bar{b}z + \bar{c} = 0$.
If $|a| \neq |b|$, one z. If $|a| = |b|$, none or infinitely many according as $a\bar{c} \neq$ or $= c\bar{b}$.

9. Use $|A|^2 = A\bar{A}$. Equality in Cauchy–Schwarz if $a_r/\bar{b}_r$ same for all r.

10. Assume $>$. Repeated bisection (C1, 59) of $[z_1, z_2]$ gives contradiction.

10 (b)

1. If im $a > 0$, then $|z-a| < |z-\bar{a}| \Leftrightarrow$ im $z > 0$.

2. (i) Circles on diameters $(-3, -\frac{1}{3})$ and $(0, \frac{4}{3})$.

3. From §10.5 (8), $w-1 = k\dfrac{z-a}{1-\bar{a}z}$. Values for $z = 0, 1$ give $k = -1$, $a = -\frac{1}{2}$.

4. By continuity at frontier, w is real if z is real, Use §10.5 (2) with $0, 1, \infty$. For last part, put $z = i$.

5. If two fixed points α, β, then $\dfrac{z_n - \alpha}{z_n - \beta} = k\dfrac{z_{n-1} - \alpha}{z_{n-1} - \beta}$.

If $|k| < 1$, $z_n \to \alpha$; if $|k| > 1$, $z_n \to \beta$. If $|k| = 1$, no limit.
If $\alpha = \beta$, $z_n \to \alpha$.

6. $\dfrac{w-1}{w+1} = \left(\dfrac{z-1}{z+1}\right)^3$. $|w| < 1$ gives Ph $\dfrac{w-1}{w+1}$ in $\left[\dfrac{\pi}{2}, \dfrac{3\pi}{2}\right]$.

z- locus is three lunes, vertices at ± 1 and angles in ranges $\left[\dfrac{\pi}{6}, \dfrac{\pi}{2}\right]$, $\left[\dfrac{5\pi}{6}, \dfrac{7\pi}{6}\right]$,

$\left[\dfrac{3\pi}{2}, \dfrac{11\pi}{6}\right]$.

7. $\dfrac{w-1}{w+1} = \left(\dfrac{z-4}{z-2}\right)^2$. See **6**.

8. Fixed points are $e^{\pi i/n}$, $e^{-\pi i/n}$. Mapping is

$$\frac{w - e^{\pi i/n}}{w - e^{-\pi i/n}} = e^{2\pi i/n}\frac{z - e^{\pi i/n}}{z - e^{-\pi i/n}}.$$

Regions re $z_1 > \frac{1}{2}$ and $|z_2 - 1| > 1$.

9. $z_1 = z^2$, $z_2 = \dfrac{1+z_1}{1-z_1}$, $z_3 = z_2^2$, $w = \dfrac{z_3 - i}{z_3 + i}$.

10. Given R, there exists n_0 such that $|z_n| > R$ for all $n > n_0$.

10 (c)

1. (i) and (ii). Every value of either side is a value of the other. Principal values are equal in (i), not always in (ii).

2. $\exp(-2m+\frac{1}{2})\pi$.

3. (i) $|z| = k$ go into confocal ellipses; ph $z = \alpha$ into confocal hyperbolas. For (ii), combine (i) with $z = \exp \zeta$.

4. $iz \exp(-iz)$.

5. Strip $\frac{1}{2}(2m-1)\pi < x \leqslant \frac{1}{2}(2m+1)\pi$.

6. If $0 \leqslant r < 1$,

$$\log(1+z) = \tfrac{1}{2}\log(1+2r\cos\theta+r^2)+i\arctan\left(\frac{r\sin\theta}{1+r\cos\theta}\right).$$

Equate real parts in $\Sigma(-1)^{n-1}z^n/n = \log(1+z)$.

Let $r \to 1-$. First series $= \frac{1}{2}\log(4\cos^2\frac{1}{2}\theta)$ for $-\pi < \theta < \pi$. Second series: sum $= $ re $\log(1+e^{i\theta}\cos\theta) = \frac{1}{2}\log(1+3\cos^2\theta)$.

7. Let $z_r = x_r+iy_r$ $(r = 1, ..., n)$ be zeros of p. $x_r \geqslant 0$. $\dfrac{p'(z)}{p(z)} = \displaystyle\sum_1^n \frac{1}{z-z_r}$.

Suppose $p'(z) = 0$ for $z = x+iy$, where $x < 0$.

Then $0 = \Sigma\dfrac{\bar{z} - \bar{z}_r}{|z-z_r|^2}$.

Real part of right side < 0. Contradiction.

8. Absolute convergence if $x > 1$. Uniform convergence for $x \geqslant 1 + \delta$.

9. (i) If $z = z_0$, $|z_0| < 1$, take δ with $|z_0| < 1 - \delta < 1$. u.c. in $|z| \leqslant 1 - \delta$ by M-test.

    (ii) if re $w_0 < -\delta < 0$, then $|z^n n^{w-1}| \leqslant n^{-1-\delta}$.

    (iii) Dirichlet's test.

10. Let $\sum_0^\infty A_n(\theta) = \sum_0^\infty (a_n \cos n\theta + b_n \sin n\theta)$ be the Fourier series of $f(\theta)$. Then the Poisson integral (theorem 9.73) is

$$\sum_0^\infty A_n(\theta) r^n = \text{re} \, \Sigma(a_n - ib_n) z^n = \text{re} \, \phi(z), \text{ say,}$$

where ϕ is regular for $|z| < 1$.

$$2 - \sum_1^\infty \frac{(-1)^n r^n \cos n\theta}{n^2 - \frac{1}{4}}.$$

11 (a)

1. $\pi i R^2$, $-\pi R^2$, 0, 0.

2. $\frac{3}{2}(1+i)$. From primitives $\sin z$, e^z.

3. $2\pi i$, $-8i/3$, 0, -4.

4. All. $(1-z)^{-1}$.

5. Yes, yes.

6. Z. Z with segment $[-2i, 2i]$ deleted. A region containing no circuit which encircles a point $(\pm n + \frac{1}{2})\pi$, e.g. the strip $-\frac{1}{2}\pi < \text{re } z < \frac{1}{2}\pi$.

7. $R = \text{re} \int_\gamma (u - iv)(u_x + iv_x)(dx + idy) = \int (uu_x + vv_x) \, dx + \int (-uv_x + vu_x) \, dy$.

By C–R equations, latter $\int = \int (uu_y + vv_y) \, dy$. So $R = \frac{1}{2}[u^2 + v^2]_\gamma = 0$.

8. Theorem 11.11. Since $|e^{iz}| = e^{-y} \leqslant 1$ on γ, $\int$ is $O(1/R)$.

    Integration by parts into

$$\left[\frac{e^{iz}}{iz^2}\right]_\gamma - \int_\gamma \frac{2e^{iz}}{iz^3} \, dz$$

gives each term $O(1/R^2)$, a smaller bound.

9. $f(z) = \{k + \epsilon(z)\}/(z - a)$, where $|\epsilon(z)| < \epsilon$ if $\delta < \delta_0$ and $\alpha \leqslant \text{ph}(z - a) \leqslant \beta$.

$$\int_\gamma f = \int_\alpha^\beta \{k + \epsilon(z)\} i \, d\theta \to ik(\beta - \alpha) \quad \text{as} \quad \delta \to 0.$$

11. $\displaystyle\int_{-\infty}^\infty \frac{i \, dy}{1 + iy} = \int_{-\infty}^\infty \frac{y + i}{1 + y^2} \, dy = \pi i$.

12. Call the integrals I, J and the inequality to be proved $|S| \leqslant \pi T$. Use

$$|f(e^{i\theta})|^2 = f(e^{i\theta}) \overline{f(e^{i\theta})};$$

$$\int_{-\pi}^{\pi} \theta e^{ip\theta} d\theta = \frac{2\pi(-1)^p}{ip} \text{ (integer } p \neq 0), \quad = 0 \ (p = 0);$$

$$\int_{-\pi}^{\pi} e^{ip\theta} d\theta = 0 \ (p \neq 0), \quad 2\pi \ (p = 0).$$

Then
$$I = \sum_{m, n=1}^{N} (-1)^{m+n} a_m \bar{a}_n \int_{-\pi}^{\pi} \theta e^{(m-n)i\theta} d\theta = -2\pi i S,$$

$$J = 2\pi T,$$

$$|I| \leqslant \pi J.$$

Hence $|S| \leqslant \pi T$.

13. Let $z = z(t)$ for $a \leqslant t \leqslant b$ be a representation of γ,

$$\text{r.h.s.} = \int_a^b g\{f(z(t))\} f'\{z(t)\} z'(t) dt$$

$$= \int_a^b g(w) \frac{dw}{dt} dt = \text{l.h.s.}$$

11 (b)

1. Quadrisect a triangle by joining mid-points of sides. Simple polygon can be split into triangles by diagonals *inside* it—proof by induction on the number of sides.

2. Given ϵ, take R_0 a square, centre ζ, side 2δ so small that $|f(z)| \leqslant \epsilon/\delta$ for z on ∂R_0.

By theorem 11.11, $\left| \int_{\partial R_0} f \right| \leqslant 8\epsilon$.

3. Continuity of f' is assumed.

4. Along one side $a(1-i)$, $a(1+i)$ of the square

$$\int \frac{dz}{z} = \int_{-a}^a \frac{i \, dy}{a+iy} = \int_{-a}^a \frac{y+ia}{a^2+y^2} dy = \tfrac{1}{2}\pi i.$$

11 (c)

1. $2\pi i a^4$ (theorem 11.81), $2\pi i e^a/3!$ (theorem 11.91).

2. arc tan z ($\pm n\pi$).

3. $2\pi i \{n(\gamma, 0) - \tfrac{1}{2} n(\gamma, -1) - \tfrac{1}{2} n(\gamma, 1)\}$. Possible values are $\pm 2\pi i$, $\pm \pi i$, 0.

4. Theorem 11.91.

$$\sum_0^\infty \left(\frac{x^n}{n!}\right)^2 = \frac{1}{2\pi i} \int_\gamma \left(\sum_0^\infty \frac{x^n}{n! z^n} \right) \frac{\exp zx}{z} dz,$$

since, for any given x, the Σ is u.c. on γ. Let γ be $z = e^{i\theta}$ ($0 \leqslant \theta \leqslant 2\pi$).

5. Theorem 11.81. The $\leqslant$ from §10.5 (8).

6. Integrand is $\left\{\dfrac{1}{z-a}+\dfrac{\bar a}{R^2-z\bar a}\right\}f(z)$. $R^2/\bar a$ is outside C.

7. If γ is $\zeta = re^{i\theta}$, $f'(0) = \dfrac{1}{2\pi i}\displaystyle\int_\gamma \dfrac{f(\zeta)}{\zeta^2}\,d\zeta = \dfrac{1}{2\pi r}\int_0^{2\pi}(u+iv)e^{-i\theta}\,d\theta$.

Also $\displaystyle\int_0^{2\pi}(u-iv)e^{-i\theta}\,d\theta = $ conjugate of $\displaystyle\int_0^{2\pi}(u+iv)e^{i\theta}\,d\theta = 0$.

8. (i) $f(z) = \dfrac{1}{z}\displaystyle\int_\gamma \left(\dfrac{1}{\zeta-z}-\dfrac{1}{\zeta}\right)d\zeta$.

If $|z| < 1$, $f(z) = 0$.   If $|z| > 1$, $f(z) = -2\pi i/z$.

(ii) $f(z) = \dfrac{1}{2}\displaystyle\int_\gamma \dfrac{\zeta+\bar\zeta}{\zeta-z}\,d\zeta$, and, on γ, $\bar\zeta = 1/\zeta$.

9. Integrate along semicircle $\zeta = Xe^{i\theta}$ $(0 \leqslant \theta \leqslant \pi)$ and its diameter. By theorem 11.81 and continuity of integrand for im $\zeta \geqslant 0$.

$$f(z)-0 = \dfrac{1}{2\pi i}\int\left(\dfrac{1}{\zeta-z}-\dfrac{1}{\zeta-\bar z}\right)f(\zeta)\,d\zeta.$$

10. If a rectangle R in $D_1 \cup D_2$ is divided by l into two rectangles R_1, R_2, then, by theorem 11.42,

$$\int_{\partial R} = \int_{\partial R_1} + \int_{\partial R_2} = 0.$$

11. $f_2 = f_1$ on l. $\dfrac{f_2(z+h)-f_2(z)}{h} = $ conjugate of $\dfrac{f_1(\bar z+\bar h)-f_1(\bar z)}{\bar h}$.

Since $f_1'(\bar z)$ exists, $f_2'(z)$ exists.

12 (a)

1. Branch point is $z = -1$ (if k is not an integer).

Principal branch f, defined to be 1 for $z = 0$, is regular and has Taylor's series in $|z| < 1$. From p. 354 and theorem 12.11,

$$f^{(n)}(z) = k(k-1)\dots(k-n+1)(1+z)^{k-n}$$

and $$(1+z)^k = 1+kz+\dots+\dfrac{k(k-1)\dots(k-n+1)}{n!}z^n+\dots.$$

2. $\tfrac{1}{2}z^2-\tfrac{1}{12}z^4+\tfrac{1}{45}z^6\dots$; $R = \tfrac{1}{2}\pi$.

3. Method of (say) C1, ex. 6(d), 6, namely:
If $w = (\text{arc sin } z)^2$, then $(1-z^2)w'' - zw' = 2$.
From Leibniz's theorem,

$$(1-z^2)w^{(n+2)} - (2n+1)zw^{(n+1)} - n^2w^{(n)} = 0.$$

Putting $z = 0$, we have coefficients. $R = 1$.

4. $z+\tfrac{1}{3}z^3+\tfrac{2}{15}z^5+\dots$; $R = \tfrac{1}{2}\pi$.

5. $1-\tfrac{1}{4}z^2-\tfrac{1}{96}z^4\dots$; $R = \tfrac{1}{2}\pi$.

6. $\frac{1}{2} - \frac{1}{4}z + \frac{1}{48}z^3\ldots$; $R = 2\pi$.

7. $\sum\limits_0^\infty (-1)^n z^{2n+1}/n!\,(2n+1)!\,(2n+1)$, all z.

8. $\Sigma d_n z^n$, where d_n = number of divisors of n, including 1 and n. $R = 1$.

9. $\cot z = -\tan(z - \frac{1}{2}\pi)$ say.

10. 2^k, $2^{1/k}$, 1, 1.

11. f, regular at 0 and $= 0$ at $\frac{1}{2}, \frac{1}{4}, \frac{1}{6}, \ldots$, $\equiv 0$.   (i), (ii), (v) No.

12. $|a_n| \leqslant M/r^n$ gives

$$|f(z) - a_0| \leqslant \sum_1^\infty |a_n| \cdot |z|^n \leqslant M\Sigma(|z|/r)^n = M|z|/(r - |z|).$$

No zero if this $< |a_0|$.

13. If $f \not\equiv 0$, $\exists \zeta$ in D with $f(\zeta) \neq 0$. This $\neq$ holds in a disc $B(\zeta; r)$. $fg \equiv 0 \Rightarrow g \equiv 0$ in $B(\zeta; r)$ and, by theorem 12.12, in D.

14. Use $|A|^2 = A\bar{A}$. By uniform convergence of the series,

$$\frac{1}{2\pi} \int_0^{2\pi} |f(re^{i\theta})|^2 d\theta = \frac{1}{2\pi} \int_0^{2\pi} \sum_{m,n} a_m \bar{a}_n r^{m+n} e^{i(m-n)\theta}\, d\theta = \sum_n a_n \bar{a}_n r^{2n},$$

all terms with $m \neq n$ vanishing.

15. By **14**, $a_n = 0$ if $n \neq k$.

16. If $p_k(z) = \sum\limits_0^k b_n z^n$, the given $\int$ is

$$\sum_0^k |a_n - b_n|^2 + \sum_{k+1}^\infty |a_n|^2.$$

17. Let γ be the circle $z = Re^{i\theta}$ ($0 \leqslant \theta \leqslant 2\pi$). By theorem 11.81,

$$\int_\gamma \frac{1}{a-b} \left\{ \frac{f(z)}{z-a} - \frac{f(z)}{z-b} \right\} dz = \frac{2\pi i}{a-b} \{f(a) - f(b)\}.$$

If $|f(z)| \leqslant M$, theorem $11.11 \Rightarrow \left| \int_\gamma \right| \leqslant \dfrac{M\pi R}{|a-b|(R-|a|)(R-|b|)} \to 0$.

18. (i) In theorem 12.12, $|a_n| = o(r^{1-n})$. True.
(ii) f is a polynomial of degree at most k.

19. (ii) If $g(z) = f(z) - 8i$, then $1/g$ is regular and $|1/g| \leqslant 1$.
If, a, b, c being suitable constants, $au + bv - c \geqslant 0$ [or $\leqslant 0$], then f is constant.

20. Take f^2.

21. $1/|f(z) - w_0| \leqslant 1/\delta$.

22. f/g is constant.

12 (b)

1. For $1 < |z| < 2$, binomial expansions give $-\frac{1}{2}\sum_{0}^{\infty}\left(\frac{z}{2}\right)^n - \sum_{1}^{\infty}\left(\frac{1}{z}\right)^n$.

2. For $|z| < 1$, $1 - z + z^3 - z^4 + z^6 - \ldots$

For $|z| > 1$, $\frac{1}{z^2} - \frac{1}{z^3} + \frac{1}{z^5} - \frac{1}{z^6} + \frac{1}{z^8} - \ldots$

4. For $|z| > 0$, taking $\int$ round $z = e^{i\theta}$,

$$J_n(u) = \frac{1}{2\pi i}\int_{-\pi}^{\pi}\exp(iu\sin\theta)\,\frac{id\theta}{e^{ni\theta}}.$$

To obtain the series for J_n multiply the series for $\exp\frac{1}{2}uz$ and $\exp(-u/2z)$.

5. Either branch is one-valued in $|z| < 1$ (or in $|z| > 2$), hence expansible in Taylor (or Laurent) series. Cannot expand in $1 < |z| < 2$.

7. From §10.1, the Jacobian $u_x v_y - u_y v_x = |f'|^2$.

Area $= \int_{r_1}^{r_2} r\,dr \int_{0}^{2\pi} |f'(re^{i\theta})|^2 d\theta$.   See 12 (a), 14.

12 (c)

0. $f(z) = z^4 - 3z^3 + 7z^2 - 9z + 6 = 0$. We first look for real roots. $f'(x) = 0$ if and only if $x = 1$. $f(1) = 2$, minimum. No real roots.

To find the number of zeros of f in the first quadrant, let z describe the simple circuit (positively oriented) composed of (1) real axis 0 to R (R large), (2) quadrant $z = Re^{i\theta}$ ($0 \leqslant \theta \leqslant \frac{1}{2}\pi$), (3) imaginary axis from iR to 0. Call the corresponding paths of $w\,(=f(z))$, $\Gamma_1, \Gamma_2, \Gamma_3$. We need to find the index of $w = 0$ for $\Gamma_1 + \Gamma_2 + \Gamma_3$. Γ_1 is part of the real axis and at $f(R)$ we can take $\mathrm{Ph}\,w = 0$. On Γ_2, $w = R^4 e^{4i\theta}\{1 + O(1/R)\}$. At $f(iR)$, $\mathrm{Ph}\,w$ is $2\pi\{1 + O(1/R)\}$. To trace the change in $\mathrm{Ph}\,w$ on Γ_3, we have $w = u + iv = y^4 - 7y^2 + 6 + i(3y^3 - 9y)$.

Mark the crossings with the imaginary axis $u = 0$. First is at $y = \sqrt{6}$, second when $y = 1$. At former $v > 0$, at latter $v < 0$; u changes sign from $+$ to $-$ at former, from $-$ to $+$ at latter. When w returns to $f(0)$, the value of $\mathrm{Ph}\,w$ is 4π. The index of 0 for $\Gamma_1 + \Gamma_2 + \Gamma_3$ is 2. Two roots in first quadrant. By conjugates, two in fourth.

1. By theorem 12.5 integrals are $-\frac{1}{3}\pi i$, $(-1)^n 2\pi i/(n-1)!$, πi.

2. Take $a \neq b$. $c = a - b$. Residue at a is coefficient of $1/\zeta$ in Laurent expansion about $\zeta = 0$ of $\zeta^{-n}(\zeta + c)^{-n}$. This is k_n/c^{2n-1}, k_n being the relevant binomial coefficient. Residue at b is $k_n/(-c)^{2n-1}$.

3. Follow method detailed in **0**. Rouché (theorem 12.62) with $f = 5$ shows no zeros in $|z| < \frac{1}{2}$.

4. 5. 8. As **3**.

6. On $|z| = 1 + \delta$, $|e^{z-1}| < |z^n|$ if δ small. Rouché shows n roots in $|z| < 1 + \delta$. There is one root $z = e^{i\theta}$ with $\theta = 0$ on $|z| = 1$.

7. $w = \sinh \pi z$ gives

$$u = \sinh \pi x \cos \pi y, \quad v = \cosh \pi x \sin \pi y.$$

The four lines $y = \pm \frac{1}{2}$, $x = 0$, $x = X$ bound a rectangle.

As z describes its positively oriented boundary, w describes Γ composed of right half of the ellipse

$$\frac{u^2}{\sinh^2 X} + \frac{v^2}{\cosh^2 X} = 1$$

from $-i \cosh \pi X$ to $+i \cosh \pi X$ and the line segment iv from $v = \cosh \pi X$ to $-\cosh \pi X$. If re $w_0 > 0$ and X large, index of w_0 for Γ is 1. Hence $w = w_0$ exactly once in the strip.

9.
$$f(z) = z^m \left(\frac{z-a}{1-\bar{a}z} \right)^n$$

has m zeros at 0 and n at a. $|f| = 1$ on $|z| = 1$ from §10.5 (8). Take in Rouché $g = -a$.

10. Method of **7**.

11. Put ζ for $1/z$ and $p_n(\zeta) = \sum_0^n \zeta^r/r!$

Given R, ϵ, there is N with $|p_n(\zeta) - \exp \zeta| < \epsilon$ on $|\zeta| = R$ for $n \geqslant N$.

Take $\epsilon < \inf |\exp(Re^{i\theta})|$ for $0 \leqslant \theta \leqslant 2\pi$. By Rouché, $\exp \zeta$ and $p_n(\zeta)$ have same number of zeros in $|\zeta| < R$. $\exp \zeta$ has no zeros. All zeros of p_n have $|\zeta| \geqslant R$.

12 (*d*)

1. None (removable at 1, -1).

2. 3. Simple poles at $(-1)^{\frac{1}{3}}$. At ∞, **2** has pole, **3** essential singularity.

4. Simple poles at $\frac{1}{2}(2n+1)\pi$. Limit point is ∞, singularity.

7. Essential singularity at 1.

8. Simple poles at $2n\pi i$ ($n \neq 0$). Remember ∞.

9. $\cos z = a + ib$ gives $\cos x \cosh y = a$, $\sin x \sinh y = -b$.

Continue either (i) as in 12(*c*), **7** or (ii) directly (long) $\cosh y = A_1 \pm A_2$,

where $\quad A_1 = \frac{1}{2}\sqrt{\{(a+1)^2 + b^2\}}, \quad A_2 = \frac{1}{2}\sqrt{\{(a-1)^2 + b^2\}}.$

Then $\cos x = A_1 - A_2$ and $\cosh y = A_1 + A_2$. Solve these.

13 (*a*)

1. (i)–(iii) $|z| < 1$ and $|z| > 1$.
   (iv) (vii) None.
   (viii) Strip $|y| < \log 2$.

2. (i) $|z| < 1$.   (ii) Z.   (iii) $x > 0$.
   (iv) $x < 2$.   (v) $x > 0$.   (vi) Z, punctured at 0.
   (vii), (viii) None.

3. $\phi_n = (g+f)+(g-f)\chi_n,$

where $2\chi_n(z) = (z^n-1)/(z^n+1).$

4. M-test and theorem 13.11. Simple poles at $\pm ni$ and $\pm i/n$ $(n \neq 1)$, double poles at $\pm i$.

Singularities at 0 and ∞, being limit points of poles.

5. Analogue of corollary 1 of theorem 13.11.

6. By theorem 13.11, any function continuous, but not regular, in D is an example, e.g. $|z|$ in any D.

13 (*b*)

1. Find, if possible, z for which all terms in the Taylor series have the same phase.
  (i) $z = -ir$ gives e^r, (ii) $\sinh r$, (iii) $(\sinh \sqrt{r})/\sqrt{r}$.

2. $1/f$ in theorem 13.22.

3. Theorem 12.62.

5. Let $\omega = \exp(2\pi i/n)$, where $n(\beta-\alpha) > 2\pi$. Apply theorem 13.22 (corollary) to $f(z)f(\omega z)...f(\omega^{n-1}z)$ and then use exercise 12(*a*), 13.

6. Write $M = M(r)$, $M_1 = M(r_1)$, $M_2 = M(r_2)$; p, q integers.

$$r^p M^q \leqslant \max(r_1^p M_1^q, r_2^p M_2^q).$$

$r_1^p M_1^q \leqslant r_2^p M_2^q$ if and only if

(1) $\dfrac{p}{q} \geqslant \dfrac{\log(M_1/M_2)}{\log(r_2/r_1)}.$

With such p, q we have $r^p M^q \leqslant r_2^p M_2^q$, i.e.

(2) $\log M \leqslant \dfrac{p}{q}\log\dfrac{r_2}{r}+\log M_2.$

The infimum of the p/q satisfying (1) is

$$\frac{\log(M_1/M_2)}{\log(r_2/r_1)}.$$

(2) holds if p/q is replaced by this infimum.

7. If $f(z) = \sum\limits_0^\infty a_n z^n$, then $I_2(r) = \sum\limits_0^\infty |a_n|^2 r^{2n}$.

  (i) $I_2'(r) \geqslant 0$.
  (ii) For theorem 13.22, suppose $f(z) = \Sigma a_n(z-z_0)^n$ is regular in $B(z_0, R)$ and $|f|$ has a maximum at z_0.
  Then, if r is small enough,

$$|a_0|^2 = |f(z_0)|^2 \geqslant \Sigma |a_n|^2 r^{2n}.$$

Hence $a_n = 0$ for $n \geqslant 1$.

(iii) Write $u = \log r$ and $J(u) = I_2(e^u) = \Sigma |a_n|^2 e^{2nu}$

$$\frac{d^2}{du^2} \log J_2 = \frac{J_2 J'' - J'^2}{J^2}.$$

Numerator is

$$(\Sigma |a_n|^2 e^{2nu})(\Sigma |a_n|^2 4n^2 e^{2nu}) - (\Sigma |a_n|^2 2n e^{2nu})^2 \geqslant 0 \quad \text{(Schwarz)}.$$

13 (c)

1. The last three. $\tan \frac{1}{2}\pi z$ has poles when z is an odd integer.

2. $|f(r \exp(2\pi i p / 2^r))| \to \infty$ as $r \to 1-$.

3. By theorems 12.13, 13.13,

$$\int_0^\infty e^{-(x+iy)t^2} dt = \frac{1}{2} \left\{ \frac{\pi}{x+iy} \right\}^{\frac{1}{2}} \quad \text{(principal value)}.$$

Write $r^2 = x^2 + y^2$. $\operatorname{re}(\text{r.h.s.}) = \frac{1}{2} \left\{ \frac{\pi(x+r)}{2r^2} \right\}^{\frac{1}{2}}$. $\qquad \int_0^\infty e^{-iyt^2} dt$ converges.

Analogue for integrals of Abel's limit theorem gives

$$\int_0^\infty \cos yt^2 \, dt = \lim_{x \to 0} \frac{1}{2} \left\{ \frac{\pi(x+r)}{2r^2} \right\}^{\frac{1}{2}} = \frac{1}{2} \sqrt{\frac{\pi}{2y}}.$$

4. If false, there is ρ for which $\sum_{\nu=0}^\infty \frac{f^\nu(\rho)}{\nu!} (z-\rho)^\nu$ converges for $z = 1 + \delta > 1$. The series is

$$\sum_{\nu=0}^\infty \frac{(z-\rho)^\nu}{\nu!} \sum_{n=\nu}^\infty n(n-1)...(n-\nu+1) a_n \rho^{n-\nu}.$$

Terms are positive. $\sum_n \sum_\nu$ is $\sum_{n=0}^\infty a_n \{(z-\rho)+\rho\}^n$, which therefore converges for $z = 1 + \delta$. This contradicts $R = 1$.

5. No generality is lost by operating only with discs having rational centres and radii. $\quad z^\alpha$ with irrational α, $\log z$.

6. First part from theorem 13.11.

If $x > 1$, $\qquad (1 - 2^{1-z}) \zeta(z) = \sum_1^\infty (-1)^{n-1} n^{-z}$.

As $z \to 1$, $\qquad \phi(z) \dfrac{z-1}{1 - 2^{1-z}} \to \log 2 \dfrac{1}{\log 2} = 1$.

7. If δ is small enough, the rectangle with corners $-\delta \pm i\delta$, $1 + \delta \pm i\delta$ is in D. Let $\delta < b - a$.

Let C_n be the disc $B(\frac{1}{2}n\delta; \delta)$ for $n = 0, 1, \dots$.

Let $a < \frac{1}{2}k\delta < b$. f and all its derivatives are real at $x_k = \frac{1}{2}k\delta$, $y = 0$. From Taylor's series in C_k, f and all its derivatives are real at x_{k-1} and so, successively, at $x_{k-2}, \dots, x_0$.

Counterexample, $a = \frac{1}{2}$, $b = \frac{3}{4}$, $f(z) =$ value of $(z - \frac{1}{4})^{\frac{1}{2}}$ which is positive for $\frac{1}{2} \leqslant x \leqslant \frac{3}{4}$, $y = 0$.

$D = $ a simply-connected region containing $z = 0$ and segment $[\frac{1}{2}, \frac{3}{4}]$ of $y = 0$ and excluding $z = \frac{1}{4}$.

14 (a)

1. Integrate round boundary of sector $|z| \leqslant R$, $0 \leqslant \mathrm{ph}\, z \leqslant \frac{1}{4}\pi$.

2. $(\pi/n)\operatorname{cosec}(\pi/n)$. Without evaluation,

$$0 < 1 - \int_0^1 \frac{dx}{1+x^n} = \int_0^1 \frac{x^n\,dx}{1+x^n} < \int_0^1 x^n\,dx = \frac{1}{n+1},$$

and

$$0 < \int_1^\infty \frac{dx}{1+x^n} < \int_1^\infty \frac{dx}{x^n} = \frac{1}{n-1}.$$

3. Validity for $a = b$ can be proved from theorem 8.73.
Write

$$\phi(a, b) = \int_{-\infty}^\infty \frac{dx}{(x^2+a^2)^2(x^2+b^2)}.$$

Suppose a is fixed and (say) $\frac{1}{2}a \leqslant b \leqslant \frac{3}{2}a$. Integrand $\leqslant 1/(x^2+a^2)^2(x^2+\frac{1}{4}a^2)$ and $\int_{-\infty}^\infty$ of this exists. Hence the integral $\phi(a, b)$ converges uniformly for $\frac{1}{2}a \leqslant b \leqslant \frac{3}{2}a$ and so is continuous for $b = a$.

Alternatively (and more simply) we may note that (for fixed a), $\phi(a, b)$ decreases as b increases. Hence $\phi(a, a-) \geqslant \phi(a, a) \geqslant \phi(a, a+)$. As $b \to a$, the first and last of these have the same limit $3\pi/8a^5$.

4. If $k < 0$, value is $-\frac{1}{2}\pi e^{ka}\cos ka$. If $k = 0$, $\int = 0$.

5. §11.2, example 3. Differentiate the factor $1/z$.

$$\int \frac{e^{iz}}{z}\,dz = \left[\frac{e^{iz}}{iz}\right] + \int \frac{e^{iz}}{iz^2}\,dz = O\left(\frac{1}{R}\right) + \int O\left(\frac{1}{R^2}\right)dz.$$

6. $\int_0^\infty$ as in example 2. To prove $\int_X^\infty \frac{\cos ax}{x^2}\,dx = O\left(\frac{1}{X^2}\right)$,

$$\int_X^{X_1} = \left[\frac{\sin ax}{ax^2}\right]_X^{X_1} + 2\int_X^{X_1} \frac{\sin ax}{ax^3}\,dx.$$

8. Vertices ab, aB, Ab, AB.

9. If γ is $z = e^{i\theta}$ $(0 \leqslant \theta \leqslant 2\pi)$,

$$\int = \frac{i}{4}\int_\gamma \frac{(z^2-1)^2\,dz}{z^2(z-k)(kz-1)}.$$

Poles inside γ at 0, k.

If $|k| > 1$, put $k = \frac{1}{l}$. $\int = \frac{\pi}{k^2}$. Examine $k = \pm 1$.

10. Method of **9.** $2\pi/n!$, 0.

12. Rectangle of example 3, indented at 0, $2\pi i$.

14. Example 4 (method (i)).

$$\int_0^\infty \frac{(\log x)^3 - (\log x + 2\pi i)^3}{1+x^2}\, dx = 2\pi i \{\text{res}\,(\tfrac{1}{2}\pi i) + \text{res}\,(\tfrac{3}{2}\pi i)\} = \tfrac{13}{4}\pi^4 i.$$

16. Sector $|z| \leqslant R,\, 0 \leqslant \text{ph}\, z \leqslant \tfrac{1}{4}\pi$.

The integral for $1/\sqrt{x}$ states that it is its own Fourier transform if the definition of §9.9 is widened to admit unbounded integrands. ($1/\sqrt{x}$ in $x > 0$.)

17. By theorem 8.75, $\phi'(x) = \int_0^\infty \exp(-\tfrac{1}{2}u^2)(-u \sin ux)\, du$. Integrate by parts.

18. For the latter, integrate $\sin zu/(e^{az}-1)$, where $a = \sqrt{(2\pi)}$, round the rectangle $0,\, X,\, X+2\pi i/a,\, 2\pi i/a$. Let $X \to \infty$.

19. From theorem 9.95, if $\displaystyle\int_{-\infty}^\infty |g(t)|\, dt$ exists and g satisfies conditions for the convergence of its Fourier series at t, then

$$\tfrac{1}{2}\{g(t+)+g(t-)\} = \frac{1}{2\pi} \lim_{U\to\infty} \int_{-U}^{U} du \int_{-\infty}^\infty g(\tau) \cos u(t-\tau)\, d\tau.$$

Since $\sin u(t-\tau)$ is an odd function of u, the right-hand side is

$$\frac{1}{2\pi} \lim_{U\to\infty} \int_{-U}^{U} du \int_{-\infty}^\infty g(\tau)\, e^{iu(t-\tau)}\, d\tau.$$

Put $g(t) = e^{-at}f(t)$ if $t \geqslant 0$ and $g(t) = 0$ if $t < 0$. Then, when $t \geqslant 0$,

$$\tfrac{1}{2}\{f(t+)+f(t-)\} = \frac{1}{2\pi} \lim_{U\to\infty} \int_{-U}^{U} e^{(a+iu)t} du \int_0^\infty f(\tau)\, e^{-(a+iu)\tau}\, d\tau.$$

The inner integral is $F(a+iu)$.

14 (b)

1. 2. 3. Method of §§ 14.2, 14.3, or derive from results obtained there; e.g. $z - \tfrac{1}{2}$ for z in § 14.2 gives

$$\sum_{-\infty}^\infty \frac{1}{\{z-(n+\tfrac{1}{2})\}^2} = \frac{\pi^2}{\cos^2 \pi z}.$$

Integrate, as in § 14.3, to deduce **2**.

4. Poles of $\dfrac{1}{8z^3-1}$ are at $z = \tfrac{1}{2}(1$ or ω or $\omega^2)$ where $\omega = e^{2\pi i/3}$; residues respectively $\tfrac{1}{6}(1$ or ω or $\omega^2)$. Take, as in § 14.2,

$$\int_{\partial C_n} \frac{\pi \cot \pi \zeta}{8\zeta^3-1}\, d\zeta$$

and we have

$$\sum_{r=1}^\infty \frac{1}{64r^6-1} = \frac{1}{2} - \frac{\pi}{12}(\omega \cot \tfrac{1}{2}\pi\omega + \omega^2 \cot \tfrac{1}{2}\pi\omega^2). \quad \text{Simplify.}$$

8. (i) $\tfrac{1}{2}z \coth \tfrac{1}{2}z$ (which $\to 1$ as $z \to 0$) has Taylor's series in $|z| < R$, where R is distance of nearest singularity, namely $\pm 2\pi i$.

(iii) Equate coefficients of z^{2n} in

$$\cosh \tfrac{1}{2}z = (2/z) \sinh \tfrac{1}{2}z \Sigma B_n z^n / n!$$

i.e. in

$$\sum_0^\infty \frac{z^{2n}}{2^{2n}(2n)!} = \left\{ \sum_0^\infty \frac{z^{2r}}{2^{2r}(2r+1)!} \right\} \left\{ \sum_0^\infty \frac{B_{2r}}{(2r)!} z^{2r} \right\}.$$

14 (c)

10. $\left\{ \begin{array}{c} \\ \end{array} \right\}$ on right $= \displaystyle\sum_{n=1}^\infty \sum_{m=1}^\infty a_{mn}$, where $a_{mn} = \dfrac{(-1)^{n-1}}{n} z^{mn}$.

Now $\displaystyle\sum_{n=1}^\infty |a_{mn}| = -\log(1 - |z|^m) < \dfrac{|z|^m}{1 - |z|^m} < 2|z|^m$ if $m > m_0$.

Hence $\displaystyle\sum_{m=1}^\infty \sum_{n=1}^\infty |a_{mn}|$ converges if $|z| < 1$.

By theorem 4.74,

$$\sum_{n=1}^\infty \sum_{m=1}^\infty a_{mn} = \sum_{m=1}^\infty \sum_{n=1}^\infty a_{mn} = \sum_{m=1}^\infty \log(1 + z^m).$$

Take exponential of each side.

11. $\cos \pi z = \displaystyle\prod_{n=1}^\infty \left\{ 1 - \dfrac{z^2}{(n - \tfrac{1}{2})^2} \right\}.$

12. (i) False. $a_n = 1/n \log n.$

    (ii) False. $a_n = \dfrac{(-1)^{n-1}}{\sqrt{n}} + \dfrac{1}{n}, \quad b_n = \dfrac{(-1)^{n-1}}{\sqrt{n}}.$

    (iii) True. $\displaystyle\prod_1^n$ tends to a finite limit, zero or non-zero according as Σa_n^2 diverges or converges.

13. (i) If p_k is the kth prime,

$$\prod_{k=1}^m \left(1 - \frac{1}{p_k^z} \right)^{-1} = \prod_{k=1}^m \left(\sum_{r=0}^\infty p_k^{-rz} \right) = \Sigma^{(m)} n^{-z},$$

where $\Sigma^{(m)}$ means summation over all n not divisible by $p_{m+1}, \dots$.
    As $m \to \infty$, every integer n appears in $\Sigma^{(m)}$ and r.h.s. $\to (z)$.

    (ii) If $x > 1$, Σp^{-z} (a sub-series of Σn^{-z}) converges absolutely. Hence product converges, and so $\zeta(z) \neq 0$.

    (iii) In (i) write $z = 1$. As $m \to \infty$, $\Sigma^{(m)} n^{-1} \to \infty$. Therefore $\Pi (1 - p^{-1})$ diverges to 0. Then (theorem 14.41) $\Sigma(1/p)$ diverges.

14. Let g be a function with simple zeros at the a_n. Then, for suitable q_n

$$\sum_1^\infty g(z) \frac{b_n \exp\{q_n(z-a_n)\}}{g'(a_n)(z-a_n)}$$

converges. $q_n =$ real multiple of $\bar{a}_n$.

15. (i) Theorem 14.51.

(ii) As for s'_N/s_N in theorem 14.52.

(iii) Theorem 13.11.

(iv) (a) $\wp(-z)$ is got from $\wp(z)$ by rearranging terms of an absolutely convergent series.

(b) Similar.

(c) $\wp'(z+2\omega_1) = -2\Sigma(z-\Omega_{m,n}+2\omega_1)^{-3}$. The sets of points $\Omega_{m,n}-2\omega_1$ and $\Omega_{m,n}$ are the same. The series for $\wp'(z+2\omega_1)$ is a rearrangement of that for $\wp'(z)$.

(d) Integrating (c), $\wp(z+2\omega_1) = \wp(z)+A$. $z = -\omega_1$ gives $A = 0$.

(v) $\wp(z)-z^{-2} = \Sigma'\left\{\dfrac{1}{(z-\Omega_{m,n})^2}-\dfrac{1}{\Omega_{m,n}^2}\right\}$ (even function)

$$= \Sigma'\left\{\frac{3z^2}{\Omega_{m,n}^4}+\frac{5z^4}{\Omega_{m,n}^6}+\ldots\right\} \quad \text{if } |z| \text{ small enough}$$

$$= a_2 z^2+a_3 z^4+O(|z|).$$

Thus

$$\{\wp(z)\}^3 = \frac{1}{z^6}+\frac{3a_2}{z^2}+3a_3+O(|z|),$$

$$\{\wp'(z)\}^2 = \frac{4}{z^6}-\frac{8a_2}{z^2}-16a_3+O(|z|).$$

Hence $\{\wp'(z)\}^2-4\{\wp(z)\}^3+20a_2\,\wp(z)+28a_3 = O(|z|).$

Expression on left is an elliptic function, regular for all z and 0 for $z = 0$, hence $\equiv 0$ (Liouville).

The coefficients g_2, g_3 are $60\Sigma'\Omega_{m,n}^{-4}$, $140\Sigma'\Omega_{m,n}^{-6}$.

14 (d)

1 to **5** follow from §14.6 (5), e.g.

2. If $\omega = e^{2\pi i/3}$ $\displaystyle\prod_1^\infty \left(1+\frac{1}{n^3}\right) = \Pi\frac{(n+1)(n+\omega)(n+\omega^2)}{n^3} = \frac{1}{\Gamma(\omega+1)\Gamma(-\omega)}$

$$= \frac{\sin(-\pi\omega)}{\pi} = \frac{\cosh\frac{1}{2}\pi\sqrt{3}}{\pi}.$$

5. Product is

$$\Pi\left(1-\frac{z}{2n-1}\right)\left(1+\frac{z}{2n}\right) = \Pi\frac{(n-\frac{1}{2}-\frac{1}{2}z)(n+\frac{1}{2}z)}{(n-\frac{1}{2})n}.$$

The rearranged product is

$$\Pi\frac{(4n-z-3)(4n-z-1)(2n+z)}{(4n-3)(4n-1)2n} = \frac{\Gamma(\frac{1}{4})\Gamma(\frac{3}{4})}{\Gamma(\frac{1}{4}-\frac{1}{4}z)\Gamma(\frac{3}{4}-\frac{1}{4}z)\Gamma(1+\frac{1}{2}z)},$$

which simplifies to $\dfrac{\sqrt{\pi}}{2^{z/2}\Gamma(\frac{1}{2}-\frac{1}{2}z)\Gamma(1+\frac{1}{2}z)}.$

6. Pole at $z = -n$ $(n = 1, 2, \ldots)$.

$$\text{Residue} = \lim_{z \to -n} (n+z)\Gamma(1+z) = \lim \frac{(n+z)\Gamma(n+z)}{(1+z)(2+z)\ldots(n-1+z)} = \frac{(-1)^{n-1}}{(n-1)!}.$$

14 (e)

1. (i) $\dfrac{\Gamma(a)\Gamma(b)}{\Gamma(a+b)} \dfrac{1}{(1+k)^a k^b}$.   (ii) $\dfrac{\Gamma(a)\Gamma(b)}{\Gamma(a+b)} 2^{a+b-2}$.

2. (i) $\dfrac{c^2}{4\sqrt{(2\pi)}}$,   (ii) $\dfrac{c^4 - 8\pi^2}{4c^2\sqrt{(2\pi)}}$, where $c = \Gamma(\tfrac{1}{4})$.

3. Integrate $e^{-z}z^{a-1}$ round boundary of $\delta \leqslant |z| \leqslant R$, $0 \leqslant \text{ph } z \leqslant \tfrac{1}{2}\pi$.

4. Theorem 14.72.

5. Integrate $(z+z^{-1})^a z^{b-1}$ round the boundary of $\delta \leqslant |z| \leqslant 1$, $-\tfrac{1}{2}\pi \leqslant \text{ph } z \leqslant \tfrac{1}{2}\pi$.

6. Integrand is $t^{z-1}\sum_1^\infty e^{-nt}$. Justify inversion of $\int$ and Σ.

REFERENCES

Ahlfors, L. V., *Complex Analysis* (New York: McGraw-Hill, 1966).

Banach, S., *Théorie des Opérations Linéaires* (Monografje Matematyczne, Warsaw, 1932).

Bartle, R. G., *The Elements of Integration* (New York: Wiley, 1966).

Bourbaki, N., *Eléments de Mathématique* (Paris: Hermann, 1939–   ).

Buck, R. C. (editor), *Studies in Modern Analysis* (New York: Prentice-Hall, 1962).

Burkill, J. C., *The Lebesgue Integral* (Cambridge, 1951).

A First Course in Mathematical Analysis (Cambridge, 1962).

Carslaw, H. S., *An Introduction to the Theory of Fourier's Series and Integrals* (London: Macmillan, 1930).

Cartan, H., *Analytic Functions* (New York: Addison Wesley, 1963).

Cohen, L. W. & Ehrlich, G., *The Structure of the Real Number System* (New York: Van Nostrand, 1963).

Dieudonné, J., *Foundations of Modern Analysis* (London and New York: Academic Press, 1960).

Flanders, H., *Differential Forms with Applications to the Physical Sciences* (London and New York: Academic Press, 1963).

Halmos, P. R., *Naive Set Theory* (New York: Van Nostrand, 1960).

Hancock, H., *Maxima and Minima* (New York: Dover, 1960).

Hardy, G. H., *Divergent Series* (Oxford, 1949).

Hardy, G. H., Littlewood, J. E. & Pólya, G., *Inequalities* (Cambridge, 1934).

Jeffreys, H. & B., *Methods of Mathematical Physics* (Cambridge, 1962).

Landau, E., *Foundations of Analysis* (New York: Chelsea, 1951).

Mirsky, L., *An Introduction to Linear Algebra* (Oxford, 1955).

Newman, M. H. A., *Elements of the Topology of Plane Sets of Points* (Cambridge, 1951).

Rotman, B. & Kneebone, G. T., *Theory of Sets and Transfinite Numbers* (London: Oldbourne, 1966).

Thurston, H. A., *The Number System* (London: Blackie, 1956).

INDEX